MAVA Math: Enhanced Skills Solutions

Marla Weiss

MAVA Books and Education Company
www.mavabooks.com

authorHOUSE

AuthorHouse™
1663 Liberty Drive
Bloomington, IN 47403
www.authorhouse.com
Phone: 1 (800) 839-8640

Published by AuthorHouse 12/08/2015

ISBN: 978-1-5049-6790-7 (sc)

CONTENTS

NOTES

Books By Marla Weiss

Available at AuthorHouse and most online bookstores

FICTION

School Scandalle

School Scoundrelle

MATH WORKBOOKS

MAVA Math: Number Sense

MAVA Math: Number Sense Solutions

MAVA Math: Grade Reviews

MAVA Math: Grade Reviews Solutions

MAVA Math: Middle Reviews

MAVA Math: Middle Reviews Solutions

MAVA Math: Enhanced Skills

MAVA Math: Enhanced Skills Solutions

NONFICTION (available 2016)

How To Finally Fix Math Education

Available at www.terrapinlogo.com

COMPUTER PROGRAMMING WORKBOOKS

Go, Logo!

Go, Logo! Solutions

An Important Message To Teachers, Parents, and Students

What are Enhanced Skills?
All *MAVA Math* workbooks have two-word titles: *Number Sense*, *Grade Reviews*, and *Middle Reviews*. *Enhanced Skills* keeps that format while indicating that the combined lessons are of a more advanced nature, suitable for students who wish to study math in depth or to enter math competitions.

What is this book's curriculum?
The detailed specificity of the Contents acts as a curriculum guide. This book includes a blend of basic skills as well as enriched material.

What is the grade level of this book?
While this book aims at grades six through eight, some students may begin earlier or later.

How many years should students take to complete this book?
Mathematically gifted students may complete the book in grades 6 and 7, finishing any remaining pages concurrently with Honors Algebra I in 8th grade. Other students could complete the book in 3 years.

Why do topics appear in alphabetical order?
Presenting the topics in alphabetical order offers maximum flexibility for teachers and students who may construct their own desired progression of lessons. Alphabetical order also permits students to easily supplement classroom work or contest preparation.

Does this book use a developmental approach?
Yes. Within each of the 79 topics, lesson 1 precedes lesson 2 developmentally. With AGE PROBLEMS for example, students should learn how to solve problems linearly prior to constructing charts. However, for the topics with many lessons, slight reordering may make sense for some teachers and students.

Are these topics and problems all that exist in advanced middle school math?
No. While this book is very comprehensive, another 426 pages still would not exhaust the rich variety of possible exercises.

Does this book's curriculum vary dramatically in any way from those typically seen?
Yes. This book's breadth, depth, and quantity of problems are not commonly found in one comprehensive volume.

Was this book field-tested?
Yes. Many problems in this book were used as part of comprehensive worksheets written by Marla Weiss for classroom settings. This material, unified for the first time in this book, yielded students who loved math, performed high on standardized tests, and earned countless awards at math competitions.

Does MAVA Math offer another middle school level workbook?
Yes. *MAVA Math: Middle Reviews*, intended to supplement the daily textbook, is designed for all middle school students. The accompanying answer book is *MAVA Math: Middle Reviews Solutions*.

May this book be used as the sole textbook?
Yes. This book in draft form was used as the sole textbook, supplemented with contest materials, for grades 6 and 7. Those students excelled in math.

Should certain pages of certain topics be done before others?
Yes. For example, students should know DIVISIBILITY 1 (rules) before FRACTIONS 1 (simplifying) or FRACTIONS 2 (multiplying). This book trusts teachers to order math.

Does this book prepare for future math instruction?
Yes. This book provides a solid foundation for high school math, including algebra skills, algebra word problems, plane geometry, and space geometry.

Should students use a calculator with this book?
Students should use a calculator only when absolutely essential. This book encourages practice of number sense throughout, not only on the pages titled MENTAL MATH.

What are the different types of MENTAL MATH?
Mental math may mean: 1. responding to an oral question; 2. doing all work in one's head for a written problem; and 3. doing minimal written work for a written problem.

Does this book cover a complete course in Algebra I?
No. This book covers most topics in pre-algebra and some topics in algebra but is not in any way intended to cover a full Algebra I course.

Why do some answers abbreviate words?
Math class is not an opportunity to teach language arts, whether spelling, handwriting, or composition. Attaching verbal skills to the study of math slows students both in daily work and annual progress. Ideally, students will learn to express themselves mathematically. However, mathematicians write in a specific style, totally different from English essays. Most abbreviations used in this book are found on pages 427–428.

Why do some geometry problems use upper case as well as lower case abbreviations?
When a problem involves two circles, use capital R for the radius of the larger one and lowercase r for the radius of the smaller one. Moreover, use B for the larger base of a trapezoid and b for the smaller one. Clarity of notation leads to improved focus and accurate answers.

Why are the answers to some measurement problems a number without a label following?
Consider the question: How far do you live from school? The answer could be 2 miles or 2 turtle steps. A label is needed for accuracy. Now consider the question: How many miles do you live from school? The answer may be 2 without ambiguity because a label, namely miles, is built into the question. Requiring a label at all times is unnecessary.

Should all improper fractions be converted to mixed numbers?
No. The term "improper fraction" is a misnomer. Converting from a fraction greater than one to a mixed number is a valuable skill, but it need not always be done. For example, as a solution to an equation, the fraction is better because it may be substituted directly to check its validity. However, measurements are best as mixed numbers. For example, one and three fourths cups flour is usually more helpful than seven fourths cups.

How does a student best learn problem solving?
While the ultimate goal is to solve a variety of problems in random order, students need to learn one problem type at a time. After mastery, then problems may be jumbled.

Can cumulative review be built into this book?
Yes. Students do not have to complete a page once starting it. Returning to a page at a later time builds in cumulative review.

Why do some problems have charts and diagrams pre-drawn while others do not?
At the advanced level of this book, students need to be able to draw their own diagrams and construct their own charts. However, in some problems, a diagram shows information that is not in the text.

Which are more valuable–fractions or decimals?
Decimal math may be easily done on a calculator. Fractions are more important in higher math. For example, a student who does not understand how to add 1/2 + 1/3 cannot possibly add 1/x + 1/y (diagonal fraction lines for ease of typing only). Furthermore, fractions yield an exact answer when decimals sometimes yield an approximation.

Do some problems or skills have more than one method of solution?
Yes. To find the perimeter of a rectangle, should one add the length and width and then double, or should one double each measurement and then add? To find the slope of a line given 2 points, which point should be considered the 1st point and which the 2nd? Students should understand both methods and decide based on the numbers in the problems, always seeking fast and accurate calculation. The inherent richness and beauty of math yield multiple approaches to many problems. PERCENTS 5 (x is y% of z) is just one of many pages that show multiple methods. The WLOG method is another example.

Why do some of the 79 topics have "see" following them?
Due to the richness of math, placing a concept or skill into a category may be difficult. For example, integers may be even or odd, and equations may contain exponents. Similarly, a problem about the percent change in the area of a rectangle covers three topics. The placement is often arbitrary. Thus, the "see" references guide the user.

Why does this book use two different fonts?
All problems appear in the Arial font. All work, answers, and comments appear in the Chalkboard font. Final answers are in Chalkboard Bold.

Why are negative, opposite, and subtraction signs all represented by the same symbol?
Some math texts use both a smaller, higher line and a longer, mid-level line. Because all three signs operate equivalently, this book uses just the longer, mid-level line for simplicity.

Why are the decimal points bold?
Some students do not see decimal points in normal font. Similarly, some students do not write decimal points darkly enough. A happy medium exists between a light dot and a wart.

How can one learn more about various problem types and solution methods?
The website www.mavabooks.com offers short videos giving instruction on various topics. More will be added as time permits.

MAVA Math: Enhanced Skills Solutions Copyright © 2015 Marla Weiss

Why are equilateral triangles also labeled isosceles?
The definition of isosceles triangle is a three-sided polygon with at least 2 congruent sides. Therefore, an equilateral triangle is isosceles, but the converse is not true.

Why is functional notation used in situations that do not involve functions?
Functional notation is precise and concise. For example, P(even) is neater than writing "probability of tossing an even." Similarly, GCF(35, 49) is neater than writing the "greatest common factor of 35 and 49."

What does the word "unit" mean?
Unit is a general term. Regarding distance, "unit" may mean many different measurements such as inches, feet, miles, or centimeters. Understanding that the label must be square units for area and cubic units for volume is more important than what the actual unit is. By using the generic "unit," students may focus on specific skills.

Why does this book list over 500 vocabulary words?
Students cannot do math problems without understanding the words contained therein. Unfortunately, many math words have multiple meanings. Consider base–e.g., base of a triangle, base two arithmetic, and a number (base) raised to a power (exponent). Students learn math vocabulary when they continually hear the words used correctly.

May parents help with the pages?
Students who receive continual math help from their parents often show less growth than students who learn to work independently. Moreover, most parents have forgotten math or don't know the best ways to approach many problems. Parents should only monitor a child's work, determining weak areas needing further help.

May students and teachers write diagonal fraction lines?
Never! For correct fraction work, students must clearly see numerators and denominators, only accomplished by writing horizontal fraction lines. This book occasionally uses diagonal fraction lines for ease of typing only.

Should students memorize all of the Pythagorean triples and prime numbers at the end of this book?
No. The lists are for reference. However, math memorization should not stop with the four whole number operation facts. Further memorization, which will speed work, may include primes, Pythagorean triples, formulae, perfect squares and cubes, and square roots of non-squares (root 48 is 4 root 3).

What happens if MAVA Math: Enhanced Skills or MAVA Math: Enhanced Skills Solutions contains an error?
Both books were thoroughly proofed. However, any needed corrections will be posted on www.mavabooks.com. Please send a concise and precise e-mail to info@mavabooks.com if a correction does not address your concern.

Should math be fun?
Of course, math should be fun. However, teaching math solely as a game does not lead to growth. Students who study rigorous math truly learn math. Understanding in turn leads to natural enjoyment. Competence is pleasurable.

MAVA Math: Enhanced Skills Solutions Copyright © 2015 Marla Weiss

Absolute Value 1

Evaluate, working down. Show one line before the answer.

1. $|3 - 7| - 3 + 4 - |-2 - 6| - |1 - 5| - 5$

   $4 + 1 - 8 - 4 - 5$
   -12

2. $|-9 - 11| - (-2)^2 + |-11 + 9| - |3 - 9|$

   $20 - 4 + 2 - 6$
   12

3. $|1 - 9 - 7 + 1| - 2 |8 - 9| - 5 + 2 |8 - 9|$

   $14 - 2 - 5 + 2$
   9

4. $|2 - 5| + 2^5 - 3 |2 - 5 - 5 + 2| - |-4|$

   $3 + 32 - 18 - 4$
   13

5. $|-3 - 4 - 1| + |5 - 6 + 2 - 3| - |4 - 7|$

   $8 + 2 - 3$
   7

6. $|-4 + 8 - 4 - 8| - |-8 - 4| - (4)(-8) - 4$

   $8 - 12 + 32 - 4$
   24

7. $|-8 + 9 - 5 - 2| - |-7 - 9| - 2 |-7 + 5|$

   $6 - 16 - 4$
   -14

8. $5^2 - |2 - 5| - |5 - 2| + 2^5 - |-5| + (-2)$

   $25 - 3 - 3 + 32 - 5 - 2$
   44

9. $|-9 + 8 - 5 - 4| - |-6 - 9| - |(-8)(2)|$

   $10 - 15 - 16$
   -21

10. $|-1 - 6 + 5| - 4 |-2 - 4 + 3| - |-6|$

    $2 - 12 - 6$
    -16

11. $|6 - 11| - 1^{16} - |6 - 7 + 2| + |7 - 16|$

    $5 - 1 - 1 + 9$
    12

12. $- |3 - 10| - |4 - 9| + |-5 - 3| - |-3|$

    $-7 - 5 + 8 - 3$
    -7

13. $3^4 - |-3 - 3| + |3 - 6| - 3^3 - |-6|$

    $81 - 6 + 3 - 27 - 6$
    45

14. $|-8 - 9 - 3| - |- 4 + 11 - 2| - 10$

    $20 - 5 - 10$
    5

15. $- |-2 - 5| + |-3 - 4| - |-6 - 1| + 1$

    $-7 + 7 - 7 + 1$
    -6

16. $|3 - 4| + 4^3 - |4 - 3 - 4 + 3| - 32$

    $1 + 64 - 0 - 32$
    33

Absolute Value 2

Solve by mental math.

1. $\|x\| = 6$ **6, −6**	17. $\|x + 3\| = 8$ **5, −11**	33. $\|9 − x\| = \|−2\|$ **7, 11**
2. $\|x\| = −6$ **No Sol**	18. $\|x + 1\| = 7$ **6, −8**	34. $\|x + 4\| = \|−15\|$ **11, −19**
3. $\|−6\| = x$ **6**	19. $\|x − 5\| = −1$ **No Sol**	35. $\|4 + x\| = \|−6\|$ **2, −10**
4. $\|6\| = −x$ **−6**	20. $\|x − 1\| = 10$ **11, −9**	36. $\|x − 1\| + 5 = 10$ **6, −4**
5. $\|−x\| = 6$ **6, −6**	21. $\|3 − x\| = 4$ **7, −1**	37. $\|2x + 4\| = \|−15 + 3\|$ **4, −8**
6. $x = \|−5\|$ **5**	22. $\|x + 5\| = 8$ **3, −13**	38. $\|2x + 3\| = \|−17 + 4\|$ **5, −8**
7. $\|x\| = −5$ **No Sol**	23. $\|x − 4\| = 8$ **12, −4**	39. $\|x + 7\| = \|−13\|$ **6, −20**
8. $\|−x\| = 5$ **5, −5**	24. $\|x − 5\| = 17$ **22, −12**	40. $\|x + 7\| + 5 = 16$ **4, −18**
9. $\|5\| = −x$ **−5**	25. $\|x + 7\| = 14$ **7, −21**	41. $\|x − 8\| + 2 = 14$ **20, −4**
10. $\|5\| = x$ **5**	26. $\|9 + x\| = 20$ **11, −29**	42. $\|x − 6\| − 7 = 8$ **21, −9**
11. $\|x\| = 5$ **5, −5**	27. $\|6 − x\| = −2$ **No Sol**	43. $\|x + 3\| − 4 = 11$ **12, −18**
12. $\|x\| = 7$ **7, −7**	28. $\|x + 6\| = 15$ **9, −21**	44. $3x + 5 = \|−19 + 5\|$ **3**
13. $\|x\| = −7$ **No Sol**	29. $\|2 − x\| = 11$ **13, −9**	45. $\|x + 7\| − 7 = 7$ **7, −21**
14. $x = \|7\|$ **7**	30. $\|x + 2\| = 9$ **7, −11**	46. $\|5x + 5\| = \|−11 − 4\|$ **2, −4**
15. $\|−7\| = x$ **7**	31. $\|6 + x\| = 12$ **6, −18**	47. $\|2x + 4\| = \|−13 + 3\|$ **3, −7**
16. $\|7\| = −x$ **−7**	32. $\|8 − x\| = 13$ **21, −5**	48. $7x − 5 = \|−21 − 2\|$ **4**

Absolute Value 3

Find the least value of x such that:	Find the greatest value of x such that:	Find the number of integers x that satisfy:	Find the sum of the integral solutions.
1. $\|x + 1\| \le 6$ **−7**	17. $\|x - 2\| \le 8$ **10**	33. $\|x + 3\| \le 10$ −13 to 7 **21**	49. $\|x + 2\| < 5$ −6, . . . , 2 **−18**
2. $\|x - 8\| \le 11$ **−3**	18. $\|x + 2\| \le 6$ **4**	34. $\|x - 1\| \le 4$ −3 to 5 **9**	50. $\|x - 1\| \le 4$ −3, . . . , 5 **9**
3. $\|x - 4\| \le 10$ **−6**	19. $\|2x + 9\| \le 17$ **4**	35. $\|4x + 1\| \le 9$ −2 to 2 **5**	51. $\|x + 3\| < 7$ −9, . . . , 3 **−39**
4. $\|x + 5\| \le 12$ **−17**	20. $\|4x + 6\| \le 13$ **7/4**	36. $\|5x - 4\| \le 14$ −2 to 3 **6**	52. $\|x - 4\| \le 6$ −2, . . . , 10 **52**
5. $\|x + 7\| \le 8$ **−15**	21. $\|3x - 2\| \le 20$ **22/3**	37. $\|x - 6\| \le 2$ 4 to 8 **5**	53. $\|2x + 2\| < 8$ −4, . . . , 2 **−7**
6. $\|2x + 7\| \le 11$ **−9**	22. $\|x + 7\| \le 11$ **4**	38. $\|6x - 1\| \le 7$ −1, 0, 1 **3**	54. $\|x + 5\| \le 5$ −10, . . . , 0 **−55**
7. $\|x + 9\| \le 15$ **−24**	23. $\|2x - 9\| \le 5$ **7**	39. $\|x + 9\| \le 11$ −20 to 2 **23**	55. $\|x - 6\| < 10$ −3, . . . , 15 **114**
8. $\|x - 6\| \le 13$ **−7**	24. $\|5x - 2\| \le 12$ **14/5**	40. $\|2x - 4\| \le 5$ 0 to 4 **5**	56. $\|3x + 1\| < 11$ −3, . . . , 3 **0**
9. $\|3x + 4\| \le 20$ **−8**	25. $\|2x + 4\| \le 16$ **6**	41. $\|8x - 1\| \le 3$ 0 **1**	57. $\|3x - 3\| \le 9$ −2, . . . , 4 **7**
10. $\|x - 9\| \le 16$ **−7**	26. $\|7x - 3\| \le 9$ **12/7**	42. $\|x + 4\| \le 12$ −16 to 8 **25**	58. $\|4x + 3\| \le 15$ −4, . . . , 3 **−4**
11. $\|2x + 1\| \le 15$ **−8**	27. $\|2x + 5\| \le 14$ **9/2**	43. $\|3x - 4\| \le 11$ −2 to 5 **8**	59. $\|x - 4\| < 5$ 0, . . . , 8 **36**
12. $\|x + 2\| \le 15$ **−17**	28. $\|11x - 5\| \le 7$ **12/11**	44. $\|x - 3\| \le 8$ −5 to 11 **17**	60. $\|x + 1\| < 3$ −3, . . . , 1 **−5**
13. $\|4x - 3\| \le 23$ **−5**	29. $\|6x + 6\| \le 19$ **13/6**	45. $\|3x - 2\| \le 5$ −1 to 2 **4**	61. $\|2x + 1\| \le 9$ −5, . . . , 4 **−5**
14. $\|x - 3\| \le 17$ **−14**	30. $\|8x - 9\| \le 21$ **15/4**	46. $\|x + 3\| \le 7$ −10 to 4 **15**	62. $\|10x + 3\| < 15$ −1, 0, 1 **0**
15. $\|5x + 4\| \le 21$ **−5**	31. $\|9x + 8\| \le 1$ **−7/9**	47. $\|2x - 1\| \le 7$ −3 to 4 **8**	63. $\|4x + 6\| \le 10$ −4, . . . , 1 **−9**
16. $\|6x - 5\| \le 29$ **−4**	32. $\|2x - 3\| \le 0$ **3/2**	48. $\|x + 5\| \le 6$ −11 to 1 **13**	64. $\|3x + 2\| < 8$ −3, . . . , 1 **−5**

Absolute Value 4

Solve.

1. $| 3x - 7 | = 20$

$3x - 7 = 20$ OR $3x - 7 = -20$
$3x = 27$ $\qquad 3x = -13$
$x = \mathbf{9}$ $\qquad x = \dfrac{-13}{3}$

2. $2 | 8 + 5x | = 66$

$5x + 8 = 33$ OR $5x + 8 = -33$
$5x = 25$ $\qquad 5x = -41$
$x = \mathbf{5}$ $\qquad x = \dfrac{-41}{5}$

3. $5 | 9x - 1 | = -10$

No Solution
nonnegative ≠ negative

4. $| 7x + 9 | = 30$

$7x + 9 = 30$ OR $7x + 9 = -30$
$7x = 21$ $\qquad 7x = -39$
$x = \mathbf{3}$ $\qquad x = \dfrac{-39}{7}$

5. $| -4x - 12 | = 48$

$-4x - 12 = 48$ OR $-4x - 12 = -48$
$-4x = 60$ $\qquad -4x = -36$
$x = \mathbf{-15}$ $\qquad x = \mathbf{9}$

6. $6 | -5x + 8 | = 72$

$-5x + 8 = 12$ OR $-5x + 8 = -12$
$-5x = 4$ $\qquad -5x = -20$
$x = \dfrac{-4}{5}$ $\qquad x = \mathbf{4}$

7. $| 14x + 7 | = 49$

$14x + 7 = 49$ OR $14x + 7 = -49$
$14x = 42$ $\qquad 14x = -56$
$x = \mathbf{3}$ $\qquad x = \mathbf{-4}$

8. $-7 | 8x - 5 | = -84$

$8x - 5 = 12$ OR $8x - 5 = -12$
$8x = 17$ $\qquad 8x = -7$
$x = \dfrac{17}{8}$ $\qquad x = \dfrac{-7}{8}$

9. $| -6 - 11x | = 60$

$-6 - 11x = 60$ OR $-6 - 11x = -60$
$-11x = 66$ $\qquad -11x = -54$
$x = \mathbf{-6}$ $\qquad x = \dfrac{54}{11}$

10. $-5 | 10x - 8 | = -75$

$10x - 8 = 15$ OR $10x - 8 = -15$
$10x = 23$ $\qquad 10x = -7$
$x = \dfrac{23}{10}$ $\qquad x = \dfrac{-7}{10}$

11. $6 | 13 - 5x | = 0$

$13 - 5x = 0$
$5x = 13$
$x = \dfrac{13}{5}$

12. $| 20x - 9 | = 51$

$20x - 9 = 51$ OR $20x - 9 = -51$
$20x = 60$ $\qquad 20x = -42$
$x = \mathbf{3}$ $\qquad x = \dfrac{-21}{10}$

13. $| 11x - 2 | = 20$

$11x - 2 = 20$ OR $11x - 2 = -20$
$11x = 22$ $\qquad 11x = -18$
$x = \mathbf{2}$ $\qquad x = \dfrac{-18}{11}$

14. $- | 10x + 11 | = -39$

$10x + 11 = 39$ OR $10x + 11 = -39$
$10x = 28$ $\qquad 10x = -50$
$x = \dfrac{14}{5}$ $\qquad x = \mathbf{-5}$

15. $| 12x + 5 | = 51$

$12x + 5 = 51$ OR $12x + 5 = -51$
$12x = 46$ $\qquad 12x = -56$
$x = \dfrac{23}{6}$ $\qquad x = \dfrac{-14}{3}$

16. $8 | 2x - 13 | = 72$

$2x - 13 = 9$ OR $2x - 13 = -9$
$2x = 22$ $\qquad 2x = 4$
$x = \mathbf{11}$ $\qquad x = \mathbf{2}$

Age Problems 1

Answer by making a vertical timeline. Distinguish between the relative (write on the left) and absolute (write on the right) ages.

1. Freddie is 10 years older than Frenchie who is 6 years younger than Frannie. If Frannie is 17 now, how old is Freddie?

   4 (Freddie **21**
   Frannie 17
   6 (Frenchie 11

5. Barry is 5 years older than Cary who is 3 years older than Harry. Mary, now 24, is 6 years younger than Barry. Find the sum of their ages.

   5 (Barry 30
   Cary 25
   1 (Mary 24
   2 (Harry 22   **101**

2. Katie is 6 years older than Kathy who is 2 years older than Cathy. Katy, who is 19, is five years older than Cathy. How old is Katie?

   3 (Katie **22**
   Katy 19
   3 (Kathy 16
   2 (Cathy 14

6. Joe is 9 years older than Jo who is 4 years younger than Joan. If Josie is the oldest by 5 years and is now 22, how old is Jo?

   5 (Josie 22
   Joe 17
   5 (Joan 12
   4 (Jo **8**

3. Shana is 9 years older than Jana who is 5 years younger than Dana. If Lana is the oldest by one year and is 24 now, how old is Jana?

   1 (Lana 24
   Shana 23
   4 (Dana 19
   5 (Jana **14**

7. Jan is 20 years older than Jim who is 4 years younger than Jon. Jen, who is 52, is 4 years older than Jon. How old is Jan?

   12 (Jan **64**
   Jen 52
   4 (Jon 48
   4 (Jim 44

4. TJ is 2 years older than PJ, and PJ is 6 years younger than RJ. If RJ is 20, how old are the other boys?

   4 (RJ 20
   TJ **16**
   2 (PJ **14**

8. Hank is younger than Hal by 6 years but older than Henry by 7 years. If Hank is now 20, find the sum of their ages.

   6 (Hal 26
   Hank 20
   7 (Henry 13
   59

Age Problems 2

Answer by making a chart.

1. Three years ago Shauna was one year younger than Liz will be in 4 years. If Shauna is 18 now, how old is Liz?

	−3	now	+4
S	15	18	22
L		**12**	16

2. Hal's age in 30 years will be half a century. Six years ago, Bo was one year younger than Hal is now. How old will Bo be in 5 years?

	−6	now	+5	+30
H	14	20	25	50
B	19	25	**30**	

3. Sarah's age is the average of Paul and Matt's. If Paul was 8 two years ago and Matt will be 20 in six years, how old is Sarah now?

	−2	now	+6
S		**12**	
P	8	10	16
M	12	14	20

4. Five years ago Karen was the same age as Mike now. Marc is now twice Mike's age. If Karen will be 18 in 3 years, how old was Marc 5 years ago?

	−5	now	+3
K	10	15	18
Mi	5	10	13
Ma	**15**	20	

5. Winston 2 years ago was as old as James is now. If Winston is now 16, how old will James be in five years?

	−2	now	+5
W	14	16	21
J		14	**19**

6. Eli is 11 years old. When Eli is 23, the sum of his age and his brother's will be 53. How old is his brother now?

	now	+12
E	11	23
Br	**18**	30

7. PJ's age is one-fifth greater than Ian's. Four years ago Ian's age was the least two-digit prime number. How old will PJ be in 7 years?

	−4	now	+7
PJ		18	**25**
I	11	15	22

8. On the first day of 6th grade, Joy was one year younger than Vicki. Four years later Vicki was 16. How old was Joy when she started 2nd grade?

	2nd	6th	+4
J	**7**	11	
V	8	12	16

Angles 1

Find the complement of the angle in degrees by mental math.

Find the supplement of the angle in degrees by mental math.

1. 70°	**20**	17. 64°	**26**	33. 130°	**50**	49. 70°	**110**
2. 8°	**82**	18. 11°	**79**	34. 46°	**134**	50. 31°	**149**
3. 35°	**55**	19. 66°	**24**	35. y°	**180−y**	51. 87°	**93**
4. 29°	**61**	20. 42°	**48**	36. 51°	**129**	52. 5g°	**180−5g**
5. 89.5°	**0.5**	21. 29.8°	**60.2**	37. 92°	**88**	53. 111°	**69**
6. 60°	**30**	22. 14.7°	**75.3**	38. 99.5°	**80.5**	54. 14.7°	**165.3**
7. x°	**90−x**	23. m°	**90−m**	39. 73°	**107**	55. 154.6°	**25.4**
8. 23.4°	**66.6**	24. 61°	**29**	40. 24°	**156**	56. 9x°	**180−9x**
9. 17°	**73**	25. 75.5°	**14.5**	41. 152°	**28**	57. 53°	**127**
10. 5°	**85**	26. 22°	**68**	42. 3w°	**180−3w**	58. 178.3°	**1.7**
11. 13°	**77**	27. 41°	**49**	43. 106°	**74**	59. 1.5°	**178.5**
12. 54.9°	**35.1**	28. 39.6°	**50.4**	44. 43°	**137**	60. 99.3°	**80.7**
13. 72°	**18**	29. 3y°	**90−3y**	45. 145°	**35**	61. 4h°	**180−4h**
14. 54°	**36**	30. 21.5°	**68.5**	46. 68°	**112**	62. 138°	**42**
15. 33°	**57**	31. 50°	**40**	47. 19°	**161**	63. 103°	**77**
16. 2s°	**90−2s**	32. 30.1°	**59.9**	48. 54.5°	**125.5**	64. 16.9°	**163.1**

Angles 2

Find the angle measure in degrees.		*Answer as indicated.*	
1. Find the supplement of the complement of 25°.	C(25) = 65 S(65) = **115**	12. Two supplementary angles are in the ratio 7:2. Find their product.	180/9=20 (140)(40)= **5600**
2. Find the complement of the supplement of 127°.	S(127) = 53 C(53) = **37**	13. Two supplementary angles are in the ratio 5:4. Find their positive difference.	180/9=20 100−80= **20**
3. Find the complement of the supplement of 110°.	S(110) = 70 C(70) = **20**	14. Two complementary angles are in the ratio 9:1. Find their positive difference.	90/10=9 81−9= **72**
4. Find the supplement of the complement of 42°.	C(42) = 48 S(48) = **132**	15. Two complementary angles are in the ratio 2:1. Find their positive difference.	90/3=30 60−30= **30**
5. Find the complement of the supplement of 154°.	S(154) = 26 C(26) = **64**	16. Two supplementary angles are in the ratio 7:5. Find their positive difference.	180/12=15 105−75= **30**
6. Find the supplement of the complement of 69°.	C(69) = 21 S(21) = **159**	17. Two supplementary angles are in the ratio 8:7. Find the square of their difference.	180/15=12 12(8−7)=12 **144**
7. Find the supplement of the complement of 51°.	C(51) = 39 S(39) = **141**	18. Two complementary angles are in the ratio 3:7. Find their positive difference.	90/10=9 63−27= **36**
8. Find the complement of the supplement of 163°.	S(163) = 17 C(17) = **73**	19. Two complementary angles are in the ratio 5:1. Find their positive difference.	90/6=15 75−15= **60**
9. Find the complement of the supplement of 145°.	S(145) = 35 C(35) = **55**	20. Two complementary angles are in the ratio 11:7. Find their product.	90/18=5 (55)(35)= **1925**
10. Find the supplement of the complement of 78°.	C(78) = 12 S(12) = **168**	21. Two supplementary angles are in the ratio 8:1. Find their product.	180/9=20 (160)(20)= **3200**
11. Find the complement of the supplement of 172°.	S(172) = 8 C(8) = **82**	22. Two complementary angles are in the ratio 17:13. Find the supplement of the greater angle.	90/30=3 S(51)= **129**

Angles 3

Label the angles in the diagram alternate exterior (AE), alternate interior (AI), corresponding (C), supplementary (S), or vertical (V).

1. ∠3 and ∠6 **V**	14. ∠5 and ∠8 **AE**	27. ∠2, ∠3, and ∠4 **S**
2. ∠1 and ∠4 **V**	15. ∠11 and ∠12 **S**	28. (∠5 & ∠6) and ∠14 **C**
3. ∠12 and ∠13 **S**	16. ∠13 and ∠14 **S**	29. ∠1, ∠2, and ∠6 **S**
4. ∠2 and ∠8 **C**	17. ∠1 and ∠11 **C**	30. (∠3 & ∠4) and ∠9 **C**
5. ∠11 and ∠14 **S**	18. ∠2 and ∠10 **AI**	31. (∠1 & ∠6) and ∠9 **AE**
6. ∠2 and ∠5 **V**	19. ∠1 and ∠13 **AE**	32. ∠1, ∠5, and ∠6 **S**
7. ∠4 and ∠13 **C**	20. ∠7 and ∠10 **S**	33. (∠2 & ∠3) and ∠14 **AI**
8. ∠8 and ∠10 **V**	21. ∠7 and ∠9 **V**	34. (∠1 & ∠6) and ∠7 **C**
9. ∠5 and ∠10 **C**	22. ∠9 and ∠10 **S**	35. ∠1, ∠2, and ∠3 **S**
10. ∠4 and ∠11 **AI**	23. ∠3, ∠4, and ∠5 **S**	36. (∠5 & ∠6) and ∠12 **AE**
11. ∠12 and ∠14 **V**	24. (∠2 & ∠3) and ∠12 **C**	37. (∠2 & ∠3) and (∠5 & ∠6) **V**
12. ∠11 and ∠13 **V**	25. ∠8 and ∠9 **S**	38. (∠3 & ∠4) and ∠7 **AI**
13. ∠7 and ∠8 **S**	26. ∠4, ∠5, and ∠6 **S**	39. (∠1 & ∠2) and (∠4 & ∠5) **V**

Angles 4

Solve algebraically. | *Answer as indicated. NTS*

---

1. What are the measures of two supplementary angles, the greater of which measures 4 times the lesser?

$180 - L = 4L$ | $L = \mathbf{36}$
$180 = 5L$ | $G = \mathbf{144}$

8. Find x + y.

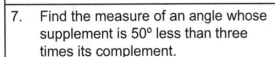

$4y = 180$
$y = 45$
$x = 45$ (vert)
$x + y = \mathbf{90}$

---

2. The sum of the measures of the complement and supplement of an angle is 230°. Find the angle.

$90 - A + 180 - A = 230$
$40 = 2A$
$A = \mathbf{20}$

9. Find x.

$supp(135) = 45$
$45 + 45 = 90$
$supp(45) = 135$
$x = \mathbf{135}$
(corr)

---

3. The complement of an angle is 36° greater than twice the angle. Find the supplement of the angle.

$90 - A = 2A + 36$ | $A = 18$
$54 = 3A$ | $supp(A) = \mathbf{162}$

10. Find y.

$3x - 35 = 2x$
$x = 35$
$35 + y = 90$
$y = \mathbf{55}$

---

4. The sum of triple the complement of an angle and twice its supplement is 455°. Find the angle.

$3(90 - A) + 2(180 - A) = 455$ | $175 = 5A$
$270 - 3A + 360 - 2A = 455$ | $A = \mathbf{35}$

11. Find x.

$6y + 6 = 180$
$6y = 174$
$y = 29$
$x = \mathbf{29}$
(corr)

---

5. The sum of one-half of the supplement of an angle and the complement of the angle is 135°. Find the supplement.

$(180 - A)/2 + (90 - A) = 135$
$180 - A + 180 - 2A = 270$ | $A = 30$
$90 = 3A$ | $supp(A) = \mathbf{150}$

12. Find y.

$5x - 10 = 90$
$5x = 100$
$x = 20$
$y = \mathbf{70}$
(vert)

---

6. Find the greater of two supplementary angles if one is 30° more than twice the other.

$180 - L = 30 + 2L$ | $L = 50$
$150 = 3L$ | $G = \mathbf{130}$

13. $y + z = 185°$
Find x.

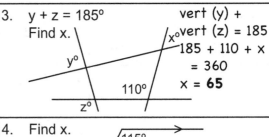

vert (y) +
vert (z) = 185
$185 + 110 + x$
$= 360$
$x = \mathbf{65}$

---

7. Find the measure of an angle whose supplement is 50° less than three times its complement.

$180 - A = 3(90 - A) - 50$ | $2A = 40$
$180 - A = 270 - 3A - 50$ | $A = \mathbf{20}$

14. Find x.

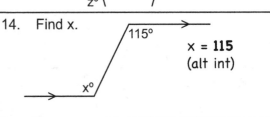

$x = \mathbf{115}$
(alt int)

Angles 5

Find the missing angle measures using the diagram. Each problem is independent. NTS

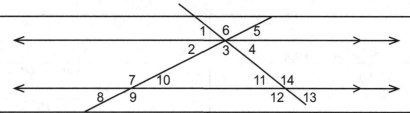

1. If m∠3 = 42º and m∠10 = 56º,
   then m∠11 = **82** º.

   3, 10, 11 triangle.

2. If m∠3 = 38º and m∠8 = 45º,
   then m∠11 = **97** º.

   8 & 10 vertical. 3, 10, 11 triangle.

3. If m∠11 = 25º and m∠10 = 55º,
   then m∠6 = **100** º.

   3, 10, 11 triangle. 3 & 6 vertical.

4. If m∠13 = 39º,
   then m∠11 = **39** º.

   11 & 13 vertical.

5. If m∠7 = 125º and m∠1 = 40º,
   then m∠6 = **85** º.

   1 & 6 corresponding to 7.

6. If m∠11 = 26º,
   then m∠4 = **26** º.

   4 & 11 alternate interior.

7. If m∠7 = 110º and m∠14 = 120º,
   then m∠3 = **50** º.

   7 & 10 supp. 11 & 14 supp. 3, 10, 11 triangle.

8. If m∠12 = 106º and m∠7 = 99º,
   then m∠3 = **25** º.

   11 & 12 supp. 7 & 10 supp. 3, 10, 11 triangle.

9. If m∠5 = 65º,
   then m∠10 = **65** º.

   5 & 10 corresponding.

10. If m∠5 = 62º and m∠12 = 116º,
    then m∠6 = **54** º.

    12 & 14 vertical. 14 corresp to 5 & 6 tog.

11. If m∠1 = 35º and m∠3 = 55º,
    then m∠5 = **90** º.

    3 & 6 vertical. 1, 5, & 6 supp.

12. If m∠8 = 42º and m∠1 = 33º,
    then m∠6 = **105** º.

    8 & 2 corresp. 2 & 5 vertical. 1, 5, & 6 supp.

13. If m∠9 = 108º and m∠6 = 44º,
    then m∠13 = **64** º.

    9 & 10 supp. 6 & 3 vert. Triangle. 11 & 13 vert.

14. If m∠8 = 44º and m∠13 = 63º,
    then m∠6 = **73** º.

    8 & 10 vert. 13 & 11 vert. Triangle. 3 & 6 vert.

15. If m∠13 = 81º and m∠6 = 36º,
    then m∠2 = **63** º.

    13 & 4 corr. 6 & 3 vert. 2, 3, & 4 supp.

16. If m∠13 = 76º and m∠3 = 41º,
    then m∠9 = **117** º.

    11 & 13 vert. 3, 10 & 11 triangle. 10 & 9 supp.

17. If m∠9 = 111º and m∠6 = 22º,
    then m∠14 = **91** º.

    9 & 7 vert. 7 corresp to 1 & 6. 1 & 4 vert.
    4 & 14 same side interior.

18. If m∠9 = 103º and m∠1 = 35º,
    then m∠3 = **68** º.

    1 & 4 vert. 9 & 7 vert. 7 & 2 same side
    interior. 2, 3, & 4 supp.

Angles 6

Find x in degrees in the crook diagram. Assume parallel lines. NTS

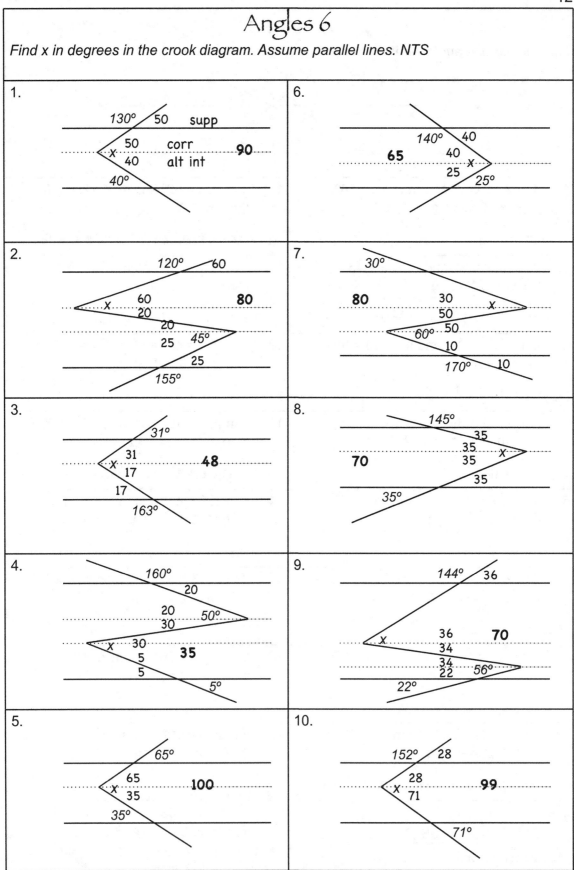

1. 130° / 50 supp; 50 corr x / 40 alt int **90**; 40°

2. 120° / 60; 60 x / 20; 20; 25 45°; 25; 155° **80**

3. 31°; 31 x / 17; 17; 163° **48**

4. 160° / 20; 20 50°; 30; 30 x / 5; 5; 5° **35**

5. 65°; 65 x / 35; 35° **100**

6. 140° / 40; 65 40 x; 25 25°

7. 30°; 80 30 x; 50; 60° 50; 10; 170° 10

8. 145° / 35; 35 x; 70 35; 35; 35°

9. 144° / 36; x 36 **70**; 34; 34 56°; 22 22°

10. 152° / 28; 28 x / 71 **99**; 71°

MAVA Math: Enhanced Skills Solutions Copyright © 2015 Marla Weiss

Area 1

Find the area in square units. Assume semicircles. One box equals one unit.

1. 16 x 18 − 123 = 288 − 123 = **165**

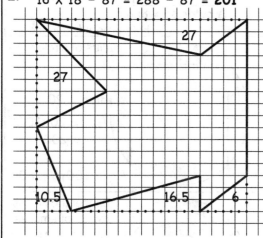

4. 15x16−47.5−12.5π = **192.5 − 12.5π**

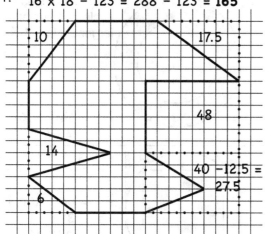

2. 16 x 18 − 87 = 288 − 87 = **201**

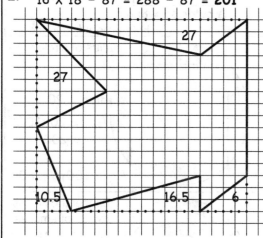

5. 16 x 19 − 73.5 = 304 − 73.5 = **230.5**

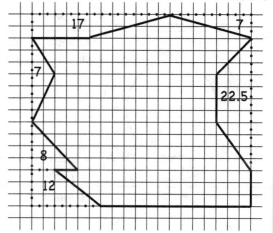

3. 18x16−82.5−4.5π = **205.5 − 4.5π**

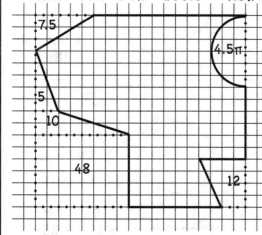

6. 18x17−98.5−4.5π = **207.5 − 4.5π**

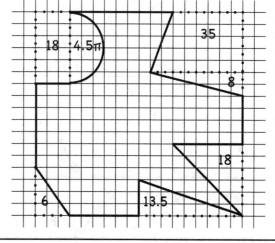

Area 2

Find the shaded area in square units. Assume right angles and tangency. Answer in terms of π. NTS

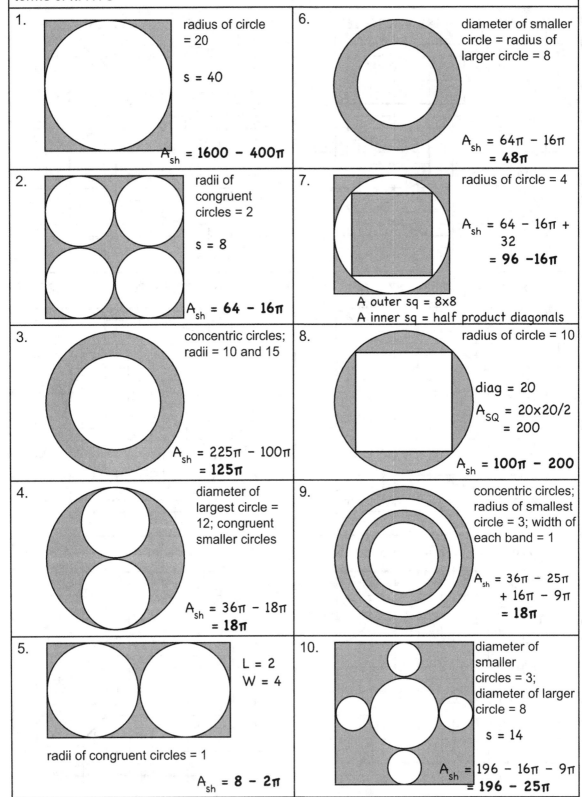

1. radius of circle = 20

   s = 40

   A_{sh} = **1600 − 400π**

2. radii of congruent circles = 2

   s = 8

   A_{sh} = **64 − 16π**

3. concentric circles; radii = 10 and 15

   A_{sh} = 225π − 100π = **125π**

4. diameter of largest circle = 12; congruent smaller circles

   A_{sh} = 36π − 18π = **18π**

5. L = 2
   W = 4

   radii of congruent circles = 1

   A_{sh} = **8 − 2π**

6. diameter of smaller circle = radius of larger circle = 8

   A_{sh} = 64π − 16π = **48π**

7. radius of circle = 4

   A_{sh} = 64 − 16π + 32 = **96 − 16π**

   A outer sq = 8x8
   A inner sq = half product diagonals

8. radius of circle = 10

   diag = 20
   A_{SQ} = 20x20/2 = 200

   A_{sh} = **100π − 200**

9. concentric circles; radius of smallest circle = 3; width of each band = 1

   A_{sh} = 36π − 25π + 16π − 9π = **18π**

10. diameter of smaller circles = 3; diameter of larger circle = 8

    s = 14

    A_{sh} = 196 − 16π − 9π = **196 − 25π**

Area 3

Answer as indicated. All structures (photos, ponds, etc.) are rectangles. All borders are uniform width. All linear dimensions are units, and areas are square units.

1. A picture 7 by 11 has a border of size 3 all around. Find the area of the border.

   $13 \times 17 - 7 \times 11 =$
   $221 - 77 =$
   144

6. A photo 5 longer than it is wide has a border of area 100 and width 2. Find the dimensions of the photo.

   $(W+4)(W+9) = W(W+5) + 100$
   $W^2 + 13W + 36 = W^2 + 5W + 100$
   $8W = 64$
   $W = 8$ **8 by 13**

2. A photo 6 by 12 has a border of size 1.5 all around. Find the ratio of areas, border to photo.

   $9 \times 15 - 6 \times 12 =$ $\dfrac{63}{72}$ $\dfrac{7}{8}$
   $135 - 72 =$
   63

7. A pool twice as long as it is wide has a border of area 940 and width 5. Find the dimensions of the pool.

   $(W)(2W) + 940 = (W + 10)(2W + 10)$
   $2W^2 + 940 = 2W^2 + 30W + 100$
   $30W = 840$
   $W = 28$ **28 by 56**

3. A pond 12 by 15 has a border of size 2.5 all around. Find the ratio of areas, pond to border.

   $17 \times 20 - 12 \times 15 =$ $\dfrac{180}{160}$ $\dfrac{9}{8}$
   $340 - 180 =$
   160

8. A garden 6 wider than it is long has a border of area 156 and width 3. Find the dimensions of the garden.

   $(L)(L+6) + 156 = (L + 6)(L + 12)$
   $L^2 + 6L + 156 = L^2 + 18L + 72$
   $84 = 12L$
   $L = 7$ **7 by 13**

4. A pool 25 by 40 has a walkway all around. Find the width of the walkway if the total area is 1750.

   $1750 = 35 \times 50$
   $\quad = (25 + 10) \times (40 + 10)$
   width = **5**

9. A picture 4 longer than it is wide has a border of area 200 and width 2. Find the dimensions of the picture.

   $(W)(W+4) + 200 = (W + 4)(W + 8)$
   $W^2 + 4W + 200 = W^2 + 12W + 32$
   $168 = 8W$
   $W = 21$ **21 by 25**

5. A garden 18 by 20 has a border of size 6 all around. Find the ratio of areas, garden to border.

   $30 \times 32 - 18 \times 20 =$ $\dfrac{360}{600}$ $\dfrac{3}{5}$
   $960 - 360 =$
   600

10. A lawn with area 192 is triple as wide as it is long. Find the width of the lawn's border if the ratio of the areas, border to total, is 5:8.

   $L(3L) = 192$ | $\dfrac{B}{B + 192} = \dfrac{5}{8}$ | total = 512
   $(L)(L) = 64$ | | $= 16 \times 32$
   $L=8$ $W=24$ | $3B = 192 \times 5$ |
   | $B = 64 \times 5 = 320$ | **4**

Area 4

Find the area in square units of the rectangle bounded by:	*Find the area in square units of the triangle bounded by both axes and the line:*
1. y = –3, x = –1, x = 9, and y = –11 W = 10 H = 8 A = **80**	9. y = –x + 4 (0, 4) (–4, 0) A = 4 × 4 / 2 = **8**
2. x = 4, x = –5, y = 9, and y = –6 W = 9 H = 15 A = **135**	10. y = –2x – 6 (0, –6) (–3, 0) A = 3 × 6 / 2 = **9**
3. x = –6, y = –2, x = 5, and y = –15 W = 11 H = 13 A = **143**	11. y = 2x – 8 (0, –8) (4, 0) A = 8 × 4 / 2 = **16**
4. y = –2, y = 13, x = –17, and x = –11 W = 6 H = 15 A = **90**	12. y = 3x + 3 (0, 3) (–1, 0) A = 1 × 3 / 2 = **1.5**
5. x = 9, x = –3, y = 11, and y = –2 W = 12 H = 13 A = **156**	13. y = –x – 9 (0, –9) (–9, 0) A = 9 × 9 / 2 = **40.5**
6. x = –12, y = 8, x = –2, and y = –11 W = 10 H = 19 A = **190**	14. y = 4x + 8 (0, 8) (–2, 0) A = 2 × 8 / 2 = **8**
7. y = 1, y = –15, x = –19, and x = 11 W = 30 H = 16 A = **480**	15. y = 2x – 5 (0, –5) (5/2, 0) A = (5/2) × 5 / 2 = 25/4 = **6.25**
8. y = 6, x = –10, x = 4, and y = –8 W = 14 H = 14 A = **196**	16. y = –6x – 12 (0, –12) (–2, 0) A = 2 × 12 / 2 = **12**

Area 5

Find the area in square units using Pick's Formula: A = (B/2) + I − 1.
B = # border points; I = # interior points.

1.
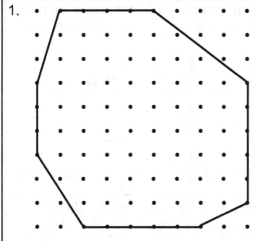

21/2 + 60 − 1 = **69.5**

4.
24/2 + 53 − 1 = **64**

2.
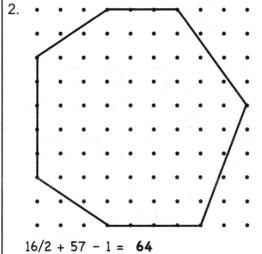

16/2 + 57 − 1 = **64**

5.
19/2 + 56 − 1 = **64.5**

3.
24/2 + 52 − 1 = **63**

6.
6/2 + 50 − 1 = **52**

Area 6

Find the specified area in square units.

1. Find the area of the parallelogram bounded by $x = 0$, $x = 4$, $y = 2x - 4$, and $y = 2x + 5$.

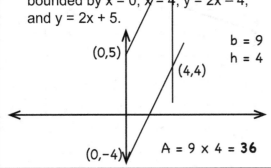

(4,13)

(0,5)

(4,4)

b = 9
h = 4

(0,−4)

A = 9 × 4 = **36**

2. Find the area of the parallelogram bounded by $y = 5$, $y = -6$, $y = 3x - 5$, and $y = 3x + 7$.

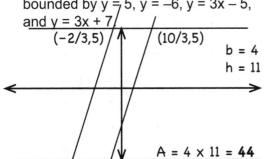

(−2/3,5)   (10/3,5)

b = 4
h = 11

A = 4 × 11 = **44**

3. Find the area of the parallelogram bounded by $y = 5$, $y = x$, $y = -4$, and $y = x + 5$.

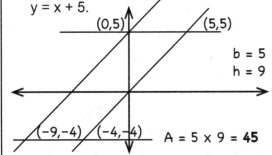

(0,5)   (5,5)

b = 5
h = 9

(−9,−4) (−4,−4)

A = 5 × 9 = **45**

4. Find the area of the parallelogram bounded by $x = -2$, $x = 3$, $y = 2x - 2$, and $y = 2x + 3$.

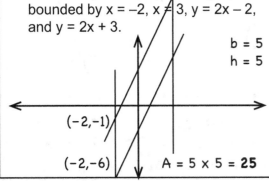

b = 5
h = 5

(−2,−1)

(−2,−6)

A = 5 × 5 = **25**

5. Find the area of the trapezoid bounded by $x = 6$, $y = 4$, $y = 2x$, and the x-axis.

B = 6
b = 4
M = 5
h = 4

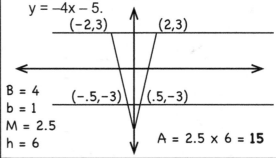

(2,4)   (6,4)

(0,0)   (6,0)

A = 5 × 4 = **20**

6. Find the area of the trapezoid bounded by $x = 6$, $y = 8$, $y = -2x + 3$, and $x = -2$.

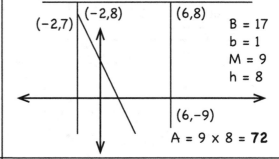

(−2,7) (−2,8)   (6,8)

B = 17
b = 1
M = 9
h = 8

(6,−9)

A = 9 × 8 = **72**

7. Find the area of the trapezoid bounded by $y = 3$, $y = -3$, $y = 4x - 5$, and $y = -4x - 5$.

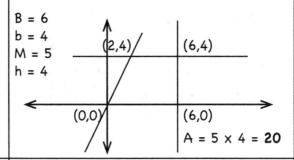

(−2,3)   (2,3)

B = 4
b = 1
M = 2.5
h = 6

(−.5,−3) (.5,−3)

A = 2.5 × 6 = **15**

8. Find the area of the trapezoid bounded by $x = 4$, $x = -2$, $y = (1/2)x - 4$, and $y = -x + 4$.

(−2,6)

(4,0)

(4,−2)

B = 11
b = 2
M = 6.5
h = 6

(−2,−5)

A = 6.5 × 6 = **39**

Averages 1

Find the average by mental math. When summing, group to make multiples of 10.

1. 5, 825, 75, 95, 105, 155 $\dfrac{1260}{6} = \mathbf{210}$	12. 45, 15, 75, 95, 65, 45, 55, 85 $\dfrac{480}{8} = \mathbf{60}$
2. 99, 53, 101, 50, 147 $\dfrac{450}{5} = \mathbf{90}$	13. 32.4, 11.8, 88, 48.6, 19.2 $\dfrac{200}{5} = \mathbf{40}$
3. 56, 420, 91, 60, 39, 60, 44 $\dfrac{770}{7} = \mathbf{110}$	14. 161, 87, 113, 158, 142, 39, 70 $\dfrac{770}{7} = \mathbf{110}$
4. 30, 19, 19, 11, 31, 51, 89, 70 $\dfrac{320}{8} = \mathbf{40}$	15. 13.7, 12.1, 38.9, 37.3, 15.5, 2.5 $\dfrac{120}{6} = \mathbf{20}$
5. 75, 76, 75, 76, 48 $\dfrac{350}{5} = \mathbf{70}$	16. 40, 30, 20, 10, 60, 70, 80, 90 $\dfrac{400}{8} = \mathbf{50}$
6. 56, 34, 72, 38, 91, 29, 170 $\dfrac{490}{7} = \mathbf{70}$	17. 14, 99, 86, 11, 105, 75, 30 $\dfrac{420}{7} = \mathbf{60}$
7. 109, 850, 391, 200, 150, 100 $\dfrac{1800}{6} = \mathbf{300}$	18. 38, 99, 51, 77, 23, 72 $\dfrac{360}{6} = \mathbf{60}$
8. 100, 2000, 450, 950, 300, 400 $\dfrac{4200}{6} = \mathbf{700}$	19. 14, 19, 96, 21, 88, 32, 80 $\dfrac{350}{7} = \mathbf{50}$
9. 820, 60, 20, 900, 200 $\dfrac{2000}{5} = \mathbf{400}$	20. 115, 698, 173, 227, 302, 285 $\dfrac{1800}{6} = \mathbf{300}$
10. 13, 87, 54, 46, 72, 28 $\dfrac{300}{6} = \mathbf{50}$	21. 50, 75, 325, 150, 450, 225, 125 $\dfrac{1400}{7} = \mathbf{200}$
11. 90, 125, 300, 110, 375, 800 $\dfrac{1800}{6} = \mathbf{300}$	22. 17.8, 28, 49.2, 43.5, 22.5, 61 $\dfrac{222}{6} = \mathbf{37}$

20

Averages 2

Find the average using the arithmetic sequence method.

1.	12, 14, 16, 18, 20		12.	102, 104, 106, 108, 110, 112
	n odd; avg is the middle — **16**			n even; avg of two in middle — **107**
2.	1, 7, 13, 19, 25, 31, 37, 43		13.	1.4, 3.2, 5.0, 6.8
	d = 6; 6/2 = 3; 19 + 3 — **22**			8.2/2 — **4.1**
3.	91, 88, 85, 97, 94		14.	25, 32, 39, 46, 53, . . . , 81
	85, 88, 91, 94, 97 — **91**			60, 67, 74 n=9; avg is 5th — **53**
4.	12, 15, 18, 21, . . . , 36		15.	84, 92, 86, 90, 82, 88
	24, 27, 30, 33 n=9; avg is 5th — **24**			82, 84, 86, 88, 90, 92 — **87**
5.	4, 9, 14, 19, 24, 29		16.	979, 981, 983, 985, 987, 989
	d = 5; 5/2 = 2.5; 14 + 2.5 — **16.5**			n even; avg of two in middle — **984**
6.	1, 3, 5, 7, 9, 11, 13, 15, 17, 19		17.	3.5, 6.5, 9.5, 12.5, 15.5, 18.5
	n even; avg of two in middle — **10**			d = 3; 3/2 = 1.5; 9.5 + 1.5 — **11**
7.	117, 128, 139, 150		18.	17, 43, 69, 82, 30, 56
	d = 11; 11/2 = 5.5; 128 + 5.5 — **133.5**			17, 30, 43, 56, 69, 82 — **49.5**
8.	20, 22, 24, 26, . . . , 38		19.	150, 300, 50, 200, 100, 350, 250
	2x10, . . . , 2x19; n = 10 — **29**			50, 100, 150, 200, 250, 300, 350 — **200**
9.	26, 10, 6, 22, 18, 14		20.	4, 9, 14, . . . , 44
	6, 10, 14, 18, 22, 26 — **16**			4x1, . . . , 4x11; n = 11 — **24**
10.	20, 5, 25, 15, 10, 30		21.	40, 90, 60, 50, 30, 70, 80, 20
	5, 10, 15, 20, 25, 30 — **17.5**			20, 30, 40, 50, 60, 70, 80, 90 — **55**
11.	29, 35, 41, 47, 53, . . . , 77		22.	6.1, 9.1, 12.1, 15.1, 18.1, 21.1
	59, 65, 71 n=9; avg is 5th — **53**			d = 3; 3/2 = 1.5; 12.1 + 1.5 — **13.6**

MAVA Math: Enhanced Skills Solutions   Copyright © 2015 Marla Weiss

Averages 3

Find the average using the rightmost digits method.

Add positives above and negatives below the "keep" number. The keep number may vary.

1. 821, 826, 829, 827, 825, 822 Keep 820. 30 ÷ 6 = 5. 820 + 5 **825**	12. 11, 17, 10, 14, 13, 18, 19, 17, 16 Keep 10. 45 ÷ 9 = 5. 10 + 5 **15**
2. 621, 632, 602, 614 Keep 600. 69 ÷ 4 = 17.25 **617.25**	13. 34, 31, 35, 33, 33, 38 Keep 30. 24 ÷ 6 = 4. 30 + 4 **34**
3. 91, 97, 94, 98, 95 Keep 90. 25 ÷ 5 = 5. 90 + 5 **95**	14. 15, 19, 12, 13, 17 Keep 10. 26 ÷ 5 = 5.2. 10 + 5.2 **15.2**
4. 738, 738, 738, 738, 738, 750 Keep 740. 0 ÷ 6 = 0. 740 + 0 **740**	15. 3401, 3407, 3409, 3410, 3405 Keep 3400. 32 ÷ 5 = 6.4 **3406.4**
5. 29,325; 29,331; 29,326; 29,322 Keep 29,300. 104 ÷ 4 = 26 **29,326**	16. 103, 108, 107, 102, 110 Keep 100. 30 ÷ 5 = 6. 100 + 6 **106**
6. 1202, 1209, 1217, 1211 Keep 1200. 39 ÷ 4 = 9.75 **1209.75**	17. 5407, 5406, 5411, 5404 Keep 5400. 28 ÷ 4 = 7 **5407**
7. 780, 780, 786, 784, 785 Keep 780. 15 ÷ 5 = 3. 780 + 3 **783**	18. 1363, 1369, 1357, 1373, 1367, 1361 Keep 1360. 30 ÷ 6 = 5 **1365**
8. 80, 80, 81, 80 Keep 80. 1 ÷ 4 = .25. 80 + .25 **80.25**	19. 731, 735, 739, 731 Keep 730. 16 ÷ 4 = 4. 730 + 4 **734**
9. 521, 567, 555, 525 Keep 500. 168 ÷ 4 = 42 **542**	20. 1212, 1234, 1240, 1251, 1231 Keep 1230. 18 ÷ 5 = 3.6 **1233.6**
10. 29, 25, 24, 26 Keep 20. 24 ÷ 4 = 6. 20 + 6 **26**	21. 414, 412, 419, 411 Keep 410. 16 ÷ 4 = 4. 410 + 4 **414**
11. 818, 820, 819, 815, 825, 835 Keep 820. 12 ÷ 6 = 2. 820 + 2 **822**	22. 327, 335, 322, 321, 320 Keep 320. 25 ÷ 5 = 5. 320 + 5 **325**

Averages 4

Find the sum of the numbers.   The average times n is the sum.

1. Four numbers average to 14.	**56**	17. Six numbers average to 25.	**150**
2. Nine numbers average to 8.	**72**	18. Thirty numbers average to 13.	**390**
3. Twenty numbers average to 35.	**700**	19. Thirty numbers average to 80.	**2400**
4. Twelve numbers average to 12.	**144**	20. Eleven numbers average to 75.	**825**
5. Forty numbers average to 15.	**600**	21. Eight numbers average to 61.	**488**
6. Eight numbers average to 70.	**560**	22. Fifty numbers average to 12.	**600**
7. Twenty-five numbers average to 40.	**1000**	23. Eleven numbers average to 60.	**660**
8. Seven numbers average to 12.	**84**	24. Five numbers average to 71.	**355**
9. Five numbers average to 41.	**205**	25. Twenty numbers average to 62.	**1240**
10. Fifty numbers average to 31.	**1550**	26. Seven numbers average to 111.	**777**
11. Nine numbers average to 81.	**729**	27. Twelve numbers average to 110.	**1320**
12. Eleven numbers average to 9.	**99**	28. Fifteen numbers average to 40.	**600**
13. Four numbers average to 70.	**280**	29. Nine numbers average to 31.	**279**
14. Eighty numbers average to 51.	**4080**	30. Twenty numbers average to 41.	**820**
15. Thirty numbers average to 32.	**960**	31. Sixteen numbers average to 100.	**1600**
16. Fifteen numbers average to 11.	**165**	32. Ten numbers average to 450.	**4500**

MAVA Math: Enhanced Skills Solutions   Copyright © 2015 Marla Weiss

Averages 5

Answer as indicated using the umbrella method.

The word "umbrella" derives from the arc drawn.

---

1. Three numbers average to 26. If two of the numbers are 20 and 22, find the third number.

sum = 26 x 3 = 78

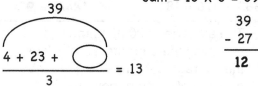

$\dfrac{20 + 22 + \bigcirc}{3} = 26$

```
   78
 - 42
 ----
   36
```

---

6. Two tests average to 85. What score on a third test would raise the average to 90?

sum = 90 x 3 = 270
85 x 2 = 170

```
   270
 - 170
 -----
   100
```

---

2. The average of four test scores is 78. If three of the scores are 56, 70, and 92, find the fourth score.

sum = 78 x 4 = 312
56 + 70 + 92 = 218

```
   312
 - 218
 -----
    94
```

---

7. Four test grades average to 78. What score on the next test would raise the average to 80?

sum = 80 x 5 = 400
78 x 4 = 312

```
   400
 - 312
 -----
    88
```

---

3. Three numbers average to 13. If two of the numbers are 4 and 23, find the third number.

sum = 13 x 3 = 39

$\dfrac{4 + 23 + \bigcirc}{3} = 13$

```
   39
 - 27
 ----
   12
```

---

8. The average of four test scores is 82. Three of the scores are 70, 85, and 90. Find the fourth score.

sum = 82 x 4 = 328
70 + 85 + 90 = 245

```
   328
 - 245
 -----
    83
```

---

4. The average of 3 numbers is 67. Two of the numbers are 56 and 91. Find the 3rd number.

sum = 67 x 3 = 201

$\dfrac{56 + 91 + \bigcirc}{3} = 67$

```
   201
 - 147
 -----
    54
```

---

9. Four numbers average to 85. If three of the numbers are 70, 72, and 88, find the fourth number.

sum = 85 x 4 = 340
70 + 72 + 88 = 230

```
   340
 - 230
 -----
   110
```

---

5. If five test grades are 55, 56, 57, 58, and 59, what score on another test would raise the average to 60?

sum = 60 x 6 = 360
55 + 56 + 57 + 58 + 59 = 285

```
   360
 - 285
 -----
    75
```

---

10. Test grades are 58, 60, 55, 50, and 55. What grade on the next test raises the average to 60?

sum = 60 x 6 = 360
58 + 60 + 55 + 50 + 55 = 278

```
   360
 - 278
 -----
    82
```

Averages 6

Answer as indicated.

1. The average of 5 numbers is 85. Eliminating the least, the average is 91. Find the least number.

   sum = 85 x 5 = 425
   new sum = 91 x 4 = 364
   425 − 364 = **61**

2. The average of 8 numbers is 12. Subtract 3 from 2 of the numbers. What is the new average?

   sum = 12 x 8 = 96
   new sum = 96 − 6 = 90
   new average = 90/8 = **11.25**

3. The average of 7 numbers is 105. Eliminating the least, the average is 120. Find the least number.

   sum = 105 x 7 = 735
   new sum = 120 x 6 = 720
   735 − 720 = **15**

4. The average of 5 tests is 80. If the lowest grade is dropped, 86 is the new average. Find the lowest grade.

   sum = 80 x 5 = 400
   new sum = 86 x 4 = 344
   400 − 344 = **56**

5. The average of 9 numbers is 44. Subtract 6 from 3 of the numbers. What is the new average?

   sum = 44 x 9 = 396
   new sum = 396 − 18 = 378
   new average = 378/9 = **42**

6. The average of 6 numbers is 312. Eliminating the greatest, the average is 280. Find the greatest number.

   sum = 312 x 6 = 1872
   new sum = 280 x 5 = 1400
   1872 − 1400 = **472**

7. The average of 10 numbers is 84. After removing the 2 greatest and 2 least, the average is 72. Find the average of those removed.

   sum = 84 x 10 = 840
   new sum = 72 x 6 = 432
   840 − 432 = 408   408/4 = **102**

8. The average of 7 tests is 68. After dropping the 2 lowest grades, the average is 78. Find the average of the 2 grades dropped.

   sum = 68 x 7 = 476
   new sum = 78 x 5 = 390
   476 − 390 = 86   86/2 = **43**

9. The average of 4 test scores was 90. After removing two identical scores, the average was 88. What was the score removed twice?

   sum = 90 x 4 = 360
   new sum = 88 x 2 = 176
   360 − 176 = 184   184/2 = **92**

10. The average of 20 numbers is 90. By removing the 3 greatest and 3 least, the average becomes 60. Find the average of those removed.

    sum = 90 x 20 = 1800
    new sum = 60 x 14 = 840
    1800 − 840 = 960   960/6 = **160**

11. The average of 6 tests (range 0–100) is 70. If the first 4 tests average to 60, what is the lowest possible score on the last 2 tests?

    sum = 70 x 6 = 420
    420 = 60 x 4 + max + min
    420 = 240 + 100 + **80**

12. The average of 8 tests (range 0–100) is 72. If the first 6 tests average to 64, what is the lowest possible score on the last 2 tests?

    sum = 72 x 8 = 576
    576 = 64 x 6 + max + min
    576 = 384 + 100 + **92**

Bases 1

Count by writing the specified numerals.

From right to left, place values are 1s, the base, the base squared, the base cubed, etc.

By convention base eleven uses T for the digit ten instead of A like base sixteen.

1. zero to ten in base four

0, 1, 2, 3, 10, 11, 12, 13, 20, 21, 22

2. three to thirteen in base five

3, 4, 10, 11, 12, 13, 14, 20, 21, 22, 23

3. five to fifteen in base seven

5, 6, 10, 11, 12, 13, 14, 15, 16, 20, 21

4. ten to nineteen in base six

14, 15, 20, 21, 22, 23, 24, 25, 30, 31

5. fifteen to twenty-two in base eight

17, 20, 21, 22, 23, 24, 25, 26

6. twelve to twenty-one in base sixteen

C, D, E, F, 10, 11, 12, 13, 14, 15

7. eight to nineteen in base eleven

8, 9, T, 10, 11, 12, 13, 14, 15, 16, 17, 18

8. twelve to seventeen in base two

1100, 1101, 1110, 1111, 10000, 10001

9. eight to nineteen in base nine

8, 10, 11, 12, 13, 14, 15, 16, 17, 18, 20, 21

10. nine to twenty in base twelve

9, T, E, 10, 11, 12, 13, 14, 15, 16, 17, 18

11. three to twelve in base three

10, 11, 12, 20, 21, 22, 100, 101, 102, 110

12. thirty-two to forty in base six

52, 53, 54, 55, 100, 101, 102, 103, 104

13. twenty-two to thirty in base twelve

1T, 1E, 20, 21, 22, 23, 24, 25, 26

14. four to eleven in base two

100, 101, 110, 111, 1000, 1001, 1010, 1011

15. fifty-two to fifty-nine in base seven

103, 104, 105, 106, 110, 111, 112, 113

16. twenty to twenty-six in base five

40, 41, 42, 43, 44, 100, 101

17. thirty to thirty-eight in base eleven

28, 29, 2T, 30, 31, 32, 33, 34, 35

18. ten to eighteen in base three

101, 102, 110, 111, 112, 120, 121, 122, 200

19. eleven to nineteen in base four

23, 30, 31, 32, 33, 100, 101, 102, 103

20. thirty to forty in base sixteen

1E, 1F, 20, 21, 22, 23, 24, 25, 26, 27, 28

21. sixty-two to seventy in base eight

76, 77, 100, 101, 102, 103, 104, 105, 106

22. fifty-five to sixty-four in base nine

61, 62, 63, 64, 65, 66, 67, 68, 70, 71

Bases 2

Write the numeral one before.		*Write the numeral one after.*	
1. E_{twelve}	T_{twelve}	17. 3133_{four}	3200_{four}
2. 4000_{eight}	3777_{eight}	18. 589_{eleven}	$58T_{eleven}$
3. $13_{thirteen}$	$12_{thirteen}$	19. 366_{seven}	400_{seven}
4. $39C_{sixteen}$	$39B_{sixteen}$	20. 199_{twelve}	$19T_{twelve}$
5. 101100_{two}	101011_{two}	21. $13_{thirteen}$	$14_{thirteen}$
6. $3T0_{eleven}$	$39T_{eleven}$	22. $1E_{twelve}$	20_{twelve}
7. 333_{four}	332_{four}	23. 555_{six}	1000_{six}
8. 560_{seven}	556_{seven}	24. 477_{eight}	500_{eight}
9. 1022_{three}	1021_{three}	25. TTT_{eleven}	1000_{eleven}
10. 800_{nine}	788_{nine}	26. $2B5B_{sixteen}$	$2B5C_{sixteen}$
11. 20000_{four}	13333_{four}	27. 212_{three}	220_{three}
12. $635A_{fourteen}$	$6359_{fourteen}$	28. 10111_{two}	11000_{two}
13. 1300_{five}	1244_{five}	29. $43EF_{sixteen}$	$43F0_{sixteen}$
14. ET_{twelve}	$E9_{twelve}$	30. $7FF_{sixteen}$	$800_{sixteen}$
15. 310_{eight}	307_{eight}	31. $E_{fifteen}$	$10_{fifteen}$
16. $9TT_{eleven}$	$9T9_{eleven}$	32. $5CD_{fourteen}$	$5D0_{fourteen}$

Bases 3

Convert to base ten.

1. 343_{five}	3 x 25 =	75	
	4 x 5 =	20	
	3 x 1 =	3	
		98	

8. 354_{seven}	3 x 49 =	147
	5 x 7 =	35
	4 x 1 =	4
		186

2. 2385_{nine}	2 x 729 =	1458
	3 x 81 =	243
	8 x 9 =	72
	5 x 1 =	5
		1778

9. 1011011_{two}	1 x 64 =	64
	1 x 16 =	16
	1 x 8 =	8
	1 x 2 =	2
	1 x 1 =	1
		91

3. $A2B_{sixteen}$	10 x 256 =	2560
	2 x 16 =	32
	11 x 1 =	11
		2603

10. $T3E_{twelve}$	10 x 144 =	1440
	3 x 12 =	36
	11 x 1 =	11
		1487

4. 22122_{three}	2 x 81 =	162
	2 x 27 =	54
	1 x 9 =	9
	2 x 3 =	6
	2 x 1 =	2
		233

11. $3T8_{eleven}$	3 x 121 =	363
	10 x 11 =	110
	8 x 1 =	8
		481

5. 562_{eight}	5 x 64 =	320
	6 x 8 =	48
	2 x 1 =	2
		370

12. 2451_{six}	2 x 216	432
	4 x 36 =	144
	5 x 6 =	30
	1 x 1 =	1
		607

6. $13A_{fifteen}$	1 x 225 =	225
	3 x 15 =	45
	10 x 1 =	10
		280

13. $25B_{thirteen}$	2 x 169 =	338
	5 x 13 =	65
	11 x 1 =	11
		414

7. 2032_{four}	2 x 64 =	128
	0 x 16 =	0
	3 x 4 =	12
	2 x 1 =	2
		142

14. $1AC_{fourteen}$	1 x 196 =	196
	10 x 14 =	140
	12 x 1 =	12
		348

Bases 4

Convert from base ten to the specified base.

1. $190 = ?_{five}$ **1230**$_{five}$

1	2	3	0
125s	25s	5s	1s

```
  190
 -125
 ----
   65
  -50
 ----
   15
```

8. $762 = ?_{four}$ **23322**$_{four}$

2	3	3	2	2
256s	64s	16s	4s	1s

```
  762
 -512
 ----
  250
 -192
 ----
   58
  -48
```

2. $668 = ?_{eight}$ **1234**$_{eight}$

1	2	3	4
512s	64s	8s	1s

```
  668
 -512
 ----
  156
 -128
 ----
   28
  -24
```

9. $279 = ?_{eleven}$ **234**$_{eleven}$

2	3	4
121s	11s	1s

```
  279
 -242
 ----
   37
  -33
 ----
    4
```

3. $94 = ?_{three}$ **10111**$_{three}$

1	0	1	1	1
81s	27s	9s	3s	1s

```
   94
  -81
 ----
   13
   -9
 ----
    4
```

10. $1326 = ?_{twelve}$ **926**$_{twelve}$

9	2	6
144s	12s	1s

```
  1326
 -1296
 -----
    30
   -24
 -----
     6
```

4. $600 = ?_{seven}$ **1515**$_{seven}$

1	5	1	5
343s	49s	7s	1s

```
  600
 -343
 ----
  257
 -245
 ----
   12
```

11. $1043 = ?_{fourteen}$ **547**$_{fourteen}$

5	4	7
196s	14s	1s

```
  1043
  -980
 -----
    63
   -56
 -----
     7
```

5. $696 = ?_{nine}$ **853**$_{nine}$

8	5	3
81s	9s	1s

```
  696
 -648
 ----
   48
  -45
 ----
    3
```

12. $267 = ?_{six}$ **1123**$_{six}$

1	1	2	3
216s	36s	6s	1s

```
  267
 -216
 ----
   51
  -36
 ----
   15
```

6. $1234 = ?_{five}$ **14414**$_{five}$

1	4	4	1	4
625s	125s	25s	5s	1s

```
  1234
  -625
 -----
   609
  -500
 -----
   109
  -100
```

13. $729 = ?_{fifteen}$ **339**$_{fifteen}$

3	3	9
225s	15s	1s

```
  729
 -675
 ----
   54
  -45
 ----
    9
```

7. $41 = ?_{two}$ **101001**$_{two}$

1	0	1	0	0	1
32s	16s	8s	4s	2s	1s

```
   41
  -32
 ----
    9
   -8
 ----
    1
```

14. $1952 = ?_{sixteen}$ **7A0**$_{sixteen}$

7	A	0
256s	16s	1s

```
  1952
 -1792
 -----
   160
  -160
 -----
     0
```

Bases 5

Find the missing base b.

1. $102_b = 291_{ten}$	$100_b = 289$ b = **seventeen**	12. $101_b = 145_{ten}$ $100_b = 144$ b = **twelve**
2. $11111_b = 31_{ten}$	$11110_b = 30$ b = **two**	13. $403_b = 103_{ten}$ $400_b = 100$ b = **five**
3. $222_b = 114_{ten}$	$220_b = 112$ $110_b = 56$ b = **seven**	14. $111_b = 13_{ten}$ $110_b = 12$ b = **three**
4. $117_b = 79_{ten}$	$110_b = 72$ b = **eight**	15. $505_b = 325_{ten}$ $500_b = 320$ b = **eight**
5. $204_b = 166_{ten}$	$200_b = 162$ $100_b = 81$ b = **nine**	16. $1001_b = 126_{ten}$ $1000_b = 125$ b = **five**
6. $105_b = 41_{ten}$	$100_b = 36$ b = **six**	17. $1005_b = 1336_{ten}$ $1000_b = 1331$ b = **eleven**
7. $103_b = 19_{ten}$	$100_b = 16$ b = **four**	18. $111_b = 133_{ten}$ $110_b = 132$ b = **eleven**
8. $907_b = 1096_{ten}$	$900_b = 1089$ $100_b = 121$ b = **eleven**	19. $22_b = 34_{ten}$ $20_b = 32$ $10_b = 16$ b = **sixteen**
9. $333_b = 63_{ten}$	$330_b = 60$ b = **four**	20. $10002_b = 83_{ten}$ $10000_b = 81$ b = **three**
10. $402_b = 786_{ten}$	$400_b = 784$ $100_b = 196$ b = **fourteen**	21. $1004_b = 220_{ten}$ $1000_b = 216$ b = **six**
11. $1003_b = 732_{ten}$	$1000_b = 729$ b = **nine**	22. $111_b = 31_{ten}$ $110_b = 30$ b = **five**

MAVA Math: Enhanced Skills Solutions   Copyright © 2015 Marla Weiss

Bases 6

Convert among bases two, eight, and sixteen.

---

1. $111011_{two} = ?_{eight}$

111　011　　　　　　　73_{eight}

12. $111011_{two} = ?_{sixteen}$

11　1011　　　　　　　$3B_{sixteen}$

---

2. $777_{eight} = ?_{two}$

7　7　7　　　　　　　111111111_{two}

13. $10000001_{two} = ?_{sixteen}$

1000　0001　　　　　　$81_{sixteen}$

---

3. $567_{eight} = ?_{sixteen}$

101　110　111
1　0111　0111　　　　$177_{sixteen}$

14. $FF_{sixteen} = ?_{two}$

F　F　　　　　　　11111111_{two}

---

4. $314_{eight} = ?_{sixteen}$

11　001　100
1100　1100　　　　　$CC_{sixteen}$

15. $4AB_{sixteen} = ?_{eight}$

100　1010　1011
10　010　101　011　　2253_{eight}

---

5. $1234_{sixteen} = ?_{eight}$

1　0010　0011　0100
1　001　000　110　100　11064_{eight}

16. $C32_{sixteen} = ?_{eight}$

1100　0011　0010
110　000　110　010　6062_{eight}

---

6. $5732_{eight} = ?_{sixteen}$

101　111　011　010
1011　1101　1010　　$BDA_{sixteen}$

17. $4521_{eight} = ?_{sixteen}$

100　101　010　001
1001　0101　0001　　$951_{sixteen}$

---

7. $1011101101_{two} = ?_{sixteen}$

10　1110　1101　　　$2ED_{sixteen}$

18. $D7A_{sixteen} = ?_{eight}$

1101　0111　1010
110　101　111　010　6572_{eight}

---

8. $1010101_{two} = ?_{eight}$

1　010　101　　　　　125_{eight}

19. $10101011_{two} = ?_{eight}$

10　101　011　　　　　253_{eight}

---

9. $1010101_{two} = ?_{sixteen}$

101　0101　　　　　　$55_{sixteen}$

20. $10101011_{two} = ?_{sixteen}$

1010　1011　　　　　　$AB_{sixteen}$

---

10. $ABC_{sixteen} = ?_{two}$

1010　1011　1100　　101010111100_{two}

21. $2176_{eight} = ?_{two}$

10　001　111　110　10001111110_{two}

---

11. $9DC_{sixteen} = ?_{eight}$

1001　1101　1100
100　111　011　100　4734_{eight}

22. $EA1_{sixteen} = ?_{two}$

1110　1010　0001　111010100001_{two}

Bases 7

Operate in the given base.

1.
$$2135_{six}$$
$$+ 1424_{six}$$
$$\mathbf{4003_{six}}$$

8.
$$5273_{eight}$$
$$+ 4365_{eight}$$
$$\mathbf{11660_{eight}}$$

15.
$$21012_{three}$$
$$+ 22102_{three}$$
$$\mathbf{120121_{three}}$$

22.
$$3462_{seven}$$
$$+ 2536_{seven}$$
$$\mathbf{6331_{seven}}$$

2.
$$34_{five}$$
$$\times 12_{five}$$
$$123$$
$$340$$
$$\mathbf{1013}\ \mathbf{five}$$

9.
$$22_{three}$$
$$\times 22_{three}$$
$$121$$
$$1210$$
$$\mathbf{2101}\ \mathbf{three}$$

16.
$$133_{four}$$
$$\times 23_{four}$$
$$1131$$
$$3320$$
$$\mathbf{11111}\ \mathbf{four}$$

23.
$$101_{two}$$
$$\times 11_{two}$$
$$101$$
$$1010$$
$$\mathbf{1111}\ \mathbf{two}$$

3.
$$8341_{nine}$$
$$- 5278_{nine}$$
$$\mathbf{3052_{nine}}$$

10.
$$8375_{eleven}$$
$$- 38T6_{eleven}$$
$$\mathbf{457T_{eleven}}$$

17.
$$3254_{eight}$$
$$- 1657_{eight}$$
$$\mathbf{1375_{eight}}$$

24.
$$2210_{three}$$
$$- 1012_{three}$$
$$\mathbf{1121_{three}}$$

4.
$$1001011_{two}$$
$$+ 10111_{two}$$
$$\mathbf{1100010_{two}}$$

11.
$$7235_{twelve}$$
$$+ 8967_{twelve}$$
$$\mathbf{13ET0_{twelve}}$$

18.
$$10332_{four}$$
$$+ 31023_{four}$$
$$\mathbf{102021_{four}}$$

25.
$$12343_{five}$$
$$+ 23342_{five}$$
$$\mathbf{41240_{five}}$$

5.
$$53_{seven}$$
$$\times 34_{seven}$$
$$305$$
$$2220$$
$$\mathbf{2525}\ \mathbf{seven}$$

12.
$$45_{six}$$
$$\times 23_{six}$$
$$223$$
$$1340$$
$$\mathbf{2003}\ \mathbf{six}$$

19.
$$475_{nine}$$
$$\times 23_{nine}$$
$$1546$$
$$10610$$
$$\mathbf{12256}\ \mathbf{nine}$$

26.
$$394_{eleven}$$
$$\times 53_{eleven}$$
$$1061$$
$$18290$$
$$\mathbf{19341}\ \mathbf{eleven}$$

6.
$$2131_{four}$$
$$- 1333_{four}$$
$$\mathbf{132_{four}}$$

13.
$$3245_{seven}$$
$$- 1266_{seven}$$
$$\mathbf{1646_{seven}}$$

20.
$$418B_{sixteen}$$
$$- 23AD_{sixteen}$$
$$\mathbf{1DDE_{sixteen}}$$

27.
$$7654_{eight}$$
$$- 2356_{eight}$$
$$\mathbf{5276_{eight}}$$

7.
$$4323_{five}$$
$$+ 1443_{five}$$
$$\mathbf{11321_{five}}$$

14.
$$2465_{nine}$$
$$+ 8574_{nine}$$
$$\mathbf{12150_{nine}}$$

21.
$$56T9_{eleven}$$
$$+ 4T78_{eleven}$$
$$\mathbf{T676_{eleven}}$$

28.
$$1543_{six}$$
$$+ 3354_{six}$$
$$\mathbf{5341_{six}}$$

Bases 8

Similar concept to Bases 6.

Convert between the bases by using the given as one less than 1 in the next place.	*Convert between the bases without passing through base ten.*
1. $22222_{three} = ?_{ten}$ 22222 is 1 less than 243 base ten **242**	9. $475_{nine} = ?_{three}$ $9 = 3^2$ 4 = 11 7 = 21 5 = 12 **112112**
2. $55_{six} = ?_{two}$ 55 is 1 less than 36 base ten 35 = 32 + 3 **100011**	10. $149_{twenty\text{-}five} = ?_{five}$ $25 = 5^2$ 1 = 1 4 = 04 9 = 14 **10414**
3. $1111111_{two} = ?_{five}$ 1111111 is 1 less than 128 base ten 127 = 125 + 2 **1002**	11. $320112_{four} = ?_{sixteen}$ $16 = 4^2$ 32 = 14 01 = 1 12 = 6 **E16**
4. $888_{nine} = ?_{ten}$ 888 is 1 less than 729 base ten **728**	12. $9A52_{sixteen} = ?_{four}$ $16 = 4^2$ 9 = 21 A = 22 5 = 11 2 = 02 **21221102**
5. $666_{seven} = ?_{ten}$ 666 is 1 less than 343 base ten **342**	13. $1211201_{three} = ?_{nine}$ $9 = 3^2$ 1 = 1 21 = 7 12 = 5 01 = 1 **1751**
6. $4444_{five} = ?_{ten}$ 4444 is 1 less than 625 base ten **624**	14. $2863_{nine} = ?_{three}$ $9 = 3^2$ 2 = 2 8 = 22 6 = 20 3 = 10 **2222010**
7. $TTT_{eleven} = ?_{ten}$ TTT is 1 less than 1331 base ten **1330**	15. $233211_{five} = ?_{twenty\text{-}five}$ $25 = 5^2$ 23 = 13 32 = 17 11 = 6 **DH6**
8. $EE_{twelve} = ?_{five}$ EE is 1 less than 144 base ten 143 = 125 + 15 + 3 **1033**	16. $A98_{twenty\text{-}seven} = ?_{three}$ $27 = 3^2$ A = 101 9 = 100 8 = 022 **101100022**

Boundary 1

Answer as indicated.

1. $-3 \leq x \leq 12$
   $4 \leq y \leq 15$
   Find the least value of xy.

   $(-3)(15) = $ **−45**

8. $-3 \leq x \leq 15$
   $-15 \leq y \leq -2$
   Find the greatest value of x − y.

   $15 - -15 = $ **30**

2. $-10 \leq x \leq 30$
   $-5 \leq y \leq 15$
   Find the greatest value of x − y.

   $30 - -5 = $ **35**

9. $2 \leq x \leq 25$
   $-5 \leq y \leq 30$
   Find the least value of xy.

   $(25)(-5) = $ **−125**

3. $-10 \leq x \leq 11$
   $-50 \leq y \leq -5$
   A = greatest value of x − y
   B = least value of x + y
   Find A − B. $A = 11 - -50 = 61$
   $B = -10 + -50 = -60$ **121**

10. $-11 \leq x \leq 11$
    $-12 \leq y \leq -10$
    A = greatest value of x − y
    B = least value of xy
    Find A − B. $A = 11 - -12 = 23$
    $B = (11)(-12) = -132$ **155**

4. $6 \leq x \leq 24$
   $-11 \leq y \leq -2$
   A = least value of x − y
   B = greatest value of x + y
   Find A + B. $A = 6 - -2 = 8$
   $B = 24 + -2 = 22$ **30**

11. $-20 \leq x \leq -6$
    $-30 \leq y \leq -15$
    A = greatest value of x − y
    B = least value of x + y
    Find A − B. $A = -6 + -15 = -21$
    $B = -20 + -30 = -50$ **29**

5. $-20 \leq x \leq -8$
   $-15 \leq y \leq -6$
   Find the greatest value of xy.

   $(-20)(-15) = $ **300**

12. $-6 \leq x \leq 9$
    $4 \leq y \leq 20$
    Find the least value of xy.

    $(-6)(20) = $ **−120**

6. $-25 \leq x \leq 20$
   $-50 \leq y \leq -25$
   A = greatest value of x + y
   B = least value of y − x
   Find A + B. $A = 20 + -25 = -5$
   $B = -50 - 20 = -70$ **−75**

13. $-10 \leq x \leq -2$
    $2 \leq y \leq 10$
    A = greatest value of x + y
    B = least value of x − y
    Find AB. $A = -2 + 10 = 8$
    $B = -10 - 10 = -20$ **−160**

7. $-30 \leq x \leq 70$
   $-10 \leq y \leq 20$
   A = least value of x + y
   B = greatest value of x − y
   Find AB. $A = -30 + -10 = -40$
   $B = 70 - -10 = 80$ **−3200**

14. $4 \leq x \leq 35$
    $11 \leq y \leq 20$
    A = greatest value of x − y
    B = least value of x + y
    Find B − A. $A = 35 - 11 = 24$
    $B = 4 + 11 = 15$ **−9**

34

Boundary 2

Answer as indicated.

1.	The perimeter of a rectangle is 30. The width is less than 12. Find the range of values for its length. L + W = 15 W < 12 **3 < L < 15**	8.	The measure of an obtuse angle is 2x − 30. Find the range of values for x. 90 < 2x − 30 < 180 120 < 2x < 210 **60 < x < 105**
2.	The perimeter of a rectangle is 50. The length is greater than 10. Find the range of values for its width. L + W = 25 L > 10 **0 < W < 15**	9.	The perimeter of a rectangle is 90. The length is greater than 15. Find the range of values for its width. L + W = 45 L > 15 **0 < W < 30**
3.	The measure of an obtuse angle is 3x − 15. Find the range of values for x. 90 < 3x − 15 < 180 105 < 3x < 195 **35 < x < 65**	10.	The area of a rectangle is 144. Find its least and greatest perimeter if all sides are whole. LW = 144 least P = 12 x 4 = **48** greatest P = 2(144 + 1) = **290**
4.	The measure of an acute angle is 2x − 12. Find the range of values for x. 0 < 2x − 12 < 90 12 < 2x < 102 **6 < x < 51**	11.	The perimeter of a rectangle is 144. Find its least and greatest area if all sides are whole. L + W = 72 least A = 1 x 71 = **71** greatest A = 36 x 36 = **1296**
5.	The area of a rectangle is 100. Find its least and greatest perimeter if all sides are whole. LW = 100 least P = 10 x 4 = **40** greatest P = 2(100 + 1) = **202**	12.	The perimeter of a rectangle is 70. The width is less than 25. Find the range of values for its length. L + W = 35 W < 25 **10 < L < 35**
6.	The perimeter of a rectangle is 100. Find its least and greatest area if all sides are whole. L + W = 50 least A = 1 x 49 = **49** greatest A = 25 x 25 = **625**	13.	The measure of an acute angle is 4x − 6. Find the range of values for x. 0 < 4x − 6 < 90 6 < 4x < 96 **1.5 < x < 24**
7.	A triangle has two vertices on a circle and one vertex at the center. If the central angle is less than 30°, find the range of values for the other 2 angles of the isosceles triangle. **75 < A < 90**	14.	A triangle has two vertices on a circle and one vertex at the center. If the central angle is less than 56°, find the range of values for the other 2 angles of the isosceles triangle. **62 < A < 90**

MAVA Math: Enhanced Skills Solutions*   Copyright © 2015 Marla Weiss

Calendar 1

Find the day of the week.   SMTWRFS

1. 358 days from today if today is Friday 7 divides 357 Friday + 1　　　　**SAT**	11. the 1st of the month if the 27th is Friday 　　　　27 is Fri 　　　　6 is Fri　　　**SUN**
2. 285 days ago if today is Saturday 7 divides 280 Saturday − 5　　**MON**	12. the 4th of the month if the 29th is Monday 　　　　29 is Mon 　　　　1 is Mon　　**THURS**
3. 421 days ago if today is Wednesday 7 divides 420 Wednesday − 1　**TUES**	13. the 31st of the month if the 4th is Friday 　　　　4 is Fri 　　　　25 is Fri　　**THURS**
4. 80 days from today if today is Monday 7 divides 77 Monday + 3　　**THURS**	14. the 3rd of the month if the 30th is Tuesday 　　　　30 is Tues 　　　　2 is Tues　　**WED**
5. 86 days ago if today is Thursday 7 divides 84 Thursday − 2　　**TUES**	15. the 27th of the month if the 3rd is Thursday 　　　　3 is Thurs 　　　　24 is Thurs　**SUN**
6. 290 days from today if today is Friday 7 divides 287 Friday + 3　　**MON**	16. the 4th of the month if the 26th is Monday 　　　　26 is Mon 　　　　5 is Mon　　**SUN**
7. 216 days from today if 3 days ago was Tuesday −3 is Tues; tod is Fri; 7 divides 210; Friday + 6　**THURS**	17. the 28th of the month if the 1st is Sunday 　　　　1 is Sun 　　　　29 is Sun　　**SAT**
8. 149 days before yesterday if tomorrow is Sunday tom is Sun; tod is Sat; yest is Fri 7 divides 147; Friday − 2　**WED**	18. the 30th of the month if the 5th is Saturday 　　　　5 is Sat 　　　　26 is Sat　　**WED**
9. 496 days from tomorrow if yesterday was Thursday yest is Thurs; tod is Fri; tom is Sat 7 divides 490; Saturday + 6　**FRI**	19. the 2nd of the month if the 31st is Wednesday 　　　　31 is Wed 　　　　3 is Wed　　**TUES**
10. 723 days from today if today is Friday 7 divides 721 Friday + 2　　**SUN**	20. the 29th of the month if the 5th is Monday 　　　　5 is Mon 　　　　26 is Mon　　**THURS**

Calendar 2

Find the ones digit of the product of the calendar numbers marked by dots.	*Find the positive difference of two products: the • numbers and the ø numbers.*

1.

Su	M	T	W	R	F	Sa
				•		
				•	15	
		•				
					•	

6x3x8x9 ends in **6**

5.

Su	M	T	W	R	F	Sa
		ø		•		
		•		ø		

$x(x+23) -$ $(x+2)(x+21)$

42

2.

Su	M	T	W	R	F	Sa
	•	•				
				•		
					•	
24		•				

4x5 ends in **0**

6.

Su	M	T	W	R	F	Sa
ø					•	
•					ø	

$x(x+12) -$ $(x+5)(x+7)$

35

3.

Su	M	T	W	R	F	Sa
		•		•		•
	15				•	
•						•

9x1x3x9x 1x7 ends in **1**

7.

Su	M	T	W	R	F	Sa
ø						•
•						ø

$x(x+20) -$ $(x+6)(x+14)$

84

4.

Su	M	T	W	R	F	Sa
		•	•			•
			17	•		
					•	
	•					

8x9x2x8x 6x7 ends in **4**

8.

Su	M	T	W	R	F	Sa
		ø				•
	•					ø

$x(x+25) -$ $(x+4)(x+21)$

84

| A century year has to be divisible by 400 to be leap. *Find the day of the week, observing leap years.* | Calendar 3 | Non-leap year is 52 weeks + 1 day. Leap year is 52 weeks + 2 days. |

1. June 1, 2005 was a Wednesday.
   June 1, 2015 was a _____.

   8 non-leap years
   2 leap years (08, 12)
   advance 12 days   **Monday**

8. January 4, 2005 was a Tuesday.
   January 4, 2007 was a _____.

   2 non-leap years
   advance 2 days   **Thursday**

2. July 14, 2014 was a Monday.
   July 14, 2009 was a _____.

   4 non-leap years
   1 leap year (12)
   backup 6 days   **Tuesday**

9. May 9, 2005 was a Mondday.
   May 9, 2008 was a _____.

   2 non-leap years
   1 leap year (08)
   advance 4 days   **Friday**

3. January 1, 1991 was a Tuesday.
   January 1, 2000 was a _____.

   7 non-leap years
   2 leap years (92, 96)
   advance 11 days   **Saturday**

10. December 12, 2004 was a Sunday.
    December 12, 2009 was a _____.

    4 non-leap years
    1 leap year (08)
    advance 6 days   **Saturday**

4. January 1, 1900 was a Monday.
   January 1, 1904 was a _____.

   1900 not leap
   2/29/1904 didn't occur yet
   advance 4 days   **Friday**

11. April 11, 2008 was a Friday.
    April 11, 2014 was a _____.

    5 non-leap years
    1 leap year (12)
    advance 7 days   **Friday**

5. February 2, 1992 was a Sunday.
   February 2, 1998 was a _____.

   4 non-leap years
   2 leap years (92, 96)
   advance 8 days   **Monday**

12. February 1, 2009 was a Sunday.
    February 1, 2012 was a _____.

    3 non-leap years
    2/29/2012 didn't occur yet
    advance 3 days   **Wednesday**

6. March 9, 1998 was a Monday.
   March 9, 2005 was a _____.

   5 non-leap years
   2 leap years (00, 04)
   advance 9 days   **Wednesday**

13. January 1, 2015 was a Thursday.
    January 1, 2006 was a _____.

    7 non-leap years
    2 leap years (08, 12)
    backup 11 days   **Sunday**

7. November 30, 2004 was a Tuesday.
   November 30, 2014 was a _____.

   8 non-leap years
   2 leap years (08, 12)
   advance 12 days   **Sunday**

14. July 4, 2014 was a Friday.
    July 4, 2006 was a _____.

    6 non-leap years
    2 leap years (08, 12)
    backup 10 days   **Tuesday**

Calendar 4

Answer as indicated.

1. What day is the 420,716th day from a Friday?

    7 divides 420,714
    420,716/7 has Remainder 2
    2 more days past full cycles
    Sunday

2. If this month is April, what will be the month 1995 months from today?

    div by 12 is div by 3 and 4
    12 divides 1992
    1995 − 1992 = 3
    advance 3 months from April   **July**

3. If a full moon occurs once every 28 days, at most how many full moons can occur in one calendar year?

    assume full moon on day 1
    div by 28 = div by 4 and 7
    28 divides 364; 364/28 = 13
    14th full moon on day 365   **14**

4. A fictitious calendar year has 14 months with even numbered months having 30 days and odd numbered months having 32 days. Find the number of days in the year.

    30 x 7 = 210
    32 x 7 = 224   **434**

5. A month with 5 Thursdays could start on which days of the week?

    Thursday (1, 8, 15, 22, 29)
    Wednesday (2, 9, 16, 23, 30)
    Tuesday (3, 10, 17, 24, 31)

6. A calendar year has 13 months with prime-numbered months having 30 days, composite-numbered months have 25 days, and the 1st month having 35 days. Find the number of days in the year.

    30 x 6 = 180
    25 x 6 = 150   **365**

7. Suppose a calendar with 363 days has 11 months, each with 11 weeks. 438 of these days is how many years-months-weeks?

    363/11 = 33 days per month   **1 year**
    33/11 = 3 days per week   **2 months**
    438 = 363 + 66 + 9   **3 weeks**

8. Find the date of the first Monday in a month given the sum of the dates of all Mondays in the month is 80.

    80/5 = 16 = average of arithmetic sequence of dates
    2, 9, 16, 23, 30

9. A fictitious calendar year has 56 weeks with 8 days per week. Find the absolute difference of the number of days of this calendar and a standard one if neither has a leap year.

    56 x 8 = 448
    448 − 365 = **83**

10. In what month does the 311th day of the year occur?

    estimate 300/30 = 10
    10 full months is end of October
    November

11. Find the least possible sum of the dates of all Mondays in any month.

    least is February in non-leap year
    1 + 8 + 15 + 22 = **46**

12. Find the greatest possible sum of the dates of all Mondays in any month.

    greatest (5 Mondays) =
    31 + 24 + 17 + 10 + 3 = **85**

Ciphers 1

Decipher the addition problem. Each letter is a different digit.

1.
```
   A          1
+ BB       + 99
 AYY        100
```
A = 1 or else only 2 digits in answer. B = 9 or else BB is YB. Y = 0 because 9+1=10.

9.
```
  ABA         505
  ABA         505
+ ABA       + 505
PAPA         1515
```
From 1s, A = 0 or 5. But A ≠ 0 (3-dgit no.). A = 5. P = 1. 3B + 1 = 1. B = 0.

2.
```
  ABA        161
+ BBA      + 661
  EYY        822
```
Y = 2A, so Y is even. Y = 2B, so no carry from ones column. Try A = 1, B = 6. Works. (A = 2, B = 7 No, etc.)

10.
```
  CBA         371
  CBA         371
+ CBA       + 371
 AAAC        1113
```
A = 1 or 2 in answer. If A = 2, C = 6 by 1s. In 10s, B = 4 making answer 1926. Fails. So, A = 1, C = 3, B = 7.

3.
```
  CBA        497
+ CCA      + 447
  BCC        944
```
C = 2A, so C is even. B ≠ 0, C ≠ 0 by 3-digit nos. For tens, B = 9 and ones carry 1. 9 = 4+4+1, so C = 4. A = 7.

11.
```
  CBA         471
  CBA         471
+ CBA       + 471
 ACAP        1413
```
A = 1 or 2 in answer. 1s do not carry. C = 4 or 9 with carry 2. B = 7, 8, 9 to carry 2. B = 7. A = 1. P = 3. C = 4.

4.
```
  AAB        661
+ PAB      + 261
  EPP        922
```
P = 2B, so P is even. A, P, E ≠ 0 by 3-digit nos. Try B = 1, P = 2, A = 6. Need low nos so hundreds do not carry.

12.
```
   ABB        144
+ KPPB     + 9884
 AYYDP       10028
```
A = 1 in answer. Most carried is 1, so K = 9, Y = 0. In 100s 1 + P = 0. P = 8. In 1s B ≠ 9, so B = 4. D = 2.

5.
```
  CAB        497
+ CAB      + 497
  AAC        994
```
C = 2B, so C is even. A, C ≠ 0 by 3-digit nos. For 10s, A = 9, ones and tens carry. Then C = 4, B = 7.

13.
```
  ABBC       2557
+ KPBB     + 8155
 PSCPA      10712
```
P = 1 (carrying from 2 addends). B = 0 or 5 in 10s. But B ≠ 0 in 1s. C = 7 (5+1+carry 1) in 100s. A = 2 in 1s. K = 8. S = 0.

6.
```
  ABB        266
+ BCB      + 656
  EAA        922
```
A = 2B, so A is even. By 1s & 10s, ones carry and C is 1 less than B. A ≠ 0, so B ≠ 5. Try B = 6. C = 5, A = 2, E = 9.

14.
```
  ABCD       6201
+ APPD     + 6441
 DBAPB      12642
```
D = 1 (carrying from 2 addends). B = 2 in 1s. A + A = 12 in 1000s. A = 6. C = 0 in 10s. P = 4 since no carrying to 100s.

7.
```
  BAB        272
+ CAB      + 472
  ACC        744
```
B, C, A ≠ 0. Answer is not 4-digit, so try low nos. B = 1 fails. B = 2, C = 4, A = 7.

15.
```
  ABPB       1898
+ PCCK     + 9220
 AAAAB      11118
```
A = 1 in 10,000s. K = 0 in 1s. In 1000s 1 + P + 1 (carry) = 1. P = 9. In 10s 9 + C = 1. C = 2. In 100s B + 2 + 1 = 1. B = 8.

8.
```
  ABA        565
+ ABY      + 561
 YYEB       1126
```
Y = 1, max carried from 2 addends. In 100s, A = 5. In 1s, B = 6. In 10s, E = 2.

16.
```
  ABCD       8594
+ PBCD     + 1594
 PSPAA      10188
```
P = 1. 1s cannot carry to have AA in answer. A = 8 for 1000s to carry. D = 4. C = 9. B = 5. S = 0.

MAVA Math: Enhanced Skills Solutions   Copyright © 2015 Marla Weiss

Ciphers 2

Decipher the multiplication problem. Each letter is a different nonzero digit. The letter O is not used. A letter cannot be a digit already appearing in the problem.

1. BB 55 66 x _B_ x _5_ x _6_ A9B 275 396 B x B = B implies B = 1, 5, or 6. B ≠ 1 because of hundreds place in aswer.		**A = 3** **B = 6**	

1.		
BB 55 66	**A = 3**	8.

#	problem	result	#	problem	result

1.
```
  BB        55        66      A = 3
x  B     x  5     x  6       B = 6
A9B       275       396
```
B x B = B implies B = 1, 5, or 6.
B ≠ 1 because of hundreds place in aswer.

8.
```
 AAB       226       776      A = 7
x  B     x  6     x  6       B = 6
CB5B      1356      4656      C = 4
```
B = 1, 5,   no   yes
or 6
B ≠ 1, 5

2.
```
  A2        32        82      A = 8
x 7A     x 73     x 78
6396       96       656
          224       574
                   6396
```
A = 3 or 8

9.
```
 A2C       A21       721      A = 7
x   2    x   2    x   2      C = 1
CPP2      1442      1442      P = 4
```
C = 1 max   P = 4   A = 7
carrying

3.
```
 AB        A7        83       A = 8
x  B     x  7     x  3       B = 3
249       249       249
```
B = 3 or 7   7x=20   yes
           no sol

10.
```
 AB2       162                A = 1
x  B     x  6                B = 6
972       972
```
B = 1 or 6   yes
1 x 1 ≠ 7

4.
```
 AB        A1        89       A = 8
x  B     x  1     x  9       B = 9
A01       A01       801
```
B = 1 or 9   Ax1 not   yes
           2-digit

11.
```
 A6                       46     A = 4
x 5A                    x 54
2A8A                     184
                         230
                        2484
```
A ≥ 4 to have 2
A ≠ 5, 6
A < 7

5.
```
 BBB       666       555      B = 5
x  B     x  6     x  5
277B                2775
```
B = 1, 5,   too big   yes
or 6

12.
```
 AAA       999                A = 9
x   3    x   3                B = 2
BAAC      2997                C = 7
```
B = 1 or 2
A = 4 to 9

6.
```
 10A       106       105      A = 5
x   A    x   6    x   5
A2A       636       525
```
A = 5 or 6   no   yes

13.
```
 BBB       555                A = 1
x   2    x   2                B = 5
AAAC      1110                C = 0
```
A = 1 max   yes
carrying

7.
```
 1A1       181       121      A = 2
x   A    x   8    x   2
A4A                 242
```
A = 2 or 8   4-digit   yes
           product

14.
```
 ABC       A1C       816      A = 8
x   2    x   2    x   2      B = 1
BCD2      16D2      1632      C = 6
                             D = 3
```
B = 1 max   C = 6   A = 8
carrying   D = 3

Circles 1

Complete the chart for each circle: radius, diameter, circumference, and area.

#	RAD	DIAM	CIRC	AREA	#	RAD	DIAM	CIRC	AREA
1.	5	10	10π	25π	20.	80	160	160π	6400π
2.	15	30	30π	225π	21.	$\dfrac{11}{2}$	11	11π	$\dfrac{121\pi}{4}$
3.	50	100	100π	2500π	22.	25	50	50π	625π
4.	$\dfrac{9}{2}$	9	9π	$\dfrac{81\pi}{4}$	23.	$\dfrac{10\sqrt{\pi}}{\pi}$	$\dfrac{20\sqrt{\pi}}{\pi}$	$20\sqrt{\pi}$	100
5.	3	6	6π	9π	24.	30	90	90π	900π
6.	18	36	36π	324π	25.	12	24	24π	144π
7.	8	16	16π	64π	26.	20	40	40π	400π
8.	4	8	8π	16π	27.	$\dfrac{5}{2}$	5	5π	$\dfrac{25\pi}{4}$
9.	60	120	120π	3600π	28.	$\dfrac{50}{\pi}$	$\dfrac{100}{\pi}$	100	$\dfrac{2500}{\pi}$
10.	$\dfrac{1}{2}$	1	π	$\dfrac{\pi}{4}$	29.	$\dfrac{7}{2}$	7	7π	$\dfrac{49\pi}{4}$
11.	1	2	2π	π	30.	$\dfrac{6\sqrt{\pi}}{\pi}$	$\dfrac{12\sqrt{\pi}}{\pi}$	$12\sqrt{\pi}$	36
12.	19	38	38π	361π	31.	11	22	22π	121π
13.	6	12	12π	36π	32.	$\dfrac{13}{2}$	13	13π	$\dfrac{169\pi}{4}$
14.	70	140	140π	4900π	33.	π	2π	$2\pi^2$	π^3
15.	16	32	32π	256π	34.	9	18	18π	81π
16.	2	4	4π	4π	35.	$\dfrac{\pi}{2}$	π	π^2	$\dfrac{\pi^3}{4}$
17.	7	14	14π	49π	36.	$\dfrac{25}{2}$	25	25π	$\dfrac{625\pi}{4}$
18.	14	28	28π	196π	37.	17	34	34π	289π
19.	10	20	20π	100π	38.	$\dfrac{10}{\pi}$	$\dfrac{20}{\pi}$	20	$\dfrac{100}{\pi}$

Circles 2

Answer as indicated.

---

1. The ratio of the circumferences of two circles is 9:16. Find the ratio of their areas.

   d:D = 9:16
   r:R = 4.5:8 = 9:16
   a:A = **81:256**

6. Three circles are pairwise externally tangent. The triangle formed by joining their centers has sides 8, 9, and 10. Find the three diameters.

   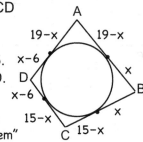

   10−(8−x) = x+2
   x+(x+2) = 9
   x = 3.5
   diams = 2x, 16−2x, 2x+4
   diams = **7, 9, 11**

---

2. A circle with circumference 2 has the same area as a square. Find the perimeter of the square.

   πd = 2   A circ = 1/π
   d = 2/π   A sq = 1/π
   r = 1/π   s = √π/π   **4√π / π**

7. Quadrilateral ABCD is circumscribed about a circle. AB = 19. BC = 15. AD = 13. Find CD.

   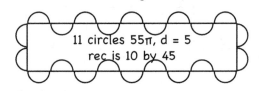

   CD=(x−6)+(15−x)
   CD = **9**
   "ice cream cone theorem"

---

3. Points A, B, C, D, E, and F in order are on a circle of area 81π, dividing it into 6 congruent arcs. Find the length of the arc AEC.

   A = 81π   18π/6 = 3π
   r = 9   3π × 4 = **12π**
   d = 18
   C = 18π

8. Congruent semicircles comprise the curved path of length 55π. Find the area of the rectangle as shown. **450**

   11 circles 55π, d = 5
   rec is 10 by 45

---

4. Four concentric circles have radii 2, 4, 6, and 8. Find the sum of the areas of the outermost and innermost rings.

   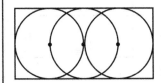

   inner ring A =
   16π−4π = 12π
   outer ring A =
   64π−36π = 28π
   28π + 12π = **40π**

9. A circle with circumference 30 has two overlapping arcs of lengths 15 and 25. Find the positive difference of the greatest and least possible overlap of the arcs.

   greatest overlap = 15
   least overlap = 10
   diff = 15 − 10 = **5**

---

5. Find the area of the rectangle if the congruent circles with centers shown have diameter 10.

   r = 5
   L = 10
   W = 20
   A = **200**

10. Find the area of circle O if the smaller square has area 11.

   A sq = 11
   s sq = r circ = √11
   A circ = **11π**

MAVA Math: Enhanced Skills Solutions Copyright © 2015 Marla Weiss

Circles 3

Calculate for the sector. CENTRAL ANGLE	FRACTION OF CIRCLE	RADIUS	AREA OF SECTOR	LENGTH OF ARC	PERIMETER OF SECTOR
1. 45°	$\frac{1}{8}$	4	2π	π	8 + π
2. 90°	$\frac{1}{4}$	8	16π	4π	16 + 4π
3. 120°	$\frac{1}{3}$	12	48π	8π	24 + 8π
4. 80°	$\frac{2}{9}$	9	18π	4π	18 + 4π
5. 100°	$\frac{5}{18}$	18	90π	10π	36 + 10π
6. 180°	$\frac{1}{2}$	30	450π	30π	60 + 30π
7. 20°	$\frac{1}{18}$	36	72π	4π	72 + 4π
8. 60°	$\frac{1}{6}$	18	54π	6π	36 + 6π
9. 135°	$\frac{3}{8}$	32	384π	24π	64 + 24π
10. 150°	$\frac{5}{12}$	24	240π	20π	48 + 20π
11. 30°	$\frac{1}{12}$	24	48π	4π	48 + 4π
12. 50°	$\frac{5}{36}$	72	720π	20π	144 + 20π
13. 10°	$\frac{1}{36}$	36	36π	2π	72 + 2π
14. 160°	$\frac{4}{9}$	45	900π	40π	90 + 40π
15. 40°	$\frac{1}{9}$	27	81π	6π	54 + 6π
16. 30°	$\frac{1}{12}$	36	108π	6π	72 + 6π
17. 140°	$\frac{7}{18}$	54	1134π	70π	108 + 70π
18. 240°	$\frac{2}{3}$	15	150π	20π	30 + 20π
19. 225°	$\frac{5}{8}$	32	640π	40π	64 + 40π

Circles 4

Answer as indicated.

1. $\overline{AD}$ is a diameter. Find the sum of the measures of the 4 angles of the triangles not opposite the diameter.

shape is quad = 360
opposite diam = 90 each
360 − 180 = **180**

6.  O is the center of the circle with radius 8. Find the area of the circle not inside the △.

A = **64π − 32**

2. 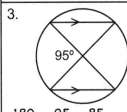 $\overline{AD}$ is a diameter. AD = 85. AB = 13. CD = 40. Find AC + BD.

△ABD and △ACD right because of diameter
BD = 84. AC = 75. **159**

7. O is the center of the circle with radius 6. Find the area of the circle not inside either △.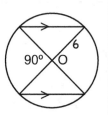

A△ = 18
A = **36π − 36**

3. 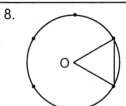 Each △ has one vertex at the center of the circle. Find the sum of the measures of the 4 noncentral angles of the △s.

180 − 95 = 85
180 − 85 = 95
95 × 2 = **190**

8. The 6 points are equally spaced on the circle with center O. Find the measures of the 3 angles of the △.

central angle = 360/6 = 60
2 radii make △ isosceles
60, 60, 60

4. 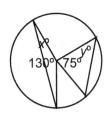 Each △ has one vertex at the center of the circle. Find x + y.
2 radii make △s isosceles
(180 − 130)/2 = 25
(180 − 75)/2 = 52.5
52.5 + 25 = **77.5**

9. Each vertex of the △ is a center of a circle. Find the ratio of the sum of the 3 circumferences to the perimeter of the △.

$\dfrac{2a\pi + 2b\pi + 2c\pi}{2a + 2b + 2c}$ = **π:1**

5. 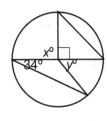 Each △ has one vertex at the center of the circle. Find x + y.
2 radii make △ isosceles
180 − 34 − 34 = 112
112 + 90 + x + y = 360
x + y = **158**

10. 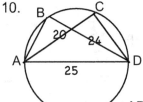 $\overline{AD}$ is a diameter. AD = 25. BD = 24. AC = 20. Find AB + CD.

△ABD and △ACD right because of diameter
AB = 7. CD = 15. **22**

Circles 5

Find the area in square units.

A square is a rhombus. Its area is half the product of the diagonals.

---

1. Find the area of a circle inscribed in a square of area 36.

$s = 6$
$r = 3$

$A_{circ} = \mathbf{9\pi}$

7. Find the area of a square inscribed in a circle with area 9π.

$A_{circ} = 9\pi$
$r = 3$
$diag = 6$
$A_{SQ} = 6\times6/2 = \mathbf{18}$

---

2. Find the area of a circle inscribed in a square with side 5.

$s = 5$
$r = 5/2$

$A_{circ} = \dfrac{\mathbf{25\pi}}{\mathbf{4}}$

8. Find the area of a square inscribed in a circle with radius 2.

$r = 2$
$diag = 4$
$A_{SQ} = 4\times4/2 = \mathbf{8}$

---

3. Find the area of a circle inscribed in a square of area 10.

$A = 10$
$s = \sqrt{10}$
$r = (\sqrt{10})/2$

$A_{circ} = \dfrac{\mathbf{5\pi}}{\mathbf{2}}$

9. Find the area of a square inscribed in a circle with area 20π.

$A_{circ} = \mathbf{20\pi}$
$r = 2\sqrt{5}$
$diag = 4\sqrt{5}$
$A_{SQ} = (4\sqrt{5})(4\sqrt{5})/2 = \mathbf{40}$

---

4. Find the area of a square circumscribed about a circle with radius 7.5.

$r = 7.5$
$s = 15$

$A_{SQ} = \mathbf{225}$

10. Find the area of a circle circumscribed about a square with area 1.

$A_{SQ} = 1$
$s = 1$
$diag = d = \sqrt{2}$
$A_{circ} = (\sqrt{2})(\sqrt{2})\pi/4 = \dfrac{\mathbf{\pi}}{\mathbf{2}}$

---

5. Find the area of a square circumscribed about a circle with circumference 12π.

$C = 12\pi$
$d = 12$

$A_{SQ} = \mathbf{144}$

11. Find the area of a circle circumscribed about a square with side 5.

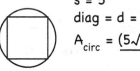

$s = 5$
$diag = d = 5\sqrt{2}$
$A_{circ} = \dfrac{(5\sqrt{2})(5\sqrt{2})\pi}{4} = \dfrac{\mathbf{25\pi}}{\mathbf{2}}$

---

6. Find the area of a square circumscribed about a circle with area 49π.

$A_{circ} = 49\pi$
$r = 7$
$d = 14$

$A_{SQ} = \mathbf{196}$

12. Find the area of a circle circumscribed about a square with area 10.

$A_{SQ} = 10$
$s = \sqrt{10}$
$diag = d = \sqrt{20} = 2\sqrt{5}$
$A_{circ} = (\sqrt{5})(\sqrt{5})\pi = \mathbf{5\pi}$

Circles 6

Answer for the inscribed polygon.	Find the length in units of the common external tangent of the two circles.

---

1. An equilateral triangle is inscribed in circle with radius 6. Find the perimeter of the triangle.

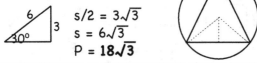

$s/2 = 3\sqrt{3}$
$s = 6\sqrt{3}$
P = 18√3

7. radii 2 and 10, centers 17 apart

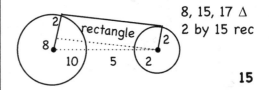

8, 15, 17 △
2 by 15 rec

15

---

2. A regular hexagon is inscribed in circle with circumference 18π. Find area of the hexagon. d = 18, r = 9

$\dfrac{6 \times 81\sqrt{3}}{4}$ $\dfrac{243\sqrt{3}}{2}$

8. radii 8 and 18, externally tangent to each other

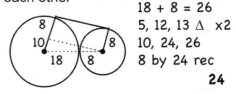

18 + 8 = 26
5, 12, 13 △ ×2
10, 24, 26
8 by 24 rec

24

---

3. A regular octagon is inscribed in circle. Find least interior angle of the triangle formed by connecting 3 adjacent vertices.

135, 22.5, **22.5**

ext = 360/8 = 45
int = 180 − 45 = 135

9. radii 3 and 14, centers 61 apart

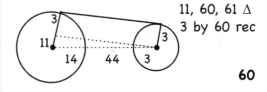

11, 60, 61 △
3 by 60 rec

60

---

4. A regular hexagon is inscribed in circle with circumference 16π. Find perimeter of the hexagon.

d = 16
r = 8 = side hex
P = 6 × 8 = **48**

10. radii 3 and 10, centers 25 apart

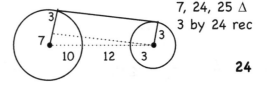

7, 24, 25 △
3 by 24 rec

24

---

5. A regular pentagon is inscribed in circle. Find least interior angle of the triangle formed by connecting 3 adjacent vertices.

108, 36, **36**

ext = 360/5 = 72
int = 180 − 72 = 108

11. radii 9 and 16, externally tangent to each other

16 + 9 = 25
7, 24, 25 △
9 by 24 rec

24

---

6. An equilateral triangle is inscribed in circle with radius 8. Find the area of the triangle.

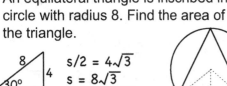

$s/2 = 4\sqrt{3}$
$s = 8\sqrt{3}$
A = 48√3

12. radii 2 and 8, externally tangent to each other

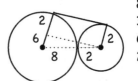

8 + 2 = 10
3, 4, 5 △ ×2
6, 8, 10
2 by 8 rec

8

Clocks 1

Convert units.

1.	.5 day =	**12**	hours	20.	1/3 minute =	**20**	seconds	
2.	5/6 hour =	**50**	minutes	21.	.15 hour =	**9**	minutes	
3.	.25 day =	**6**	hours	22.	1.1 minute =	**66**	seconds	
4.	.25 hour =	**15**	minutes	23.	.4 hour =	**24**	minutes	
5.	.1 hour =	**6**	minutes	24.	.75 hour =	**45**	minutes	
6.	.4 minutes =	**24**	seconds	25.	2/3 hour =	**40**	minutes	
7.	1/3 hour =	**20**	minutes	26.	.45 hour =	**27**	minutes	
8.	.85 hour =	**51**	minutes	27.	.8 hour =	**48**	minutes	
9.	5/6 day =	**20**	hours	28.	1/6 day =	**240**	minutes	
10.	.6 hour =	**36**	minutes	29.	.9 hour =	**54**	minutes	
11.	1/12 day =	**120**	minutes	30.	.25 year =	**3**	months	
12.	5/12 minute =	**25**	seconds	31.	.7 hour =	**42**	minutes	
13.	.1 hour =	**360**	seconds	32.	7/8 day =	**21**	hours	
14.	3/5 hour =	**36**	minutes	33.	.15 minute =	**9**	seconds	
15.	.75 minute =	**45**	seconds	34.	.05 hour =	**3**	minutes	
16.	.3 hour =	**18**	minutes	35.	7/12 minute =	**35**	seconds	
17.	.125 day =	**3**	hours	36.	7/15 hour =	**1680**	seconds	
18.	.2 hour =	**12**	minutes	37.	11/30 hour =	**22**	minutes	
19.	4/15 minute =	**16**	seconds	38.	11/12 day =	**1320**	minutes	

Clocks 2

Answer as indicated.

1. What time is 3540 seconds before 2:57 PM?

   round 3540 to 3600
   3600/60 = 60 min = 1 hour
   2:57 – 1 hr + 1 min   **1:58 PM**

8. If a clock stopped running 313 minutes after 3:17 PM, at what time did it stop?
   3:17 + 13 min = 3:30
   300/60 = 5 hr
   3:30 + 5 hr = 8:30   **8:30 PM**

2. If Eli goes to sleep at 9:30 PM and sleeps for 10 hours and 30 minutes, when will he wake up?

   9:30 to 12:00--2.5
   12:00 to 8:00--8   **8:00 AM**

9. The time on a 12-hour clock is 9:00 AM. Find the time after the minute hand goes around 3 times.

   1 circle of min hand in 1 hr
   3 circles of min hand in 3 hrs **12:00 PM**

3. What time is 10 hours and 58 minutes after 3:19 PM?

   round 58 min to 1 hr
   3:19 + 11 hrs = 2:19
   subtract 2 min   **2:17 AM**

10. What time is 2460 seconds before 4:39 PM?

    2460/60 = 246/6 = 41
    4:39 – 39 min = 4:00
    4:00 – 2 min   **3:58 PM**

4. The time on a 12-hour clock is 10:00 AM. Find the time after the minute hand goes around 4 times.

   1 circle of min hand in 1 hr
   4 circles of min hand in 4 hrs **2:00 PM**

11. What time is 8 hours and 52 minutes after 1:37 PM?

    round 52 min to 1 hr
    1:37 + 9 hrs = 10:37
    subtract 8 min   **10:29 PM**

5. The number of minutes in 5 hours is the same as the number of hours in how many days?

   5 x 60 = 24 x D
   5 x 5 x 12 = 12 x 2 x D   **12.5**

12. If a clock stopped running 257 minutes after 1:54 PM, at what time did it stop?

    round 257 to 240
    240/60 = 4 hr
    1:54 + 4 hr + 17 min   **6:11 PM**

6. What time is 335 minutes after 2:55 PM?

   335 = 300 + 35
   300 = 5 hrs
   2:55 + 5 hrs = 7:55   **8:30 PM**
   +5, + 30

13. What time is 285 minutes after 4:25 PM?

    round 285 to 300
    300/60 = 5 hr
    4:25 + 5 hr – 15 min   **9:10 PM**

7. Find the time 281 hours ago if the present time is 7:00 AM.
   281 = 288 – 7
   288/12 = 24
   7:00 backed up 24 hrs is 7:00 **2:00 PM**
   7 AM + 7 hrs

14. On any day, 8:39 AM is how many minutes before 3:26 PM?

    round 3:26 to 3:39
    8:39 AM to 3:39 PM is 7 hrs
    7 x 60 – 13   **407**

Clocks 3

Operate as indicated. Regroup/simplify answers.

1. 13 days 8 hours 9 min 55 sec + 5 days 16 hours 54 min 7 sec 18 days 24 hours 63 min 62 sec 18 days 25 hrs 4 min 2 sec **19 days 1 hr 4 min 2 sec**	**9.** 6 days 30 hrs 66 min 67 sec 7 days 7 hours 7 min 7 sec − 5 days 7 hours 9 min 11 sec **1 day 23 hr 57 min 56 sec**
2. 12 days 29 hours 68 min 63 sec 13 days 6 hours 9 min 3 sec − 5 days 16 hours 54 min 5 sec **7 days 13 hr 14 min 58 sec**	**10.** 17 days 43 hrs 99 min 61 sec 18 days 19 hrs 99 min 61 sec 18 days 20 hours 40 min 1 sec − 10 days 22 hours 50 min 9 sec **7 days 21 hrs 49 min 52 sec**
3. (9) (5 days 7 hrs 7 min 7 sec) 0 wk 45 d 63 h 63 m 63 s 6 wk 3 d 63 h 64 m 3 s 6 wk 3 d 64 h 4 m 3 s **6 weeks 5 days 16 hrs 4 min 3 sec**	**11.** $\frac{1}{4}$ (5 days 9 hours 8 minutes) (1/4) 4 days 33 hrs 8 min (1/4) 4 days 32 hrs 68 min **1 day 8 hrs 17 min**
4. $\frac{1}{3}$ (7 days 8 hrs 9 min 42 sec) (1/3) 6 days 32 hrs 9 min 42 sec (1/3) 6 days 30 hrs 129 min 42 sec **2 days 10 hrs 43 min 14 sec**	**12.** (4) (2 days 7 hr 20 min 40 sec) 8 days 28 hr 80 min 80 sec 8 days 28 hr 81 min 20 sec 8 days 29 hr 21 min 20 sec 9 days 5 hr 21 min 20 sec **1 wk 2 days 5 hr 21 min 20 sec**
5. (10 hrs 22 min 40 sec) ÷ 4 (8 hrs 142 min 40 sec)/4 (8 hrs 140 min 160 sec)/4 **2 hrs 35 min 40 sec**	**13.** (6 hrs 30 min 16 sec) ÷ 4 (4 hrs 150 min 16 sec) ÷ 4 (4 hrs 148 min 136 sec) ÷ 4 **1 hr 37 min 34 sec**
6. 8 hours 59 min 60 sec 9 hours − 5 hours 5 minutes 5 seconds **3 hours 54 min 55 sec**	**14.** 4 days 10 hours 22 min 32 sec + 3 days 16 hours 49 min 31 sec 7 days 26 hours 71 min 63 sec 7 days 27 hrs 12 min 3 sec **8 days 3 hrs 12 min 3 sec**
7. (11 hrs 31 min 20 sec) ÷ 5 (10 hrs 91 min 20 sec) ÷ 5 (10 hrs 90 min 80 sec) ÷ 5 **2 hrs 18 min 16 sec**	**15.** $\frac{1}{6}$ (8 days 8 hours 6 minutes) (1/6) 6 days 56 hrs 6 min (1/6) 6 days 54 hrs 126 min **1 day 9 hrs 21 min**
8. (13 hrs 16 min 18 sec) ÷ 6 (12 hrs 76 min 18 sec) ÷ 6 (12 hrs 72 min 258 sec) ÷ 6 **2 hrs 12 min 43 sec**	**16.** (14 hrs 10 min 15 sec) ÷ 3 (12 hrs 130 min 15 sec) ÷ 3 (12 hrs 129 min 75 sec) ÷ 3 **4 hrs 43 min 25 sec**

30° is between 2 adjacent numbers.

Clocks 4

Find the measure in degrees of the angle formed by the hands of a clock at the given time.

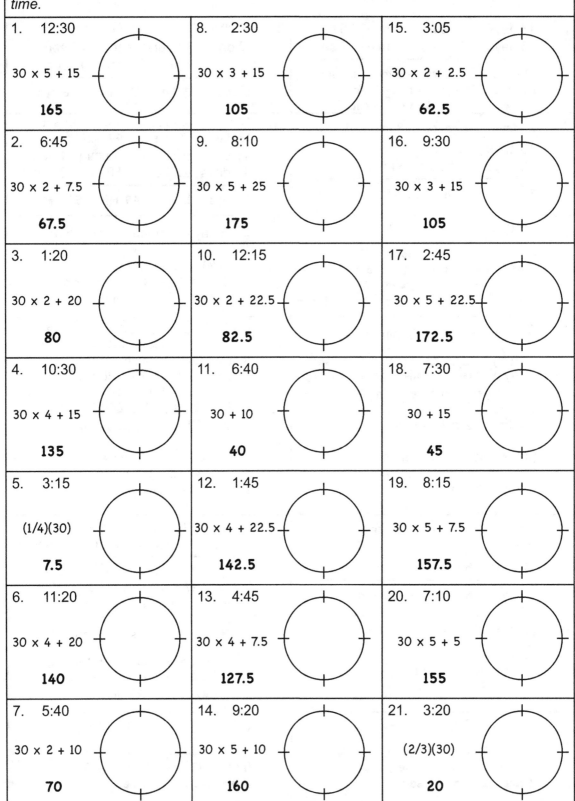

1. 12:30

30 x 5 + 15

165

2. 6:45

30 x 2 + 7.5

67.5

3. 1:20

30 x 2 + 20

80

4. 10:30

30 x 4 + 15

135

5. 3:15

(1/4)(30)

7.5

6. 11:20

30 x 4 + 20

140

7. 5:40

30 x 2 + 10

70

8. 2:30

30 x 3 + 15

105

9. 8:10

30 x 5 + 25

175

10. 12:15

30 x 2 + 22.5

82.5

11. 6:40

30 + 10

40

12. 1:45

30 x 4 + 22.5

142.5

13. 4:45

30 x 4 + 7.5

127.5

14. 9:20

30 x 5 + 10

160

15. 3:05

30 x 2 + 2.5

62.5

16. 9:30

30 x 3 + 15

105

17. 2:45

30 x 5 + 22.5

172.5

18. 7:30

30 + 15

45

19. 8:15

30 x 5 + 7.5

157.5

20. 7:10

30 x 5 + 5

155

21. 3:20

(2/3)(30)

20

Clocks 5

Find the fractional part.

	Find the time on the nonstandard clock, disregarding AM and PM.	

1. One and one half hours are what fraction of the time between 12 PM on Monday and 12 AM on Saturday of the same week?

$$\frac{1.5}{4.5 \times 24} \qquad \frac{1}{72}$$

2. Two and one third hours are what fraction of the time between 12 PM on Tuesday and 12 AM on Friday of the same week?

$$\frac{7}{3} \div (2.5 \times 24) \qquad \frac{7}{3 \times 60} \qquad \frac{7}{180}$$

3. Two hours and fifteen minutes are what fraction of the time between 8 PM on Wednesday and 11 AM on Friday of the same week?

$$\frac{9}{4} \div (24 + 15) \qquad \frac{9}{4 \times 39} \qquad \frac{3}{52}$$

4. Three and one third hours are what fraction of the time between noon on Wednesday and noon on Saturday of the same week?

$$\frac{10}{3} \div (3 \times 24) \qquad \frac{10}{9 \times 24} \qquad \frac{5}{108}$$

5. One hour and fifty minutes are what fraction of the time between 10:00 AM Monday and 7:00 PM Tuesday of the same week?

$$\frac{11}{6} \div (24 + 9) \qquad \frac{11}{6 \times 33} \qquad \frac{1}{18}$$

6. Three hours and twelve minutes are what fraction of the time between noon Sunday and 4:00 PM Tuesday of the same week?

$$\frac{16}{5} \div (48 + 4) \qquad \frac{16}{5 \times 52} \qquad \frac{4}{65}$$

7. Four hours and ten minutes are what fraction of the time between 9:00 AM Sunday and 1:00 PM Thursday of the same week?

$$\frac{25}{6} \div (96 + 4) \qquad \frac{25}{6 \times 100} \qquad \frac{1}{24}$$

8. 8-hour clock
start at 7:00
22 hours later

7 + 22 = 29
29 − 24 = 5   **5:00**

15. 4-hour clock
start at 2:00
19 hours later

2 + 19 = 21
21 − 20 = 1   **1:00**

9. 13-hour clock
start at 11:00
56 hours later

11 + 56 = 67
67 − 65 = 2   **2:00**

16. 9-hour clock
start at 1:00
70 hours later

1 + 70 = 71
71 − 63 = 8   **8:00**

10. 7-hour clock
start at 4:00
17 hours later

4 + 17 = 21
21 − 21 = 0   **7:00**

17. 20-hour clock
start at 15:00
87 hours later

15 + 87 = 102
102 − 100 = 2   **2:00**

11. 15-hour clock
start at 14:00
68 hours later

14 + 68 = 82
82 − 75 = 7   **7:00**

18. 14-hour clock
start at 12:00
30 hours later

12 + 30 = 42
42 − 42 = 0   **14:00**

12. 10-hour clock
start at 9:00
35 hours later

9 + 35 = 44
44 − 40 = 4   **4:00**

19. 16-hour clock
start at 12:00
39 hours later

12 + 39 = 51
51 − 48 = 3   **3:00**

13. 6-hour clock
start at 5:00
79 hours later

5 + 79 = 84
84 − 84 = 0   **6:00**

20. 17-hour clock
start at 8:00
29 hours later

8 + 29 = 37
37 − 34 = 3   **3:00**

14. 11-hour clock
start at 2:00
37 hours later

2 + 37 = 39
39 − 33 = 6   **6:00**

21. 5-hour clock
start at 3:00
44 hours later

3 + 44 = 47
47 − 45 = 2   **2:00**

Clocks 6

Answer as indicated for the broken clock.

1. A clock that uniformly loses 5 minutes every 12 hours was correctly set at noon on January 1. Find the time on the clock when the correct time was noon on January 5 of the same year.

12p Jan 1 to 12p Jan 5 is 8 units of 12 hr. 8x5 = 40 min lost

11:20 AM

6. A broken clock runs at the normal rate but backward. A 2nd broken clock runs forward at twice the normal rate. Both start at 12. Find the correct time when the clocks next show the same time.

Br B	Br F	Corr
12	12	12
11	2	1
10	4	2
9	6	3
8	8	4

4:00

2. A clock is correctly set at 4:00 PM, but it loses 4 minutes every hour. What is the correct time when the clock reads 8:00 AM the next day?

4p one day to 8a next day is 16 hrs. 16x4 = 64 min 64 = 60 + 4 add to 8am

9:04 AM

7. A clock is correctly set at 2:00 AM, but it loses 5 minutes every hour. What is the correct time when the clock reads 9:00 PM the next day?

2a one day to 9p next day is 43 hrs. 43x5 = 215 min 215 = 180 + 35 add to 9pm

12:35 AM

3. A 12-hour clock loses 5 minutes each day. Find the number of days for the clock to first return to the correct time.

5 min	1 day
60 min	12 days
12 hrs	144 days

144 days

8. A 12-hour clock loses 15 minutes each day. Find the number of days for the clock to first return to the correct time.

15 min	1 day
60 min	4 days
12 hrs	48 days

48 days

4. The hands on a broken 12-hour clock moved counterclockwise. Find the number to which the minute hand pointed 15 minutes before the hour hand pointed to 4.

if min hand at 3, counterclockwise 15 min lands on 12 for 4:00

3

9. The hands on a broken 12-hour clock moved counterclockwise. Find the number to which the minute hand pointed 20 minutes before the hour hand pointed to 5.

if min hand at 4, counterclockwise 20 min lands on 12 for 5:00

4

5. Two clocks show the correct time at 2:00. One runs backward and the other forward, both at the normal rate. When will both clocks next show the same time?

correct	broken
2:00	2:00
5:00	11:00
7:00	9:00
8:00	8:00

8:00

10. A clock that uniformly loses 6 minutes every 24 hours was set correctly at noon on May 1. Find the time on the clock when the correct time was 4:00 AM on May 6 of the same year.

12p May 1 to May 5 is 4 days. plus 16 hrs to May 6 = 2/3 day. (2/3)(6) = 4 4 × 6 + 4 = 28 min lost

3:32 AM

Combinations 1

Calculate.

1. C(5, 2) $\dfrac{5!}{2!\ 3!}$ $\dfrac{4 \times 5}{2}$ 2×5 **10**

2. C(8, 3) $\dfrac{8!}{3!\ 5!}$ $\dfrac{6 \times 7 \times 8}{2 \times 3}$ 7×8 **56**

3. C(7, 4) $\dfrac{7!}{4!\ 3!}$ $\dfrac{5 \times 6 \times 7}{2 \times 3}$ 5×7 **35**

4. C(10, 5) $\dfrac{10!}{5!\ 5!}$ $\dfrac{6 \times 7 \times 8 \times 9 \times 10}{2 \times 3 \times 4 \times 5}$ 7×9×4 **252**

5. C(6, 2) $\dfrac{6!}{2!\ 4!}$ $\dfrac{5 \times 6}{2}$ 5×3 **15**

6. C(12, 4) $\dfrac{12!}{4!\ 8!}$ $\dfrac{9 \times 10 \times 11 \times 12}{2 \times 3 \times 4}$ 9×5×11 **495**

7. C(9, 6) $\dfrac{9!}{6!\ 3!}$ $\dfrac{7 \times 8 \times 9}{2 \times 3}$ 7×4×3 **84**

8. C(11, 9) $\dfrac{11!}{9!\ 2!}$ $\dfrac{10 \times 11}{2}$ 5×11 **55**

9. C(4, 3) $\dfrac{4!}{3!\ 1!}$ **4**

10. C(10, 6) $\dfrac{10!}{6!\ 4!}$ $\dfrac{7 \times 8 \times 9 \times 10}{2 \times 3 \times 4}$ 7×3×10 **210**

11. C(9, 7) $\dfrac{9!}{7!\ 2!}$ $\dfrac{8 \times 9}{2}$ 4×9 **36**

12. C(15, 14) $\dfrac{15!}{14!\ 1!}$ **15**

13. C(20, 2) $\dfrac{20!}{2!\ 18!}$ $\dfrac{19 \times 20}{2}$ 19×10 **190**

14. C(11, 8) $\dfrac{11!}{8!\ 3!}$ $\dfrac{9 \times 10 \times 11}{2 \times 3}$ 3×5×11 **165**

15. C(8, 6) $\dfrac{8!}{6!\ 2!}$ $\dfrac{7 \times 8}{2}$ 7×4 **28**

16. C(7, 5) $\dfrac{7!}{5!\ 2!}$ $\dfrac{6 \times 7}{2}$ 3×7 **21**

17. C(9, 5) $\dfrac{9!}{5!\ 4!}$ $\dfrac{6 \times 7 \times 8 \times 9}{2 \times 3 \times 4}$ 7×2×9 **126**

18. C(12, 9) $\dfrac{12!}{9!\ 3!}$ $\dfrac{10 \times 11 \times 12}{2 \times 3}$ 5×11×4 **220**

19. C(10, 7) $\dfrac{10!}{7!\ 3!}$ $\dfrac{8 \times 9 \times 10}{2 \times 3}$ 4×3×10 **120**

20. C(20, 3) $\dfrac{20!}{3!\ 17!}$ $\dfrac{18 \times 19 \times 20}{2 \times 3}$ 6×19×10 **1140**

21. C(8, 5) $\dfrac{8!}{5!\ 3!}$ $\dfrac{6 \times 7 \times 8}{2 \times 3}$ 7×8 **56**

22. C(13, 11) $\dfrac{13!}{11!\ 2!}$ $\dfrac{12 \times 13}{2}$ 13×6 **78**

Combinations 2

In #9, divide by 2 because cannot shake hands with oneself.

Answer as indicated. *Answer as indicated.*

---

1. Nineteen telephones may be paired for conversations in how many ways?

$C(19, 2)$ $\dfrac{19!}{17!\ 2!}$ $\dfrac{18 \times 19}{2}$ 9×19 **171**

9. How many handshakes occur among 31 people if everyone shakes hands once?

$\dfrac{31 \cdot 30}{2}$ or $C(31, 2)$ **465**

---

2. How many ways can a doubles team be selected from 8 players?

$C(8, 2)$ $\dfrac{8!}{6!\ 2!}$ $\dfrac{7 \times 8}{2}$ 7×4 **28**

10. How many gifts are exchanged among 10 people if everyone exchanges gifts with every other person?

$10 \cdot 9$ This is multiplication principle but here to contrast with #9.   **90**

---

3. How many ways can a club of 6 members choose a committee of 3 people?

$C(6, 3)$ $\dfrac{6!}{3!\ 3!}$ $\dfrac{4 \times 5 \times 6}{2 \times 3}$ **20**

11. How many rolls occur if each of 13 children sitting in a circle rolls a ball to each other child?

$13 \cdot 12$ **156**

---

4. How many ways can 3 people be chosen from 4 couples if all are equally eligible?

$C(8, 3)$ $\dfrac{8!}{5!\ 3!}$ $\dfrac{6 \times 7 \times 8}{2 \times 3}$ **56**

12. How many cards are used if all 20 students in a class exchange valentines.

$20 \cdot 19$ **380**

---

5. How many ways can a reader select 3 books from 5 different books?

$C(5, 3)$ $\dfrac{5!}{3!\ 2!}$ $\dfrac{4 \times 5}{2}$ 2×5 **10**

13. Find the least number of line segments to connect 12 points on a circle.

$\dfrac{12 \cdot 11}{2}$ or $C(12, 2)$ **66**

---

6. How many pairs of friends may be chosen from among 6 friends?

$C(6, 2)$ $\dfrac{6!}{4!\ 2!}$ $\dfrac{5 \times 6}{2}$ 5×3 **15**

14. In a league of 16 teams, how many games are played matching each team with every other team once?

$\dfrac{16 \cdot 15}{2}$ or $C(16, 2)$ **120**

---

7. How many triangles may be formed selecting 3 of 7 points on a circle as the vertices?

$C(7, 3)$ $\dfrac{7!}{4!\ 3!}$ $\dfrac{5 \times 6 \times 7}{2 \times 3}$ **35**

15. In a league of 14 teams, how many games are played matching each team with every other team twice?

$\dfrac{14 \cdot 13 \cdot 2}{2}$ or $C(14, 2) \cdot 2$ **182**

---

8. How many quadrilaterals may be formed selecting 4 of 6 points on a circle as the vertices?

$C(6, 4)$ $\dfrac{6!}{4!\ 2!}$ $\dfrac{5 \times 6}{2}$ **15**

16. How many ways may 2 books be paired from a collection of 20 books?

$\dfrac{20 \cdot 19}{2}$ or $C(20, 2)$ **190**

MAVA Math: Enhanced Skills Solutions   Copyright © 2015 Marla Weiss

Cones 1

Find the volume of the cone in cubic units.	*Find the radius of the cone in units.*
1. radius = 10, height = 12 $V = \dfrac{\pi \cdot 10^2 \cdot 12}{3}$ **400π**	9. volume = 30π, height = 10 $30\pi = \dfrac{\pi \cdot r^2 \cdot 10}{3}$ **r = 3**
2. radius = 6, slant height = 10 h = 8 $V = \dfrac{\pi \cdot 6^2 \cdot 8}{3}$ **96π**	10. volume = 132π, height = 11 $132\pi = \dfrac{\pi \cdot r^2 \cdot 11}{3}$ **r = 6**
3. height = 10, slant height = 26 r = 24 $V = \dfrac{\pi \cdot 24^2 \cdot 10}{3}$ **1920π**	11. volume = 15π, height = 5 $15\pi = \dfrac{\pi \cdot r^2 \cdot 5}{3}$ **r = 3**
4. radius = 15, altitude = 7 $V = \dfrac{\pi \cdot 15^2 \cdot 7}{3}$ **525π**	12. volume = 90π, height = 10 $90\pi = \dfrac{\pi \cdot r^2 \cdot 10}{3}$ **r = 3√3**
5. diameter = 14, slant height = 25 r = 7 h = 24 $V = \dfrac{\pi \cdot 7^2 \cdot 24}{3}$ **392π**	13. volume = 135π, height = 5 $135\pi = \dfrac{\pi \cdot r^2 \cdot 5}{3}$ **r = 9**
6. diameter = 12, altitude = 10 $V = \dfrac{\pi \cdot 6^2 \cdot 10}{3}$ **120π**	14. volume = 576π, height = 12 $576\pi = \dfrac{\pi \cdot r^2 \cdot 12}{3}$ **r = 12**
7. height = 40, slant height = 41 r = 9 $V = \dfrac{\pi \cdot 9^2 \cdot 40}{3}$ **1080π**	15. volume = 72π, height = 6 $72\pi = \dfrac{\pi \cdot r^2 \cdot 6}{3}$ **r = 6**
8. height = 60, slant height = 61 r = 11 $V = \dfrac{\pi \cdot 11^2 \cdot 60}{3}$ **2420π**	16. volume = 80π, height = 12 $80\pi = \dfrac{\pi \cdot r^2 \cdot 12}{3}$ **r = 2√5**

Cones 2

Answer as indicated. Measurements are units, square units, and cubic units.

1. A cone "just fits" in a cylinder with diameter 10 and height 12. Find the volume between the two solids.

   $Vcyl = \pi \cdot 5^2 \cdot 12 = 300\pi$
   $Vcone = 1/3\ Vcyl = 100\pi$
   $300\pi - 100\pi = 200\pi$

   200π

5. If the radius of a cone is tripled and the height is doubled, find the percent increase in the volume.

   $v = (1/3)\pi \cdot r^2 \cdot h$
   $V = (1/3)\pi \cdot (3r)^2 \cdot 2h = 18v$

   $\dfrac{ch}{orig} = \dfrac{17v}{v} = \dfrac{17}{1}$ **1700%**

2. Two upright cones with heights 6 and 12 share a base with radius 8. Find the volume remaining when the smaller cone is removed from the larger cone.

   $h = 6, r = 8$ $H = 12, R = 8$
   $v = (1/3)\pi \cdot 8^2 \cdot 6$ $V = (1/3)\pi \cdot 8^2 \cdot 12$
   $v = 2 \cdot 64\pi$ $V = 4 \cdot 64\pi$

   $4 \cdot 64\pi - 2 \cdot 64\pi = 2 \cdot 64\pi =$ **128π**

6. The height of a cone is reduced by 1/3, and the radius is increased by 1/2. The volume of the new cone is what fraction of the volume of the original cone?

   $h = 30, r = 10$ $H = 20, R = 15$
   $v = (1/3)\pi \cdot 10^2 \cdot 30$ $V = (1/3)\pi \cdot 15^2 \cdot 20$
   $v = 1000\pi$ $V = 1500\pi$

   $\dfrac{1500\pi}{1000\pi} = \dfrac{\textbf{3}}{\textbf{2}}$

3. A cylinder with height 10 has a hemisphere with equal circumference attached at one end and a cone with altitude 16 and slant height 20 attached at the other end. Find the total volume.

   cone: 3-4-5 to 12-16-20 $\triangle$
   radius = 12

   $Vcyl = \pi \cdot 12^2 \cdot 10 = 1440\pi$
   $Vcone = (1/3)\pi \cdot 12^2 \cdot 16 = 768\pi$
   $Vhemi = (1/2) \cdot (4/3)\pi \cdot 12^3 = 1152\pi$ **3360π**

7. Two upright cones with heights 3 and 9 share a base with radius 6. Find the volume remaining when the smaller cone is removed from the larger cone.

   $h = 3, r = 6$ $H = 9, R = 6$
   $v = (1/3)\pi \cdot 6^2 \cdot 3$ $V = (1/3)\pi \cdot 6^2 \cdot 9$
   $v = 36\pi$ $V = 108\pi$

   $108\pi - 36\pi =$ **72π**

4. A cylinder with height 12 has a hemisphere attached at one end and a cone at the other, base to base. The cone has height 4 and slant height 5. Find the total surface area.

   cone: 3-4-5 $\triangle$; r = 3; d = 6
   lat SA cyl = $\pi \cdot 6 \cdot 12 = 72\pi$
   lat SA cone = $\pi \cdot 3 \cdot 5 = 15\pi$
   SA hemi = $(1/2) \cdot 4\pi \cdot 3^2 = 18\pi$
   $72\pi + 15\pi + 18\pi = 105\pi$ **105π**

8. A sphere of ice cream has radius 2. If the ice cream sits on an ice cream cone with the same radius and melts into the cone without dripping, find the height of the cone to hold all of the ice cream.

   $(4/3)\pi \cdot 2^3 = (1/3)\pi \cdot 2^2 \cdot h$
   $4 \cdot 8 = 4 \cdot h$
   $h = 8$
   8

Cones 3

A right triangle with one vertex at the origin is rotated about an axis. Find the volume in cubic units of the resulting cone with the given specifications.

1. vertices = (12,0) and (12, 5)
    rotated about *x*-axis

    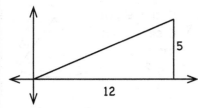

    $V = \dfrac{25\pi \times 12}{3} = 100\pi$

5. vertices = (0,10) and (6,10)
    rotated about *y*-axis

    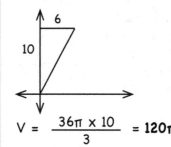

    $V = \dfrac{36\pi \times 10}{3} = 120\pi$

2. vertices = (18,–5) and (0, –5)
    rotated about *y*-axis

    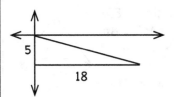

    $V = \dfrac{18 \times 18\pi \times 5}{3} = 540\pi$

6. legs = 9
    rotated about *x*-axis

    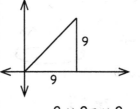

    $V = \dfrac{9 \times 9\pi \times 9}{3} = 243\pi$

3. one vertex = (15,0); one leg = 6
    rotated about *x*-axis

    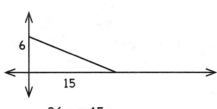

    $V = \dfrac{36\pi \times 15}{3} = 180\pi$

7. legs = 30
    rotated about *y*-axis

    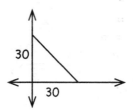

    $V = \dfrac{30 \times 30\pi \times 30}{3} = 9000\pi$

4. one vertex = (0,9); hypotenuse 15
    rotated about *y*-axis

    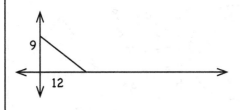

    $V = \dfrac{12 \times 12\pi \times 9}{3} = 432\pi$

8. vertices = (21,7) and (21, 0)
    rotated about *x*-axis

    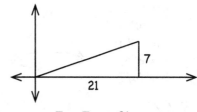

    $V = \dfrac{7 \times 7\pi \times 21}{3} = 343\pi$

radius of circle is slant
height of cone

Cones 4

Cut a circular paper into congruent sectors. Form one sector into a cone with no overlap.
Find the volume of the cone given two facts.

1. radius of circular paper = 9
    number of congruent sectors = 3

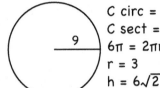

C circ = 18π
C sect = 6π
6π = 2πr
r = 3
h = $6\sqrt{2}$

V = 9π · $6\sqrt{2}$ / 3
V = $18π\sqrt{2}$

5. radius of circular paper = 10
    number of congruent sectors = 5

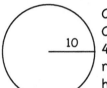

C circ = 20π
C sect = 4π
4π = 2πr
r = 2
h = $4\sqrt{6}$

V = 4π · $4\sqrt{6}$ / 3 = $\dfrac{16π\sqrt{6}}{3}$

2. radius of circular paper = 8
    number of congruent sectors = 4

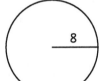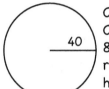

C circ = 16π
C sect = 4π
4π = 2πr
r = 2
h = $2\sqrt{15}$

V = 4π · $2\sqrt{15}$ / 3 = $\dfrac{8π\sqrt{15}}{3}$

6. radius of circular paper = 40
    number of congruent sectors = 10

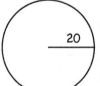

C circ = 80π
C sect = 8π
8π = 2πr
r = 4
h = $12\sqrt{11}$

V = 16π · $12\sqrt{11}$ / 3
V = $64π\sqrt{11}$

3. radius of circular paper = 18
    number of congruent sectors = 6

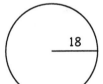

C circ = 36π
C sect = 6π
6π = 2πr
r = 3
h = $3\sqrt{35}$

V = 9π · $3\sqrt{35}$ / 3
V = $9π\sqrt{35}$

7. radius of circular paper = 20
    number of congruent sectors = 5

C circ = 40π
C sect = 8π
8π = 2πr
r = 4
h = $8\sqrt{6}$

V = 16π · $8\sqrt{6}$ / 3 = $\dfrac{128π\sqrt{6}}{3}$

4. radius of circular paper = 16
    number of congruent sectors = 8

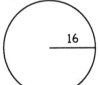

C circ = 32π
C sect = 4π
4π = 2πr
r = 2
h = $6\sqrt{7}$

V = 4π · $6\sqrt{7}$ / 3
V = $8π\sqrt{7}$

8. radius of circular paper = 24
    number of congruent sectors = 6

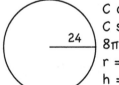

C circ = 48π
C sect = 8π
8π = 2πr
r = 4
h = $4\sqrt{35}$

V = 16π · $4\sqrt{35}$ / 3 = $\dfrac{64π\sqrt{35}}{3}$

lateral area = "πrl"	Cones 5

Find the surface area of the cone in square units.

The altitude need not be a whole number.

1. radius = 10, slant height = 12 SA = π · 10 · 12 + π · 10² **220π**	10. altitude = 8, slant height = 10 3, 4, 5 × 2; R = 6 SA = π · 6 · 10 + π · 6² SA = 60π + 36π = 96π **96π**
2. radius = 6, slant height = 11 SA = π · 6 · 11 + π · 6² **102π**	11. altitude = 24, slant height = 25 7, 24, 25; R = 7 SA = π · 7 · 25 + π · 7² SA = 175π + 49π = 224π **224π**
3. diameter = 10, slant height = 13 SA = π · 5 · 13 + π · 5² **90π**	12. altitude = 30, slant height = 34 8, 15, 17 × 2; R = 16 SA = π · 16 · 34 + π · 16² SA = 16π (34 + 16) = π · 16 · 50 **800π**
4. radius = 15, slant height = 7 SA = π · 15 · 7 + π · 15² **330π**	13. altitude = 60, slant height = 61 11, 60, 61; R = 11 SA = π · 11 · 61 + π · 11² SA = 11π (61 + 11) = π · 11 · 72 **792π**
5. diameter = 14, slant height = 25 SA = π · 7 · 25 + π · 7² **224π**	14. altitude = 15, slant height = 17 8, 15, 17; R = 8 SA = π · 8 · 17 + π · 8² SA = 8π (17 + 8) = π · 8 · 25 **200π**
6. diameter = 18, slant height = 11 SA = π · 9 · 11 + π · 9² **180π**	15. altitude = 14, slant height = 50 7, 24, 25 × 2; R = 48 SA = π · 48 · 50 + π · 48² SA = 2400π + 2304π = 4704π **4704π**
7. radius = 20, slant height = 21 SA = π · 20 · 21 + π · 20² **820π**	16. altitude = 21, slant height = 29 20, 21, 29; R = 20 SA = π · 20 · 29 + π · 20² SA = 20π (29 + 20) = π · 49 · 20 **980π**
8. diameter = 16, slant height = 17 SA = π · 8 · 17 + π · 8² **200π**	17. altitude = 9, slant height = 15 3, 4, 5 × 3; R = 12 SA = π · 12 · 15 + π · 12² SA = 180π + 144π **324π**
9. radius = 11, slant height = 18 SA = π · 11 · 18 + π · 11² **319π**	18. altitude = 24, slant height = 26 5, 12, 13 × 2; R = 10 SA = π · 10 · 26 + π · 10² SA = 260π + 100π **360π**

Cones 6

Find the volume of the cone frustrum in cubic units.

1. upper frustrum base radius = 5
   lower frustrum base radius = 15
   frustrum height = 24

   Or 5 is 1/3 of 15
   12 is 1/3 of 36

   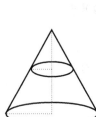

   $$\frac{h}{5} = \frac{h+24}{15}$$

   $3h = h+24$
   $2h = 24$
   $h = 12$

   V = 15x15πx36/3 = 2700π
   v = 5x5πx12/3 = 100π
   2700π − 100π = **2600π**

5. larger cone height = 45
   smaller cone height = 15
   frustrum slant height = 34

   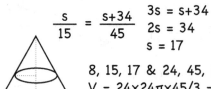

   $$\frac{s}{15} = \frac{s+34}{45}$$

   $3s = s+34$
   $2s = 34$
   $s = 17$

   8, 15, 17 & 24, 45, 51 tri
   V = 24x24πx45/3 = 8640π
   v = 8x8πx15/3 = 320π
   8640π − 320π = **8320π**

2. upper frustrum base radius = 9
   lower frustrum base radius = 12
   frustrum slant height = 5

   $$\frac{s}{9} = \frac{s+5}{12}$$

   $12s = 9s+45$
   $3s = 45$
   $s = 15$

   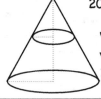

   9, 12, 15 & 12, 16, 20 tri
   V = 12x12πx16/3 = 768π
   v = 9x9πx12/3 = 324π
   768π − 324π = **444π**

6. larger cone base area = 36
   larger cone altitude = 18
   frustrum altitude = 6

   $$\frac{12\text{x}12}{18\text{x}18} = \frac{a}{36}$$

   $9a = 36\text{x}4$
   $a = 16$

   V = 36x18/3 = 216
   v = 16x12/3 = 64
   216 − 64 = **152**

3. larger cone height = 40
   smaller cone height = 20
   frustrum slant height = 29

   20 is 1/2 of 40
   29 is 1/2 of 58
   20, 21, 29 & 40, 42, 58 tri

   V = 42x42x40/3 = 23,520π
   v = 21x21πx20/3 = 2940π
   23,520π−2940π = **20,580π**

7. upper cone height = 12
   frustrum height = 8
   frustrum slant height = 10

   $$\frac{12}{20} = \frac{s}{s+10}$$

   $5s = 3s+30$
   $2s = 30$
   $s = 15$

   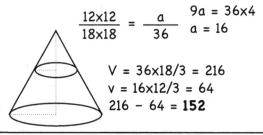

   9, 12, 15 & 15, 20, 25 tri
   V = 15x15πx20/3 = 1500π
   v = 9x9πx12/3 = 324π
   1500π − 324π = **1176π**

4. upper frustrum base radius = 6
   lower frustrum base radius = 9
   frustrum slant height = 5

   $9s = 6s+30$

   $$\frac{s}{6} = \frac{s+5}{9}$$

   $3s = 30$
   $s = 10$

   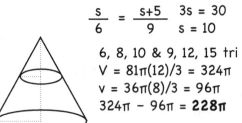

   6, 8, 10 & 9, 12, 15 tri
   V = 81π(12)/3 = 324π
   v = 36π(8)/3 = 96π
   324π − 96π = **228π**

8. upper frustrum base area = 4
   lower frustrum base area = 16
   smaller cone altitude = 6

   $$\frac{4}{16} = \frac{6\text{x}6}{H\text{x}H}$$ $H = 12$

   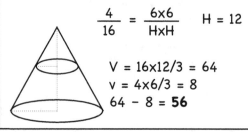

   V = 16x12/3 = 64
   v = 4x6/3 = 8
   64 − 8 = **56**

Coordinate Plane 1

Answer as indicated. All rectangles and squares are parallel to the axes. Linear measurements are units; area measurements are square units

---

1. Rectangle ABCD has vertices A(2,–3) and C(–1,4). Find vertices B and D.

    B = (–1, –3)
    D = (2, 4)   or vice versa

---

8. Find the coordinates of the point equidistant from points A(7, 4), B(–1, 4), C(–1, –2), and D(7, –2).

    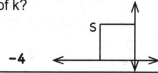
    (–1,4)   (7,4)   W = 8, H = 6
                          7–4 = 3, 4–3 = 1
    (–1,–2)   (7,–2) **(3, 1)**

---

2. Rectangle EFGH has vertices E(–5,8) and G(9,–7). Find vertices F and H.

    F = (9, 8)
    H = (–5, –7)   or vice versa

---

9. ORST is a square in Quadrant II. Point O is the origin. If point S is (k, 4), what is the value of k?

    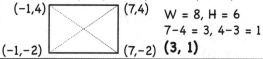
    –4

---

3. Find the area of a rectangle with two vertices (9,2) and (–1, –13).

    B = 10
    H = 15
    area = **150**

---

10. OEFG is a square in Quadrant III. Point O is the origin. If point F is (–6, k), what is the value of k?

    –6

---

4. Find the area of a rectangle with two vertices (7,10) and (–4, –2).

    B = 11
    H = 12
    area = **132**

---

11. A square with coordinates A(–6,–3), B(–6, 6), C(3, 6), and D(3,–3) is translated 3 units in the positive y direction and then reflected across the x-axis. Find the new coordinates of C.   **(3, –9)**

---

5. Find the area and perimeter of the rectangle with vertices A(9,2), B(9,–6), C(–5, –6), and D(–5, 2).

    area = 14 x 8 = **112**
    perimeter = 2 x 22 = **44**

---

12. Find the 2 missing coordinates of the rectangle.

    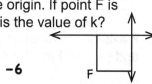
    (–x, –y)   **(x, –y)**
    (–x, y)   (x, y)

---

6. A square with coordinates A(2,–1), B(7, –1), C(7, 4), and D(2,4) is translated 4 units in the positive x direction and then reflected across the y-axis. Find the new coordinates of C.   **(–11, 4)**

---

13. Find the coordinates of the point equidistant from points A(9, 6), B(–2, 6), C(–2, –3), and D(9, –3).

    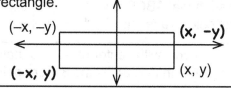
    (–2,6)   (9,6)   W = 11, H = 9
                          (3.5, 1.5)
    (–2,–3)   (9,–3)

---

7. OABC is a square in Quadrant I with coordinates O at the origin and A(3a, 0). Find the coordinates of B and C.

    B = (3a, 3a)
    C = (0, 3a)

---

14. Find the area and perimeter of the rectangle with vertices A(8,7), B(8,–8), C(–7, –8), and D(–7, 7).

    area = 15 x 15 = **225**
    perimeter = 2 x 30 = **60**

MAVA Math: Enhanced Skills Solutions   Copyright © 2015 Marla Weiss

Coordinate Plane 2

Answer as indicated.

1. A circle with center at the origin passes through (0,7). Name 3 more points on the circle.

(0, –7)   r = 7
(7, 0)
(–7, 0)

7. Name a point on an axis that is twice as far from (2,3) as it is from (1,2).

(0, 1)

2. Points A (2,1), B(8,1), and C(x,y) are such that the area of △ABC is 21. Find the 2 possible values for y.

AB = 6
h = 7
1 + 7 = 8
1 – 7 = –6   **8 and –6**

8. A circle with center (0,0) has a diameter with one endpoint (5,6). Find the other endpoint.

(–5, –6)

3. A rhombus with 2 sides parallel to the x-axis has vertices at (6, 15), (1, 3), and (14, 3). Find the 4th vertex.

14 – 1 = 13
6 + 13 = 19
(19, 15)

9. Travel from E to C on rectangle ABCD. How much farther is the trip through A and D than the trip through B?

A   E(5,7)   B

D(2,1)   C(12,1)
3 + 10 – 7 = **6**

4. AB = 9, CD = 9
D(10,6)

Find the coordinates of D such that ABCD is an isosceles trapezoid.

C(1,6) D
B(–3,2)
slope $\overline{BC}$
= slope $\overline{AD}$
A(–3,–7) = 1

10. The area of the largest rectangle is 6 times the area of the smallest. Find the coordinates of A.   **(18, 8)**

A   (24,8)
(36)(8) = (6w)(8)
w = 6
24 – 6 = 18
(–12,0)

5. A circle with center at (–2, –5) passes through (–2, 3). Name 3 more points on the circle.

(–2, –13)   r = 8
(6, –5)
(–10, –5)

11. Points A (6,2), B(10,2), and C(x,y) are such that the area of △ABC is 12. Find the 2 possible values for y.

AB = 4
h = 6
2 + 6 = 8
2 – 6 = –4   **8 and –4**

6. $\overline{AB}$ is a diameter of the circle with center at the origin. Find the coordinates of B.

(–3, –3)

A(3,3)
O
B

12. Find the area of the triangle with coordinates (1,0), (7,0), and (9,8).

b = 6
h = 8
A = **24**

Coordinate Plane 3

Find the area in square units of the convex or concave polygon determined by the coordinates.

1.   (0,0), (0,4), (9,8), (18,8), (18,4), (9,0)

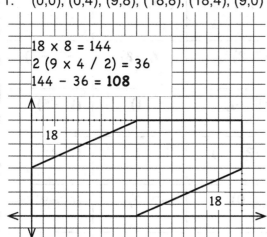

18 x 8 = 144
2 (9 x 4 / 2) = 36
144 − 36 = **108**

18

18

4.   (0,1), (2,8), (8,4), (8,1), (2, −1)

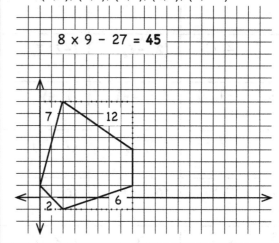

8 x 9 − 27 = **45**

7   12

.2   6

2.   (−5,7), (−2,3), (6,−1)

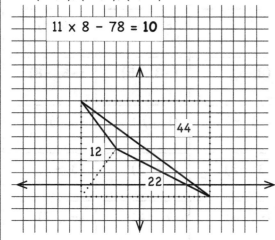

11 x 8 − 78 = **10**

44

12

22

5.   (8,−2), (4,−5), (1,3), (1,6), (10,8)

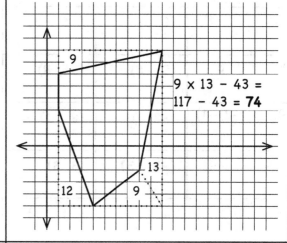

9

9 x 13 − 43 =
117 − 43 = **74**

13

12   9

3.   (9,2), (5,4), (2,8), (−6,3), (−3,−3)

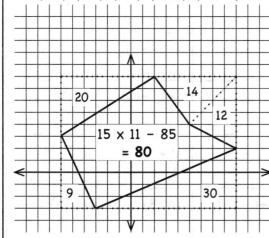

20   14

12

15 x 11 − 85
= **80**

9   30

6.   (−9,1), (−5,3), (6,1), (1,−7), (−7,−4)

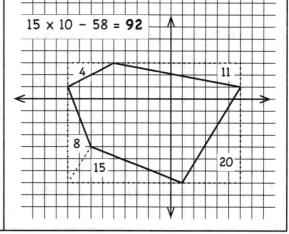

15 x 10 − 58 = **92**

4   11

8

15   20

Find the reflection of each point across the:	x-axis.	y-axis.	line y = x.	line y = –x.	origin.
1. (1, 4)	(1, –4)	(–1, 4)	(4, 1)	(–4, –1)	(–1, –4)
2. (2, 7)	(2, –7)	(–2, 7)	(7, 2)	(–7, –2)	(–2, –7)
3. (–1, 5)	(–1, –5)	(1, 5)	(5, –1)	(–5, 1)	(1, –5)
4. (–7, –2)	(–7, 2)	(7, –2)	(–2, –7)	(2, 7)	(7, 2)
5. (0, 9)	(0, –9)	(0, 9)	(9, 0)	(–9, 0)	(0, –9)
6. (8, 0)	(8, 0)	(–8, 0)	(0, 8)	(0, –8)	(–8, 0)
7. (6, –3)	(6, 3)	(–6, –3)	(–3, 6)	(3, –6)	(–6, 3)
8. (5, 8)	(5, –8)	(–5, 8)	(8, 5)	(–8, –5)	(–5, –8)
9. (–2, –1)	(–2, 1)	(2, –1)	(–1, –2)	(1, 2)	(2, 1)
10. (–3, 3)	(–3, –3)	(3, 3)	(3, –3)	(–3, 3)	(3, –3)
11. (4, –7)	(4, 7)	(–4, –7)	(–7, 4)	(7, –4)	(–4, 7)
12. (3, –4)	(3, 4)	(–3, –4)	(–4, 3)	(4, –3)	(–3, 4)
13. (9, 9)	(9, –9)	(–9, 9)	(9, 9)	(–9, –9)	(–9, –9)
14. (10, 1)	(10, –1)	(–10, 1)	(1, 10)	(–1, –10)	(–10, –1)
15. (11, 2)	(11, –2)	(–11, 2)	(2, 11)	(–2, –11)	(–11, –2)
16. (–5, –6)	(–5, 6)	(5, –6)	(–6, –5)	(6, 5)	(5, 6)

Coordinate Plane 4

Counting 1

Answer by drawing a Venn diagram.

1. Of 83 houses, 37 need new roofs and 46 need new shrubs. If 19 need both, how many need neither?

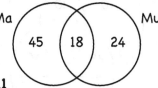

83 – 37 – 27 = **19**

2. Of 98 students, 63 take math, 42 take music, and 18 take both math and music. How many students take neither?

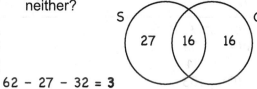

98 – 63 – 24 = **11**

3. Of 62 girls, 43 are wearing sweaters, 32 are wearing coats, and 16 are wearing both. How many are wearing neither?

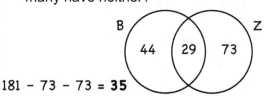

62 – 27 – 32 = **3**

4. Of 181 dresses, 73 have buttons, 102 have zippers, and 29 have both. How many have neither?

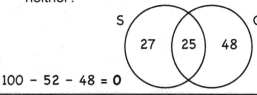

181 – 73 – 73 = **35**

5. Of 100 students, 52 take Spanish, 73 take German, and 25 take both languages. How many students take neither?

100 – 52 – 48 = **0**

6. Show the relationship among the factors of 18, 30, and 45.

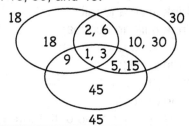

7. Show the relationship among the factors of 20, 50, and 60.

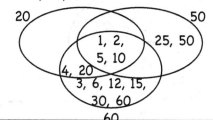

8. Show the relationship among the factors of 24, 40, and 42.

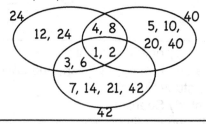

9. Show the relationship among the factors of 16, 28, and 35.

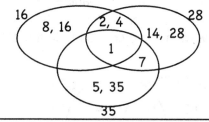

10. Show the relationship among the infinite sets N, W, Z, Q, and R. Write the digit 0 and "irrationals" in the correct band.

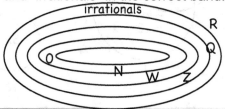

MAVA Math: Enhanced Skills Solutions Copyright © 2015 Marla Weiss

Counting 2

Answer by drawing a Venn diagram.

1. A survey of 300 children found:
45 drink juice, milk, and cocoa;
60 drink juice and milk;
70 drink milk and cocoa;
46 drink juice and cocoa;
160 drink milk;
75 drink juice; and
110 drink cocoa.
How many drink none of the three?

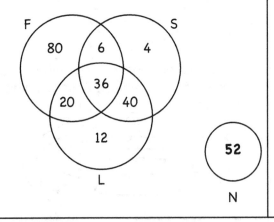

3. For sets A, B, and C:
n(A) = 100
n(B) = 80
n(C) = 60
n(A AND B) = 32
n(B AND C) = 28
n(A AND C) = 40
n(A AND B AND C) = 12
n(B′) = 74
Find n(A OR B OR C)′.

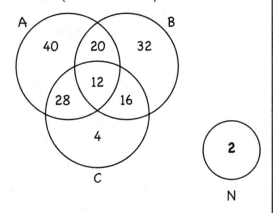

2. A survey of 250 students found:
36 study French, Spanish, and Latin;
142 study French;
86 study Spanish;
108 study Latin;
56 study French and Latin;
76 study Latin and Spanish;
42 study French and Spanish;
How many study none of the three?

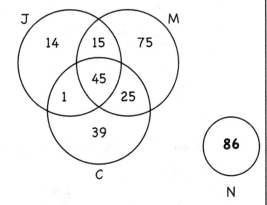

4. For sets A, B, and C:
n(A) = 109
n(B) = 111
n(C) = 153
n(A AND B) = 51
n(B AND C) = 63
n(A AND C) = 39
n(A AND B AND C) = 21
n(A′) = 133
Find n(A OR B OR C)′.

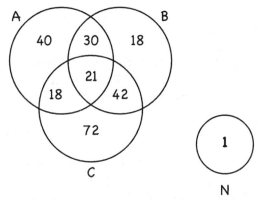

Counting 3

Answer by constructing a crosstabulation chart.

1. 200 people, half of whom were children, were asked if they liked ice skating. 80 replied yes, including 50% of the 60 men. Twelve of the women replied no. How many children said that they did not like ice skating?

	M	W	C	TOT
Y	30	28	22	80
N	30	12	**78**	120
TOT	60	40	100	200

2. 250 people, one fifth of whom were children, were asked if they used mouthwash. 160 said yes, including 90% of the 100 women. 75% of the men answered no. How many of the children used mouthwash?

	M	W	C	TOT
Y	25	90	**45**	160
N	75	10	5	90
TOT	100	100	50	250

3. In a group of 200 adults, 58% are women, 30% are Republicans, and 30% are Independents. Of the men, 1/3 are Republicans and 32 are Democrats. How many women are Independents?

	D	R	I	TOT
M	32	28	24	84
W	48	32	**36**	116
TOT	80	60	60	200

4. 500 people, of whom 1/2 were children and 1/5 were men, were asked if they liked chewing gum. 60% said yes, including 12% of the men. 16% of the women said no. How many children liked gum?

	M	W	C	TOT
Y	12	126	**162**	300
N	88	24	88	200
TOT	100	150	250	500

5. A yes/no question was asked of 300 people. Of the 150 men, 2/3 said yes. 136 people, including 39 of the women, said no. If 25% of the yes answers were women, how many children were in the group?

	M	W	C	TOT
Y	100	41	23	164
N	50	39	47	136
TOT	150	80	**70**	300

6. Of 220 stamps in a collection, 3/11 were U.S. stamps and 50% were cancelled. If 60% of the U.S. stamps were cancelled, how many of the stamps were not U.S. and not cancelled?

	Canc	Not	TOT
US	36	24	60
Not	74	**86**	160
TOT	110	110	220

Counting 4

Use a variable in the Venn to answer.

1. Of 90 students, 40 study French, 60 study Spanish, and 5 study neither. How many study both?

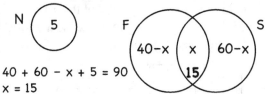

40 + 60 − x + 5 = 90
x = 15

2. Of 130 girls, 72 wore coats, 64 wore sweaters, and 4 wore neither. How many wore both?

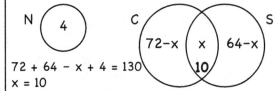

72 + 64 − x + 4 = 130
x = 10

3. Of 160 houses, 65 need new roofs, 108 need new landscaping, and 12 need neither. How many need both?

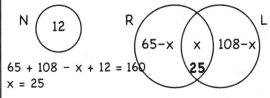

65 + 108 − x + 12 = 160
x = 25

4. Of 135 children, 95 played tennis, 50 played soccer, and 15 played neither. How any played both?

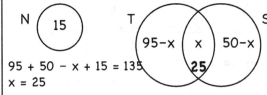

95 + 50 − x + 15 = 135
x = 25

5. Of 136 students, 42 play brass instruments, 95 play string instruments, and 7 play neither. How many play both?

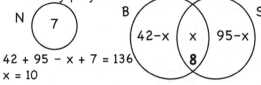

42 + 95 − x + 7 = 136
x = 10

6. Of 73 houses, 28 need new roofs and 36 need new shrubs. If 17 need neither, how many need only a roof?

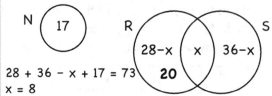

28 + 36 − x + 17 = 73
x = 8

7. Of 91 students, 54 take math, 27 take music, and 13 take neither subject. How many students take only math?

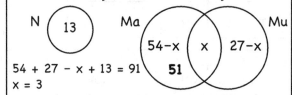

54 + 27 − x + 13 = 91
x = 3

8. Of 124 girls, 78 are wore sweaters, 63 wore coats, and 25 wore neither. How many wore only a coat?

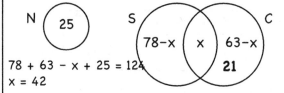

78 + 63 − x + 25 = 124
x = 42

9. Of 106 dresses, 76 have buttons, 79 have zippers, and 9 have neither. How many have only buttons?

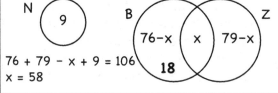

76 + 79 − x + 9 = 106
x = 58

10. Of 104 students, 72 take Spanish, 55 take German, and 18 take neither language. How many students take only German?

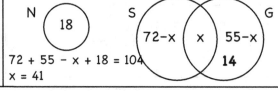

72 + 55 − x + 18 = 104
x = 41

Counting 5

Count the perfect squares between the two given numbers.	Count the numbers with the specified digit sum.
1. between 101 and 10,010 $10^2 = 100$ $11 \longrightarrow 100$ $100^2 = 10,000$ $100 - 11 + 1 = \mathbf{90}$	9. How many 2-digit numbers have a perfect square digit sum? 10, 13, 31, 22, 40, 90, perf sq = 1, 4, 9, 16 81, 18, 72, 27, 63, 36, **17** 54, 45, 97, 79, 88
2. between 890 and 2520 $30^2 = 900$ $30 \longrightarrow 50$ $50^2 = 2500$ $50 - 30 + 1 = \mathbf{21}$	10. How many 3-digit numbers have an odd 2-digit perfect square digit sum? sum = 25 997, 979, 799, 988, 898, 889 **6**
3. between 601 and 3601 $24^2 = 576$ $25 \longrightarrow 60$ $60^2 = 3600$ $60 - 25 + 1 = \mathbf{36}$	11. How many 2-digit numbers have a perfect cube digit sum? 10, 80, 71, 17, 62, 26, perf cu = 1, 8 35, 53, 44 **9**
4. between 50 and 8150 $8^2 = 64$ $64 \longrightarrow 90$ $90^2 = 8100$ $90 - 8 + 1 = \mathbf{83}$	12. How many 3-digit numbers have an odd perfect cube digit sum? sum = 1 or 27 100, 999 **2**
5. between 401 and 40,001 $20^2 = 400$ $21 \longrightarrow 200$ $200^2 = 40,000$ $200 - 21 + 1 =$ **180**	13. How many 2-digit numbers have a 1-digit prime digit sum? sum = 2, 3, 5, or 7 20, 11, 30, 21, 12, 50, 41, 14, 32, 23, 70, 61, 16, 52, 25, 43, 34 **17**
6. between 899 to 250,001 $30^2 = 900$ $30 \longrightarrow 500$ $500^2 = 250,000$ $500 - 30 + 1 =$ **471**	14. How many 2-digit numbers have a 2-digit prime digit sum? sum = 11, 13, or 17 92, 83, 74, 65, 56, 47, 38, 29, 94, 85, 76, 67, 58, 49, 98, 89 **16**
7. between 48 and 490,048 $7^2 = 49$ $7 \longrightarrow 700$ $700^2 = 490,000$ $700 - 7 + 1 =$ **694**	15. How many 4-digit numbers have an even prime digit sum? sum = 2 2000, 1100, 1010, 1001 **4**
8. between 220 and 920 $15^2 = 225$ $15 \longrightarrow 30$ $30^2 = 900$ $30 - 15 + 1 = \mathbf{16}$	16. How many whole numbers from 100 to 400 have a digit sum of 17? 197, 188, 179, 296, 287, 278, 269, 395, 386, 377, 368, 359 **12**

Counting 6

How many triangles? Labeling vertices with letters may help.

1.

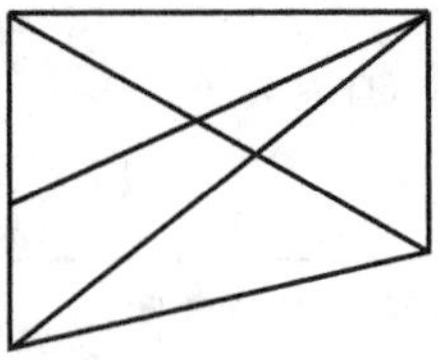

6--0
5--0
4--1
3--2
2--6
1--6

15

2.

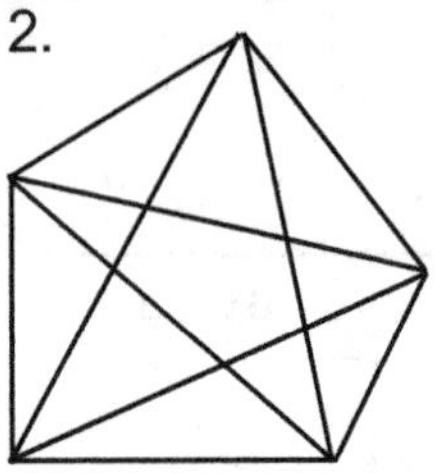

6-11--0
5--5
4--0
3--10
2--10
1--10

35

3.

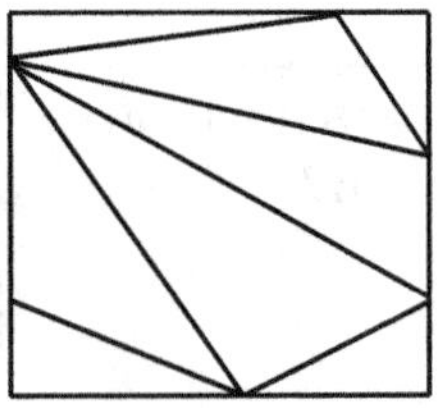

3-8--0
2--1
1--8

9

4.

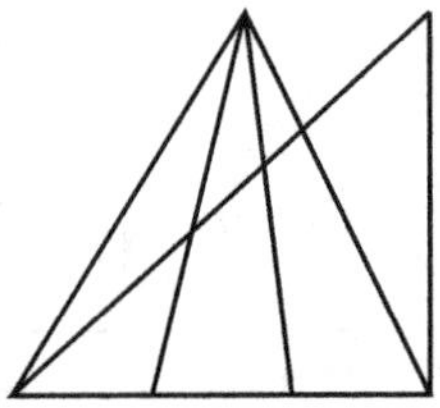

7--0
6--1
5--0
4--3
3--2
2--6
1--7

19

5.

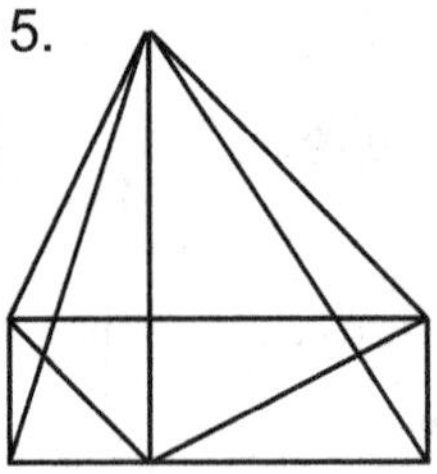

7-12--0
6--1
5--0
4--4
3--4
2--13
1--11

33

6.

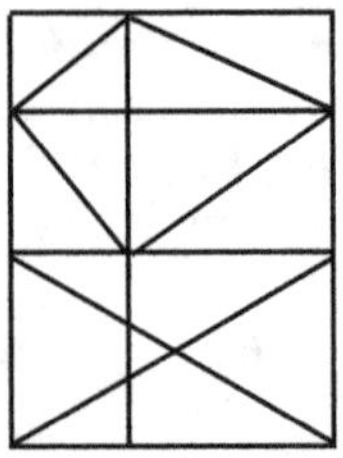

3-12--0
2--6
1--10

16

How many squares?
How many rectangles (includes squares)?

7.

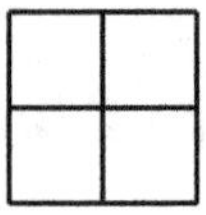

sq =
1 + 4 = **5**

rec =
$(1 + 2)^2 = 3^2 =$ **9**

8.

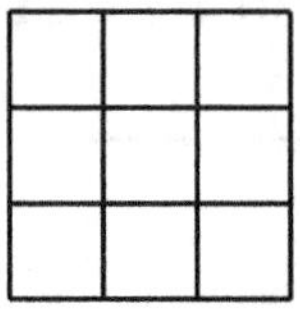

sq =
1 + 4 + 9 = **14**

rec =
$(1 + 2 + 3)^2 = 6^2 =$ **36**

9.

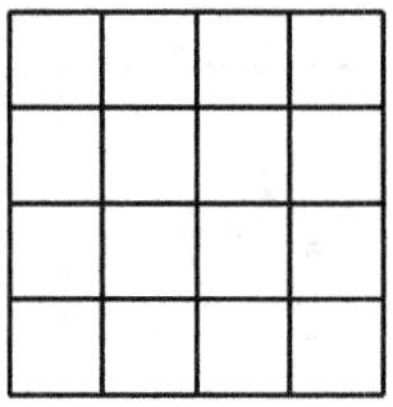

sq =
1 + 4 + 9 + 16 = **30**

rec =
$(1 + 2 + 3 + 4)^2 = 10^2$
= **100**

10.

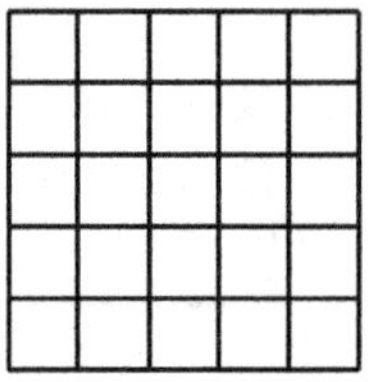

sq =
1 + 4 + 9 + 16 + 25 = **55**

rec =
$(1 + 2 + 3 + 4 + 5)^2 =$
$15^2 =$ **225**

11.

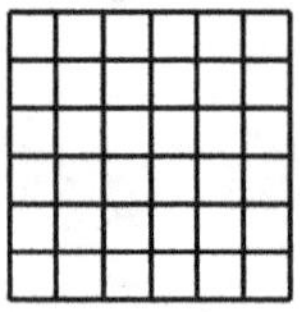

sq =
1 + 4 + 9 + 16 + 25 + 36
= **91**

rec =
(1 + 2 + 3 + 4 + 5 +
$6)^2 = 21^2 =$ **441**

12.

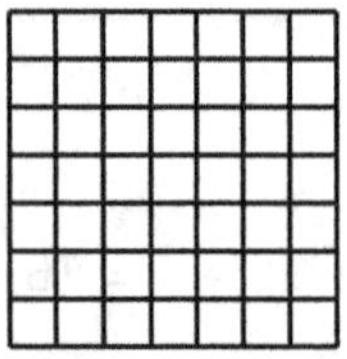

sq =
1 + 4 + 9 + 16 + 25 + 36
+ 49 = **140**

rec =
(1 + 2 + 3 + 4 + 5 + 6
$+ 7)^2 = 28^2 =$ **784**

 Copyright © 2015 Marla Weiss

71

Counting 7

Count using the Multiplication Principle.	*Count the multiples.*
1. How many 3-digit numbers with no 0s and at least one 4 are divisible by 5 ? 445 counted twice. Minus 1. $\underset{4}{1}\ \underset{}{9}\ \underset{5}{1}$ OR $\underset{}{9}\ \underset{4}{1}\ \underset{5}{1}$ **17**	8. How many numbers from 1 to 100 inclusive are multiples of 2 or 3? 100/2 = 50 100/3 = 33 50 + 33 − 16 = **67** 100/6 = 16
2. How many 3-digit numbers with different digits are divisible by 10? $\underline{9}\ \ \underline{8}\ \ \underline{1}$ **72**	9. How many numbers from 1 to 100 inclusive are multiples of 3 or 4? 100/3 = 33 100/4 = 25 33 + 25 − 8 = **50** 100/12 = 8
3. How many 3-digit numbers have the tens digit one greater than the ones digit? $\underline{9}\ \ \underline{9}\ \ \underline{1}$ **81**	10. How many numbers from 1 to 100 inclusive are multiples of 3 or 5? 100/3 = 33 100/5 = 20 33 + 20 − 6 = **47** 100/15 = 6
4. How many 3-digit numbers have the hundreds digit one greater than the tens digit? $\underline{9}\ \ \underline{1}\ \ \underline{10}$ **90**	11. How many numbers from 1 to 200 inclusive are multiples of 6 or 8? 200/6 = 33 200/8 = 25 33 + 25 − 8 = **50** 200/24 = 8
5. How many 3-digit numbers with no 0s and different digits are divisible by 2? $\underline{8}\ \ \underline{7}\ \ \underline{4}$ **224**	12. How many numbers from 1 to 100 inclusive are multiples of 2, 3, or 5? 100/2 = 50 100/6 = 16 50 + 33 + 20 − 100/3 = 33 100/10 = 10 (16 + 10 + 6) + 3 100/5 = 20 100/15 = 6 = **74** 100/30 = 3
6. How many 3-digit even numbers have all prime digits? $\underline{4}\ \ \underline{4}\ \ \underline{1}$ **16**	13. How many numbers from 1 to 100 inclusive are multiples of 2, 3, or 7? 100/2 = 50 100/6 = 16 50 + 33 + 14 − 100/3 = 33 100/14 = 7 (16 + 7 + 4) + 2 100/7 = 14 100/21 = 4 = **72** 100/42 = 2
7. How many 3-digit even numbers have exactly 3 different prime digits? $\underline{3}\ \ \underline{2}\ \ \underline{1}$ **6**	14. How many numbers from 1 to 200 inclusive are multiples of 3, 5, or 7? 200/3 = 66 200/15 = 13 66 + 40 + 28 − 200/5 = 40 200/21 = 9 (13 + 9 + 5) + 1 200/7 = 28 200/35 = 5 = **108** 200/105 = 1

MAVA Math: Enhanced Skills Solutions Copyright © 2015 Marla Weiss

Counting 8

Count the number of paths from A to B moving only right and/or up.

1.

6.

2.

7.

3.

8.

4.

9.

5.

10.

Cubes 1

How many cubes with the given edge can maximally fill a rectangular box with the specified unit measurements (or be cut from a block with those measurements)?

1. cube edge 2
   box (block) 8 by 10 by 12

   $\dfrac{8 \times 10 \times 12}{2 \times 2 \times 2}$ 4 × 5 × 6 = **120**

9. cube edge 11
   box (block) 55 by 22 by 66

   $\dfrac{55 \times 22 \times 66}{11 \times 11 \times 11}$ 5 × 2 × 6 = **60**

2. cube edge 4
   box (block) 36 by 24 by 20

   $\dfrac{36 \times 24 \times 20}{4 \times 4 \times 4}$ 9 × 6 × 5 = **270**

10. cube edge 8
    box (block) 32 by 16 by 64

    $\dfrac{32 \times 16 \times 64}{8 \times 8 \times 8}$ 4 × 2 × 8 = **64**

3. cube edge 3
   box (block) 36 by 18 by 15

   $\dfrac{36 \times 18 \times 15}{3 \times 3 \times 3}$ 12 × 6 × 5 = **360**

11. cube edge 5
    box (block) 15 by 35 by 50

    $\dfrac{15 \times 35 \times 50}{5 \times 5 \times 5}$ 3 × 7 × 10 = **210**

4. cube edge 5
   box (block) 45 by 30 by 25

   $\dfrac{45 \times 30 \times 25}{5 \times 5 \times 5}$ 9 × 6 × 5 = **270**

12. cube edge 6
    box (block) 24 by 30 by 66

    $\dfrac{24 \times 30 \times 66}{6 \times 6 \times 6}$ 4 × 5 × 11 = **220**

5. cube edge 12
   box (block) 24 by 48 by 60

   $\dfrac{24 \times 48 \times 60}{12 \times 12 \times 12}$ 2 × 4 × 5 = **40**

13. cube edge 9
    box (block) 24 by 48 by 60

    $\dfrac{18 \times 36 \times 63}{9 \times 9 \times 9}$ 2 × 4 × 7 = **56**

6. cube edge 6
   box (block) 36 by 18 by 60

   $\dfrac{36 \times 18 \times 60}{6 \times 6 \times 6}$ 6 × 3 × 10 = **180**

14. cube edge 13
    box (block) 26 by 52 by 65

    $\dfrac{26 \times 52 \times 65}{13 \times 13 \times 13}$ 2 × 4 × 5 = **40**

7. cube edge 3
   box (block) 6 by 6 by 9

   $\dfrac{6 \times 6 \times 9}{3 \times 3 \times 3}$ 2 × 2 × 3 = **12**

15. cube edge 8
    box (block) 24 by 40 by 72

    $\dfrac{24 \times 40 \times 72}{8 \times 8 \times 8}$ 3 × 5 × 9 = **135**

8. cube edge 7
   box (block) 14 by 42 by 56

   $\dfrac{14 \times 42 \times 56}{7 \times 7 \times 7}$ 2 × 6 × 8 = **96**

16. cube edge 15
    box (block) 45 by 75 by 90

    $\dfrac{45 \times 75 \times 90}{15 \times 15 \times 15}$ 3 × 5 × 6 = **90**

Cubes 2

Answer as indicated. Measurements are in units, square units, and cubic units.

1. Find the volume of a cube with surface area 384.	SA = 384 AF = 64 E = 8 V = 8^3 = **512**	9. Find the volume of a cube with surface area 150.	SA = 150 AF = 25 E = 5 V = 5^3 = **125**
2. Find the surface area of a cube with volume 1000.	V = 1000 E = 10 AF = 100 SA = **600**	10. Find the surface area of a cube with volume 1728.	V = 1728 E = 12 AF = 144 SA = **864**
3. Find the volume of a cube with surface area 216.	SA = 216 AF = 36 E = 6 V = **216**	11. If the volume and surface area of a cube are equal, find its edge.	$6e^2 = e^3$ E = **6**
4. Find the volume of a cube with surface area 18.	SA = 18 AF = 3 E = $\sqrt{3}$ V = **$3\sqrt{3}$**	12. Find the volume of a cube with surface area 726.	SA = 726 AF = 121 E = 11 V = **1331**
5. Find the sum of the edges of a cube with volume 729.	V = 729 E = 9 SE = **108**	13. Find the volume of a cube with surface area 30.	SA = 30 AF = 5 E = $\sqrt{5}$ V = **$5\sqrt{5}$**
6. Find the volume of a cube if the sum of its edges is 36.	SE = 36 E = 3 V = **27**	14. Find the sum of the edges of a cube with surface area 294.	SA = 294 AF = 49 E = 7 SE = **84**
7. Find the surface area of a cube with volume 343.	V = 343 E = 7 AF = 49 SA = **294**	15. Find the sum of the edges of a cube with surface area 96.	SA = 96 AF = 16 E = 4 SE = **48**
8. A cube has surface area 600. Find the sum of its edges.	SA = 600 AF = 100 E = 10 SE = **120**	16. Find the surface area of a cube if the sum of its edges is 156.	SE = 156 E = 13 AF = 169 SA = **1014**

Cubes 3

Answer as indicated.

---

1. Find the percent change in the surface area of a cube if the edge is increased by 50%.

Let e = 10	New e = 15	$\dfrac{ch}{orig} = \dfrac{750}{600}$
Af = 100	Af = 225	$= \dfrac{125}{100}$ **125% I**
SA = 600	SA = 1350	

7. Find the percent change in the diagonal of a cube if the volume is increased by 700%.

Let v = 8	V = 64	$\dfrac{ch}{orig} = \dfrac{2\sqrt{3}}{2\sqrt{3}}$
e = 2	E = 4	$= 1$ **100% I**
diag = $2\sqrt{3}$	diag = $4\sqrt{3}$	

---

2. Find the percent change in the surface area of a cube if the edge is decreased by 50%.

Let e = 10	New e = 5	$\dfrac{ch}{orig} = \dfrac{450}{600}$
Af = 100	Af = 25	$= \dfrac{75}{100}$ **75% D**
SA = 600	SA = 150	

8. Find the percent increase in each edge of a cube if the surface area increased by 21%.

$SA = 6e^2$

$(1.21)6e^2 = (6)(1.1e)(1.1e)$

1.1 is **10% I**

---

3. Find the percent change in the volume of a cube if the edge is increased by 50%.

Let e = 2	$\dfrac{ch}{orig} = \dfrac{19}{8} = 2\dfrac{3}{8}$
v = 8	
E = 3	**237.5% I**
V = 27	

9. Find the percent change in the surface area of a cube if the edge is increased by 20%.

Let e = 10	New e = 12	$\dfrac{ch}{orig} = \dfrac{264}{600}$
Af = 100	Af = 144	$= \dfrac{44}{100}$ **44% I**
SA = 600	SA = 864	

---

4. Find the percent change in the volume of a cube if the edge is decreased by 50%.

Let E = 2	$\dfrac{ch}{orig} = \dfrac{7}{8}$
V = 8	
e = 1	**87.5% D**
v = 1	

10. Find the percent change in the volume of a cube if the edge is increased by 20%.

Let e = 10	$\dfrac{ch}{orig} = \dfrac{728}{1000} = \dfrac{72.8}{100}$
V = 1000	
E = 12	**72.8% I**
v = 1728	

---

5. Find the ratio of the volumes of a cube to that of the cube with edge increased by 25%.

Let e = 8	$\dfrac{v}{V} = \dfrac{4 \times 4 \times 4}{5 \times 5 \times 5}$
v = 8×8×8	
E = 10	$= \dfrac{64}{125}$
V = 10×10×10	

11. Find the percent change in the surface area of a cube if the edge is decreased by 20%.

Let e = 10	New e = 8	$\dfrac{ch}{orig} = \dfrac{216}{600}$
Af = 100	Af = 64	$= \dfrac{36}{100}$ **36% D**
SA = 600	SA = 384	

---

6. Find the ratio of the surface areas of a cube to that of the cube with edge decreased by 33 1/3%.

Let E = 9	New e = 6	$\dfrac{SA}{sa} = \dfrac{3 \times 3}{2 \times 2}$
Af = 9×9	Af = 6×6	$= \dfrac{9}{4}$
SA = 9×9×6	SA = 6×6×6	

12. Find the percent change in the volume of a cube if the edge is decreased by 20%.

Let E = 10	$\dfrac{ch}{orig} = \dfrac{488}{1000} = \dfrac{48.8}{100}$
V = 1000	
e = 8	**48.8% D**
v = 512	

Cubes 4

A solid "just fits" in a cube. They have congruent heights. Find the volume inside the cube but outside the solid. Also find the ratio of the volumes, solid to cube.

1. cylinder inside cube with edge 10

Vol cyl = (5x5)π(10) = 250π

Vol cube = 10x10x10 = 1000

Vol gap = **1000 – 250π** $\frac{\pi}{4}$

---

7. sphere inside cube with edge 5

Vol sph = 4π(5x5x5)/(8x3) = 125π/6

Vol cube = 5x5x5 = 125

Vol gap = **125 –** $\frac{125\pi}{6}$ $\frac{\pi}{6}$

---

2. square pyramid inside cube with volume 216

Vol sq pyr = (6x6x6)/3 = 72

Vol cube = 216

Vol gap = 216 – 72 = **144** $\frac{1}{3}$

---

8. cone inside cube with face area 4

Vol cone = (1x1)π(2)/3 = 2π/3

Vol cube = 2x2x2 = 8

Vol gap = **8 –** $\frac{2\pi}{3}$ $\frac{\pi}{12}$

---

3. sphere inside cube with face area 324

Vol sph = 4π(9x9x9)/3 = 972π

Vol cube = 18x18x18 = 5832

Vol gap = **5832 – 972π** $\frac{\pi}{6}$

---

9. cylinder inside cube with edge 7

Vol cyl = (7x7)π(7)/4 = 343π/4

Vol cube = 7x7x7 = 343

Vol gap = **343 –** $\frac{343\pi}{4}$ $\frac{\pi}{4}$

---

4. cone inside cube with surface area 384

Vol cone = (4x4)π(8)/3 = 128π/3

Vol cube = 8x8x8 = 512

Vol gap = **512 –** $\frac{128\pi}{3}$ $\frac{\pi}{12}$

---

10. square pyramid inside cube with face diagonal 3√2

Vol sq pyr = (3x3x3)/3 = 9

Vol cube = 3x3x3 = 27

Vol gap = 27 – 9 = **18** $\frac{1}{3}$

---

5. square pyramid inside cube with volume 729

Vol sq pyr = (9x9x9)/3 = 243

Vol cube = 729

Vol gap = 729 – 243 = **486** $\frac{1}{3}$

---

11. cylinder with radius 3 inside cube

Vol cyl = (3x3)π(6) = 54π

Vol cube = 6x6x6 = 216

Vol gap = **216 – 54π** $\frac{\pi}{4}$

---

6. sphere inside cube with face area 16

Vol sph = 4π(2x2x2)/3 = 32π/3

Vol cube = 4x4x4 = 64

Vol gap = **64 –** $\frac{32\pi}{3}$ $\frac{\pi}{6}$

---

12. cone inside cube with face area 144

Vol cone = (6x6)π(12)/3 = 144π

Vol cube = 12x12x12 = 1728

Vol gap = 1728 – 144 = **1584** $\frac{\pi}{12}$

Cubes 5

Find the volume in cubic units and total surface area in square units of glued solids comprised of unit cubes. Front faces are shaded only to help visualization.

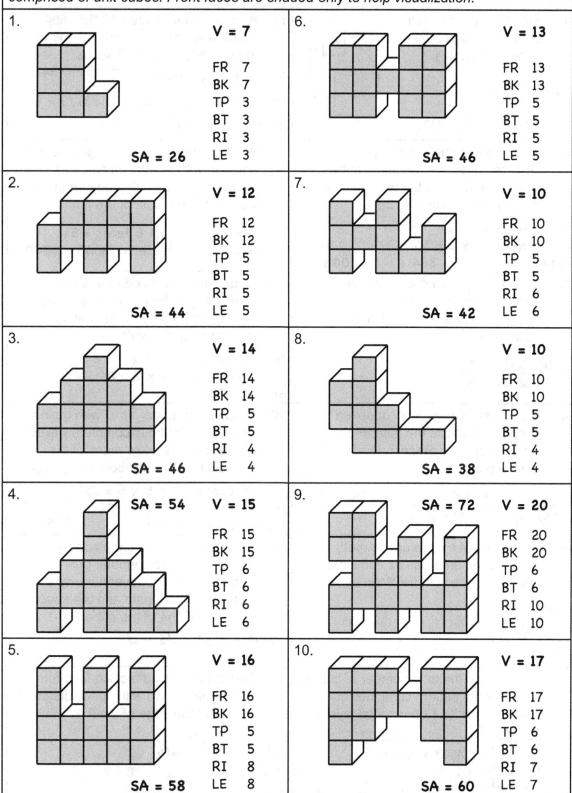

1.
V = 7

FR 7
BK 7
TP 3
BT 3
RI 3
SA = 26   LE 3

6.
V = 13

FR 13
BK 13
TP 5
BT 5
RI 5
SA = 46   LE 5

2.
V = 12

FR 12
BK 12
TP 5
BT 5
RI 5
SA = 44   LE 5

7.
V = 10

FR 10
BK 10
TP 5
BT 5
RI 6
SA = 42   LE 6

3.
V = 14

FR 14
BK 14
TP 5
BT 5
RI 4
SA = 46   LE 4

8.
V = 10

FR 10
BK 10
TP 5
BT 5
RI 4
SA = 38   LE 4

4.
SA = 54   V = 15

FR 15
BK 15
TP 6
BT 6
RI 6
LE 6

9.
SA = 72   V = 20

FR 20
BK 20
TP 6
BT 6
RI 10
LE 10

5.
V = 16

FR 16
BK 16
TP 5
BT 5
RI 8
SA = 58   LE 8

10.
V = 17

FR 17
BK 17
TP 6
BT 6
RI 7
SA = 60   LE 7

MAVA Math: Enhanced Skills Solutions   Copyright © 2015 Marla Weiss

Cubes 6

Answer as indicated.

1. Find the surface area: cube with edge 2 on cube with edge 4.

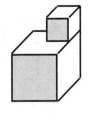

SA = 16 x 6 = 96
sa = 4 x 6 = 24
96 + 24 – 4 – 4 = **112**

7. A cube with face area 12 "just fits" inside a sphere. Find the volume of the gap between the 2 solids.

e = $2\sqrt{3}$ Vol cube = $24\sqrt{3}$ **$36\pi - 24\sqrt{3}$**

sp diag cube = diam sph = $(2\sqrt{3})\sqrt{3}$ = 6

Vol sph = $4\pi(3\times3\times3)/3$ = 36π

2. A cube with edge 6 is cut, perfectly centered, from the top into a cube with edge 12. Find the total surface area of the resulting figure.

SA = 144 x 6 = 864
A sm face = 6 x 6 = 36
864 – 36 + 5x36 = 864 + 144 = **1008**

8. A cubical box with edge 1 yard is to be covered with fabric that costs $10 per square foot. Assuming no overlap and waste, find the cost of the fabric.

e = 3 feet | A 6 faces = 54
A face = 9 sq ft | cost = 54x10 = **$540**

3. Fix one edge of a cube, double a second, and triple the third. Find the ratios of volumes: original cube to new prism.

$$\frac{(e)(e)(e)}{(e)(2e)(3e)} \quad \frac{1}{6}$$

9. A cubical box with edge 4 is filled completely with unit cubes and then closed. How many of the unit cubes touch the cardboard?

total # cubes = 4 x 4 x 4 = 64
center core = 2 x 2 x 2 = 8
64 – 8 = **56**

4. Fix one edge of a cube, double a second, and triple the third. Find the ratios of surface areas: original cube to new prism.

$$\frac{6(e)(e)}{2(e)(2e)+ 2(e)(3e) + 2(2e)(3e)} \quad \frac{6}{22} \quad \frac{3}{11}$$

10. A topless cardboard box with base 4 by 5 and height 6 is completely filled with unit cubes. How many of the cubes touch the cardboard?   Center of top also does not touch.

total # cubes = 4 x 5 x 6 = 120
center core = 2 x 3 x 4 = 24
120 – 24 – (2x3) = **90**

5. Five faces of a cube are painted. The painted area equals 245 square units. Find the volume of the cube.

$5e^2$ = 245
e^2 = 49
e = 7   V = **343**

11. When 2 unit cubes are placed face-to-face to form a rectangular prism, what is the difference between the space diagonal of the prism and the space diagonal of one of the cubes?

$\sqrt{4+1+1} - \sqrt{3}$ = $\sqrt{6} - \sqrt{3}$

6. Find the total number of cubes if the pattern continues downward for 12 horizontal layers.

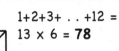

1+2+3+ . . +12 =
13 x 6 = **78**

12. Two cubes placed face-to-face form a rectangular prism with space diagonal $2\sqrt{6}$. Find the edge of a cube.

$\sqrt{e^2+e^2+4e^2}$ = $2\sqrt{6}$
$e\sqrt{6}$ = $2\sqrt{6}$ **2**

Cubes 7

Answer as indicated for the painted cubes.

1. A cube with edge 8 is totally painted and then cut into cubes with edge 2. How many of the smaller cubes have paint on exactly 3 faces?

8 (only the corners)

5. A cube with edge 4 made of 64 unit cubes is totally painted and then disassembled. For a unit cube selected randomly, find the probability it will have each number of faces painted.

0 faces	$\frac{8}{64}$	$\frac{1}{8}$
1 face	$\frac{24}{64}$	$\frac{3}{8}$
2 faces	$\frac{24}{64}$	$\frac{3}{8}$
3 faces	$\frac{8}{64}$	$\frac{1}{8}$

2. A cube with edge 5 is totally painted and then cut into 125 unit cubes. How many of the small cubes have:

0 faces painted? $3\times3\times3 = $ **27** (core)

1 face painted? $3\times3 = 9$ on each face $9\times6 = $ **54**

2 faces painted? $3\times12 = $ **36** (3 middle-most along each edge)

3 faces painted? **8** (corners)

6. A large cube, formed by gluing unit cubes, is dipped in paint. Then it is separated into the original unit cubes, 216 of which have no painted faces. How many unit cubes were used to form the large cube?

no painted faces = center core
$6 \times 6 \times 6 = 216$
edge $= 6 + 2 = 8$
cubes $= 8 \times 8 \times 8 = $ **512**

3. A large cube, formed by gluing unit cubes, is dipped in paint and then separated into the original unit cubes, of which 294 have paint on exactly one face. Find the volume of the large cube.

$294/6 = 49$ per face (in the center)
$7\times7 = 49$
$e = 1 + 7 + 1 = 9$
$V = 9\times9\times9 = $ **729**

7. A 4-inch cube is painted and then cut into one-inch cubes. Find the probability that a cube selected and placed randomly will have the top face painted.

$$\frac{0}{6}\cdot\frac{1}{8} + \frac{1}{6}\cdot\frac{3}{8} + \frac{2}{6}\cdot\frac{3}{8} + \frac{3}{6}\cdot\frac{1}{8}$$

$$\frac{1}{16} + \frac{2}{16} + \frac{1}{16} \qquad \frac{1}{4}$$

4. A cube with edge n is totally painted and then cut into n^3 unit cubes. How many of the small cubes have:
0 faces painted?

0 faces painted? $(n-2)^3$

1 face painted? $6(n-2)^2$

2 faces painted? $12(n-2)$

3 faces painted? **8**

8. A cube with edge 3 made of 27 unit cubes is totally painted and then disassembled. For a unit cube selected randomly, find the probability it will have each number of faces painted.

0 faces	$\frac{1}{27}$	
1 face	$\frac{6}{27}$	$\frac{2}{9}$
2 faces	$\frac{12}{27}$	$\frac{4}{9}$
3 faces	$\frac{8}{27}$	

MAVA Math: Enhanced Skills Solutions Copyright © 2015 Marla Weiss

Cubes 8

Complete the chart.

	EDGE	VOLUME	AREA FACE	FACE DIAG	SPACE DIAG	SURFACE AREA	SUM EDGES
1.	12	1728	144	$12\sqrt{2}$	$12\sqrt{3}$	864	144
2.	7	343	49	$7\sqrt{2}$	$7\sqrt{3}$	294	84
3.	$\sqrt{3}$	$3\sqrt{3}$	3	$\sqrt{6}$	3	18	$12\sqrt{3}$
4.	$\sqrt{30}$	$30\sqrt{30}$	30	$2\sqrt{15}$	$3\sqrt{10}$	180	$12\sqrt{30}$
5.	$\sqrt{2}$	$2\sqrt{2}$	2	2	$\sqrt{6}$	12	$12\sqrt{2}$
6.	$3\sqrt{2}$	$54\sqrt{2}$	18	6	$3\sqrt{6}$	108	$36\sqrt{2}$
7.	$\sqrt{15}$	$15\sqrt{15}$	15	$\sqrt{30}$	$3\sqrt{5}$	90	$12\sqrt{15}$
8.	$2\sqrt{2}$	$16\sqrt{2}$	8	4	$2\sqrt{6}$	48	$24\sqrt{2}$
9.	1	1	1	$\sqrt{2}$	$\sqrt{3}$	6	12
10.	$\sqrt{6}$	$6\sqrt{6}$	6	$2\sqrt{3}$	$3\sqrt{2}$	36	$12\sqrt{6}$
11.	16	4096	256	$16\sqrt{2}$	$16\sqrt{3}$	1536	192
12.	$2\sqrt{3}$	$24\sqrt{3}$	12	$2\sqrt{6}$	6	72	$24\sqrt{3}$
13.	9	729	81	$9\sqrt{2}$	$9\sqrt{3}$	486	108
14.	$5\sqrt{3}$	$375\sqrt{3}$	75	$5\sqrt{6}$	15	450	$60\sqrt{3}$
15.	$3\sqrt{3}$	$81\sqrt{3}$	27	$3\sqrt{6}$	9	162	$36\sqrt{3}$
16.	$\sqrt{10}$	$10\sqrt{10}$	10	$2\sqrt{5}$	$\sqrt{30}$	60	$12\sqrt{10}$
17.	$\sqrt{42}$	$42\sqrt{42}$	42	$2\sqrt{21}$	$3\sqrt{14}$	252	$12\sqrt{42}$
18.	$\sqrt{66}$	$66\sqrt{66}$	66	$2\sqrt{33}$	$3\sqrt{22}$	396	$12\sqrt{66}$
19.	$\sqrt{14}$	$14\sqrt{14}$	14	$2\sqrt{7}$	$\sqrt{42}$	84	$12\sqrt{14}$

Cylinders 1

Find the volume in cubic units of a cylinder with the given unit dimensions. Answer in terms of π.

1. height 12, radius 6 $V = \pi \cdot 6^2 \cdot 12 = \mathbf{432\pi}$ $360 + 72$	10. height 5, radius 2 $V = \pi \cdot 2^2 \cdot 5 = \mathbf{20\pi}$
2. height 30, diameter 18 $r = 9$ $V = \pi \cdot 9^2 \cdot 30 = \mathbf{2430\pi}$	11. diameter 8, height 20 $r = 4$ $V = \pi \cdot 4^2 \cdot 20 = \mathbf{320\pi}$

3.   diameter 10, height triple the radius
                           $r = 5$
 $V = \pi \cdot 5^2 \cdot 15 = \mathbf{375\pi}$ $h = 15$
 $250 + 125$
	12. radius 3, height triple the diameter
	                           $d = 6$
	 $V = \pi \cdot 3^2 \cdot 18 = \mathbf{162\pi}$ $h = 18$
	 $90 + 72$

4. radius 12, height 11 $V = \pi \cdot 12^2 \cdot 11 = \mathbf{1584\pi}$ $144 \cdot 11$	13. radius 13, height 3 $V = \pi \cdot 13^2 \cdot 3 = \mathbf{507\pi}$ $169 \cdot 3$
5. radius 6, height half the radius $h = 3$ $V = \pi \cdot 6^2 \cdot 3 = \mathbf{108\pi}$	14. radius 11, height double the diameter $d = 22$ $V = \pi \cdot 11^2 \cdot 44 = \mathbf{5324\pi}$ $h = 44$ $1331 \cdot 4$
6. radius 8, height 22 $V = \pi \cdot 8^2 \cdot 22 = \mathbf{1408\pi}$ $1280 + 128$	15. diameter 28, height 10.5 $r = 14$ $V = \pi \cdot 14^2 \cdot 10.5 = \mathbf{2058\pi}$ $196 \cdot 10.5 = 1960 + 98$
7. height = 40, diameter = 18 $r = 9$ $V = \pi \cdot 9^2 \cdot 40 = \mathbf{3240\pi}$ $81 \cdot 40$	16. diameter 8, height 11.5 $r = 4$ $V = \pi \cdot 4^2 \cdot 11.5 = \mathbf{184\pi}$ $16 \cdot 11.5 = 176 + 8$
8. height = 60, diameter = 22 $r = 11$ $V = \pi \cdot 11^2 \cdot 60 = \mathbf{7260\pi}$ $121 \cdot 60$	17. radius 16, height 7 $V = \pi \cdot 16^2 \cdot 7 = \mathbf{1792\pi}$ $256 \cdot 7 = 1750 + 42$
9. diameter 14, height 13 $r = 7$ $V = \pi \cdot 7^2 \cdot 13 = \mathbf{637\pi}$ $490 + 147$	18. height = 3, diameter = 5 $r = 2.5$ $V = \pi \cdot 2.5^2 \cdot 3 = \mathbf{18.75\pi}$ $6.25 \cdot 3$

MAVA Math: Enhanced Skills Solutions   Copyright © 2015 Marla Weiss

Cylinders 2

Find the surface area in square units of the cylinder with the given unit dimensions. Answer in terms of π.

1. height 12, radius 6 $\quad d = 12$

$SA = \pi \cdot 6^2 \cdot 2 + \pi \cdot 12 \cdot 12 = \mathbf{216\pi}$
$\qquad 72 + 144$

9. height 5, radius 2 $\quad d = 4$

$SA = \pi \cdot 2^2 \cdot 2 + \pi \cdot 4 \cdot 5 = \mathbf{28\pi}$
$\qquad 8 + 20$

2. height 20, diameter 20 $\quad r = 10$

$SA = \pi \cdot 10^2 \cdot 2 + \pi \cdot 20 \cdot 20 = \mathbf{600\pi}$
$\qquad 200 + 400$

10. diameter 8, height 20 $\quad r = 4$

$SA = \pi \cdot 4^2 \cdot 2 + \pi \cdot 8 \cdot 20 = \mathbf{192\pi}$
$\qquad 32 + 160$

3. diameter 22, height equals the radius

$SA = \pi \cdot 11^2 \cdot 2 + \pi \cdot 22 \cdot 11 = \mathbf{484\pi}$
$\qquad 242 + 242$ $\quad r = 11$
$\qquad\qquad\qquad h = 11$

11. radius 3, height is twice the diameter

$SA = \pi \cdot 3^2 \cdot 2 + \pi \cdot 6 \cdot 12 = \mathbf{90\pi}$
$\qquad 18 + 72$ $\quad d = 6$
$\qquad\qquad\qquad h = 12$

4. radius 9, height 11 $\quad d = 18$

$SA = \pi \cdot 9^2 \cdot 2 + \pi \cdot 18 \cdot 11 = \mathbf{360\pi}$
$\qquad 162 + 198$

12. radius 3, height 3 $\quad d = 6$

$SA = \pi \cdot 3^2 \cdot 2 + \pi \cdot 6 \cdot 3 = \mathbf{36\pi}$
$\qquad 18 + 18$

5. radius 8, height equals half the radius

$SA = \pi \cdot 8^2 \cdot 2 + \pi \cdot 16 \cdot 4 = \mathbf{192\pi}$
$\qquad 128 + 64$ $\quad d = 16$
$\qquad\qquad\qquad h = 4$

13. radius 15, height is 4/5 the diameter

$SA = \pi \cdot 15^2 \cdot 2 + \pi \cdot 30 \cdot 24 = \mathbf{1170\pi}$
$\qquad 450 + 720$ $\quad d = 30$
$\qquad\qquad\qquad h = 24$

6. radius 10, height 40 $\quad d = 20$

$SA = \pi \cdot 10^2 \cdot 2 + \pi \cdot 20 \cdot 40 = \mathbf{1000\pi}$
$\qquad 200 + 800$

14. radius 1.2, height 10 $\quad d = 2.4$

$SA = \pi \cdot 1.2^2 \cdot 2 + \pi \cdot 2.4 \cdot 10 = \mathbf{26.88\pi}$
$\qquad 2.88 + 24$

7. radius 5, height 7 $\quad d = 10$

$SA = \pi \cdot 5^2 \cdot 2 + \pi \cdot 10 \cdot 7 = \mathbf{120\pi}$
$\qquad 50 + 70$

15. diameter 5, height 9 $\quad r = 2.5$

$SA = \pi \cdot 2.5^2 \cdot 2 + \pi \cdot 5 \cdot 9 = \mathbf{57.5\pi}$
$\qquad 12.5 + 45$

8. height 30, diameter 3 $\quad r = 1.5$

$SA = \pi \cdot 1.5^2 \cdot 2 + \pi \cdot 3 \cdot 30 = \mathbf{94.5\pi}$
$\qquad 4.5 + 90$

16. height 16, radius 20 $\quad d = 40$

$SA = \pi \cdot 20^2 \cdot 2 + \pi \cdot 40 \cdot 16 = \mathbf{1440\pi}$
$\qquad 800 + 640$

Cylinders 3

Find the volume in cubic units of the cylinder formed by rotating the rectangle with the given vertices around the specified line.

1. vertices: (0,0), (0,4), (5,4), (5,0)
   line: y-axis

   r = 5
   h = 4
   V = πr²h
   V = 25 x 4π
   V = 100π

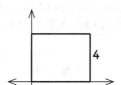

6. vertices: (1,0), (1,6), (6,6), (6,0)
   line: x-axis

   r = 6
   h = 5
   V = πr²h
   V = 36 x 5π
   V = 180π

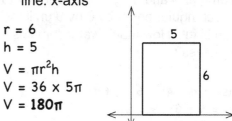

2. vertices: (0,0), (0,8), (6,8), (6,0)
   line: x-axis

   r = 8
   h = 6
   V = πr²h
   V = 64 x 6π
   V = 384π

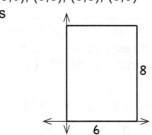

7. vertices: (2,–3), (2,5), (–1,5), (–1,–3)
   line: x = 2

   r = 3
   h = 8
   V = πr²h
   V = 9 x 8π
   V = 72π

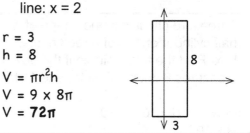

3. vertices: (3,0), (3,9), (7,9), (7,0)
   line: x = 3

   r = 4
   h = 9
   V = πr²h
   V = 16 x 9π
   V = 144π

8. vertices: (0,0), (0,7), (8,7), (8,0)
   line: x = 4

   r = 4
   h = 7
   V = πr²h
   V = 16 x 7π
   V = 112π

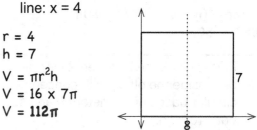

4. vertices: (0,2), (0,6), (5,6), (5,2)
   line: y = 2

   r = 4
   h = 5
   V = πr²h
   V = 16 x 5π
   V = 80π

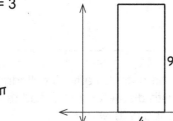

9. vertices: (1,4), (1,–6), (3,–6), (3,4)
   line: y = –1

   r = 5
   h = 2
   V = πr²h
   V = 25 x 2π
   V = 50π

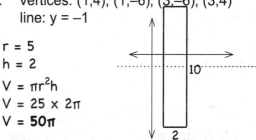

5. vertices: (–3,–2), (–3,4), (4,4), (4,–2)
   line: x = –3

   r = 7
   h = 6
   V = πr²h
   V = 49 x 6π
   V = 294π

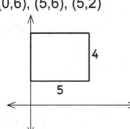

10. vertices: (0,0), (0,6), (3,6), (3,0)
    line: y-axis

    r = 3
    h = 6
    V = πr²h
    V = 9 x 6π
    V = 54π

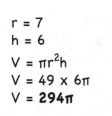

Cylinders 4

Answer as indicated. Linear measurements are in units. Volume answers are in cubic units; surface area answers are in square units.

1. Water from a full cylindrical glass with diameter 8 and height 6 is poured into a rectangular pan 4 by 6 by 9 until the pan is full. How much water remains in the glass?

   V glass = $\pi \cdot 4^2 \cdot 6 = 96\pi$
   V pan = $4 \cdot 6 \cdot 9 = 216$
   $96\pi - 216$

5. A wedge is cut at a 60° central angle from a cylindrical block of cheese. Find the volume of the wedge if the cylinder has radius 8 and height 6.

   V cyl = $\pi \cdot 8^2 \cdot 6$
   60° = 1/6 of 360°
   64π

2. A treasure chest is in the shape of a half cylinder on top of a rectangular box. Find the total volume if the box measures 8 deep, 10 wide, and 11 tall.

   V box = $8 \cdot 11 \cdot 10 = 880$
   cyl H = 10
   cyl D = 8
   V top = $(\pi \cdot 4^2 \cdot 10)/2 = 80\pi$
   $880 + 80\pi$

6. Find the remaining volume of a cube with edge 8 after a cylindrical hole with diameter 2 is drilled through the cube.

   V cube = $8 \cdot 8 \cdot 8 = 512$
   V cyl = $\pi \cdot 1^2 \cdot 8 = 8\pi$
   $512 - 8\pi$

3. Find the volume between 2 cylinders, one inside the other, with concentric circular bases of diameters 9 and 12, both 10 high.

   V out cyl = $\pi \cdot 6^2 \cdot 10 = 36\pi \cdot 10$
   V in cyl = $\pi \cdot 3^2 \cdot 10 = 9\pi \cdot 10$
   V shell = $10(36\pi - 9\pi) = 27\pi \cdot 10$
   270π

7. Find the total surface area (inside and outside) after a cylindrical hole with diameter 4 is drilled through a cube with edge 10.

   SA cube = $10 \cdot 10 \cdot 6 = 600$
   minus top & bot holes = $2(\pi \cdot 2^2) = 8\pi$
   plus lateral cyl = Ch = $\pi dh = 40\pi$
   $600 + 32\pi$

4. A cylinder has diameter 16 and height 40. A square prism with equal height to the cylinder also has equal volume. Find the side of the base of the prism.

   V cyl = $\pi \cdot 8^2 \cdot 40 = 64\pi \cdot 40$
   V prism = $s^2 \cdot 40$
   $s^2 = 64\pi$
   $s = 8\sqrt{\pi}$

8. Find the volume of a cylindrical shell with height 40, outside diameter 10, and inside diameter 8.

   V out cyl = $\pi \cdot 5^2 \cdot 40 = 25\pi \cdot 40$
   V in cyl = $\pi \cdot 4^2 \cdot 40 = 16\pi \cdot 40$
   V shell = $40(25\pi - 16\pi) = 9\pi \cdot 40$
   360π

Decimals 1

Solve by clearing the decimal points.

1. $.9x = 81$ $9x = 810$ $x = \mathbf{90}$	12. $700 = 3.5x$ $7x = 1400$ $x = \mathbf{200}$	23. $2850 = 9.5x$ $19x = 5700$ $x = \mathbf{300}$
2. $.12x = 3$ $12x = 300$ $x = \mathbf{25}$	13. $.4x = 500$ $4x = 5000$ $x = \mathbf{1250}$	24. $.17x = 51$ $17x = 5100$ $x = \mathbf{300}$
3. $7.5 = 50x$ $15 = 100x$ $x = \mathbf{.15}$	14. $3.4x = 20.4$ $34x = 204$ $x = \mathbf{6}$	25. $7.25x = 58$ $29x = 232$ $x = \mathbf{8}$
4. $12 = .3x$ $120 = 3x$ $x = \mathbf{40}$	15. $2.6 = .52x$ $260 = 52x$ $x = \mathbf{5}$	26. $26.5 = 53x$ $53 = 106x$ $x = \mathbf{.5}$
5. $.15x = 9$ $15x = 900$ $x = \mathbf{60}$	16. $.06x = 19.2$ $6x = 1920$ $x = \mathbf{320}$	27. $8.25 = 25x$ $33 = 100x$ $x = \mathbf{.33}$
6. $.6x = 14.4$ $6x = 144$ $x = \mathbf{24}$	17. $1.2x = 60$ $12x = 600$ $x = \mathbf{50}$	28. $9.2 = 20x$ $46 = 100x$ $x = \mathbf{.46}$
7. $.07x = 5.6$ $7x = 560$ $x = \mathbf{80}$	18. $.008x = 0.5$ $8x = 500$ $x = \mathbf{62.5}$	29. $41.5 = 5x$ $83 = 10x$ $x = \mathbf{8.3}$
8. $.2x = 13$ $2x = 130$ $x = \mathbf{65}$	19. $18.5 = 50x$ $37 = 100x$ $x = \mathbf{.37}$	30. $242 = 5.5x$ $11x = 484$ $x = \mathbf{44}$
9. $.025x = 3$ $5x = 600$ $x = \mathbf{120}$	20. $3.25 = 25x$ $13 = 100x$ $x = \mathbf{.13}$	31. $16.5 = 66x$ $33 = 66 \cdot 2x$ $x = \mathbf{.25}$
10. $4800 = 2.4x$ $24x = 48,000$ $x = \mathbf{2000}$	21. $11.2 = 20x$ $56 = 100x$ $x = \mathbf{.56}$	32. $3.25x = 39$ $13x = 39 \cdot 4$ $x = \mathbf{12}$
11. $.18x = 5.4$ $18x = 540$ $x = \mathbf{30}$	22. $23.5 = 5x$ $47 = 10x$ $x = \mathbf{4.7}$	33. $.11x = 264$ $11x = 24 \cdot 11 \cdot 100$ $x = \mathbf{2400}$

Decimals 2

Write the exact decimal conversion.

1. $\dfrac{7}{9}$ $.\overline{7}$	17. $\dfrac{5}{9}$ $.\overline{5}$	33. $\dfrac{5}{11}$ $\dfrac{45}{99}$ $.\overline{45}$
2. $\dfrac{7}{99}$ $.\overline{07}$	18. $\dfrac{9}{11}$ $\dfrac{81}{99}$ $.\overline{81}$	34. $\dfrac{14}{33}$ $\dfrac{42}{99}$ $.\overline{42}$
3. $\dfrac{7}{999}$ $.\overline{007}$	19. $\dfrac{17}{33}$ $\dfrac{51}{99}$ $.\overline{51}$	35. $\dfrac{53}{99}$ $.\overline{53}$
4. $\dfrac{7}{9999}$ $.\overline{0007}$	20. $\dfrac{4}{45}$ $\dfrac{8}{90}$ $.0\overline{8}$	36. $\dfrac{1}{225}$ $\dfrac{4}{900}$ $.00\overline{4}$
5. $\dfrac{7}{90}$ $.0\overline{7}$	21. $\dfrac{23}{99}$ $.\overline{23}$	37. $\dfrac{1}{180}$ $\dfrac{5}{900}$ $.00\overline{5}$
6. $\dfrac{7}{900}$ $.00\overline{7}$	22. $\dfrac{35}{99}$ $.\overline{35}$	38. $\dfrac{2}{495}$ $\dfrac{4}{990}$ $.0\overline{04}$
7. $\dfrac{7}{990}$ $.0\overline{07}$	23. $\dfrac{47}{990}$ $.0\overline{47}$	39. $\dfrac{31}{330}$ $\dfrac{93}{990}$ $.0\overline{93}$
8. $\dfrac{5}{33}$ $\dfrac{15}{99}$ $.\overline{15}$	24. $\dfrac{83}{99}$ $.\overline{83}$	40. $\dfrac{1}{11}$ $\dfrac{9}{99}$ $.\overline{09}$
9. $\dfrac{1}{45}$ $\dfrac{2}{90}$ $.0\overline{2}$	25. $\dfrac{1}{300}$ $.00\overline{3}$	41. $\dfrac{25}{33}$ $\dfrac{75}{99}$ $.\overline{75}$
10. $\dfrac{3}{11}$ $\dfrac{27}{99}$ $.\overline{27}$	26. $\dfrac{19}{330}$ $\dfrac{57}{990}$ $.0\overline{57}$	42. $\dfrac{32}{999}$ $.\overline{032}$
11. $\dfrac{13}{99}$ $.\overline{13}$	27. $\dfrac{6}{11}$ $.\overline{54}$	43. $\dfrac{25}{9999}$ $.\overline{0025}$
12. $\dfrac{1}{90}$ $.0\overline{1}$	28. $\dfrac{1}{55}$ $\dfrac{2}{110}$ $\dfrac{18}{990}$ $.0\overline{18}$	44. $\dfrac{13}{33}$ $\dfrac{39}{99}$ $.\overline{39}$
13. $\dfrac{4}{9}$ $.\overline{4}$	29. $\dfrac{127}{9999}$ $.\overline{0127}$	45. $\dfrac{2}{45}$ $\dfrac{4}{90}$ $.0\overline{4}$
14. $\dfrac{571}{999}$ $.\overline{571}$	30. $\dfrac{25}{999}$ $.\overline{025}$	46. $\dfrac{591}{9999}$ $.\overline{0591}$
15. $\dfrac{29}{990}$ $.0\overline{29}$	31. $\dfrac{8}{9}$ $.\overline{8}$	47. $\dfrac{1}{15}$ $\dfrac{6}{90}$ $.0\overline{6}$
16. $\dfrac{4}{55}$ $\dfrac{8}{110}$ $\dfrac{72}{990}$ $.0\overline{72}$	32. $\dfrac{8}{33}$ $\dfrac{24}{99}$ $.\overline{24}$	48. $\dfrac{67}{99}$ $.\overline{67}$

Decimals 3

Express as a simplified fraction.					Operate on the decimals using mental math.			

Express as a simplified fraction.

1. $.28\overline{1}$ $2\dfrac{81}{99}$ $2\dfrac{9}{11}$ $\dfrac{31}{11}$ $\dfrac{\textbf{31}}{\textbf{110}}$

2. $.8\overline{24}$ $8\dfrac{24}{99}$ $8\dfrac{8}{33}$ $\dfrac{272}{33}$ $\dfrac{272}{330}$ $\dfrac{\textbf{136}}{\textbf{165}}$

3. $.5\overline{72}$ $5\dfrac{72}{99}$ $5\dfrac{8}{11}$ $\dfrac{63}{11}$ $\dfrac{\textbf{63}}{\textbf{110}}$

4. $.3\overline{14}$ $3\dfrac{14}{99}$ $\dfrac{311}{99}$ $\dfrac{\textbf{311}}{\textbf{990}}$

5. $.3\overline{27}$ $3\dfrac{27}{99}$ $3\dfrac{3}{11}$ $\dfrac{36}{11}$ $\dfrac{36}{110}$ $\dfrac{\textbf{18}}{\textbf{55}}$

6. $.1\overline{25}$ $1\dfrac{25}{99}$ $\dfrac{124}{99}$ $\dfrac{124}{990}$ $\dfrac{\textbf{62}}{\textbf{495}}$

7. $.9\overline{18}$ $9\dfrac{18}{99}$ $9\dfrac{2}{11}$ $\dfrac{101}{11}$ $\dfrac{\textbf{101}}{\textbf{110}}$

8. $.2\overline{63}$ $2\dfrac{63}{99}$ $2\dfrac{7}{11}$ $\dfrac{29}{11}$ $\dfrac{\textbf{29}}{\textbf{110}}$

9. $.6\overline{15}$ $6\dfrac{15}{99}$ $6\dfrac{5}{33}$ $\dfrac{203}{33}$ $\dfrac{\textbf{203}}{\textbf{330}}$

10. $.5\overline{09}$ $5\dfrac{9}{99}$ $5\dfrac{1}{11}$ $\dfrac{56}{11}$ $\dfrac{56}{110}$ $\dfrac{\textbf{28}}{\textbf{55}}$

11. $.3\overline{21}$ $3\dfrac{21}{99}$ $3\dfrac{7}{33}$ $\dfrac{106}{33}$ $\dfrac{106}{330}$ $\dfrac{\textbf{53}}{\textbf{165}}$

12. $.2\overline{54}$ $2\dfrac{54}{99}$ $2\dfrac{6}{11}$ $\dfrac{28}{11}$ $\dfrac{28}{110}$ $\dfrac{\textbf{14}}{\textbf{55}}$

13. $.1\overline{36}$ $1\dfrac{36}{99}$ $1\dfrac{4}{11}$ $\dfrac{15}{11}$ $\dfrac{15}{110}$ $\dfrac{\textbf{3}}{\textbf{22}}$

14. $.3\overline{45}$ $3\dfrac{45}{99}$ $3\dfrac{5}{11}$ $\dfrac{38}{11}$ $\dfrac{38}{110}$ $\dfrac{\textbf{19}}{\textbf{55}}$

15. $.2\overline{75}$ $2\dfrac{75}{99}$ $2\dfrac{25}{33}$ $\dfrac{91}{33}$ $\dfrac{\textbf{91}}{\textbf{330}}$

16. $.3\overline{12}$ $3\dfrac{12}{99}$ $3\dfrac{4}{33}$ $\dfrac{103}{33}$ $\dfrac{\textbf{103}}{\textbf{330}}$

Operate on the decimals using mental math.

17. $.2 + .93$ **1.13**

18. $2 - .93$ **1.07**

19. $6 + .14$ **6.14**

20. $6 - .14$ **5.86**

21. $71 + 4.65$ **75.65**

22. $.71 + .465$ **1.175**

23. $.71 + 46.5$ **47.21**

24. $71 - .465$ **70.535**

25. $.8 + .032$ **.832**

26. $.8 - .032$ **.768**

27. $.8 \times .032$ **.0256**

28. $.8 \div .032$ **25**

29. $.032 \div .8$ **.04**

30. $12 \div .025$ **480**

31. $150.8 - 14.75$ **136.05**

32. $1.6 \div .008$ **200**

33. $4 + .25$ **4.25**

34. $4 - .25$ **3.75**

35. $4 \times .25$ **1**

36. $4 \div .25$ **16**

37. $.25 \div 4$ **.0625**

38. $.8 + .064$ **.864**

39. $.8 - .064$ **.736**

40. $.8 \times .064$ **.0512**

41. $.8 \div .064$ **12.5**

42. $.064 \div .8$ **.08**

43. $.8 - .005$ **.795**

44. $.8 \times .005$ **.004**

45. $.8 \div .005$ **160**

46. $20.9 - 5.06$ **15.84**

47. $50.6 - 2.09$ **48.51**

48. $.025 \times 3.2$ **.08**

Decimals 4

Write T if the fraction terminates as a decimal or R if it repeats. For repeating decimals, also write the denominator factor causing the repeat.

1. $\frac{7}{12}$		R	3	17. $\frac{9}{40}$		T		33. $\frac{171}{224}$		R	7
2. $\frac{11}{32}$		T		18. $\frac{143}{176}$	$\frac{13}{16}$	T		34. $\frac{10}{48}$	$\frac{5}{24}$	R	3
3. $\frac{35}{75}$	$\frac{7}{15}$	R	3	19. $\frac{17}{25}$		T		35. $\frac{353}{640}$		T	
4. $\frac{13}{50}$		T		20. $\frac{10}{51}$		R	3, 17	36. $\frac{111}{120}$	$\frac{37}{40}$	T	
5. $\frac{15}{60}$	$\frac{1}{4}$	T		21. $\frac{27}{55}$		R	11	37. $\frac{249}{512}$		T	
6. $\frac{7}{16}$		T		22. $\frac{91}{280}$	$\frac{13}{40}$	T		38. $\frac{49}{96}$		R	3
7. $\frac{77}{160}$		T		23. $\frac{53}{128}$		T		39. $\frac{251}{256}$		T	
8. $\frac{33}{125}$		T		24. $\frac{169}{375}$		R	3	40. $\frac{23}{575}$	$\frac{1}{25}$	T	
9. $\frac{121}{175}$		R	7	25. $\frac{99}{250}$		T		41. $\frac{777}{1024}$		T	
10. $\frac{5}{24}$		R	3	26. $\frac{113}{320}$		T		42. $\frac{19}{85}$		R	17
11. $\frac{21}{56}$	$\frac{3}{8}$	T		27. $\frac{73}{100}$		T		43. $\frac{12}{45}$	$\frac{4}{15}$	R	3
12. $\frac{49}{625}$		T		28. $\frac{33}{150}$	$\frac{11}{50}$	T		44. $\frac{3421}{4000}$		T	
13. $\frac{19}{80}$		T		29. $\frac{15}{64}$		T		45. $\frac{343}{750}$		R	3
14. $\frac{20}{57}$		R	3, 19	30. $\frac{24}{25}$		T		46. $\frac{169}{800}$		T	
15. $\frac{101}{680}$		R	17	31. $\frac{111}{440}$		R	11	47. $\frac{243}{700}$		R	7
16. $\frac{131}{275}$		R	11	32. $\frac{599}{600}$		R	3	48. $\frac{25}{52}$		R	13

MAVA Math: Enhanced Skills Solutions   Copyright © 2015 Marla Weiss

Distance-Rate-Time 1

Write and solve a D=RT equation. Mph is miles per hour.

1. How many minutes does Liz spend biking 4 miles at 12 mph?

D = RT
4 = 12T
T = 1/3 **20**

2. Find Dana's rate in mph to run 12.5 miles in 2.5 hours.

D = RT
12.5 = R (2.5)
R = **5**

3. How many miles does Abby drive in 80 minutes at 45 mph?

D = RT
D = 45 (4/3)
D = **60**

4. How many minutes does Jon spend driving 12 miles at 40 mph?

D = RT
12 = 40T
T = 12/40
T = 3/10 **18**

5. Find Eli's rate in mph when he runs 2 miles in 12 minutes.

D = RT
2 = R (1/5)
R = **10**

6. How many miles are covered at 15 mph for 24 minutes?

D = RT
D = 15 (2/5)
D = **6**

7. Find the rate in mph of a person who runs 4.5 miles in 45 minutes.

D = RT
4.5 = R (3/4)
R = **6**

8. What is the rate of a car in mph that travels 1 mile in 75 seconds?

D = RT
1 = R (1.25/60)
R = 60/1.25 = **48**

9. Kristen bikes 2 miles at 16 mph in how many minutes?

D = RT
2 = 16T
T = 1/8 **7.5**

10. How many miles does a bird fly at 36 mph in 25 minutes?

D = RT
D = 36 (5/12)
D = **15**

11. A plane flies 300 miles at 450 mph in how many minutes?

D = RT
300 = 450T
T = 30/45
T = 2/3 **40**

12. What is the rate in mph of a plane that flies 90 miles in 18 minutes?

D = RT
90 = R (3/10)
R = **300**

13. How many miles does a plane fly at 360 mph in 5 minutes?

D = RT
D = 360 (1/12)
D = **30**

14. Find a snail's rate in mph to crawl half a mile in 20 hours.

D = RT
0.5 = R (20)
R = **.025**

Distance-Rate-Time 2

Answer by constructing a DRT chart. Mph is miles per hour.

1. Harrison rode 60 miles at 12 mph, returning along the same route at 20 mph. What was his average rate in mph for the round trip?

	R	T	D
To	12	5	60
Fr	20	3	60
TOT	**15**	8	120

5. A plane flew 3000 miles at 200 mph and returned by the same route at 300 mph. Find the average rate in mph for the round trip.

	R	T	D
To	200	15	3000
Fr	300	10	3000
TOT	**240**	25	6000

2. David jogged 12 miles at 6 mph and then walked back on the same path at 2 mph. What was his average rate in mph for the round trip?

	R	T	D
To	6	2	12
Fr	2	6	12
TOT	**3**	8	24

6. Joslyn drove 120 miles at 30 mph and returned by the same route at 60 mph. What was her average rate for the full trip?

	R	T	D
To	30	4	120
Fr	60	2	120
TOT	**40**	6	240

3. Amy drove 180 miles at 40 mph and returned by the same route at 60 mph. What was her average rate in mph for the whole trip?

	R	T	D
To	40	4.5	180
Fr	60	3	180
TOT	**48**	7.5	360

7. Mia biked a total of 360 miles–one third at 8 mph, one half at 9 mph, and one sixth at 12 mph. What was her average rate?

	R	T	D
1/3	8	15	120
1/2	9	20	180
1/6	12	5	60
TOT	**9**	40	360

4. Sam drove at 45 mph for 4 hours and then at 60 mph for 8 hours. What was his overall rate in mph for the entire trip?

	R	T	D
To	45	4	180
Fr	60	8	480
TOT	**55**	12	660

8. Gil traveled 20 miles at 50 mph and the rest of his 100 mile trip at 40 mph. Find his overall rate in mph.

	R	T	D
To	50	$\frac{2}{5}$	20
Fr	40	2	80
TOT	**$41\frac{2}{3}$**	$2\frac{2}{5}$	100

Distance-Rate-Time 3

Answer by constructing a DRT chart before writing and solving an equation. Mph is miles per hour.

1. At noon Mo drives north at 30 mph. Jo leaves from the same place one hour later driving north 10 mph faster to catch him. When does Jo catch Mo?

	R	T	D
M	30	x	30x
J	40	x – 1	40x – 40

$30x = 40x - 40$
$40 = 10x$
$x = 4$
4:00 PM

2. Sal and Hal left from the same place driving in opposite directions. Hal drove 12 mph faster than Sal. Two hours later they were 192 miles apart. Find Hal's rate in mph.

	R	T	D
S	x	2	2x
H	x + 12	2	2x + 24

$4x + 24 = 192$
$4x = 168$
$x = 42$
Hal 54

3. Jane walked from home to school at 3 mph and got a ride home at 42 mph. Her total travel time was 1/2 hour. How many minutes was the drive home?

	R	T	D
To	3	x	3x
Fr	42	.5 – x	21 – 42x

$3x = 21 - 42x$
$45x = 21$
$15x = 7$
$x = 7/15$

$(7/15)(60) = 28$
2 minutes driving

4. Kyra and Lily walked toward each other from opposite ends of a 9-mile long road. Kyra walked at 5 mph; Lily walked at 4 mph. How many miles had each walked when they met?

	R	T	D
K	5	x	5x
L	4	x	4x

$4x + 5x = 9$
$9x = 9$
$x = 1$
Kyra **5**; Lily **4**

5. Eli paddled a canoe down a river at a steady rate of 6 mph. Two hours later Hudson left from the same place, driving a motorboat at 18 mph. After how long did Hudson reach Eli?

	R	T	D
E	6	x	6x
H	18	x – 2	18x–36

$6x = 18x - 36$
$36 = 12x$
$x = 3$
1 hour

6. The trip from City A to City B at 60 mph takes a half hour longer than if driving at 80 mph. How many miles is the trip?

	R	T	D
1	60	x + .5	60x+30
2	80	x	80x

$60x + 30 = 80x$
$30 = 20x$
$x = 3/2$
120

Distance-Rate-Time 4

Answer by constructing a DRT chart, if necessary, before writing and solving a system of equations. Mph is miles per hour.

1. A boat travels upstream against the current at 7 mph and downstream with the current at 15 mph. Find the rate of the current in mph.

   B + C = 15
   B − C = 7
   2B = 22
   B = 11
   Current: **4**

5. A boat travels upstream against the current at 16 mph and downstream with the current at 30 mph. Find the rate of the current in mph.

   B + C = 30
   B − C = 16
   2B = 46
   B = 23
   Current: **7**

2. A bird flies 80 miles in 5 hours with the wind but needs 20 hours for this trip against the wind. Find the rate of the wind in mph.

	R	T	D
W	16	5	80
A	4	20	80

   B + W = 16
   B − W = 4
   2B = 20
   B = 10
   Wind: **6**

6. A bird flies 1000 yards in 5 minutes with the wind, returning along the same path in 25 minutes against the wind. Find the rate of the wind in yards per minute.

	R	T	D
W	200	5	1000
A	40	25	1000

   B + W = 200
   B − W = 40
   2B = 240
   B = 120
   Wind: **80**

3. A fish swims 240 yards downstream in 12 minutes but needs 15 minutes for the same trip in the opposite direction. Find the rate of the current in yards per minute.

	R	T	D
D	20	12	240
U	16	15	240

   F + C = 20
   F − C = 16
   2F = 36
   F = 18
   Current: **2**

7. A boater travels 90 miles downstream in 5 hours. Returning upstream, the boater's rate is 10 mph. Find the rate of the current in mph.

	R	T	D
D	18	5	90
U	10	9	90

   B + C = 18
   B − C = 10
   2B = 28
   B = 14
   Current: **4**

4. A plane flight of 8400 miles takes 6 hours with a tail wind but 7 hours on the exact-path return flight with a head wind. Find the rate of the wind in mph.

	R	T	D
T	1400	6	8400
H	1200	7	8400

   P+W = 1400
   P−W = 1200
   2P = 2600
   P = 1300
   Wind: **100**

8. A plane flight of 360 miles takes 2 hours with a tail wind but 5 hours on the exact-path return flight with a head wind. Find the rate of the wind in mph.

	R	T	D
T	180	2	360
H	72	5	360

   P+W = 180
   P−W = 72
   2P = 252
   P = 126
   Wind: **54**

Divisibility 1

Check the box if the row number is divisible by the column factor. Use divisibility rules for all.

		2	3	4	5	6	8	9	10	11	12	15	18	20	22
1.	210	✔	✔	–	✔	✔	–	–	✔	–	–	✔	–	–	–
2.	264	✔	✔	✔	–	✔	✔	–	–	✔	✔	–	–	–	✔
3.	384	✔	✔	✔	–	✔	✔	–	–	–	✔	–	–	–	–
4.	396	✔	✔	✔	–	✔	–	✔	–	✔	✔	–	✔	–	✔
5.	500	✔	–	✔	✔	–	–	–	✔	–	–	–	–	✔	–
6.	528	✔	✔	✔	–	✔	✔	–	–	✔	✔	–	–	–	✔
7.	792	✔	✔	✔	–	✔	✔	✔	–	✔	✔	–	✔	–	✔
8.	900	✔	✔	✔	✔	✔	–	✔	✔	–	✔	✔	✔	✔	–
9.	968	✔	–	✔	–	–	✔	–	–	✔	–	–	–	–	✔
10.	1008	✔	✔	✔	–	✔	✔	✔	–	–	✔	–	✔	–	–
11.	2640	✔	✔	✔	✔	✔	✔	–	✔	✔	✔	✔	–	✔	✔
12.	3324	✔	✔	✔	–	✔	–	–	–	–	✔	–	–	–	–
13.	4345	–	–	–	✔	–	–	–	–	✔	–	–	–	–	–
14.	4923	–	✔	–	–	–	–	✔	–	–	–	–	–	–	–
15.	5082	✔	✔	–	–	✔	–	–	–	✔	–	–	–	–	✔
16.	5192	✔	–	✔	–	–	✔	–	–	✔	–	–	–	–	✔
17.	6336	✔	✔	✔	–	✔	✔	✔	–	✔	✔	–	✔	–	✔
18.	6435	–	✔	–	✔	–	–	✔	–	✔	–	✔	–	–	–
19.	7260	✔	✔	✔	✔	✔	–	–	✔	✔	✔	✔	–	✔	✔
20.	7920	✔	✔	✔	✔	✔	✔	✔	✔	✔	✔	✔	✔	✔	✔
21.	8140	✔	–	✔	✔	–	–	–	✔	✔	–	–	–	✔	✔

Divisibility 2

Answer as indicated using divisibility rules.

1. The four-digit number 5xx8 is divisible by 9. Find x.

13 + 5 = 18 no
13 + 14 = 27 yes   **7**

2. Find all digits d so that the five-digit number 4137d is divisible by 6.

0, 2, 4, 6, 8
ds = 15 + d   **0, 6**

3. Find the digit D such that the five-digit number D2D3D is divisible by 11.

3D − 5 = 0 or 11 or 22
3D = 27   **9**

4. Find all values of A for which the four-digit number A55A is divisible both by 3 and 11.
div by 11 for all A
10 + 2A div by 3   **1, 4, 7**

5. Find the digit d so that the six-digit number 83d152 is divisible by 9.

8 + 1 = 9; remove from ds
ds = 10 + d   **8**

6. Find the least digit x so that 5xx8 is a four-digit number divisible by 6.

already div by 2
13 + 2x div by 3   **1**

7. Find the digit x so that 273x is a four-digit number divisible by 12.

2 + 7 + 3 divisible by 3
0, 3, 6, 9   **6**
only 36 divisible by 4

8. How many numbers of the form 1DD2 are divisible by 6?
1 + 2 divisible by 3
0, 3, 6, 9   **4**
all work due to 1s digit 2

9. Find the digit x so that the four-digit number x2x3 is divisible by 11.

2x − 5 = 0 or 11 or 22
2x = 16   **8**

10. Find the digit x so that the four-digit number x5x8 is divisible by 11.

13 − 2x = 0 or 11 or 22
2x = 2   **1**

11. The five-digit number 12A3B is divisible by 36. Find A and B if A ≠ B.
"3B" div by 4; B = 2, 6
ds = 6 + A + B   **B = 2**
B = 6, A = 6 no   **A = 1**

12. Find the digit A so that the five-digit number 2A365 is divisible by 9.

6 + 3 = 9; remove from ds
ds = 7 + A   **2**

13. Find the digit d so that the five-digit number 7142d is divisible by 6.

d = 0, 2, 4, 6, 8
4 + 2 = 6; ds = 8 + d   **4**

14. Find the digit x so that the six-digit number 3x2x51 is divisible by 9.

3 + 5 + 1 = 9; remove from ds
ds = 2 + 2x   **8**

15. Find the digit x so that the 4-digit number 3xx8 is divisible by 9.

ds = 11 + 2x
9 no; 18 no; 27 yes   **8**

16. How many numbers of the form 1DD5 are divisible by 15?
all div by 5
1 + 5 = 6; ds = 2D   **4**
0, 3, 6, 9 all work

Divisibility 3

State the remainder without dividing.		State the divisibility rule for the number.
1. 712,359 ÷ 5 5 divides 712,355 **4**	17. 572,815 ÷ 9 9 divides 572,814 **1**	33. 12 if divisible by **3 and 4**
2. 163,967 ÷ 4 4 divides 163,964 **3**	18. 335,122 ÷ 6 6 divides 335,118 **4**	34. 14 if divisible by **2 and 7**
3. 362,461 ÷ 6 6 divides 362,460 **1**	19. 350,636 ÷ 14 14 divides 350,630 **6**	35. 15 if divisible by **3 and 5**
4. 427,892 ÷ 3 3 divides 427,890 **2**	20. 444,558 ÷ 9 9 divides 444,555 **3**	36. 18 if divisible by **2 and 9**
5. 874,598 ÷ 10 10 divides 874,590 **8**	21. 214,845 ÷ 12 12 divides 214,836 **9**	37. 20 if divisible by **4 and 5**
6. 836,733 ÷ 9 9 divides 836,730 **3**	22. 973,217 ÷ 10 10 divides 973,210 **7**	38. 24 if divisible by **3 and 8**
7. 146,371 ÷ 7 7 divides 146,370 **1**	23. 376,326 ÷ 8 8 divides 376,320 **6**	39. 25 if ends in **00, 25, 50, or 75**
8. 534,673 ÷ 5 5 divides 534,670 **3**	24. 174,985 ÷ 15 15 divides 174,975 **10**	40. 28 if divisible by **4 and 7**
9. 697,938 ÷ 11 11 divides 697,939 **10**	25. 964,574 ÷ 6 6 divides 964,572 **2**	41. 30 if divisible by **5 and 6**
10. 643,987 ÷ 2 2 divides 643,986 **1**	26. 371,245 ÷ 3 3 divides 371,244 **1**	42. 33 if divisible by **3 and 11**
11. 854,593 ÷ 4 4 divides 854,592 **1**	27. 246,754 ÷ 11 11 divides 246,752 **2**	43. 35 if divisible by **5 and 7**
12. 924,729 ÷ 8 8 divides 924,728 **1**	28. 745,569 ÷ 8 8 divides 745,568 **1**	44. 36 if divisible by **4 and 9**
13. 208,579 ÷ 15 15 divides 208,575 **4**	29. 316,796 ÷ 5 5 divides 316,795 **1**	45. 40 if divisible by **5 and 8**
14. 192,812 ÷ 11 11 divides 192,808 **4**	30. 210,753 ÷ 7 7 divides 210,749 **4**	46. 42 if divisible by **6 and 7**
15. 456,147 ÷ 12 12 divides 456,144 **3**	31. 497,855 ÷ 2 2 divides 497,854 **1**	47. 44 if divisible by **4 and 11**
16. 667,154 ÷ 4 4 divides 667,152 **2**	32. 495,165 ÷ 11 11 divides 495,165 **0**	48. 45 if divisible by **5 and 9**

Divisibility 4

Answer as indicated for natural numbers.

1.	When a certain number is divided by 11, the quotient is 7 and the remainder is 9. Find the number. n = (11) (7) + 9 n = **86**	9.	What is the greatest number by which every 6-digit palindrome must be divisible? ABCCBA (A+C+B) − (B+C+A) = 0 divisible by **11**
2.	What is the greatest number by which every 4-digit palindrome must be divisible? ABBA (A + B) − (B + A) = 0 divisible by **11**	10.	$6^5 \times 10^7$ can be divided by 20 how many times with no remainder? seven 5s twelve 2s = six 4s six 20s **6**
3.	What is the greatest 2-digit number by which n must be divisible if n is divisible by 2, 3, 6, and 11? Ignore 2 and 3 (factors of 6) 6 x 11 = **66**	11.	How many positive integers less than 200 are divisible by 2, 3, and 7? 42, 84, 126, 168 **4**
4.	Find n if n is greater than 1 and less than 9, and if the remainders are the same when 9 and 16 are each divided by n. divide by 7, get R2 **7**	12.	Find the least n such that n divided by 2, 3, 4, 5, and 6 has a remainder one less than the divisor. 3 x 4 x 5 = 60 60 − 1 = **59**
5.	Find the product of the thousands digit and ones digit of the greatest four-digit number divisible by 4. n = 9996 9 x 6 = **54**	13.	Find the probability that a positive integer less than 1000 selected randomly is divisible by 9. 999/9 = 111 $\frac{111}{999}$ $\frac{1}{9}$
6.	Find the least n divisible by 3, 6, 7, 8, and 9. 9 x 8 x 7 = 72 x 7 = **504**	14.	$6^6 \times 10^8$ can be divided by 30 how many times with no remainder? eight 5s, fourteen 2s, six 3s, six 30s **6**
7.	Find the probability that a positive integer less than 500 selected randomly is divisible by 7. 499/7 = 71 + R $\frac{71}{499}$	15.	Find the least n divisible by each of the first eight natural numbers. 8 x 7 x 3 x 5 = 21 x 40 = **840**
8.	How many positive integers less than 100 are divisible by 2, 3, and 5? 30, 60, 90 **3**	16.	When a certain number is divided by 9, the quotient is 12 and the remainder is 7. Find the number. n = (9) (12) + 7 n = **115**

Divisibility 5

Determine divisibility by 7 and 19 using the two rules.

1. 36,036		4. 32,604		7. 88,452		10. 35,112	
3603	3603	3260	3260	8845	8845	3511	3511
−12	+12	−8	+8	−4	+4	−4	+4
3591	3615	3252	3268	8841	8849	3507	3515
359	361	325	326	884	884	350	351
−2	+10	−4	+16	−2	+18	−14	+10
357	371	321	342	882	902	336	361
35	37	32	34	88	90	33	36
−14	+2	−2	+4	−4	+4	−12	+2
21	39	30	38	84	94	21	38
YES 7	NO 19	NO 7	YES 19	YES 7	NO 19	YES 7	YES 19

2. 29,925		5. 37,908		8. 41,895		11. 26,125	
2992	2992	3790	3790	4189	4189	2612	2612
−10	+10	−16	+16	−10	+10	−10	+10
2982	3002	3774	3806	4179	4199	2602	2622
298	300	377	380	417	419	260	262
−4	+4	−8	+12	−18	+18	−4	+4
294	304	369	392	399	437	256	266
29	30	36	39	39	43	25	26
−8	+8	−18	+4	−18	+14	−12	+12
21	38	18	43	21	57	13	38
YES 7	YES 19	NO 7	NO 19	YES 7	YES 19	NO 7	YES 19

3. 55,575		6. 75,803		9. 63,175		12. 87,516	
5557	5557	7580	7580	6317	6317	8751	8751
−10	+10	−6	+6	−10	+10	−12	+12
5547	5567	7574	7586	6307	6327	8739	8763
554	556	757	758	630	632	873	876
−14	+14	−8	+12	−14	+14	−18	+6
540	570	749	770	616	646	855	882
54	57	74	77	61	64	85	88
−0	+0	−18	+0	−12	+12	−10	+4
54	57	56	77	49	76	75	92
NO 7	YES 19	YES 7	NO 19	YES 7	YES 19	NO 7	NO 19

Divisibility 6

Answer as indicated.

---

1. How many permutations of the digits 2, 3, 5, 6 result in a 4-digit number divisible by 11?

2365	2563	3256	3652
5236	5632	6325	6523

8

---

2. How many permutations of the digits 0, 2, 3, 4, 9 result in a 5-digit number divisible by 90?

digit sum = 18
end in 0
by Mult Princ 4 x 3 x 2 x 1 x 1   **24**

---

3. How many permutations of the digits 1, 3, 5, 6 result in a 4-digit number divisible by 4?

end in 16, 36, or 56
2 permutations for each   **6**

---

4. How many permutations of the digits 0, 4, 6, 8, 9 result in a 5-digit number divisible by 8?

2 each ending in 048, 064, 096, 408, 480, 640, 840, 904, 960; 1 each ending in 608, 648, 864, 968, 984   **23**

---

5. How many permutations of the digits 1, 2, 5, 6, 7 result in a 5-digit number divisible by 6?

digit sum = 21
end in 2 or 6
by Mult Princ 4 x 3 x 2 x 1 x 2   **48**

---

6. How many permutations of the digits 1, 2, 5, 7, 9 result in a 5-digit number divisible by 25?

end in 25 or 75
6 permutations for each: 6 x 2   **12**

---

7. How many permutations of the digits 1, 2, 3, 6, 7, 9 result in a 6-digit number divisible by 11?

group 167 and 239
by Mult Princ 3 x 3 x 2 x 2 x 1 x 1
reverse even & odd locations x 2   **72**

---

8. How many permutations of the digits 1, 3, 5, 6, 9 result in a 5-digit number divisible by 6?

digit sum div by 3
end in 6
by Mult Princ 4 x 3 x 2 x 1 x 1   **24**

---

9. How many permutations of the digits 1, 2, 5, 7, 9 result in a 5-digit number divisible by 15?

digit sum div by 3
end in 5
by Mult Princ 4 x 3 x 2 x 1 x 1   **24**

---

10. How many permutations of the digits 1, 2, 4, 6, 8 result in a 5-digit number divisible by 12?

digit sum div by 3
end in 12, 16, 24, 28, 48, 64, 68, 84
by Mult Princ, 6 for each: 6 x 8   **48**

---

11. How many permutations of the digits 0, 2, 5, 8, 9 result in a 5-digit number divisible by 18?

rule for 18: div by 2 and 9
digit sum is not div by 9   **0**

---

12. How many permutations of the digits 1, 2, 3, 4, 6, 8 result in a 6-digit number divisible by 11?

group 138 and 246
by Mult Princ 3 x 3 x 2 x 2 x 1 x 1
reverse even & odd locations x 2   **72**

---

13. How many permutations of the digits 0, 2, 5, 7, 9 result in a 5-digit number divisible by 25?

end in 25, 50, or 75; not start with 0
4 permutations for 25, 75: 4 x 2
6 permutations for 50: 6 x 1   **14**

---

14. How many permutations of the digits 2, 3, 6, 7, 8 result in a 5-digit number divisible by 8?

2 for each ending in 328, 368, 632, 672, 728, 736, 832, 872   **16**

Divisibility 7

Find the least natural number with the given remainders.

Find the number described.

1. R 4 when divided by 5 **4**
R 1 when divided by 3

   4, 9, 14
   1, 4

9. between 200 and 500
divisible by 4 and 11
prime digit sum
44 × 5 = 220
220, 264, **308**, 352, 396, 440, 484

2. R 1 when divided by 2 **47**
R 2 when divided by 5
R 5 when divided by 7
all odds
2, 7, 12, 17, 22, 27, 32, 37, 42, 47
5, 12, 19, 26, 33, 40, 47

10. between 400 and 650
divisible by 4 and 9
digits in ascending order
36 × 12 = 432
432, **468**, 504, 540, 576, 612, 648

3. R 3 when divided by 5 **43**
R 3 when divided by 4
R 1 when divided by 6
3, 8, 13, 18, 23, 28, 33, 38, 43
3, 7, 11, 15, 19, 23, 27, 31, 35, 39, 43
1, 7, 13, 19, 25, 31, 37, 43

11. between 375 and 475
divisible by 2 and 7
digits form arithmetic sequence
14 × 27 = 378
378, 392, 406, **420**, 434, 448, 462

4. R 2 when divided by 4 **26**
R 2 when divided by 3
R 1 when divided by 5
2, 6, 10, 14, 18, 22, 26
2, 5, 8, 11, 14, 17, 20, 23, 26
1, 6, 11, 16, 21, 26

12. between 440 and 840
divisible by 2, 5, and 9
100s digit is 7 greater than 10s digit
90 × 5 = 450
450, 540, 630, 720, **810**

5. R 7 when divided by 11 **73**
R 3 when divided by 10
R 1 when divided by 4
7, 18, 29, 40, 51, 62, 73
3, 13, 23, . . . , 73
1, 5, 9, 13, 13 + 4 × 15 = 73

13. between 200 and 500
divisible by 2 and 11
palindrome
2-digit digit sum
242, **484**

6. R 3 when divided by 9 **39**
R 3 when divided by 6
R 4 when divided by 5
3, 12, 21, 30, 39
3, 9, 15, 21, 27, 33, 39
4, 9, 14, 19, 24, 29, 34, 39

14. between 200 and 400
divisible by 12
palindrome
144 + 60 = 204
204, 216, 228, 240, **252**

7. R 5 when divided by 10 **15**
R 1 when divided by 7
R 3 when divided by 4
5, 15, 25
1, 8, 15, 22
3, 7, 11, 15

15. between 200 and 300
divisible by 5
digit sum 9
not a perfect square
225, **270**

8. R 2 when divided by 11 **57**
R 1 when divided by 8
R 2 when divided by 5
2, 13, 24, 35, 46, 57
1, 9, 17, 25, 33, 41, 49, 57
2, 7, 12, 17, . . . , 57

16. between 400 and 600
divisible by 5 and 7
multiple of 10
has odd perfect square factor, not 1
420, **490**, 560

Divisibility 8

Answer as indicated. The variable x is any digit.

1. Use 2, 5, 8, 9, and x to form the greatest 5-digit number divisible by 6. 2 + 5 + 8 + 9 = 24 24 + 9 = 33 **99,852**	9. Find the least 3-digit number divisible by 22 subtracted from the greatest 4-digit number divisible by 14. 9996 $\underline{-\ 110}$ **9886**
2. Use 3, 4, 7, and x to form the greatest 4-digit number divisible by 12. 3 + 4 + 7 = 14 14 + 7 = 21 7, 7, 4, 3 no **7,344**	10. Find the least 5-digit number divisible by 35 plus the greatest 3-digit number divisible by 45. 10,010 $\underline{+\ 990}$ **11,000**
3. Use 2, 4, 6, and x to form the least 4-digit number divisible by 12. 2 + 4 + 6 = 12 **2,064**	11. Find the least 3-digit number divisible by 11 subtracted from the greatest 4-digit number divisible by 17. 9996 $\underline{-\ 110}$ **9886**
4. Use 2, 3, 6, 7, 8, and x to form the least 6-digit number divisible by 22. 203,687 div by 11, not 2 **203,786**	12. Find the greatest 5-digit number divisible by 15 minus the greatest 3-digit number divisible by 36. 99,990 $\underline{-\ 972}$ **99,018**
5. Use 3, 5, 6, 7, 9, and x to form the greatest 6-digit number divisible by 33. 3 + 5 + 6 + 7 + 9 = 30 3, 5, 6, 7, 9, 9 no **976,635**	13. Find the greatest 3-digit number divisible by 3, 4, and 5 plus the least 3-digit number divisible by 3, 4, and 5. div by 60 960 $\underline{+\ 120}$ **1080**
6. Use 4, 5, 8, and x to form the greatest 4-digit number divisible by 99. 4 + 5 + 8 = 17 27 no, 18 yes **8,514**	14. Find the least 4-digit number divisible by 2, 3, and 5 plus the greatest 3-digit number divisible by 2, 5, and 7. div by 30 div by 70 1020 $\underline{+\ 980}$ **2000**
7. Use 2, 4, 5, 6, and x to form the least 5-digit number divisible by 20. must end in 0 **24,560**	15. Find the least 4-digit number divisible by 3 and 10 plus the greatest 4-digit number divisible by 5 and 8. 1020 $\underline{+9960}$ **10,980**
8. Use 1, 2, 4, 5, 8, and x to form the greatest 6-digit number divisible by 36. 1 + 2 + 4 + 5 + 8 = 20 20 + 7 = 27 **875,412**	16. Find the greatest 4-digit number divisible by 5, 6, and 11 plus the least 4-digit number divisible by 5, 6, and 11. 9900 $\underline{+\ 1320}$ **11,220**

Equations 1

Solve mentally.

1.	$x - 9 = -4$ **x = 5**	17.	$4 = 6 + x$ **x = -2**	33.	$-20x = -100$ **x = 5**	49.	$x - 12 = -5$ **x = 7**
2.	$3x = -21$ **x = -7**	18.	$x - 5 = -5$ **x = 0**	34.	$x + 13 = 2$ **x = -11**	50.	$-7x = -84$ **x = 12**
3.	$x - 2 = 1$ **x = 3**	19.	$-7x = 42$ **x = -6**	35.	$4x = -16$ **x = -4**	51.	$x - 8 = 7$ **x = 15**
4.	$-5 = 5x$ **x = -1**	20.	$350 = x + 372$ **x = -22**	36.	$5x = 400$ **x = 80**	52.	$-20 = 4x$ **x = -5**
5.	$x + 8 = 12$ **x = 4**	21.	$x + 7 = -3$ **x = -10**	37.	$0 = x + 21$ **x = -21**	53.	$x + 9 = 23$ **x = 14**
6.	$-18 = x - 11$ **x = -7**	22.	$-9 = 3x$ **x = -3**	38.	$-63 = -7x$ **x = 9**	54.	$-19 = x - 13$ **x = -6**
7.	$-8x = -40$ **x = 5**	23.	$-9 = x + 3$ **x = -12**	39.	$-19 = x + 7$ **x = -26**	55.	$-9x = -63$ **x = 7**
8.	$4 = x + 7$ **x = -3**	24.	$-9 = x - 3$ **x = -6**	40.	$12 = x - 15$ **x = 27**	56.	$22 = x + 14$ **x = 8**
9.	$-7x = 56$ **x = -8**	25.	$-27 = -3x$ **x = 9**	41.	$10x = 560$ **x = 56**	57.	$-8x = 64$ **x = -8**
10.	$x + 9 = -4$ **x = -13**	26.	$x - 42 = 42$ **x = 84**	42.	$80 = x - 15$ **x = 95**	58.	$x + 11 = -3$ **x = -14**
11.	$-2x = 0$ **x = 0**	27.	$3x = 333$ **x = 111**	43.	$11x = -99$ **x = -9**	59.	$-2x = 22$ **x = -11**
12.	$56 = -8x$ **x = -7**	28.	$x + 120 = 144$ **x = 24**	44.	$x - 15 = 32$ **x = 47**	60.	$0 = -16x$ **x = 0**
13.	$0 = x + 17$ **x = -17**	29.	$-9x = -81$ **x = 9**	45.	$4x = 484$ **x = 121**	61.	$0 = x + 16$ **x = -16**
14.	$-13x = 0$ **x = 0**	30.	$3x = 45$ **x = 15**	46.	$11x = -66$ **x = -6**	62.	$-4x = 48$ **x = -12**
15.	$72 = -6x$ **x = -12**	31.	$x + 12 = 25$ **x = 13**	47.	$x - 12 = 21$ **x = 33**	63.	$75 = -15x$ **x = -5**
16.	$0 = x + 23$ **x = -23**	32.	$-9x = -72$ **x = 8**	48.	$5x = 60$ **x = 12**	64.	$70 = x + 35$ **x = 35**

Equations 2

Solve.

1. $\dfrac{3}{4}x = 36$

   $x = 36\,\dfrac{4}{3}$

   $x = 12 \cdot 4 = \mathbf{48}$

9. $\dfrac{2}{7}x = 40$

   $x = 40\,\dfrac{7}{2}$

   $x = 20 \cdot 7 = \mathbf{140}$

17. $\dfrac{-7}{16}x = 35$

    $x = 35\,\dfrac{16}{-7}$

    $x = -5 \cdot 16 = \mathbf{-80}$

2. $\dfrac{2}{5}x = 30$

   $x = 30\,\dfrac{5}{2}$

   $x = 15 \cdot 5 = \mathbf{75}$

10. $\dfrac{-4}{9}x = 68$

    $x = 68\,\dfrac{9}{-4}$

    $x = -17 \cdot 9 = \mathbf{-153}$

18. $\dfrac{4}{5}x = 60$

    $x = 60\,\dfrac{5}{4}$

    $x = 15 \cdot 5 = \mathbf{75}$

3. $\dfrac{4}{9}x = 400$

   $x = 400\,\dfrac{9}{4}$

   $x = 100 \cdot 9 = \mathbf{900}$

11. $\dfrac{3}{7}x = -24$

    $x = -24\,\dfrac{7}{3}$

    $x = -8 \cdot 7 = \mathbf{-56}$

19. $\dfrac{-9}{20}x = -63$

    $x = -63\,\dfrac{20}{-9}$

    $x = 7 \cdot 20 = \mathbf{140}$

4. $\dfrac{5}{7}x = -45$

   $x = -45\,\dfrac{7}{5}$

   $x = -9 \cdot 7 = \mathbf{-63}$

12. $\dfrac{11}{15}x = 44$

    $x = 44\,\dfrac{15}{11}$

    $x = 4 \cdot 15 = \mathbf{60}$

20. $\dfrac{7}{6}x = 49$

    $x = 49\,\dfrac{6}{7}$

    $x = 7 \cdot 6 = \mathbf{42}$

5. $\dfrac{3}{8}x = 24$

   $x = 24\,\dfrac{8}{3}$

   $x = 8 \cdot 8 = \mathbf{64}$

13. $\dfrac{-5}{80}x = -25$

    $x = -25\,\dfrac{80}{-5}$

    $x = 5 \cdot 80 = \mathbf{400}$

21. $\dfrac{-8}{13}x = 32$

    $x = 32\,\dfrac{13}{-8}$

    $x = -4 \cdot 13 = \mathbf{-52}$

6. $\dfrac{2}{3}x = -14$

   $x = -14\,\dfrac{3}{2}$

   $x = -7 \cdot 3 = \mathbf{-21}$

14. $\dfrac{9}{5}x = -81$

    $x = -81\,\dfrac{5}{9}$

    $x = -9 \cdot 5 = \mathbf{-45}$

22. $\dfrac{11}{12}x = -55$

    $x = -55\,\dfrac{12}{11}$

    $x = -5 \cdot 12 = \mathbf{-60}$

7. $\dfrac{13}{20}x = 39$

   $x = 39\,\dfrac{20}{13}$

   $x = 3 \cdot 20 = \mathbf{60}$

15. $\dfrac{3}{5}x = 48$

    $x = 48\,\dfrac{5}{3}$

    $x = 16 \cdot 5 = \mathbf{80}$

23. $\dfrac{-7}{30}x = -42$

    $x = -42\,\dfrac{30}{-7}$

    $x = 6 \cdot 30 = \mathbf{180}$

8. $\dfrac{15}{16}x = 75$

   $x = 75\,\dfrac{16}{15}$

   $x = 5 \cdot 16 = \mathbf{80}$

16. $\dfrac{12}{17}x = 60$

    $x = 60\,\dfrac{17}{12}$

    $x = 5 \cdot 17 = \mathbf{85}$

24. $\dfrac{4}{7}x = 28$

    $x = 28\,\dfrac{7}{4}$

    $x = 7 \cdot 7 = \mathbf{49}$

Equations 3

Solve. Work vertically, writing one line before the final solution.

1.	$3x + 7 = 22$ $3x = 15$ **x = 5**	12.	$-6 + 7x = 57$ $7x = 63$ **x = 9**	23.	$-4x + 5 = 53$ $-4x = 48$ **x = -12**
2.	$5x - 2 = 18$ $5x = 20$ **x = 4**	13.	$14 - 5x = -16$ $-5x = -30$ **x = 6**	24.	$13x - 2 = 37$ $13x = 39$ **x = 3**
3.	$-4x - 5 = -29$ $-4x = -24$ **x = 6**	14.	$22 + 3x = -2$ $3x = -24$ **x = -8**	25.	$-6x + 3 = 33$ $-6x = 30$ **x = -5**
4.	$6 + 4x = 14$ $4x = 8$ **x = 2**	15.	$9x + 5 = 86$ $9x = 81$ **x = 9**	26.	$5 - 3x = 38$ $-3x = 33$ **x = -11**
5.	$-6 - 3x = -12$ $-3x = -6$ **x = 2**	16.	$-49 - 8x = 23$ $-8x = 72$ **x = -9**	27.	$-3x + 2 = -19$ $-3x = -21$ **x = 7**
6.	$13x - 4 = -30$ $13x = -26$ **x = -2**	17.	$13 - 2x = 7$ $-2x = -6$ **x = 3**	28.	$-4x - 9 = 23$ $-4x = 32$ **x = -8**
7.	$4x + 2 = 14$ $4x = 12$ **x = 3**	18.	$28 - 2x = -6$ $-2x = -34$ **x = 17**	29.	$-2x + 7 = -1$ $-2x = -8$ **x = 4**
8.	$15 - 3x = 30$ $-3x = 15$ **x = -5**	19.	$-3 - 4x = -11$ $-4x = -8$ **x = 2**	30.	$-43 - 16x = 5$ $-16x = 48$ **x = -3**
9.	$7x + 2 = -47$ $7x = -49$ **x = -7**	20.	$6x - 1 = 35$ $6x = 36$ **x = 6**	31.	$-10x - 7 = 73$ $-10x = 80$ **x = -8**
10.	$4x - 20 = 8$ $4x = 28$ **x = 7**	21.	$5x - 7 = 28$ $5x = 35$ **x = 7**	32.	$9x + 11 = 47$ $9x = 36$ **x = 4**
11.	$7x + 6 = 13$ $7x = 7$ **x = 1**	22.	$-x - 1 = 1$ $-x = 2$ **x = -2**	33.	$-7x - 4 = -18$ $-7x = -14$ **x = 2**

Equations 4

Simplify by adding or subtracting like terms.

1. $12x + 9x$ **21x**	17. $7\pi + 5 + 9\pi + 12$ **16π + 17**	33. $12x - 20x + 16$ **16 – 8x**
2. $2y - 5y$ **–3y**	18. $3c - 4c + 5c - 6c$ **–2c**	34. $15x - 55x$ **–40x**
3. $4\pi + 9\pi$ **13π**	19. $2x - 5 + 10x + 15$ **12x + 10**	35. $3y - 2 - 5y + 19$ **17 – 2y**
4. $5rs + 8rs$ **13rs**	20. $-4x + 7 - 3x - 3$ **4 – 7x**	36. $4\pi + 2 + 6\pi - 2$ **10π**
5. $6x + 8x - 3x$ **11x**	21. $2x + 5 + x - 3x$ **5**	37. $3e - 7 + 7e - 2$ **10e – 9**
6. $7ab - 9ab$ **–2ab**	22. $14abc - 20abc + 1$ **1 – 6abc**	38. $3 + 2x + 2 + 7x$ **9x + 5**
7. $7x + 3 + 2x + 4$ **9x + 7**	23. $(6a - 7b) + (3a + 3b)$ **9a – 4b**	39. $13x - 30x + 30$ **30 – 17x**
8. $7u - 13u$ **–6u**	24. $8m - 2m + 5m$ **11m**	40. $9\pi + 5 + 9x + 11$ **9x + 9π + 16**
9. $9bc - 9b - 9c$ **9bc – 9b – 9c**	25. $10a + 5b + 2a + 5$ **12a + 5b + 5**	41. $4y + 8 + 3y - 11$ **7y – 3**
10. $9\pi - 20\pi + 20$ **20 – 11π**	26. $(x + 2) + (x + 2)$ **2x + 4**	42. $6e + 4w + 2e + 4$ **8e + 4w + 4**
11. $16 + 5x - 11 - 2x$ **3x + 5**	27. $3 + 9g - 8 + 11g$ **20g – 5**	43. $12\pi + 5 + 3\pi - 9$ **15π – 4**
12. $22p - 17p$ **5p**	28. $7 + 7x - 10 + 4x$ **11x – 3**	44. $-6 - 6y - 5 + 4y$ **–2y – 11**
13. $4x + 3y + 2x + y$ **6x + 4y**	29. $2ab + 3a + 4b$ **2ab + 3a + 4b**	45. $8rst - 20rst$ **–12rst**
14. $4ab - 5a - 16ab + 8a$ **3a – 12ab**	30. $13 + 4x - 9 + 7x$ **11x + 4**	46. $13 - 9x - 20 - 2x$ **–11x – 7**
15. $12xy - 30xy$ **–18xy**	31. $-7c - 14 + 9c + 15$ **2c + 1**	47. $6\pi - 16 - 7\pi + 7$ **–π – 9**
16. $5 - 5a - 15 - 15a$ **–10 – 20a**	32. $9ae - 9a - 9a + 9ae$ **18ae – 18a**	48. $(5x - 11) + (5x - 11)$ **10x – 22**

Equations 5

Write the phrase as an algebraic expression.

1.	the sum of triple a number and five $3x + 5$	14.	the number b decreased by ten $b - 10$
2.	the product of a number and six $6x$	15.	fifteen times half the number c $\dfrac{15c}{2}$
3.	eight times a number $8x$	16.	nine minus half the number e $9 - \dfrac{e}{2}$
4.	an even number decreased by four $2x - 4$	17.	half the sum of a number and four $\dfrac{x + 4}{2}$
5.	five more than twice a number $2x + 5$	18.	triple the quantity s plus two $3(s + 2)$
6.	seven less than a number $x - 7$	19.	the positive difference of six and k $|6 - k|$
7.	one fifth of an odd number $\dfrac{2x + 1}{5}$	20.	the cube of the number w w^3
8.	one third the quantity x minus one $\dfrac{x - 1}{3}$	21.	one less than triple a number $3x - 1$
9.	the number d increased by twelve $d + 12$	22.	six more than one third of a number $\dfrac{x}{3} + 6$
10.	six less than double a number $2x - 6$	23.	the product of nine and a number squared $9x^2$
11.	the sum of half a number and ten $\dfrac{x}{2} + 10$	24.	the sum of four and a number cubed $4 + x^3$
12.	the product of three numbers xyz	25.	the positive difference of p and two $|p - 2|$
13.	the square root of the sum of twice a number and four $\sqrt{2x + 4}$	26.	the square of the sum of one and a number $(x + 1)^2$

Equations 6

Express each sentence as an algebraic equation or inequality.

1.	Eight more than a number is fifty. $x + 8 = 50$	12.	One fifth a number is eighty. $\dfrac{x}{5} = 80$
2.	Twice a number is five more than the number. $2x = x + 5$	13.	Five times the sum of a number and three is sixty-two. $5(x + 3) = 62$
3.	The positive difference of a number and three is less than seventeen. $|x - 3| < 17$	14.	A number decreased by five is one fourth the number. $x - 5 = \dfrac{x}{4}$
4.	The sum of four and triple a number is twelve. $3x + 4 = 12$	15.	Nine times a number, then decreased by six, is sixteen. $9x - 6 = 16$
5.	Thirty is four more than half a number. $30 = 4 + \dfrac{x}{2}$	16.	Nine times the quantity of a number decreased by six is sixteen. $9(x - 6) = 16$
6.	Half a number is less than or equal to nineteen. $\dfrac{x}{2} \leq 19$	17.	Twice a number decreased by one is greater than triple the number. $2x - 1 > 3x$
7.	Ten less than a number is fifteen. $x - 10 = 15$	18.	Five is greater than double a number. $5 > 2x$
8.	One half the sum of a number and 1 is twenty. $\dfrac{x + 1}{2} = 20$	19.	Nine times a number is less than two. $9x < 2$
9.	One third a number increased by six is thirty. $\dfrac{x}{3} + 6 = 30$	20.	The sum of a number squared and ten is triple the number. $x^2 + 10 = 3x$
10.	Four is seven less than six times a number. $4 = 6x - 7$	21.	The positive difference of triple a number and two is the number. $|3x - 2| = x$
11.	A number squared is three more than half the number. $x^2 = 3 + \dfrac{x}{2}$	22.	A number cubed increased by eight is the number. $x^3 + 8 = x$

Equations 7

Solve, simplifying each side before combining sides.

1. $14x + 1 - x + 6 = 5 - 6x - 6 + 3x$

   $13x + 7 = -3x - 1$
   $16x = -8$
   $x = \dfrac{-1}{2}$

8. $-4x + 9 - 3x - 13 = 2x - 4 + x - 5$

   $-7x - 4 = 3x - 9$
   $5 = 10x$
   $x = \dfrac{1}{2}$

2. $-2 + 4x + 7 - 2x = -3x + 9 + 7x$

   $2x + 5 = 4x + 9$
   $-2x = 4$
   $x = -2$

9. $3x - 3 - 8x - 9 = -2x - 15 - x + 11$

   $-5x - 12 = -3x - 4$
   $-2x = 8$
   $x = -4$

3. $15 - 10x + 6 - x = 3x - 1 - 2x + 4$

   $21 - 11x = x + 3$
   $18 = 12x$
   $x = \dfrac{3}{2}$

10. $8x + 19 - 11x - 18 = 7x - 7 - 3x - 1$

   $1 - 3x = 4x - 8$
   $9 = 7x$
   $x = \dfrac{9}{7}$

4. $13x - 9 + x - 3 = -8x - 9 - 2x + 6$

   $14x - 12 = -10x - 3$
   $24x = 9$
   $x = \dfrac{3}{8}$

11. $-2x + 10 - 5x - 2 = 5x + 23 - 8x - 11$

   $8 - 7x = -3x + 12$
   $-4 = 4x$
   $x = -1$

5. $-7 + 5x - 8 + 2x = 3 - 6x - 4 + x$

   $7x - 15 = -5x - 1$
   $12x = 14$
   $x = \dfrac{7}{6}$

12. $5 + 8x + 6 - 3x = 4x - 30 - 19x + 19$

   $5x + 11 = -15x - 11$
   $20x = -22$
   $x = \dfrac{-11}{10}$

6. $7x - 7 + 5x - 2 = -7 + 4x - 2 - 9x$

   $12x - 9 = -5x - 9$
   $17x = 0$
   $x = 0$

13. $30x - 20 - 38x + 9 = 15 + 5x - 8 + x$

   $-8x - 11 = 6x + 7$
   $-18 = 14x$
   $x = \dfrac{-9}{7}$

7. $12x - 6 - 5x + 5 = 3x + 4 - 9x + 2$

   $7x - 1 = -6x + 6$
   $13x = 7$
   $x = \dfrac{7}{13}$

14. $8x - 6 - 9x - 1 = 3x - 9 - 7x + 17$

   $-x - 7 = -4x + 8$
   $3x = 15$
   $x = 5$

Equations 8

Distribute.

Distribute. Combine like terms.

1. $3(2a + 3b + 9c)$

   $6a + 9b + 27c$

2. $4(8x - 5y - 3)$

   $32x - 20y - 12$

3. $-5(6x - 2y + 8)$

   $-30x + 10y - 40$

4. $-7a(4a - 8b + 10)$

   $-28a^2 + 56ab - 70a$

5. $-2(-3w - 4x + 5xw - 5)$

   $6w + 8x - 10xw + 10$

6. $-6w(5w + 2x - 12)$

   $-30w^2 - 12wx + 72w$

7. $3ab(4b^3 + 5ab - 2a + 7b)$

   $12ab^4 + 15a^2b^2 - 6a^2b + 21ab^2$

8. $-8c(4cd - 5c + 11)$

   $-32c^2d + 40c^2 - 88c$

9. $12c(4de - 7e + 11c^3)$

   $48cde - 84ce + 132c^4$

10. $-9b^2(2b^6 - 5b^3 - 7b^2 + 3)$

    $-18b^8 + 45b^5 + 63b^4 - 27b^2$

11. $11x^3(-4x^3 - 2x^2 - 5x + 8)$

    $-44x^6 - 22x^5 - 55x^4 + 88x^3$

12. $2(x + 2y) + 5(4x + 5y)$

    $2x + 4y + 20x + 25y$
    $22x + 29y$

13. $6(4x - 2y) + 3(3x - 4y)$

    $24x - 12y + 9x - 12y$
    $33x - 24y$

14. $-7(-2a - 5b) - 6(4a + 7b)$

    $14a + 35b - 24a - 42b$
    $-10a - 7b$

15. $3(2a + 5b - c) - 2(a - 2b + 7c)$

    $6a + 15b - 3c - 2a + 4b - 14c$
    $4a + 19b - 17c$

16. $-4(3x - 2y) - 5(2x - 5y)$

    $-12x + 8y - 10x + 25y$
    $-22x + 33y$

17. $-8(6x - 2y) - 9(3x - 6y)$

    $-48x + 16y - 27x + 54y$
    $-75x + 70y$

18. $14x(2x - 5) - 5x(3x + 2)$

    $28x^2 - 70x - 15x^2 - 10x$
    $13x^2 - 80x$

19. $-3a(4a + 10) - a(2a + 9)$

    $-12a^2 - 30a - 2a^2 - 9a$
    $-14a^2 - 39a$

20. $-5e(4e + 3x) - 4e(3e - 7x)$

    $-20e^2 - 15ex - 12e^2 + 28ex$
    $-32e^2 + 13ex$

21. $15(a + 2b - 3c) - 12(2a - b + c)$

    $15a + 30b - 45c - 24a + 12b - 12c$
    $-9a + 42b - 57c$

22. $3(4w - 8y + 6) - 2(6w - 12y + 9)$

    $12w - 24y + 18 - 12w + 24y - 18$
    0

Equations 9

Write a variable expression.

1.	the value in cents of d dimes **10d**	14.	the number of days in x leap years **366x**
2.	Jan's age in 5 years if Jan is y years old now **y + 5**	15.	the next odd number after the odd number d **d + 2**
3.	the number of eggs in x dozen **12x**	16.	the number of days in h hours $\dfrac{h}{24}$
4.	the number of quarts in g gallons **4g**	17.	the number of hours in d days **24d**
5.	the number of gallons in q quarts $\dfrac{q}{4}$	18.	the value in cents of q quarters **25q**
6.	the number of hours in y minutes $\dfrac{y}{60}$	19.	the supplement of a° in degrees **180 − a**
7.	the number of minutes in y hours **60y**	20.	the money earned in dollars in d days at y dollars per day **dy**
8.	the value in cents of n nickels **5n**	21.	twice the complement of a° in degrees **2(90 − a)**
9.	the number of cups in q quarts **4q**	22.	the next even number after the odd number d **d + 1**
10.	the number of quarts in c cups $\dfrac{c}{4}$	23.	the value in cents of h half dollars and p pennies **50h + p**
11.	the number of hours in w weeks **168w**	24.	the next even number after the even number e **e + 2**
12.	the number of cents in d dollars **100d**	25.	the next natural number after the natural number n **n + 1**
13.	the sum of two consecutive whole numbers **2x + 1**	26.	the money earned in dollars in h hours at x dollars per hour **hx**

Equations 10

Solve.

1. $-5(x-6)-6 = 10-2(x-1)$

   $-5x + 30 - 6 = 10 - 2x + 2$
   $-5x + 24 = 12 - 2x$
   $12 = 3x$
   $x = 4$

2. $9-4(2x-3) = 6(x-2)+5$

   $9 - 8x + 12 = 6x - 12 + 5$
   $21 - 8x = 6x - 7$
   $28 = 14x$
   $x = 2$

3. $3x+1-2(x-6) = -3(x-1)+2$

   $3x + 1 - 2x + 12 = -3x + 3 + 2$
   $x + 13 = -3x + 5$
   $4x = -8$
   $x = -2$

4. $3-7(2x-4)+3x = -6(x-1)+5$

   $3 - 14x + 28 + 3x = -6x + 6 + 5$
   $31 - 11x = -6x + 11$
   $20 = 5x$
   $x = 4$

5. $-6(3x-2)-7-x = 10-3(x-5)+4x$

   $-18x + 12 - 7 - x = 10 - 3x + 15 + 4x$
   $-19x + 5 = x + 25$
   $-20 = 20x$
   $x = -1$

6. $7x+5(x-3)+5 = -3(3x-3)+2x$

   $7x + 5x - 15 + 5 = -9x + 9 + 2x$
   $12x - 10 = 9 - 7x$
   $19x = 19$
   $x = 1$

7. $7+6x-5(x-2) = -3x-(-5-x)$

   $7 + 6x - 5x + 10 = -3x + 5 + x$
   $x + 17 = -2x + 5$
   $3x = -12$
   $x = -4$

8. $-2(5x-3)-2x+12 = 9+6(3x-1)$

   $-10x + 6 - 2x + 12 = 9 + 18x - 6$
   $-12x + 18 = 18x + 3$
   $15 = 30x$
   $x = \dfrac{1}{2}$

9. $-3(x-5)-4x+2 = -5(x-2)-3$

   $-3x + 15 - 4x + 2 = -5x + 10 - 3$
   $-7x + 17 = -5x + 7$
   $10 = 2x$
   $x = 5$

10. $(-3x)(-5)+4(x-6) = -(x-5)+9+x$

    $15x + 4x - 24 = -x + 5 + 9 + x$
    $19x - 24 = 14$
    $19x = 38$
    $x = 2$

11. $-4(2x-4)-2x+9 = 5x-9-(x+8)$

    $-8x + 16 - 2x + 9 = 5x - 9 - x - 8$
    $-10x + 25 = 4x - 17$
    $42 = 14x$
    $x = 3$

12. $-2(4x-7)+6x+4 = 17+8(2x-1)$

    $-8x + 14 + 6x + 4 = 17 + 16x - 8$
    $-2x + 18 = 16x + 9$
    $9 = 18x$
    $x = \dfrac{1}{2}$

Equations 11

Solve.

---

1. $\overset{12}{\dfrac{3}{4}}x - 6 = 7x + \overset{12}{\dfrac{2}{3}}$

$9x - 72 = 84x + 8$
$-75x = 80$
$-15x = 16 \qquad x = \dfrac{-16}{15}$

7. $\overset{15}{\dfrac{11}{3}}x - 3 = 4 + \overset{15}{\dfrac{2}{5}}x$

$55x - 45 = 60 + 6x$
$49x = 105$
$7x = 15 \qquad x = \dfrac{15}{7}$

---

2. $\overset{12}{\dfrac{3}{4}} - 5x = 10 + \overset{12}{\dfrac{7x}{6}}$

$9 - 60x = 120 + 14x$
$-74x = 111$
$-2x = 3 \qquad x = \dfrac{-3}{2}$

8. $\overset{24x}{\dfrac{5}{8x}} + 2 = 9 - \overset{24x}{\dfrac{5}{6x}}$

$15 + (2)24x = (9)24x - 20$
$35 = (7)24x$
$5 = 24x \qquad x = \dfrac{5}{24}$

---

3. $\overset{40}{\dfrac{5}{8}} - 2x = 6 + \overset{40}{\dfrac{3x}{20}}$

$25 - 80x = 240 + 6x$
$-86x = 215$
$-2x = 5 \qquad x = \dfrac{-5}{2}$

9. $\overset{6}{\dfrac{7}{3}}x - 5 = 9x + \overset{6}{\dfrac{5}{6}}$

$14x - 30 = 54x + 5$
$-40x = 35$
$-8x = 7 \qquad x = \dfrac{-7}{8}$

---

4. $\overset{30}{\dfrac{5}{6}}x + 5 + \dfrac{2x}{3} = 3x + \dfrac{1}{2} - \overset{30}{\dfrac{4x}{5}}$

$25x + 150 + 20x = 90x + 15 - 24x$
$135 = 21x$
$45 = 7x$
$\qquad x = \dfrac{45}{7}$

10. $\overset{9x}{\dfrac{5}{9x}} + 2 = \dfrac{10}{3x} - \overset{9x}{\dfrac{1}{x}}$

$5 + 18x = 30 - 9$
$18x = 16$
$9x = 8 \qquad x = \dfrac{8}{9}$

---

5. $\overset{30}{\dfrac{2}{15}}x - 4 + \dfrac{x}{6} = 5x + \dfrac{1}{2} - \overset{30}{\dfrac{2x}{5}}$

$4x - 120 + 5x = 150x + 15 - 12x$
$9x - 120 = 138x + 15$
$-135 = 129x$
$-45 = 43x \qquad x = \dfrac{-45}{43}$

11. $\overset{28}{\dfrac{3}{4}}x - 2 = 3x + \overset{28}{\dfrac{5}{14}}$

$21x - 56 = 84x + 10$
$-63x = 66$
$-21x = 22 \qquad x = \dfrac{-22}{21}$

---

6. $\overset{40}{\dfrac{9}{10}}x + 3x + \dfrac{5}{2} = 13 + \overset{40}{\dfrac{3}{8}}x$

$36x + 120x + 100 = 520 + 15x$
$141x = 420$
$47x = 140$
$\qquad x = \dfrac{140}{47}$

12. $\overset{18x}{\dfrac{5}{6x}} + 3 = 8 - \overset{18x}{\dfrac{5}{9x}}$

$15 + (3)18x = (8)18x - 10$
$25 = (5)18x$
$5 = 18x \qquad x = \dfrac{5}{18}$

Equations 12

Solve the system by substitution.

1. $X = 9$
   $X - Y = 20$

   $9 - Y = 20$
   Y = -11
   X = 9

2. $P = 15$
   $5Q + 2P = 60$

   $5Q + 30 = 60$
   Q = 6
   P = 15

3. $9X = 63$
   $3X - 2Y = 71$

   X = 7
   $21 - 2Y = 71$
   $2Y = -50$
   Y = -25

4. $4Y + 1 = 17$
   $-11X + 5Y = 75$

   Y = 4
   $-11X + 20 = 75$
   $-11X = 55$
   X = -5

5. $M = N$
   $4M + 7N = 88$

   $11M = 88$
   M = 8
   N = 8

6. $15 - 3Y = 6$
   $X - 4Y = 38$

   Y = 3
   $X - 12 = 38$
   X = 50

7. $-6A + 5 = 17$
   $5A + 3B = 41$

   A = -2
   $-10 + 3B = 41$
   $3B = 51$
   B = 17

8. $X + Y = 5$
   $2X - 3Y = 25$

   $2(5-Y) - 3Y = 25$
   $-5Y = 15$
   Y = -3 X = 8

9. $P + 3Q = -7$
   $3P - 2Q = 1$

   $3(-7-3Q) - 2Q = 1$
   $-21 - 11Q = 1$
   Q = -2 P = -1

10. $X + Y = 9$
    $2X + 5Y = 9$

    $2(9-Y) + 5Y = 9$
    $18 + 3Y = 9$
    Y = -3 X = 12

11. $X + 2Y = 6$
    $3X - 4Y = 8$

    $3(6-2Y) - 4Y = 8$
    $18 - 10Y = 8$
    Y = 1 X = 4

12. $X - Y = -1$
    $2X - 3Y = 0$

    $2(Y-1) - 3Y = 0$
    $-Y - 2 = 0$
    Y = -2 X = -3

13. $4M - N = 2$
    $10M - 3N = 4$

    $10M - 3(4M-2) = 4$
    $6 - 2M = 4$
    M = 1 N = 2

14. $A + B = 7$
    $2A - B = 8$

    $2(7-B) - B = 8$
    $14 - 3B = 8$
    B = 2 A = 5

15. $3M + 4N = 4$
    $2M + N = 6$

    $3M + 4(6-2M) = 4$
    $24 - 5M = 4$
    M = 4 N = -2

16. $4X + 3Y = 11$
    $4X - Y = 7$

    $4X + 3(4X-7) = 11$
    $16X - 21 = 11$
    X = 2 Y = 1

17. $4P - Q = 9$
    $P - 3Q = 16$

    $4(16 + 3Q) - Q = 9$
    $64 + 11Q = 9$
    Q = -5 P = 1

18. $X + 3Y = 1$
    $-5X + Y = 11$

    $X + 3(5X+11) = 1$
    $16X + 33 = 1$
    X = -2 Y = 1

19. $3X - Y = 1$
    $9X - 2Y = 5$

    $9X - 2(3X-1) = 1$
    $3X + 2 = 5$
    X = 1 Y = 2

20. $3X + Y = 7$
    $7X - 3Y = 43$

    $7X - 3(7-3X) = 43$
    $16X - 21 = 43$
    X = 4 Y = -5

21. $X + 4Y = 10$
    $5X + 6Y = 8$

    $5(10-4Y) + 6Y = 8$
    $50 - 14Y = 8$
    Y = 3 X = -2

Equations 13

Solve the system by addition/subtraction.

1. M + N = 28
   M − N = 14

   2M   = 42
   M = 21
   N = 7

7. 3M + N = −6
   2M − N = 1

   5M   = −5
   M = −1
   N = −3

13. A + B = 38
    A − B = 20

    2A = 58
    A = 29
    B = 9

2. 7X − 3Y = 66
   2X − 3Y = 21

   5X   = 45
   X = 9
   Y = −1

8. 4X − 3Y = 26
   2X − 3Y = 16

   2X   = 10
   X = 5
   Y = −2

14. 5X − 4Y = 64
    3X − 4Y = 32

    2X   = 32
    X = 16
    Y = 4

3. 6X − 9Y = 60
   6X − 2Y = 18

   −7Y   = 42
   Y = −6
   X = 1

9. 3X − 9Y = −51
   −3X + 5Y = 15

   −4Y   = −36
   Y = 9
   X = 10

15. 5X − 7Y = 60
    5X − 6Y = 45

    −Y   = 15
    Y = −15
    X = −9

4. 9C + 3D = −3
   2C − 3D = −8

   11C   = −11
   C = −1
   D = 2

10. 5C + 6D = 37
    2C − 6D = 12

    7C   = 49
    C = 7
    D = $\frac{1}{3}$

16. 8C + 7D = −4
    2C − 7D = −6

    10C   = −10
    C = −1
    D = $\frac{4}{7}$

5. X − 4Y = 13
   X − 7Y = 70

   3Y   = −57
   Y = −19
   X = −63

11. 2X − 3Y = −15
    2X − 9Y = −33

    6Y   = 18
    Y = 3
    X = −3

17. M − 4N = 42
    M − 8N = 90

    4N   = −48
    N = −12
    M = −6

6. 2A − 6B = 11
   7A + 6B = 52

   9A   = 63
   A = 7
   B = $\frac{1}{2}$

12. 3A − B = 0
    A + B = 4

    4A   = 4
    A = 1
    B = 3

18. A − B = 4
    2A + B = 5

    3A   = 9
    A = 3
    B = −1

Equations 14

Multiply by the "FOIL" method.

1. $(x + 5)(x + 4)$

   $x^2 + 9x + 20$

2. $(x + 3)(x + 7)$

   $x^2 + 10x + 21$

3. $(x - 6)(x - 2)$

   $x^2 - 8x + 12$

4. $(x - 7)(x - 8)$

   $x^2 - 15x + 56$

5. $(x + 6)(x - 3)$

   $x^2 + 3x - 18$

6. $(x - 9)(x + 4)$

   $x^2 - 5x - 36$

7. $(x + 11)(x + 5)$

   $x^2 + 16x + 55$

8. $(x + 7)(x - 10)$

   $x^2 - 3x - 70$

9. $(x - 4)(x + 13)$

   $x^2 + 9x - 52$

10. $(x - 8)(x - 3)$

    $x^2 - 11x + 24$

11. $(x + 12)(x + 5)$

    $x^2 + 17x + 60$

12. $(2x + 9)(x + 3)$

    $2x^2 + 15x + 27$

13. $(3x + 2)(4x + 5)$

    $12x^2 + 23x + 10$

14. $(x - 5)(4x - 7)$

    $4x^2 - 27x + 35$

15. $(11x - 2)(3x - 1)$

    $33x^2 - 17x + 2$

16. $(6x - 5)(2x + 4)$

    $12x^2 + 14x - 20$

17. $(10x + 3)(x - 2)$

    $10x^2 - 17x - 6$

18. $(7x + 2)(3x + 2)$

    $21x^2 + 20x + 4$

19. $(x - 9)(8x + 4)$

    $8x^2 - 68x - 36$

20. $(9x - 4)(5x - 6)$

    $45x^2 - 74x + 24$

21. $(4x + 11)(2x - 3)$

    $8x^2 + 10x - 33$

22. $(2x - 7)(3x - 8)$

    $6x^2 - 37x + 56$

23. $(4x + y)(x + 6y)$

    $4x^2 + 25xy + 6y^2$

24. $(2x + 4y)(3x + 5y)$

    $6x^2 + 22xy + 20y^2$

25. $(8x + 3y)(7x - 2y)$

    $56x^2 + 5xy - 6y^2$

26. $(5x - 3y)(3x + 5y)$

    $15x^2 + 16xy - 15y^2$

27. $(x - 2y)(11x - 9y)$

    $11x^2 - 31xy + 18y^2$

28. $(7x - 5y)(6x - y)$

    $42x^2 - 37xy + 5y^2$

29. $(4x + 3y)(2x + 8y)$

    $8x^2 + 38xy + 24y^2$

30. $(11x + 5y)(2x - 6y)$

    $22x^2 - 56xy - 30y^2$

31. $(8x - 4y)(10x + 3y)$

    $80x^2 - 16xy - 12y^2$

32. $(2x - 7y)(9x - 8y)$

    $18x^2 - 79xy + 56y^2$

33. $(7x + 4y)(7x - 4y)$

    $49x^2 - 16y^2$

Equations 15

Make a chart using one variable before writing equations and solving.

1. Sam scored twice as many points at the Sunday game as the Friday game. Rob scored the same as Sam on Friday but half as many as Sam on Sunday. If their total points were 60, find the number Sam scored on Sunday.

	Fri	Sun
Rob	x	x
Sam	x	2x

$5x = 60$
$x = 12$
24

2. Val swam 5 times as many races as she ran. Hal ran 4 fewer races than Val swam and swam 3 more than Val ran. If they entered 35 total races, find Hal's number of running races.

	ran	swam
Val	x	5x
Hal	5x−4	x+3

$12x - 1 = 35$
$12x = 36$
$x = 3$
11

3. Pam earned half as much writing as she did teaching. Bob's teaching earnings were $3000 less than Pam's, but his writing earnings were $4000 more. If their total earnings were $85,000, find Bob's teaching earnings.

	write	teach
Pam	P	2P
Bob	P+4000	2P−3000

$6P+1000 = 85,000$
$6P = 84,000$
$P = 14,000$
$25,000

4. A bus used the same amount of gas all 7 days one week. A taxi used half as much as the bus on the weekend days and one third as much as the bus on the weekdays. If 580 gallons of gas were used in total, find the taxi's weekend gas used in gallons.

	Day	End
Bus	15x	6x
Taxi	5x	3x

$29x = 580$
$x = 20$
60

5. The number of students in Lara's BA class was 150 fewer than Greg's. Her MBA class had half as many as her BA class, while Greg's MBA class had 300 fewer than his BA class. If the total number enrolled was 980, find the number in Lara's MBA class.

	BA	MBA
Lara	2x	x
Greg	2x+150	2x−150

$7x = 980$
$x = 140$
140

6. On Tuesday Bo worked triple his Friday hours, while Jo worked double Bo's Friday hours. On Friday Jo worked half of Bo's Tuesday hours. If their total hours for the 2 days is 90, find Bo's Tuesday hours.

	Tues	Fri
Jo	2x	1.5x
Bo	3x	x

$7.5x = 90$
$x = 12$
36

7. One week, Meg walked 6 times as many miles as she ran. Peg ran twice as many miles as Meg ran but only half the number that Meg walked. Their total combined miles are 102. Find the number of miles that Peg ran.

	walk	ran
Meg	6x	x
Peg	3x	2x

$12x = 102$
$x = 8.5$
17

8. Lil baked 4 times as many cocoa cookies as sugar cookies. Jon baked half as many sugar cookies as Lil and one third as many cocoa cookies. The total number of cookies baked was 820. Find the number of sugar cookies baked.

	C	S
Lil	24S	6S
Jon	8S	3S

$41S = 820$
$S = 20$
180

MAVA Math: Enhanced Skills Solutions   Copyright © 2015 Marla Weiss

Equations 16

Answer as indicated by writing and solving equations.

---

1. There are triple as many A as C, half as many C as B, and 55 A and B combined. Find A − (B + C). **0**

A = 3C
B = 2C
A + B = 55
3C + 2C = 55
C = 11
A = 33
B = 22

5.
a + b + c + d = 500
a + b = 185
b + c = 290
a + c = 345
Find all 4 variables.

c + d = 315
a + d = 210
c − a = 105
c + a = 345
2c = 450
c = 225
a = 120
b = 65
d = 90

---

2. A total of 200 tickets were sold, some adult at $60 each and some student at $15, yielding $5700 income. How many of each type were sold?

60A + 15(200 − A) = 5700
60A + 3000 − 15A = 5700
45A = 2700
A = 60
S = 140

6. Rob's items are worth $10 each, while Sue's are worth $25 each. How many does each have if together they have 150 at a total value of $2325?

10R + 25(150 − R) = 2325
10R + 3750 − 25R = 2325
15R = 1425
R = 95
S = 55

---

3. Jo, Bo, Mo, and Ro are dividing $175,000. Mo gets half as much as Jo. Bo gets 3/4 of Jo's share. Ro gets 5,000 less than Bo. Find how much each received.

J + B + M + R = 175,000
M = J/2
B = (3/4)J
R = B − 5000
J + 2B + M = 180,000
J + 1.5J + .5J = 180,000

3J = 180,000
J = 60,000
M = 30,000
B = 45,000
R = 40,000

7. Hal's age is 9 more than half Tom's age. Bob's age is 5 more than one-third Tom's age. Hal is fourteen years older than Bob. Find their ages.

H = 9 + T/2
2H = 18 + T
B = 5 + T/3
3B = 15 + T
H = B + 14

2(B + 14) = 18 + 3B − 15
2B + 28 = 3B + 3
B = 25
H = 39
T = 60

4.
e + f + g + h = 450
e + g = 160
e + f = 90
f + g = 120
Find h.

g + h = 360
f + h = 290
g − f = 70
g + f = 120
2g = 190
g = 95
f = 25
e = 65
h = 265

8. The average of 3 numbers is 40. The first is twice the second. The third is 5 more than the first. Find the numbers.

A + B + C = 120
A = 2B
C = A + 5
A + (A/2) + A + 5 = 120
2A + A + 2A + 10 = 240
5A = 230

A = 46
B = 23
C = 51

Equations 17

Solve the system with multiplication followed by addition/subtraction.

1. 6A + 5B = 22
12A − 7B = 10

12A + 10B = 44
17B = 34
B = 2
A = 2

2. 4X + 3Y = 22
8X − 7Y = 70

8X + 6Y = 44
8X − 7Y = 70
13Y = −26
Y = −2
X = 7

3. 3X + 5Y = 65
4X − 3Y = −10

9X + 15Y = 195
20X − 15Y = −50
29X = 145
X = 5
Y = 10

4. 6X − 7Y = 50
3X + 5Y = 8

6X − 7Y = 50
6X + 10Y = 16
−17Y = 34
Y = −2
X = 6

5. 10B + 3C = 45
5B + 3C = 15

10B + 3C = 45
10B + 6C = 30
−3C = 15
C = −5
B = 6

6. 5M − 2N = 23
4M − 3N = 3

20M − 8N = 92
20M − 15N = 15
7N = 77
N = 11
M = 9

7. 9M + 5N = −3
5M − 9N = 69

81M + 45N = −27
25M − 45N = 345
106M = 318
M = 3
N = −6

8. 5X + Y = 8
3X − 4Y = 14

20X + 4Y = 32
23X = 46
X = 2
Y = −2

9. 5X + 7Y = 6
10X − 3Y = 46

10X + 14Y = 12
−17Y = 34
Y = −2
X = 4

10. 2P + 3Q = 12
3P − 4Q = 1
8P + 12Q = 48
9P − 12Q = 3
17P = 51
P = 3
Q = 2

11. 2X + 3Y = 20
5X − 6Y = 23

4X + 6Y = 40
9X = 63
X = 7
Y = 2

12. 4X + 3Y = −1
2X + 5Y = 3

4X + 10Y = 6
−7Y = −7
Y = 1
X = −1

13. 4M + 5N = 11
5M − 4N = 24

16M + 20N = 44
25M − 20N = 120
41M = 164
M = 4
N = −1

14. 5X + 3Y = 19
2X − 5Y = 20

25X + 15Y = 95
6X − 15Y = 60
31X = 155
X = 5
Y = −2

15. 6X + 7Y = 8
7X − 2Y = 50

12X + 14Y = 16
49X − 14Y = 350
61X = 366
X = 6
Y = −4

16. 5X − 3Y = 24
3X + 5Y = 28

25X − 15Y = 120
9X + 15Y = 84
34X = 204
X = 6
Y = 2

17. 2B + 3C = 12
5B + 7C = 29

10B + 15C = 60
10B + 14C = 58
C = 2
B = 3

18. 2M − 3N = −7
5M + 4N = 17

8M − 12N = −28
15M + 12N = 51
23M = 23
M = 1
N = 3

MAVA Math: Enhanced Skills Solutions Copyright © 2015 Marla Weiss

Equations 18

Solve the quadratic equation in factored form. Use mental math.

1. $(x + 7)(x - 5) = 0$	$x = -7, 5$	17. $(2x + 3)(6x + 5) = 0$	$x = \dfrac{-3}{2}, \dfrac{-5}{6}$	
2. $(x + 2)(x + 9) = 0$	$x = -2, -9$	18. $(7x + 9)(8x - 7) = 0$	$x = \dfrac{-9}{7}, \dfrac{7}{8}$	
3. $(x - 1)(x - 8) = 0$	$x = 1, 8$	19. $(7x - 4)(4x + 1) = 0$	$x = \dfrac{4}{7}, \dfrac{-1}{4}$	
4. $(x - 4)(x + 10) = 0$	$x = 4, -10$	20. $(5x - 8)(3x - 2) = 0$	$x = \dfrac{8}{5}, \dfrac{2}{3}$	
5. $(x + 3)(x - 11) = 0$	$x = -3, 11$	21. $(4x - 3)(6x + 11) = 0$	$x = \dfrac{3}{4}, \dfrac{-11}{6}$	
6. $(x + 1)(x + 4) = 0$	$x = -1, -4$	22. $(5x - 9)(9x - 7) = 0$	$x = \dfrac{9}{5}, \dfrac{7}{9}$	
7. $(x - 7)(x - 9) = 0$	$x = 7, 9$	23. $(3x + 4)(2x - 9) = 0$	$x = \dfrac{-4}{3}, \dfrac{9}{2}$	
8. $(x - 10)(x + 8) = 0$	$x = 10, -8$	24. $(9x + 5)(8x + 1) = 0$	$x = \dfrac{-5}{9}, \dfrac{-1}{8}$	
9. $(2x + 1)(7x - 2) = 0$	$x = \dfrac{-1}{2}, \dfrac{2}{7}$	25. $(8x - 5)(2x - 5) = 0$	$x = \dfrac{5}{8}, \dfrac{5}{2}$	
10. $(9x + 4)(8x + 3) = 0$	$x = \dfrac{-4}{9}, \dfrac{-3}{8}$	26. $(4x + 7)(5x - 2) = 0$	$x = \dfrac{-7}{4}, \dfrac{2}{5}$	
11. $(4x - 5)(5x - 6) = 0$	$x = \dfrac{5}{4}, \dfrac{6}{5}$	27. $(3x - 1)(10x + 3) = 0$	$x = \dfrac{1}{3}, \dfrac{-3}{10}$	
12. $(3x - 8)(6x + 7) = 0$	$x = \dfrac{8}{3}, \dfrac{-7}{6}$	28. $(7x + 3)(9x + 2) = 0$	$x = \dfrac{-3}{7}, \dfrac{-2}{9}$	
13. $(8x + 9)(3x - 10) = 0$	$x = \dfrac{-9}{8}, \dfrac{10}{3}$	29. $(11x + 4)(6x - 1) = 0$	$x = \dfrac{-4}{11}, \dfrac{1}{6}$	
14. $(6x + 1)(7x + 4) = 0$	$x = \dfrac{-1}{6}, \dfrac{-4}{7}$	30. $(9x - 5)(4x - 11) = 0$	$x = \dfrac{5}{9}, \dfrac{11}{4}$	
15. $(2x - 7)(4x - 9) = 0$	$x = \dfrac{7}{2}, \dfrac{9}{4}$	31. $(3x - 5)(12x + 5) = 0$	$x = \dfrac{5}{3}, \dfrac{-5}{12}$	
16. $(9x - 10)(5x + 8) = 0$	$x = \dfrac{10}{9}, \dfrac{-8}{5}$	32. $(10x - 1)(3x + 7) = 0$	$x = \dfrac{1}{10}, \dfrac{-7}{3}$	

Estimating 1

Estimate to the closest multiple of 5.

1.

$5 \quad 5 \quad 10$

$$\dfrac{(5.19)(4.92)}{0.51}$$

5 **50**

2.

$50 \quad 5 \quad 100$

$$\dfrac{(51.9)(4.92)}{0.51}$$

50 **500**

3.

$5 \quad 50$

$$\dfrac{(0.519)(49.2)}{5.1}$$

$5 \quad 10$ **5**

4.

$8 \quad 8 \quad 10$

$$\dfrac{(8.14)(7.88)}{0.81}$$

8 **80**

5.

$10 \quad 90$

$$\dfrac{(0.984)(88.8)}{9.1}$$

$9 \quad 10$ **10**

6.

$80 \quad 8 \quad 10$

$$\dfrac{(81.4)(7.98)}{0.81}$$

8 **800**

7.

$6 \quad 1000$

$$\dfrac{(0.614)(986.721)}{(1.93)(31.1)}$$

$2 \quad 30 \quad 10$ **10**

8.

$9 \quad 100 \quad 65$

$$\dfrac{(0.886)(99.81)(6.52)}{(5.04)(9.07)(1.32)}$$

$5 \quad 9 \quad 13 \quad 10$ **10**

9.

$90 \quad 75 \quad 1$

$$\dfrac{(89.7)(74.65)(0.11)}{(2.98)(0.317)(24.895)}$$

$3 \quad 3 \quad 25$ **30**

10.

$6 \quad 50 \quad 52$

$$\dfrac{(6.03)(0.491)(51.89)}{(4.04)(0.59)(12.98)}$$

$4 \quad 60 \quad 13$ **5**

11.

$80 \quad 72 \quad 3$

$$\dfrac{(78.791)(71.921)(0.3)}{(5.87)(7.99)(1.188)}$$

$6 \quad 8 \quad 12$ **30**

12.

$7 \quad 50 \quad 9 \quad 50$

$$\dfrac{(6.96)(49.88)(0.89)(48.9)}{(4.984)(69.9)(8.8)}$$

$5 \quad 70 \quad 9 \quad 10$ **5**

13.

$11 \quad 7 \quad 14 \quad 5$

$$\dfrac{(10.89)(6.941)(14.03)(0.497)}{(7.02)(2.16)(0.91)}$$

$7 \quad 22 \quad 1$ **35**

14.

$5 \quad 30 \quad 66 \quad 5$

$$\dfrac{(0.507)(29.82)(6.61)(4.882)}{(6.06)(0.29)(10.98)(0.991)}$$

$6 \quad 3 \quad 11 \quad 10$ **25**

15.

$9 \quad 45 \quad 11 \quad 4$

$$\dfrac{(8.896)(4.501)(10.95)(0.391)}{(11.01)(0.891)(0.402)}$$

$11 \quad 9 \quad 4$ **45**

16.

$5 \quad 100 \quad 4 \quad 10 \quad 7$

$$\dfrac{(0.516)(98.721)(3.87)(10.001)(0.686)}{(4.04)(9.9)(0.498)(6.498)(1.001)}$$

$4 \quad 10 \quad 5 \quad 7 \quad 1 \quad 10$ **10**

Estimating 2

Answer as indicated.	Estimate to the nearest multiple of 10.
1. Eli has 159 pictures to put in an album. If each page holds 15 photos, how many pages will he need? 10 pages 150 photos too few **11**	8. 4478 − 92 round up to 4480 round down to 90 4480 − 90 = **4390**
2. Weight at age one is about triple the birthweight. If a baby weighed 7 pounds 7 ounces at birth, find the weight at age one to the nearest half pound. round to 7 lb 7 oz = 7.5 **22.5**	9. 5798 + 94 round up to 5800 round down to 90 5800 + 90 = **5890**
3. Would 3 school buses, each holding 40 people plus the driver, be enough to transport 6 classes of 18 students each plus 1 teacher per class? round 18 + 1 to 20 3 × 40 = 120 is upper bound **yes**	10. 3877 + 54 round up to 3880 round down to 50 3880 + 50 = **3930**
4. Hudson drove 1496 miles during a five-day trip. About how many miles did he average each day? round 1496 to 1500 1500/5 = 300 **300**	11. 9218 − 2181 round up to 9220 round down to 2180 9220 − 2180 = **7040**
5. Harrison's annual salary is $72,250. How much does he earn each month, to the nearest ten dollars? round 72,250 to 72,240 72,240/12 = 6020 **$6020**	12. 5602 − 2818 round down to 5600 round up to 2820 5600 − 2820 = **2780**
6. A room and carpet are both 12 feet wide. The room is 11 feet long. The carpet costs $44.95 a running yard and may be cut to any length. About how much will the carpet cost? **$165** $15/ft 15 × 11 = 165	13. 72,596 ÷ 597 round up to 72,600 726 ÷ 6 = 121 round up to 600 **120** divide both by 100
7. Kyra wishes to buy a jacket that costs $59.95. If she saves $2.50 each week, in how many weeks will she have enough money for the purchase? 2 weeks $5 24 weeks $60 **24**	14. 9011 ÷ 89 round down to 9000 round up to 90 9000 ÷ 90 = **100**

Evaluating 1

Evaluate.

1.	$90 - 3xy + x^2$ for $x = 3$, $y = 6$ $90 - 54 + 9$ **45**	12.	$3xz - 2x + 4yz$ for $x = 2$, $y = 3$, $z = 5$ $30 - 4 + 60$ **86**
2.	$100 - 2(x + 4)^2$ for $x = 3$ $100 - 98$ **2**	13.	$z^2 + 2yz - 5y$ for $y = 4$, $z = -5$ $25 - 40 - 20$ **-35**
3.	$3(x - y) + 2xy$ for $x = 7$, $y = 3$ $12 + 42$ **54**	14.	$5xy(x - z)$ for $x = 2$, $y = 3$, $z = -4$ $30(6)$ **180**
4.	$3xy - 3x - 3y$ for $x = -2$, $y = 4$ $-24 + 6 - 12$ **-30**	15.	$xy + yz - xz$ for $x = 3$, $y = 4$, $z = -2$ $12 - 8 + 6$ **10**
5.	$-5x - 2x^2 + 4y$ for $x = 4$, $y = -6$ $-20 - 32 - 24$ **-76**	16.	$xy - 4x + 5z$ for $x = 6$, $y = -7$, $z = 2$ $-42 - 24 + 10$ **-56**
6.	$7x(x - y) + 3y$ for $x = 2$, $y = 5$ $-42 + 15$ **-27**	17.	$2y^2(5xy - 2x - 4)$ for $x = -2$, $y = -4$ $32(40 + 4 - 4)$ **1280**
7.	$8x - 2(2x + y)$ for $x = -2$, $y = 6$ $-16 - 4$ **-20**	18.	$8x - (xy - z)$ for $x = 5$, $y = -3$, $z = 7$ $40 - (-22)$ **62**
8.	$2xy + x^y + y^x$ for $x = 3$, $y = 4$ $24 + 81 + 64$ **169**	19.	$2z^y + 5y^2 - 2z$ for $y = 3$, $z = -5$ $-250 + 45 + 10$ **-195**
9.	$x^2 + y^3 + 4xy$ for $x = 2$, $y = 3$ $4 + 27 + 24$ **55**	20.	$\dfrac{4x}{4y - x}$ for $x = 6$, $y = 2$ $24/2$ **12**
10.	$y^2(3xy - 2y - 6)$ for $x = -2$, $y = -3$ $9 \times (18 + 6 - 6)$ **162**	21.	$\dfrac{8xy}{5(x - y)}$ for $x = 2$, $y = 10$ $160/(-40)$ **-4**
11.	$\dfrac{x - y}{x + y}$ for $x = 4$, $y = 10$ $-6/14$ $\dfrac{-3}{7}$	22.	$\dfrac{2x - 3y}{4x + 2y}$ for $x = 3$, $y = -3$ $15/6$ $\dfrac{5}{2}$

Evaluating 2

Evaluate.

1. $20 - 12xy$ for $x = \frac{3}{4}$, $y = \frac{2}{3}$ $20 - 6 = \mathbf{14}$	9. $xy + 4x - y$ for $x = \frac{3}{8}$, $y = \frac{2}{5}$ $\frac{3}{20} + \frac{3}{2} - \frac{2}{5} = \frac{3 + 30 - 8}{20} = \mathbf{\frac{5}{4}}$
2. $100x - 25(x + 1)^2$ for $x = \frac{3}{5}$ $60 - 64 = \mathbf{-4}$	10. $xy - 3x + 2y$ for $x = \frac{2}{9}$, $y = \frac{3}{4}$ $\frac{1}{6} - \frac{2}{3} + \frac{3}{2} = \frac{1 - 4 + 9}{6} = \mathbf{1}$
3. $28(y - x) + 7xy$ for $x = \frac{1}{4}$, $y = \frac{4}{7}$ $(16 - 7) + 1 = \mathbf{10}$	11. $16xy(y - x)$ for $x = \frac{3}{8}$, $y = \frac{2}{3}$ $4(\frac{16 - 9}{24}) = \mathbf{\frac{7}{6}}$
4. $27xy - 12x - 9y$ for $x = \frac{5}{6}$, $y = \frac{4}{9}$ $10 - 10 - 4 = \mathbf{-4}$	12. $6xz - 2x + 3z$ for $x = \frac{1}{6}$, $z = \frac{11}{12}$ $\frac{11}{12} - \frac{1}{3} + \frac{11}{4} = \frac{11 - 4 + 33}{12} = \mathbf{\frac{10}{3}}$
5. $-5x - 16y + 4xy$ for $x = \frac{3}{5}$, $y = \frac{5}{8}$ $-3 - 10 + 1.5 = \mathbf{-11.5}$	13. $3z - 5yz - 5y$ for $y = \frac{3}{5}$, $z = \frac{7}{9}$ $\frac{7}{3} - \frac{7}{3} - 3 = \mathbf{-3}$
6. $24x - 2(2x + y)$ for $x = \frac{5}{8}$, $y = \frac{3}{4}$ $15 - 2(2) = \mathbf{11}$	14. $3y^2(6xy + 2x)$ for $x = \frac{3}{4}$, $y = \frac{1}{3}$ $\frac{1}{3}(\frac{3}{2} + \frac{3}{2}) = \frac{1}{3} \times \frac{6}{2} = \mathbf{1}$
7. $6x(3x + 2y) + 4y$ for $x = \frac{5}{3}$, $y = \frac{5}{2}$ $10(5 + 5) + 10 = \mathbf{110}$	15. $y^2(3xy - 2x)$ for $x = \frac{5}{6}$, $y = \frac{3}{2}$ $\frac{9}{4}(\frac{15}{4} - \frac{5}{3}) = \frac{9}{4} \times \frac{25}{12} = \mathbf{\frac{75}{16}}$
8. $-18x - 7y + 21xy$ for $x = \frac{-2}{9}$, $y = \frac{3}{7}$ $4 - 3 - 2 = \mathbf{-1}$	16. $3y^3 + 5x^2 - xy$ for $x = \frac{2}{5}$, $y = \frac{1}{3}$ $\frac{1}{9} + \frac{4}{5} - \frac{2}{15} = \frac{5 + 36 - 6}{45} = \mathbf{\frac{7}{9}}$

Evens & Odds 1

Answer as indicated.

1. If 16 is written as the product of 3 positive integers, each greater than 1, how many of them are even?

2 x 2 x 4
Because 16 is a power of 2, no factors are odd.
3

2. If the sum of two consecutive odd numbers is 28, what is their product?

13 + 15 = 28
13 x 15 = 39 x 5
195

3. If x is an odd number, what is the sum of the next two odd numbers greater than 3x+1?

x odd
3x odd
3x + 1 even
(3x + 2) + (3x + 4)
6x + 6

4. Find the least positive integer n such that 3n is both even and a perfect square.

n needs a 2 for 3n to be even
n needs another 2 & 3 to make a square
n = 2 x 2 x 3
12

5. Define n*=3n if n is even and n*=2n if n is odd. Find a* + b* if a and b are the first two prime numbers.

a = 2, b = 3
a* = 6
b* = 6
6 + 6
12

6. Subtract the sum of the odd numbers 1 through 39 from the sum of the even numbers 2 through 40.

2 + 4 + . . . + 38 + 40
− 1 + 3 + . . . + 37 + 39
1 + 1 + . . . + 1 + 1
20

7. Remove 5 even numbers from the first 55 whole numbers. What percent of the remaining numbers are even?

0 – 54
27 odd, 28 even
27 odd, 23 even
23 even of 50
46%

8. Remove 9 odd numbers from the first 40 whole numbers. What percent of the odd numbers were removed?

0 – 39
20 odd, 20 even
11 odd, 20 even
9 of 20
45%

9. If x is an even number, what is the sum of the next four odd numbers greater than 2x+3?

x even
2x even
2x + 3 odd
2x + 5
2x + 7
2x + 9
2x + 11
8x + 32

10. If the product of the digits of a two-digit number is odd, is the sum of the digits even or odd?

Both digits must be odd.
Odd + odd = even
even

11. If 210 is written as the product of 3 positive integers, each greater than 1, how many of them are even?

2 x 3 x 5 x 7
The 2 pairs with 1 of the odd factors, or the 2 remains alone.
1

12. The results of 3/4 of an even number and 2/3 of the consecutive even are consecutive and descending. Find the odd number between the two evens.

4 and 6 no: 3, 4 ascend
16 and 18 no: 12, 12 equal
28, 30 yes: 21, 20 descend
29

Evens & Odds 2

Operate as indicated on natural numbers.	*Find the greatest number that must be a factor of the given sum.*

1. the difference of the 231st odd number and the 159th even number
 	
  461
– 318
  143   1st odd = 1   nth odd =
            2nd odd = 3   2n–1
            3rd odd = 5
	9. the sum of 2 consecutive odd numbers
	
	$(2n + 1) + (2n + 3)$
	$4n + 4$
	
	                                                    4

2. the sum of the 105th even number and the 175th odd number
 	
  210
+ 349
  559   1st even = 2   nth even = 2n
            2nd even = 4
            3rd even = 6
	10. the sum of 3 consecutive odd numbers
	
	$(2n + 1) + (2n + 3) + (2n + 5)$
	$6n + 9$
	
	                                                    3

3. the quotient of the 210th even number and the 95th odd number $\dfrac{420}{189}$ $\dfrac{140}{63}$ $\dfrac{\mathbf{20}}{\mathbf{9}}$	11. the sum of 4 consecutive odd numbers $(2n + 1) + (2n + 3) + (2n + 5) + (2n + 7)$ $8n + 16$ **8**
4. the product of the 55th even number and the 43rd odd number $110 \times 85 = 85 \times 11 \times 10 = 935 \times 10 =$ **9350**	12. the sum of 5 consecutive odd numbers $(2n+1) + (2n+3) + (2n+5) + (2n+7) + (2n+9)$ $10n + 25$ **5**

5. the sum of the 195th even number and the 243rd odd number
 	
  390
+ 485
  875
	13. the sum of 2 consecutive even numbers
	
	$(2n) + (2n + 2)$
	$4n + 2$
	
	                                                    2

6. the quotient of the 116th odd number and the 132nd even number $\dfrac{231}{264}$ $\dfrac{21}{24}$ $\dfrac{\mathbf{7}}{\mathbf{8}}$	14. the sum of 3 consecutive even numbers $(2n) + (2n + 2) + (2n + 4)$ $6n + 6$ **6**
7. the product of the 100th even number and the 345th odd number $200 \times 689 = 689 \times 2 \times 100 =$ $1378 \times 100 =$ **137,800**	15. the sum of 4 consecutive even numbers $(2n) + (2n + 2) + (2n + 4) + (2n + 6)$ $8n + 12$ **4**

8. the difference of the 318th even number and the 131st odd number
 	
  636
– 261
  375
	16. the sum of 5 consecutive even numbers
	
	$(2n) + (2n+2) + (2n+4) + (2n+6) + (2n+8)$
	$10n + 20$
	
	                                                    10

Exponents 1

Evaluate by mental math.

1. 2^3 **8**	17. 8^2 **64**	33. 2^7 **128**	49. 700^2 **490,000**
2. 3^2 **9**	18. 100^2 **10,000**	34. 196^1 **196**	50. 90^2 **8100**
3. 10^2 **100**	19. 3^4 **81**	35. 2^{10} **1024**	51. 70^2 **4900**
4. 40^2 **1600**	20. 2^8 **256**	36. 110^2 **12,100**	52. 2^5 **32**
5. 20^2 **400**	21. 5^3 **125**	37. 80^2 **6400**	53. 150^2 **22,500**
6. 6^2 **36**	22. 30^3 **27,000**	38. 11^3 **1331**	54. 781^0 **1**
7. 2^6 **64**	23. 60^2 **3600**	39. 3^5 **243**	55. 2^{11} **2048**
8. 12^2 **144**	24. 3^6 **729**	40. 4^4 **256**	56. 30^3 **27,000**
9. 600^2 **360,000**	25. 10^4 **10,000**	41. 1^{82} **1**	57. 2^2 **4**
10. 3^3 **27**	26. 0^{24} **0**	42. 5^4 **625**	58. 11^4 **14,641**
11. 11^2 **121**	27. 20^3 **8000**	43. 20^4 **160,000**	59. 90^2 **8100**
12. 4^2 **16**	28. 4^5 **1024**	44. 7^2 **49**	60. 50^2 **2500**
13. 15^2 **225**	29. 10^3 **1000**	45. 50^3 **125,000**	61. 56^0 **1**
14. 9^3 **729**	30. 1^{37} **1**	46. 4^6 **4096**	62. 120^2 **14,400**
15. 13^2 **169**	31. 14^2 **196**	47. 2^9 **512**	63. 2^4 **16**
16. 4^3 **64**	32. 300^2 **90,000**	48. 9^2 **81**	64. 0^{86} **0**

Exponents 2

Evaluate.

1. $.2^2$	**.04**	17. $.17^2$	**.0289**	33. $.2^7$	**.0000128**
2. 1.3^2	**1.69**	18. $.001^2$	**.000001**	34. 1.1^3	**1.331**
3. $.02^3$	**.000008**	19. 1.1^2	**1.21**	35. $.2^{10}$	**.0000001024**
4. $.05^2$	**.0025**	20. $.09^3$	**.000729**	36. $.2^4$	**.0016**
5. $.02^4$	**.00000016**	21. $.1^5$	**.00001**	37. $.8^2$	**.64**
6. $.6^2$	**.36**	22. $.5^3$	**.125**	38. $.3^4$	**.0081**
7. $.2^6$	**.000064**	23. $.3^3$	**.027**	39. $.14^2$	**.0196**
8. 1.2^2	**1.44**	24. $.6^3$	**.216**	40. $.4^4$	**.0256**
9. $.5^2$	**.25**	25. 1.6^2	**2.56**	41. $.0965^1$	**.0965**
10. $.3^3$	**.027**	26. $.018^2$	**.000324**	42. $.05^4$	**.00000625**
11. $.11^2$	**.0121**	27. $.12^2$	**.0144**	43. $.3^2$	**.09**
12. $.004^2$	**.000016**	28. 1.7^2	**2.89**	44. $.7^2$	**.49**
13. $.15^2$	**.0225**	29. $.04^3$	**.000064**	45. $.01^3$	**.000001**
14. $.01^4$	**.00000001**	30. $.1^3$	**.001**	46. $.09^2$	**.0081**
15. $.1^2$	**.01**	31. $.011^2$	**.000121**	47. $.2^3$	**.008**
16. $.011^2$	**.000121**	32. 1.4^2	**1.96**	48. $.4^2$	**.16**

Exponents 3

Simplify using rules of exponents. Answer in exponential form.

1. $3^8 \cdot 3^6$ 3^{14}
2. $7^2 \cdot 7^5$ 7^7
3. $4^{20} \div 4^{10}$ 4^{10}
4. $3^6 \div 3^3$ 3^3
5. $(3^4)^2$ 3^8
6. $(5^5)^3$ 5^{15}
7. $2^5 \cdot 2^6$ 2^{11}
8. $9^9 \div 9^5$ 9^4
9. $(2^3)^3$ 2^9
10. $17^4 \cdot 17^8$ 17^{12}
11. $7^8 \div 7^3$ 7^5
12. $16^3 \cdot 16^4$ 16^7
13. $(7^2)^5$ 7^{10}
14. $25^6 \div 25^4$ 25^2
15. $(1.5^5)^4$ 1.5^{20}
16. $1.9^3 \cdot 1.9^6$ 1.9^9

17. $4^8 \div 4^2$ 4^6
18. $6.2^6 \cdot 6.2^2$ 6.2^8
19. $7^9 \div 7^5$ 7^4
20. $67^7 \cdot 67^8$ 67^{15}
21. $(11^9)^3$ 11^{27}
22. $22^{14} \cdot 22^3$ 22^{17}
23. $25^{12} \div 25^8$ 25^4
24. $(32^5)^6$ 32^{30}
25. $12^3 \cdot 12^3$ 12^6
26. $5^2 \cdot 5^6$ 5^8
27. $26^{13} \div 26^7$ 26^6
28. $(10^8)^4$ 10^{32}
29. $14^5 \cdot 14^6$ 14^{11}
30. $(12^3)^{10}$ 12^{30}
31. $(16^7)^2$ 16^{14}
32. $19^{15} \div 19^6$ 19^9

33. $18^4 \cdot 18^9$ 18^{13}
34. $12^3 \cdot 12^{11}$ 12^{14}
35. $15^{14} \div 15^7$ 15^7
36. $(9^9)^2$ 9^{18}
37. $(6.1^5)^7$ 6.1^{35}
38. $8^{15} \div 8^9$ 8^6
39. $(2^7)^6$ 2^{42}
40. $43^9 \cdot 43^{12}$ 43^{21}
41. $6^{11} \div 6^6$ 6^5
42. $.7^4 \cdot .7^{10}$ $.7^{14}$
43. $(16^8)^2$ 16^{16}
44. $37^{29} \div 37^{15}$ 37^{14}
45. $(2.9^4)^4$ 2.9^{16}
46. $(53^4)^6$ 53^{24}
47. $1.5^9 \cdot 1.5^7$ 1.5^{16}
48. $9^{10} \div 9^4$ 9^6

49. $14^7 \cdot 14^{13}$ 14^{20}
50. $8^9 \div 8^7$ 8^2
51. $17^{13} \div 17^4$ 17^9
52. $(39^7)^4$ 39^{28}
53. $11^{11} \div 11^8$ 11^3
54. $33^9 \cdot 33^9$ 33^{18}
55. $21^{16} \div 21^2$ 21^{14}
56. $(88^9)^7$ 88^{63}
57. $71^6 \cdot 71^7$ 71^{13}
58. $8^{21} \div 8^{17}$ 8^4
59. $(64^5)^{12}$ 64^{60}
60. $92^9 \cdot 92^{21}$ 92^{30}
61. $31^{15} \div 31^8$ 31^7
62. $16^9 \div 16^7$ 16^2
63. $(35^9)^9$ 35^{81}
64. $(77^2)^{18}$ 77^{36}

Exponents 4

Simplify using rules of exponents. Assume no denominator variable is zero.

1. $x^2 \cdot x^7$	x^9	17. $wb^6 \cdot w^5b^7$	w^6b^{13}	33. $25b^9 \div (5b^2)$	$5b^7$
2. $y^{18} \div y^{12}$	y^6	18. $h^5 \cdot h^6 \cdot h^4$	h^{15}	34. $25b^9 \div (5b)^2$	b^7
3. $h^8 \div h^3$	h^5	19. $(g^2)^9$	g^{18}	35. $36c^{11} \div (3c)^2$	$4c^9$
4. $(g^4)^3$	g^{12}	20. $eg^{10} \cdot ge^{10}$	$e^{11}g^{11}$	36. $36c^{11} \div 3c^2$	$12c^9$
5. $(ac^5)^7$	a^7c^{35}	21. $ab^{12} \div b^9$	ab^3	37. $(7w^9)^2$	$49w^{18}$
6. $p^6 \cdot p^5$	p^{11}	22. $(a^4b)^4$	$a^{16}b^4$	38. $(3h^6)^3$	$27h^{18}$
7. $x^9 \div x^8$	x	23. $n^2 \cdot n^7 \cdot n^5$	n^{14}	39. $a^9 \cdot a^3 \cdot a^8$	a^{20}
8. $(xy^3)^3$	x^3y^9	24. $g^{14} \cdot g^6$	g^{20}	40. $(3c^7)^4$	$81c^{28}$
9. $m^4 \cdot m^9$	m^{13}	25. $x^2y^{13} \div y^8$	x^2y^5	41. $x^5 \cdot ax^5 \cdot bx^2$	abx^{12}
10. $s^{11} \div s^3$	s^8	26. $(w^4)^6$	w^{24}	42. $(6e^5)^2$	$36e^{10}$
11. $a^4b^3 \cdot a^5b^4$	a^9b^7	27. $eh^2 \cdot he^2 \cdot h^6$	e^3h^9	43. $(2g^4)^4$	$16g^{16}$
12. $(y^3)^5$	y^{15}	28. $(x^3)^9$	x^{27}	44. $(k^2s^4)^3$	k^6s^{12}
13. $e^9 \div e^2$	e^7	29. $(p^7)^3$	p^{21}	45. $c \cdot cn^6 \cdot nc^3$	c^5n^7
14. $(c^2d^5)^4$	c^8d^{20}	30. $g^{15} \div g^{14}$	g	46. $m^4 \cdot a^4 \cdot m^4$	a^4m^8
15. $c^2d^5 \cdot a^2d^6$	$a^2c^2d^{11}$	31. $a^3 \cdot a^8 \cdot a^6$	a^{17}	47. $y^{30} \div y^{20}$	y^{10}
16. $k^{13} \div k^5$	k^8	32. $a^2 \cdot an^7 \cdot na^5$	a^8n^8	48. $e^7c^{15} \div e^4$	$c^{15}e^3$

Exponents 5

Write as the least whole number times a power of ten.	Operate and simplify. Answer as powers of x and 2, 5, and/or 10.
1. $(4 \cdot 10^2) \cdot (5 \cdot 10^2)$ $20 \cdot 10^4$ **$2 \cdot 10^5$**	14. $(5x^2)^4(2x^3)^4(2x^6)(5x^7)$ $10^4 \cdot x^8 \cdot x^{12} \cdot 10 \cdot x^{13} =$ **10^5x^{33}**
2. $(200 \cdot 10^5) \div (5 \cdot 10^3)$ $40 \cdot 10^2$ **$4 \cdot 10^3$**	15. $(5x)(4x^2)^2(2x^5)(4x^2)^3(2x^6)$ $5 \cdot 2^4 \cdot x^5 \cdot 2^8 \cdot x^{17} =$ **$2^{12} \cdot 5 \cdot x^{22}$**
3. $(15 \cdot 10^2) \cdot (4 \cdot 10^6)$ $60 \cdot 10^8$ **$6 \cdot 10^9$**	16. $(2x^3)^2(8x^5)(2x^2)^3(4x^5)$ $2^5 \cdot x^{11} \cdot 2^5 \cdot x^{11} =$ **$2^{10} \cdot x^{22}$**
4. $(5 \cdot 10^5) \cdot (16 \cdot 10^4)$ $80 \cdot 10^9$ **$8 \cdot 10^{10}$**	17. $(2x^5)^3(5x^5)(2x^2)^3(5x)^5$ $2^3 \cdot x^{15} \cdot 5^6 \cdot 2^3 \cdot x^{16} =$ **10^6x^{31}**
5. $(80 \cdot 10^{11}) \div (8 \cdot 10^4)$ $10 \cdot 10^7$ **10^8**	18. $(10x^2)(5x^5)(2x^2)^4(25x^5)^2$ $10 \cdot x^7 \cdot 2^4 \cdot 5^4 \cdot x^{18} =$ **10^5x^{25}**
6. $(120 \cdot 10^8) \div (4 \cdot 10^3)$ $30 \cdot 10^5$ **$3 \cdot 10^6$**	19. $(2x^3)^3(8x^6)(2x)^3(5x^5)$ $2^3 \cdot x^9 \cdot 2^6 \cdot 5 \cdot x^{14} =$ **$2^9 \cdot 5 \cdot x^{23}$**
7. $(18 \cdot 10^5) \cdot (5 \cdot 10^5)$ $90 \cdot 10^{10}$ **$9 \cdot 10^{11}$**	20. $(4x^2)^2(2x^5)(5x^2)^3(5x^5)^2$ $2^5 \cdot x^9 \cdot 5^5 \cdot x^{16} =$ **10^5x^{25}**
8. $(80 \cdot 10^{12}) \div (2 \cdot 10^4)$ $40 \cdot 10^8$ **$4 \cdot 10^9$**	21. $(5x^2)^3(125x^5)(2x^2)^4(40x^5)$ $5^6 \cdot x^{11} \cdot 2^7 \cdot 5 \cdot x^{13} =$ **10^7x^{24}**
9. $(14 \cdot 10^6) \cdot (5 \cdot 10^7)$ $70 \cdot 10^{13}$ **$7 \cdot 10^{14}$**	22. $(4x^4)^2(16x^3)(4x^3)^2(8x^7)$ $2^8 \cdot x^{11} \cdot 2^7 \cdot x^{13} =$ **$2^{15} \cdot x^{24}$**
10. $(500 \cdot 10^7) \div (25 \cdot 10^6)$ $20 \cdot 10^1$ **$2 \cdot 10^2$**	23. $(25x^3)(8x^6)(10x^3)^3(50x^4)$ $2^3 \cdot 5^2 \cdot x^9 \cdot 10^4 \cdot 5 \cdot x^{13} =$ **10^7x^{22}**
11. $(15 \cdot 10^3) \cdot (8 \cdot 10^{11})$ $120 \cdot 10^{14}$ **$12 \cdot 10^{15}$**	24. $(2x^4)^4(5x^7)(2x)^5(32x^5)$ $2^4 \cdot x^{23} \cdot 2^{10} \cdot 5 \cdot x^{10} =$ **$2^{14} \cdot 5 \cdot x^{33}$**
12. $(330 \cdot 10^{11}) \div (11 \cdot 10^2)$ $30 \cdot 10^9$ **$3 \cdot 10^{10}$**	25. $(5x^5)^5(100x^5)(2x^5)^2(80x^2)$ $5^5 \cdot x^{25} \cdot 10^3 \cdot 2^5 \cdot x^{17} =$ **10^8x^{42}**
13. $(12 \cdot 10^6) \cdot (5 \cdot 10^6)$ $60 \cdot 10^{12}$ **$6 \cdot 10^{13}$**	26. $(8x^4)^2(64x^3)(2x^7)^2(16x^6)$ $2^6 \cdot x^{11} \cdot 2^6 \cdot 2^6 \cdot x^{20} =$ **$2^{18} \cdot x^{31}$**

Exponents 6

Solve for x.

1. $16^x = 2^{12}$

   $2^{4x} = 2^{12}$
   4x = 12
   x = 3

2. $32^x = 2^{10}$

   $2^{5x} = 2^{10}$
   5x = 10
   x = 2

3. $3^8 = 9^x$

   $3^8 = 3^{2x}$
   2x = 8
   x = 4

4. $4^x = 8^8$

   $2^{2x} = 2^{24}$
   2x = 24
   x = 12

5. $27^x = 3^6$

   $3^{3x} = 3^6$
   3x = 6
   x = 2

6. $16^5 = 2^x$

   $2^{20} = 2^x$
   x = 20

7. $4^x = 64^8$

   $4^x = 4^{24}$
   x = 24

8. $81^3 = 27^x$

   $3^{12} = 3^{3x}$
   3x = 12
   x = 4

9. $9^x \div 9^6 = 9^7$

   $9^{x-6} = 9^7$
   x − 6 = 7
   x = 13

10. $4^4 \cdot 4^x \cdot 4^5 = 4^9$

    $4^{x+9} = 4^9$
    x + 9 = 9
    x = 0

11. $3^2 \cdot 3^x \cdot 3^4 = 9^6$

    $3^{x+6} = 3^{12}$
    x + 6 = 12
    x = 6

12. $8^5 \cdot 8^x \div 8^7 = 8^8$

    $8^{x-2} = 8^8$
    x−2 = 8
    x = 10

13. $11^{16} \div 11^x = 11^9$

    $11^{16-x} = 11^9$
    16−x = 9
    x = 7

14. $8^6 \cdot 8^x \div 8^5 = 2^9$

    $2^{3x+3} = 2^9$
    3x + 3 = 9
    x = 2

15. $12^3 \div 2^x = 27$

    $4^3 \div 2^x = 1$
    $2^6 \div 2^x = 1$
    x = 6

16. $2^x \cdot 2^4 \cdot 2^7 = 4^6$

    $2^{11+x} = 2^{12}$
    11 + x = 12
    x = 1

17. $9^3 = 27^{2x-1}$

    $3^6 = 3^{6x-3}$
    6 = 6x − 3
    x = 1.5

18. $125^{2x-1} = 25^{2x+1}$

    $5^{6x-3} = 5^{4x+2}$
    6x − 3 = 4x + 2
    x = 2.5

19. $16^3 = 2^{2x-4}$

    $2^{12} = 2^{2x-4}$
    12 = 2x − 4
    x = 8

20. $121^{8x} = 1331^{4x+2}$

    $11^{16x} = 11^{12x+6}$
    16x = 12x + 6
    x = 1.5

21. $4^{3x} = 8^{3x+1}$

    $2^{6x} = 2^{9x+3}$
    6x = 9x + 3
    x = −1

22. $343^{2x+1} = 49^{6x}$

    $7^{6x+3} = 7^{12x}$
    6x + 3 = 12x
    x = 0.5

23. $16^{x+5} = 32^{x-2}$

    $2^{4x+20} = 2^{5x-10}$
    4x + 20 = 5x − 10
    x = 30

24. $8^5 = 32^{2x-3}$

    $2^{15} = 2^{10x-15}$
    15 = 10x − 15
    x = 3

Exponents 7

1.5 is 3/2; .6 is 3/5; etc.

Evaluate. Express as a simplified fraction.

1. 2^{-3} $\dfrac{1}{8}$	17. 9^{-3} $\dfrac{1}{729}$	33. 1.5^{-3} $\dfrac{8}{27}$	49. $(2^{-3})^{-2}$ 64
2. $(-4)^{-3}$ $\dfrac{-1}{64}$	18. 30^{-4} $\dfrac{1}{81,000}$	34. 2.5^{-4} $\dfrac{16}{625}$	50. $(9^{-1})^{-1}$ 9
3. 10^{-4} $\dfrac{1}{10,000}$	19. 15^{-2} $\dfrac{1}{225}$	35. 1.3^{-2} $\dfrac{100}{169}$	51. $(14^{-2})^{-1}$ 196
4. 3^{-5} $\dfrac{1}{243}$	20. $(-1)^{-7}$ -1	36. 1.2^{-2} $\dfrac{25}{36}$	52. $(.5^{-1})^{-2}$ $\dfrac{1}{4}$
5. 30^{-3} $\dfrac{1}{27,000}$	21. $(-2)^{-5}$ $\dfrac{-1}{32}$	37. $.6^{-4}$ $\dfrac{625}{81}$	53. $(1.5^{-1})^{-3}$ $\dfrac{27}{8}$
6. 80^{-2} $\dfrac{1}{6400}$	22. 10^{-3} $\dfrac{1}{1,000}$	38. $.3^{-3}$ $\dfrac{1000}{27}$	54. $(3.5^{-1})^{-2}$ $\dfrac{49}{4}$
7. 11^{-2} $\dfrac{1}{121}$	23. 40^{-3} $\dfrac{1}{64,000}$	39. $.25^{-3}$ 64	55. $2^{-2} \div 4^{-1}$ 1
8. 25^{-2} $\dfrac{1}{625}$	24. $(-3)^{-3}$ $\dfrac{-1}{27}$	40. $.2^{-3}$ 125	56. $4^{-4} \div 8^{-2}$ $\dfrac{1}{4}$
9. 1^{-5} 1	25. 5^{-3} $\dfrac{1}{125}$	41. 1.75^{-2} $\dfrac{16}{49}$	57. $(1^{-1} + 2^{-1})^{-2}$ $\dfrac{4}{9}$
10. 6^{-2} $\dfrac{1}{36}$	26. 6^{-3} $\dfrac{1}{216}$	42. $(-2.5)^{-3}$ $\dfrac{-8}{125}$	58. $(1^{-2} + 2^{-2})^{-2}$ $\dfrac{16}{25}$
11. 13^{-2} $\dfrac{1}{169}$	27. 2^{-4} $\dfrac{1}{16}$	43. 1.6^{-2} $\dfrac{25}{64}$	59. $(2^{-1} - 3^{-1})^{-1}$ 6
12. 4^{-4} $\dfrac{1}{256}$	28. 4^{-3} $\dfrac{1}{64}$	44. $.2^{-4}$ 625	60. $(3^{-1} - 3^{-2})^{-1}$ $\dfrac{9}{2}$
13. 20^{-3} $\dfrac{1}{8000}$	29. $(-10)^{-4}$ $\dfrac{1}{10,000}$	45. $.75^{-3}$ $\dfrac{64}{27}$	61. $4^{-2} \div 2^{-4}$ 1
14. $(-10)^{-3}$ $\dfrac{-1}{1000}$	30. 2^{-6} $\dfrac{1}{64}$	46. $(-.4)^{-3}$ $\dfrac{-125}{8}$	62. $4^{-3} \div 2^{-5}$ $\dfrac{1}{2}$
15. 5^{-4} $\dfrac{1}{625}$	31. 8^{-3} $\dfrac{1}{512}$	47. $(-1.4)^{-2}$ $\dfrac{25}{49}$	63. $(3^{-1} - 4^{-1})^{-1}$ 12
16. 12^{-2} $\dfrac{1}{144}$	32. 16^{-2} $\dfrac{1}{256}$	48. $.25^{-5}$ 1024	64. $(2.2^{-1})^{-2}$ $\dfrac{121}{25}$

Exponents 8

Evaluate.

1. $8^{\frac{1}{3}}$ **2**	17. $1331^{\frac{2}{3}}$ **121**	33. $16^{\frac{-3}{4}}$ $\dfrac{1}{8}$
2. $27^{\frac{1}{3}}$ **3**	18. $10{,}000^{\frac{3}{4}}$ **1000**	34. $729^{\frac{-2}{3}}$ $\dfrac{1}{81}$
3. $1000^{\frac{1}{3}}$ **10**	19. $64^{\frac{2}{3}}$ **16**	35. $256^{\frac{-1}{2}}$ $\dfrac{1}{16}$
4. $16^{\frac{1}{4}}$ **2**	20. $64^{\frac{3}{2}}$ **512**	36. $9^{\frac{-3}{2}}$ $\dfrac{1}{27}$
5. $4^{\frac{3}{2}}$ **8**	21. $625^{\frac{3}{4}}$ **125**	37. $8^{\frac{-2}{3}}$ $\dfrac{1}{4}$
6. $256^{\frac{5}{8}}$ **32**	22. $81^{\frac{3}{2}}$ **729**	38. $64^{\frac{-2}{3}}$ $\dfrac{1}{16}$
7. $243^{\frac{3}{5}}$ **27**	23. $16^{\frac{5}{4}}$ **32**	39. $125^{\frac{-4}{3}}$ $\dfrac{1}{625}$
8. $128^{\frac{4}{7}}$ **16**	24. $(-32)^{\frac{1}{5}}$ **-2**	40. $243^{\frac{-2}{5}}$ $\dfrac{1}{9}$
9. $1024^{\frac{3}{10}}$ **8**	25. $(-125)^{\frac{1}{3}}$ **-5**	41. $81^{\frac{-3}{4}}$ $\dfrac{1}{27}$
10. $32^{\frac{2}{5}}$ **4**	26. $(-128)^{\frac{5}{7}}$ **-32**	42. $(-128)^{\frac{-3}{7}}$ $\dfrac{-1}{8}$
11. $9^{\frac{5}{2}}$ **243**	27. $(-729)^{\frac{2}{3}}$ **81**	43. $121^{\frac{-3}{2}}$ $\dfrac{1}{1331}$
12. $64^{\frac{5}{3}}$ **1024**	28. $(-1331)^{\frac{2}{3}}$ **121**	44. $125^{\frac{-1}{3}}$ $\dfrac{1}{5}$
13. $125^{\frac{2}{3}}$ **25**	29. $(-125)^{\frac{4}{3}}$ **625**	45. $32^{\frac{-3}{5}}$ $\dfrac{1}{8}$
14. $729^{\frac{1}{3}}$ **9**	30. $27^{\frac{1}{3}+\frac{1}{3}}$ **9**	46. $(-64)^{\frac{-1}{3}}$ $\dfrac{-1}{4}$
15. $1{,}000{,}000^{\frac{2}{3}}$ **10,000**	31. $16^{\frac{1}{2}-\frac{1}{4}}$ **2**	47. $(-32)^{\frac{-4}{5}}$ $\dfrac{1}{16}$
16. $256^{\frac{1}{4}}$ **4**	32. $64^{\frac{1}{2}+\frac{1}{3}}$ **32**	48. $(-27)^{\frac{-1}{3}}$ $\dfrac{-1}{3}$

Factorials 1

Simplify.

1. $\dfrac{997!}{996!}$ **997**

12. $\dfrac{100!}{2!\,99!} + \dfrac{12!}{9!}$ 50 + 1320   **1370**

2. $\dfrac{9!}{4!\,4!}$ $\dfrac{5 \times 6 \times 7 \times 8 \times 9}{2 \times 3 \times 4}$ 5×7×2×9   **630**

13. $\dfrac{13!}{12!} + \dfrac{11!}{10!}$ 13 + 11   **24**

3. $\dfrac{7!\,7!}{3!\,10!}$ $\dfrac{4 \times 5 \times 6 \times 7}{8 \times 9 \times 10}$ $\dfrac{7}{6}$

14. $\dfrac{8!}{10!} + \dfrac{4!}{6!}$ $\dfrac{1}{90} + \dfrac{1}{30}$ $\dfrac{4}{90}$ $\dfrac{2}{45}$

4. $\dfrac{20!}{18!\,5!}$ $\dfrac{19 \times 20}{2 \times 3 \times 4 \times 5}$ $\dfrac{19}{6}$

15. $\dfrac{7!}{9!} + \dfrac{2!}{4!}$ $\dfrac{1}{72} + \dfrac{1}{12}$ $\dfrac{7}{72}$

5. $\dfrac{18!}{16!\,3!}$ $\dfrac{17 \times 18}{2 \times 3}$ 17×3   **51**

16. $\dfrac{3!}{5!} - \dfrac{3!}{6!}$ $\dfrac{1}{20} - \dfrac{1}{120}$ $\dfrac{5}{120}$ $\dfrac{1}{24}$

6. $\dfrac{40!}{38!\,3!}$ $\dfrac{39 \times 40}{2 \times 3}$ 13 × 20   **260**

17. $\dfrac{2!}{5!} + \dfrac{7!}{10!}$ $\dfrac{1}{60} + \dfrac{1}{720}$ $\dfrac{13}{720}$

7. $\dfrac{12!}{8!\,6!}$ $\dfrac{9 \times 10 \times 11 \times 12}{2 \times 3 \times 4 \times 5 \times 6}$ $\dfrac{33}{2}$

18. $\dfrac{(30 - 10)!}{17!\,2!\,3!}$ $\dfrac{18 \times 19 \times 20}{2 \times 2 \times 3}$ 3 × 19 × 10   **570**

8. $\dfrac{7!\,8!}{10!}$ $\dfrac{2 \times 3 \times 4 \times 5 \times 6 \times 7}{9 \times 10}$ **56**

19. $\dfrac{28!\,31!}{27!\,32!}$ $\dfrac{28}{32}$ $\dfrac{7}{8}$

9. $\dfrac{15!}{6!\,11!}$ $\dfrac{12 \times 13 \times 14 \times 15}{2 \times 3 \times 4 \times 5 \times 6}$ $\dfrac{91}{2}$

20. $\dfrac{27!\,38!}{25!\,40!}$ $\dfrac{26 \times 27}{39 \times 40}$ $\dfrac{9}{20}$

10. $\dfrac{5!\,6!}{10!}$ $\dfrac{2 \times 3 \times 4 \times 5}{7 \times 8 \times 9 \times 10}$ $\dfrac{1}{42}$

21. $\dfrac{(9 - 2)!\,8!}{3!\,5!\,7!}$ $\dfrac{6 \times 7 \times 8}{2 \times 3}$ **56**

11. $\dfrac{8!}{4!\,5!}$ $\dfrac{6 \times 7 \times 8}{2 \times 3 \times 4}$ 7 × 2   **14**

22. $\dfrac{15!}{7!\,9!}$ $\dfrac{10 \times 11 \times 12 \times 13 \times 14 \times 15}{2 \times 3 \times 4 \times 5 \times 6 \times 7}$ 143 × 5   **715**

Factorials 2

Same as asking for the number of factors of 10. *Find the number of rightmost zeros when written in standard form.*	*Find the number of specified factors of the given factorial.*
1. 13! No need to count 2s-- plenty are available. 5, 10 **2**	12. 10! has how many factors of 6? 3, 6, 9, 9 **4**
2. 20! 5, 10, 15, 20 **4**	13. 10! has how many factors of 12? eight 2s is four 4s four 3s **4**
3. 30! 5, 10, 15, 20, 25, 25, 30 **7**	14. 15! has how many factors of 9? 3 and 6, 9, 12 and 15 **3**
4. 36! 5, 10, 15, 20, 25, 25, 30, 35 **8**	15. 8! has how many factors of 2? 2 x 4 x 6 x 8 1 + 2 + 1 + 3 = 7 **7**
5. 41! 5, 10, 15, 20, 25, 25, 30, 35, 40 **9**	16. 30! has how many factors of 9? 3 and 6, 9, 12 and 15, 18, 21 and 24, 27, 27 and 30 **7**
6. 48! 5, 10, 15, 20, 25, 25, 30, 35, 40, 45 **10**	17. 10! has how many factors of 8? 2 and 4, 8 (6 and 10 are not enough) **2**
7. 63! 5 through 60: 12 25: +1 50: +1 **14**	18. 10! has how many factors of 20? 5, 10 **2**
8. 77! 5 through 75: 15 25: +1 50: +1 75: +1 **18**	19. 14! has how many factors of 14? 7, 14 **2**
9. 150! 5 through 150: 30 25, 50, 75, 100, 125, 150: +1 125: + 1 more **37**	20. 25! has how many factors of 15? 5, 10, 15, 20, 25, 25 3, 6, 9, 9, 12, 15 **6**
10. 250! 5 through 250: 50 25, 50, . . . , 250: 10 125, 250: 2 **62**	21. 18! has how many factors of 6? 3, 6, 9, 9, 12, 15, 18, 18 **8**
11. 500! 5 through 500: 100 25, 50, . . . , 500: 20 125, 250, 375, 500: 4 **124**	22. 100! has how many factors of 2? factors of 2 = 50; factors of 4 = 25 factors of 8 = 12; factors of 16 = 6; factors of 32 = 3; factors of 64 = 1 **97**

135

Factorials 3

Answer as indicated.

---

1. Find the ones digit of
5! + 6! + . . . + 12!.

The factorials all have factors of 2 and 5, so they all have ones digit 0.

0

7. Evaluate $\dfrac{(n+2)!}{(n-2)!}$ for n = 9.

$\dfrac{(11)!}{(7)!}$ = 8 × 9 × 10 × 11
= 72 × 11 × 10
= 792 × 10
= **7920**

---

2. Find the least whole number n such that n! + (n+1)! is a multiple of 10.

n! + (n+1)!
= n! × 1 + n! (n + 1)
= n! (1 + n + 1)
= n! (n + 2)

Need factors of 5 and 2 to be a multiple of 10.
n = **3**

8. How many odd factors does 11! have?

$11! = 2^8 \times 3^4 \times 5^2 \times 7 \times 11$
By multiplication principle, want no 2s (1 option); 0 to 4 3s (5 options); etc.
1 × 5 × 3 × 2 × 2 = **60**

---

3. Find the greatest odd factor of 6!.

3 × 5 × 3 = **45**

Find the greatest odd factor of 7!

3 × 5 × 3 × 7 = **315**

9. Find the greatest power of 6 as a factor of 12!.

excess 2s
3 in 3, 6, 9, 9, 12 6^5
12! can be divided by 6 how many times with no remainder?   **5**

---

4. Find the greatest value of n if 2^n is a factor of 100!.

multiples of 2 = 50
multiples of 4 = 25
multiples of 8 = 12
multiples of 16 = 6

multiples of 32 = 3
multiples of 64 = 1
50 + 25 + 12 + 6 + 3 + 1 = **97**

10. Find the greatest value of n if 3^n is a factor of 90!.

multiples of 3 = 30
multiples of 9 = 10
multiples of 27 = 3
multiples of 81 = 1

30 + 10 + 3 + 1 = **44**

---

5. When 20! is written in base 3, how many rightmost zeros will it have?

Need to count factors of 3. 3 is 10. 9 is 100. Etc.
3, 6, 9, 12, 15, 18
1 + 1 + 2 + 1 + 1 + 2 = **8**

11. When 30! is written in base 5, how many rightmost zeros will it have?

Need to count factors of 5. 5 is 10. 25 is 100. Etc.
5, 10, 15, 20, 25, 30
1 + 1 + 1 + 1 + 2 + 1 = **7**

---

6. a = the least whole number such that a! is a multiple of 10
b = the only whole number such that b! is the product of two primes
c = the only whole number such that c! is prime   a + b + c =
Find a + b + c.   5 + 3 + 2 = **10**

12. a = the least whole number such that a! is a multiple of 8
b = the least whole number such that b! has 4 prime factors
c = the least whole number such that c! is a multiple of 9   abc =
Find abc.   4 × 7 × 6 = **168**

MAVA Math: Enhanced Skills Solutions Copyright © 2015 Marla Weiss

Factorials 4

Write the prime factorization. Do each without referring to another answer.	*Solve for n.*
1. 5! $2 \times 3 \times 2^2 \times 5$ **$2^3 \times 3 \times 5$**	11. $\dfrac{11!}{n!} = 990$ $990 = 9 \times 10 \times 11$ n = **8**
2. 9! $2 \times 3 \times 2^2 \times 5 \times 2 \times 3 \times 7 \times 2^3 \times 3^2$ **$2^7 \times 3^4 \times 5 \times 7$**	12. $\dfrac{10!}{(n-1)!} = 720$ $720 = 8 \times 9 \times 10$ n = **8**
3. 10! $2 \times 3 \times 2^2 \times 5 \times 2 \times 3 \times 7 \times 2^3 \times 3^2 \times$ 2×5 **$2^8 \times 3^4 \times 5^2 \times 7$**	13. $\dfrac{12!}{(n+1)!} = 1320$ $1320 = 10 \times 11 \times 12$ n = **8**
4. 13! $2 \times 3 \times 2^2 \times 5 \times 2 \times 3 \times 7 \times 2^3$ $\times 3^2 \times 2 \times 5 \times 11 \times 2^2 \times 3 \times 13$ **$2^{10} \times 3^5 \times 5^2 \times 7 \times 11 \times 13$**	14. $\dfrac{n!}{4!\,(n-3)!} = \dfrac{n!}{5!\,(n-4)!}$ $(n-3)! = 5 \times (n-4)!$ n = **8**
5. 7! $2 \times 3 \times 2^2 \times 5 \times 2 \times 3 \times 7$ **$2^4 \times 3^2 \times 5 \times 7$**	15. $\dfrac{14!}{7!\,(n-4)!} = \dfrac{14!}{8!\,(n-5)!}$ $(n-4)! = 8 \times (n-5)!$ n = **12**

6. 11!
 	
$2 \times 3 \times 2^2 \times 5 \times 2 \times 3 \times 7 \times 2^3 \times 3^2 \times$
$2 \times 5 \times 11$ **$2^8 \times 3^4 \times 5^2 \times 7 \times 11$**
	16. $\dfrac{8!}{6!+6!} = 7n$ $6! + 6! = 2(6!)$
	                    $7 \times 4 = 7n$
	                    n = **4**

7. 6!
 	
$2 \times 3 \times 2^2 \times 5 \times 2 \times 3$
$2^4 \times 3^2 \times 5$
	17. $\dfrac{10!}{7!+7!} = 20n$ $7! + 7! = 2(7!)$
	                   $4 \times 9 \times 10 = 20n$
	                     n = **18**

8. 8! $2 \times 3 \times 2^2 \times 5 \times 2 \times 3 \times 7 \times 2^3$ **$2^7 \times 3^2 \times 5 \times 7$**	18. $\dfrac{n!}{6!\,(n-3)!} = \dfrac{n!}{4!\,(n-2)!}$ n − 2 = 30 $(n-2)! = 30 \times (n-3)!$ n = **32**
9. 12! $2 \times 3 \times 2^2 \times 5 \times 2 \times 3 \times 7 \times 2^3$ $\times 3^2 \times 2 \times 5 \times 11 \times 2^2 \times 3$ **$2^{10} \times 3^5 \times 5^2 \times 7 \times 11$**	19. $\dfrac{n!}{7!\,(n-5)!} = \dfrac{n!}{9!\,(n-6)!}$ n − 5 = 72 $(n-5)! = 72 \times (n-6)!$ n = **77**

10. 14! $2 \times 3 \times 2^2 \times 5 \times 2 \times 3 \times 7 \times 2^3$
$\times 3^2 \times 2 \times 5 \times 11 \times 2^2 \times 3 \times 13 \times 2 \times 7$
$2^{11} \times 3^5 \times 5^2 \times 7^2 \times 11 \times 13$
	20. $\dfrac{7!}{4!+4!} = 21n$ $4! + 4! = 2(4!)$
	                 $5 \times 3 \times 7 = 21n$
	                  n = **5**

Factors 1

List all factors in ascending order, working in from both sides.

1. 48 1, 2, 3, 4, 6, 8, 12, 16, 24, 48	12. 360 1, 2, 3, 4, 5, 6, 8, 9, 10, 12, 15, 18, 20, 24, 30, 36, 40, 45, 60, 72, 90, 120, 180, 360
2. 50 1, 2, 5, 10, 25, 50	13. 420 1, 2, 3, 4, 5, 6, 7, 10, 12, 14, 15, 20, 21, 28, 30, 35, 42, 60, 70, 84, 105, 140, 210, 420
3. 64 1, 2, 4, 8, 16, 32, 64	14. 550 1, 2, 5, 10, 11, 22, 25, 50, 55, 110, 275, 550
4. 72 1, 2, 3, 4, 6, 8, 9, 12, 18, 24, 36, 72	15. 640 1, 2, 4, 5, 8, 10, 16, 20, 32, 40, 64, 80, 128, 160, 320, 640
5. 80 1, 2, 4, 5, 8, 10, 16, 20, 40, 80	16. 750 1, 2, 3, 5, 6, 10, 15, 25 30, 50, 75, 125, 150, 250, 375, 750
6. 93 1, 3, 31, 93	17. 888 1, 2, 3, 4, 6, 8, 12, 24, 37, 74, 111, 148, 222, 296, 444, 888
7. 100 1, 2, 4, 5, 10, 20, 25, 50, 100	18. 980 1, 2, 4, 5, 7, 10, 14, 20, 28, 35, 49, 70, 98, 140, 196, 245, 490, 980
8. 120 1, 2, 3, 4, 5, 6, 8, 10, 12, 15, 20, 24, 30, 40, 60, 120	19. 1001 1, 7, 11, 13, 77, 91, 143, 1001
9. 150 1, 2, 3, 5, 6, 10, 15, 25, 30, 50, 75, 150	20. 1100 1, 2, 4, 5, 10, 11, 20, 22, 25, 44, 50, 55, 100, 110, 220, 275, 550, 1100
10. 196 1, 2, 4, 7, 14, 28, 49, 98, 196	21. 1250 1, 2, 5, 10, 25 50, 125, 250, 625, 1250
11. 275 1, 5, 11, 25, 55, 275	22. 2000 1, 2, 4, 5, 8, 10, 16, 20, 25, 40, 50, 80 100, 125, 200, 250, 400, 500, 1000, 2000

Factors 2

Find the missing number by examining and matching factors. Do not multiply.

1. $11 \cdot 86 = 22 \cdot ?$

    $11 \cdot 2 \cdot 43 = 2 \cdot 11 \cdot \mathbf{43}$

2. $12 \cdot 95 = 60 \cdot ?$

    $12 \cdot 5 \cdot 19 = 12 \cdot 5 \cdot \mathbf{19}$

3. $45 \cdot 16 = 90 \cdot ?$

    $45 \cdot 2 \cdot 8 = 45 \cdot 2 \cdot \mathbf{8}$

4. $32 \cdot 10 = 20 \cdot ?$

    $16 \cdot 2 \cdot 10 = 2 \cdot 10 \cdot \mathbf{16}$

5. $40 \cdot 25 = 20 \cdot ?$

    $20 \cdot 2 \cdot 25 = 20 \cdot \mathbf{50}$

6. $36 \cdot 22 = 18 \cdot ?$

    $18 \cdot 2 \cdot 22 = 18 \cdot \mathbf{44}$

7. $51 \cdot 62 = 34 \cdot ?$

    $17 \cdot 3 \cdot 2 \cdot 31 = 2 \cdot 17 \cdot \mathbf{93}$

8. $24 \cdot 30 = 80 \cdot ?$

    $8 \cdot 3 \cdot 3 \cdot 10 = 8 \cdot 10 \cdot \mathbf{9}$

9. $54 \cdot 16 = 72 \cdot ?$

    $9 \cdot 6 \cdot 8 \cdot 2 = 72 \cdot \mathbf{12}$

10. $22 \cdot 85 = 11 \cdot ?$

    $11 \cdot 2 \cdot 85 = 11 \cdot \mathbf{170}$

11. $14 \cdot 15 = 10 \cdot ?$

    $7 \cdot 2 \cdot 5 \cdot 3 = 10 \cdot \mathbf{21}$

12. $35 \cdot 44 = 55 \cdot ?$

    $7 \cdot 5 \cdot 11 \cdot 4 = 5 \cdot 11 \cdot \mathbf{28}$

13. $14 \cdot 51 \cdot 48 = 34 \cdot 42 \cdot ?$

    $7 \cdot 2 \cdot 17 \cdot 3 \cdot 8 \cdot 6 = 34 \cdot 42 \cdot \mathbf{24}$

14. $96 \cdot 95 \cdot 42 = 56 \cdot 60 \cdot ?$

    $12 \cdot 8 \cdot 5 \cdot 19 \cdot 7 \cdot 6 = 56 \cdot 60 \cdot \mathbf{114}$

15. $25 \cdot 12 \cdot 26 = 5 \cdot 39 \cdot ?$

    $5 \cdot 5 \cdot 4 \cdot 3 \cdot 13 \cdot 2 = 5 \cdot 39 \cdot \mathbf{40}$

16. $12 \cdot 14 \cdot 85 = 20 \cdot 21 \cdot ?$

    $4 \cdot 3 \cdot 7 \cdot 2 \cdot 17 \cdot 5 = 20 \cdot 21 \cdot \mathbf{34}$

17. $125 \cdot 14 = 50 \cdot ?$

    $25 \cdot 5 \cdot 7 \cdot 2 = 50 \cdot \mathbf{35}$

18. $21 \cdot 18 \cdot 22 = 14 \cdot 99 \cdot ?$

    $7 \cdot 3 \cdot 9 \cdot 2 \cdot 11 \cdot 2 = 14 \cdot 99 \cdot \mathbf{6}$

19. $121 \cdot 27 \cdot 35 = 33 \cdot 45 \cdot ?$

    $11 \cdot 11 \cdot 9 \cdot 3 \cdot 7 \cdot 5 = 33 \cdot 45 \cdot \mathbf{77}$

20. $72 \cdot 45 \cdot 15 = 24 \cdot 25 \cdot ?$

    $9 \cdot 8 \cdot 9 \cdot 5 \cdot 5 \cdot 3 = 24 \cdot 25 \cdot \mathbf{81}$

21. $15 \cdot 49 \cdot 65 = 21 \cdot 25 \cdot ?$

    $5 \cdot 3 \cdot 7 \cdot 7 \cdot 5 \cdot 13 = 21 \cdot 25 \cdot \mathbf{91}$

22. $34 \cdot 10 \cdot 30 = 50 \cdot 51 \cdot ?$

    $17 \cdot 2 \cdot 5 \cdot 2 \cdot 3 \cdot 10 = 50 \cdot 51 \cdot \mathbf{4}$

Factors 3

Write each counting number from 1 to 84 in the column labeled by the counting number's number of factors.

1	2	3	4	5	6	7	8	9	10	11	12
1	2	4	6	16	12	64	24	36	48		60
	3	9	8	81	18		30		80		72
	5	25	10		20		40				84
	7	49	14		28		42				
	11		15		32		54				
	13		21		44		56				
	17		22		45		66				
	19		26		50		70				
	23		27		52		78				
	29		33		63						
	31		34		68						
	37		35		75						
	41		38		76						
	43		39								
	47		46								
	53		51								
	59		55								
	61		57								
	67		58								
	71		62								
	73		65								
	79		69								
	83		74								
			77								
			82								

Factors 4

Find the number of rightmost zeros of each product without multiplying.

1. $2^8 \cdot 5^7$ 2s: 8 5s: 7 **7**	12. $30^4 \cdot 45^2 \cdot 55^2$ 0s: 4 2s: none extra **4**
2. $2^7 \cdot 25^3$ 2s: 7 5s: 6 **6**	13. $16^3 \cdot 125^3$ 2s: 12 5s: 9 **9**
3. 35! 2s: many 5s: 5, 10, 15, 20, 25, 25, 30, 35 **8**	14. 75! 2s: many 5s: 5, 10, 15, 20, 25, 25, 30, 35, 40, 45, 50, 50, 55, 60, 65, 70, 75, 75 **18**
4. $8^3 \cdot 25^4$ 2s: 9 5s: 8 **8**	15. $5! \cdot 5^3 \cdot 500$ 0s: 2 2s: 3 5s: 5 **5**
5. $4 \cdot 5 \cdot 8 \cdot 10 \cdot 14 \cdot 15$ 2s: many 5s: 1 1 1 **3**	16. $7! \cdot 8! \cdot 9! \cdot 10! \cdot 11!$ 2s: many 5s: 1 1 1 2 2 **7**
6. $320 \cdot 1250$ 0s: 2 2s: 5 5s: 3 **5**	17. $96 \cdot 675$ 2s: 5 5s: 2 **2**
7. $420 \cdot 7500$ 0s: 3 2s: 1 5s: 2 **4**	18. $60 \cdot 65 \cdot 70 \cdot 75 \cdot 80$ 0s: 3 2s: 1 3 5s: 1 2 **6**
8. $6 \cdot 12 \cdot 16 \cdot 20 \cdot 75$ 0s: 1 2s: many 5s: 2 **3**	19. $60^4 \cdot 15^3$ 0s: 4 2s: 4 5s: 3 **7**
9. $105 \cdot 625$ 2s: 0 5s: many **0**	20. $8! \cdot 35^3$ 2s: many 5s: 1 3 **4**
10. $25 \cdot 30 \cdot 35 \cdot 40 \cdot 45 \cdot 50$ 0s: 3 2s: 2 5s: 2 1 1 1 **5**	21. $100 \cdot 200 \cdot 300 \cdot 400 \cdot 500$ 0s: 10 2s: 1 2 5s: 1 **11**
11. $225 \cdot 128$ 2s: many 5s: 2 **2**	22. $2! \cdot 3! \cdot 4! \cdot 5! \cdot 6!$ 2s: many 5s: 1 1 **2**

Factors 5

Use multiplication principle to find the number of:	$2^3 \cdot 5^6 \cdot 7^4$	$2^3 \cdot 3^4 \cdot 5^2 \cdot 7^2 \cdot 11$
factors	4 x 7 x 5 = **140**	4 x 5 x 3 x 3 x 2 = **360**
prime factors	2, 5, or 7: **3**	2, 3, 5, 7, or 11: **5**
composite factors	140 − 3 − 1 = **136**	360 − 5 − 1 = **354**
even factors	3 x 7 x 5 = **105**	3 x 5 x 3 x 3 x 2 = **270**
odd factors	1 x 7 x 5 = **35**	1 x 5 x 3 x 3 x 2 = **90**
perfect square factors	2 x 4 x 3 = **24**	2 x 3 x 2 x 2 x 1 = **24**
perfect cube factors	2 x 3 x 2 = **12**	2 x 2 x 1 x 1 x 1 = **4**
factors with ones digit 0	3 x 6 x 5 = **90**	3 x 5 x 2 x 3 x 2 = **180**
factors with ones digit 5	1 x 6 x 5 = **30**	1 x 5 x 2 x 3 x 2 = **60**
multiple of 49 factors	4 x 7 x 3 = **84**	4 x 5 x 3 x 1 x 2 = **120**

	$2^4 \cdot 3^3 \cdot 5 \cdot 13^2$	$2^2 \cdot 3^5 \cdot 5^3 \cdot 7^3 \cdot 17^2$
factors	5 x 4 x 2 x 3 = **120**	3 x 6 x 4 x 4 x 3 = **864**
prime factors	2, 3, 5, or 13: **4**	2, 3, 5, 7, or 17: **5**
composite factors	120 − 4 − 1 = **115**	864 − 5 − 1 = **858**
even factors	4 x 4 x 2 x 3 = **96**	2 x 6 x 4 x 4 x 3 = **576**
odd factors	1 x 4 x 2 x 3 = **24**	864 − 576 = **288**
perfect square factors	3 x 2 x 1 x 2 = **12**	2 x 3 x 2 x 2 x 2 = **48**
perfect cube factors	2 x 2 x 1 x 1 = **4**	1 x 2 x 2 x 2 x 1 = **8**
factors with ones digit 0	4 x 4 x 1 x 3 = **48**	2 x 6 x 3 x 4 x 3 = **432**
factors with ones digit 5	1 x 4 x 1 x 3 = **12**	1 x 6 x 3 x 4 x 3 = **216**
multiple of 27 factors	5 x 1 x 2 x 3 = **30**	3 x 3 x 4 x 4 x 3 = **432**
multiple of 36 factors	3 x 2 x 2 x 3 = **36**	1 x 4 x 4 x 4 x 3 = **192**

Factors 6

Use multiplication principle to find the number of:	$6^3 \cdot 35^2 \cdot 22^4$ $2^7 \cdot 3^3 \cdot 5^2 \cdot 7^2 \cdot 11^4$	$8^2 \cdot 15^3 \cdot 49^2 \cdot 121$ $2^6 \cdot 3^3 \cdot 5^3 \cdot 7^4 \cdot 11^2$
factors	8 x 4 x 3 x 3 x 5 = **1440**	7 x 4 x 4 x 5 x 3 = **1680**
composite factors	1440 − 5 − 1 = **1434**	1680 − 5 − 1 = **1674**
even factors	7 x 4 x 3 x 3 x 5 = **1260**	6 x 4 x 4 x 5 x 3 = **1440**
odd factors	1440 − 1260 = **180**	1680 − 1440 = **240**
perfect square factors	4 x 2 x 2 x 2 x 3 = **96**	4 x 2 x 2 x 3 x 2 = **96**
perfect cube factors	3 x 2 x 1 x 1 x 2 = **12**	3 x 2 x 2 x 2 x 1 = **24**
factors with ones digit 0	7 x 4 x 2 x 3 x 5 = **840**	6 x 4 x 3 x 5 x 3 = **1080**
factors with ones digit 5	1 x 4 x 2 x 3 x 5 = **120**	1 x 4 x 3 x 5 x 3 = **180**
multiple of 33 factors	8 x 3 x 3 x 3 x 4 = **864**	7 x 3 x 4 x 5 x 2 = **840**
	$10^4 \cdot 21^3 \cdot 30^2$ $2^6 \cdot 3^5 \cdot 5^6 \cdot 7^3$	$6^4 \cdot 14 \cdot 35^2 \cdot 64$ $2^{11} \cdot 3^4 \cdot 5^2 \cdot 7^3$
factors	7 x 6 x 7 x 4 = **1176**	12 x 5 x 3 x 4 = **720**
prime factors	2, 3, 5, or 7: **4**	2, 3, 5, or 7: **4**
composite factors	1176 − 4 − 1 = **1171**	720 − 4 − 1 = **715**
even factors	6 x 6 x 7 x 4 = **1008**	11 x 5 x 3 x 4 = **660**
odd factors	1176 − 1008 = **168**	720 − 660 = **60**
perfect square factors	4 x 3 x 4 x 2 = **96**	6 x 3 x 2 x 2 = **72**
perfect cube factors	3 x 2 x 3 x 2 = **36**	4 x 2 x 1 x 2 = **16**
factors with ones digit 0	6 x 6 x 6 x 4 = **864**	11 x 5 x 2 x 4 = **440**
factors with ones digit 5	1 x 6 x 6 x 4 = **144**	1 x 5 x 2 x 4 = **40**
multiple of 21 factors	7 x 5 x 7 x 3 = **735**	12 x 4 x 3 x 3 = **432**

Factors 7

Given factors of a number, find all other factors conclusively determinable.	*Find the number of digits when written in standard form. Do not multiply.*

1. 1, 3, 7, 11 **21, 33, 77, 231**	14. $2^{10} \cdot 25^6$ 10 2s 12 5s 25 with 10 0s **12**
2. 1, 2, 11, 17 **22, 34, 187, 374**	15. $2^3 \cdot 5^6 \cdot 10^2$ 3+2 = 5 2s 6+2 = 8 5s 125 with 5 0s **8**
3. 1, 5, 11, 19 **55, 95, 209, 1045**	16. $2^5 \cdot 5^3 \cdot 20^2$ 5+4 = 9 2s 3+2 = 5 5s 16 with 5 0s **7**
4. 1, 2, 3, 4, 5 The 2 may already **6, 10, 12, 15, 20, 30, 60** be in the 4. Only have 2 distinct 2s.	17. $2^3 \cdot 5^6 \cdot 6^3$ 3+3 = 6 2s 6 5s, 3 3s 27 with 6 0s **8**
5. 1, 3, 4, 5 **2, 6, 10, 12, 15, 20, 30, 60**	18. $4^4 \cdot 5^4 \cdot 10^2$ 8+2 = 10 2s 4+2 = 6 5s 16 with 6 0s **8**
6. 1, 2, 3, 5, 7 **6, 10, 14, 15, 21, 30, 35, 42, 70, 105, 210**	19. $4^3 \cdot 5^7 \cdot 8^2$ 6+6 = 12 2s 7 5s 32 with 7 0s **9**
7. 1, 3, 9, 10 **2, 5, 6, 15, 18, 30, 45, 90**	20. $10^2 \cdot 15^2 \cdot 20^2$ 2+4 = 6 2s, 2 3s 2+2+2 = 6 5s 9 with 6 0s **7**
8. 1, 2, 4, 9, 11 **3, 6, 12, 18, 22, 33, 36, 44, 66, 99, 132, 198, 396**	21. $2^6 \cdot 5^3 \cdot 30^2$ 6+2 = 8 2s, 2 3s 3+2 = 5 5s 72 with 5 0s **7**
9. 1, 3, 5, 9 **15, 45**	22. $4^3 \cdot 8^2 \cdot 25^4$ 6+6 = 12 2s 8 5s 16 with 8 0s **10**
10. 1, 7, 9, 11 **3, 21, 33, 63, 77, 99, 231, 693**	23. $20^2 \cdot 25^3 \cdot 30^2$ 4+2 = 6 2s, 2 3s 2+6+2 = 10 5s 9x625 with 8 0s **12**
11. 1, 7, 8, 11 **2, 4, 14, 22, 28, 44, 56, 77, 88, 154, 308, 616**	24. $15^2 \cdot 20^3 \cdot 35^2$ 6 2s, 2 3s, 2 7s 2+3+2 = 7 5s 9x49x5 with 6 0s **10**
12. 1, 5, 6, 23 **2, 3, 10, 15, 30, 46, 69, 115, 138, 230, 345, 690**	25. $6^3 \cdot 8^3 \cdot 25^5$ 3+9 = 12 2s, 3 3s 10 5s 27x4 with 10 0s **13**
13. 1, 6, 10, 12 **2, 3, 4, 5, 15, 20, 30, 60**	26. $10^4 \cdot 25^3 \cdot 40^3$ 4+9 = 13 2s 4+6+3 = 13 5s 1 with 13 0s **14**

Factors 8

Answer as indicated.

1. Find the sum of all two-digit numbers with three factors. 25 + 49 = **74**	9. Find the greatest 5-digit even number such that each pair of adjacent digits is relatively prime (GCF = 1). **98978**
2. Find the sum of the greatest 2-digit number and the least 3-digit number, each with an odd number of factors. $10^2 = 100$ $9^2 = 81$ **181**	10. Find the sum of the greatest 2-digit number and the least 3-digit number, each with exactly three factors. 49 + 121 = **170**
3. Find the sum of the greatest 3-digit number and the least 4-digit number, each with exactly three factors. $37^2 = 1369$ $31^2 = 961$ **2330**	11. Find the sum of the greatest 3-digit number and the least 4-digit number, each with an odd number of factors. $32^2 = 1024$ $31^2 = 961$ **1985**
4. If ABCABC is a 6-digit number, find the greatest prime factor that ABCABC must have. (A+C+B) − (B+A+C) = 0 **11**	12. Find the sum of all the two-digit factors of 140. 70 + 35 + 28 + 20 + 14 + 10 = **177**
5. Find the sum of the two-digit numbers with 9 or 10 factors. p^8 and p^9 not 2–digit p^4q: 16x3 = 48; 16x5 = 80 p^2q^2: 4x9 = 36 **164**	13. Find the sum of all numbers less than 1000 with exactly 5 factors. p^4: 16 + 81 + 625 = **722**
6. Find the least natural number with exactly 5 factors. 2^4 = **16**	14. Find the least natural number with exactly 10 factors. p^4q: 16 × 3 = **48**
7. How many two-digit numbers do not have a factor of 17? 17x1 thru 17x5 90 − 5 = **85**	15. Find the sum of all the two-digit factors of 200. 50 + 40 + 25 + 20 + 10 = **145**
8. Find the greatest 6-digit number divisible by 4 such that each pair of adjacent digits has a GCF of 2. **868628**	16. Find the least 6-digit number divisible by 22 such that each pair of adjacent digits is relatively prime. div by 2 & 11 **111232**

Fictitious Operations 1

Evaluate the unary operations, following the definitions.

Define operation @ by @x = x² + 3.
For example, @4 = 19.

1. @7   49 + 3   **52**

2. @(–9)   81 + 3   **84**

3. @(.5)   .25 + 3   **3.25**

Define operation [ ] by [x] = x² – 1.
For example, [3] = 8.

16. [4]   16 – 1   **15**

17. [–6]   36 – 1   **35**

18. [2] + [5]   4 – 1 + 25 – 1   **27**

Define operation ∂ by ∂x = x³ – 5.
For example, ∂5 = 120.

4. ∂(2)   8 – 5   **3**

5. ∂(–4)   –64 – 5   **–69**

6. ∂(.1)   .001 – 5   **–4.999**

7. ∂(11)   1331 – 5   **1326**

Define operation ≈ ≈ by ≈x≈ = 2x + 11.
For example, ≈4≈ = 19.

19. ≈2≈   4 + 11   **15**

20. ≈(–10)≈   –20 + 11   **–9**

21. ≈(.5)≈   1 + 11   **12**

22. ≈(17.5)≈   35 + 11   **46**

Define operation ~ ~ by ~x~ = 5x + 2.
For example, ~4~ = 22.

8. ~2~   10 + 2   **12**

9. ~(–5)~   –25 + 2   **–23**

10. ~1.2~   6 + 2   **8**

11. ~(–3.1)~   –15.5 + 2   **–13.5**

Define operation ‡ ‡ by ‡x‡ = 2x² – 2.
For example, ‡3‡ = 16.

23. ‡(.5)‡   .5 – 2   **–1.5**

24. ‡(–5)‡   50 – 2   **48**

25. ‡4‡   32 – 2   **30**

26. ‡11‡   242 – 2   **240**

Define operation Ω by Ωx = (.5)x if x is
even and Ωx = 2x if x is odd.
For example, Ω4 = 2 and Ω9 = 18.

12. (Ω5)(Ω10)   10 × 5   **50**

13. Ω3 + Ω5   6 + 10   **16**

14. Ω20 ÷ Ω8   10 ÷ 4   **2.5**

15. Ω14 – Ω7   7 – 14   **–7**

Define operation ¥ by ¥x = 2x if x is odd
and ¥x = 3x if x is even. For example,
¥4 = 12 and ¥3 = 6.

27. ¥7 – ¥10   14 – 30   **–16**

28. ¥8 ÷ ¥5   24 ÷ 10   **2.4**

29. (¥11)(¥2)   22 × 6   **132**

30. ¥3 + ¥12   6 + 36   **42**

Fictitious Operations 2

Evaluate the binary operations, following the definitions.

Define operation @ by a@b = a² + 3b. For example, 3@4 = 9 + 12 = 21.			
1. 5@7	25 + 21	**46**	
2. 4@(–9)	16 – 27	**–11**	
3. 6@(.5)	36 + 1.5	**37.5**	

Define operation @ by $a@b = a^2 + 3b$. For example, $3@4 = 9 + 12 = 21$.

1. $5@7$ — $25 + 21$ — **46**
2. $4@(-9)$ — $16 - 27$ — **–11**
3. $6@(.5)$ — $36 + 1.5$ — **37.5**

Define operation [ ] by $[x, y] = |y| - |y - x|$. For example, $[3, -4] = 4 - 7 = -3$.

17. $[-2, -5]$ — $5 - 3$ — **2**
18. $[6, -3]$ — $3 - 9$ — **–6**
19. $[-7, 0]$ — $0 - 7$ — **–7**

Define operation ß by $x ß y = 2xy + x$. For example, $2 ß 3 = 12 + 2 = 14$.

4. $3 ß 8$ — $48 + 3$ — **51**
5. $-5 ß 4$ — $-40 - 5$ — **–45**
6. $6 ß (-5)$ — $-60 + 6$ — **–54**
7. $-8 ß 2$ — $-32 - 8$ — **–40**

Define operation Δ by $a Δ b = 8b ÷ a$. For example, $4 Δ 2 = 4$.

20. $4 Δ 5$ — $40 ÷ 4$ — **10**
21. $3 Δ 6$ — $48 ÷ 3$ — **16**
22. $6 Δ 3$ — $24 ÷ 6$ — **4**
23. $5 Δ 4$ — $32 ÷ 5$ — **6.4**

Define operation § by $x § y = x - y + 2$. For example, $7 § 5 = 2 + 2 = 4$.

8. $5 § 8$ — $-3 + 2$ — **–1**
9. $6 § 6$ — $0 + 2$ — **2**
10. $4 § 11$ — $-7 + 2$ — **–5**
11. $-3 § (-9)$ — $6 + 2$ — **8**

Define operation ¶ by $x ¶ y = y - x + 3$. For example, $9 ¶ 2 = -7 + 3 = -4$.

24. $8 ¶ 5$ — $-3 + 3$ — **0**
25. $12 ¶ 12$ — $0 + 3$ — **3**
26. $4 ¶ 10.5$ — $6.5 + 3$ — **9.5**
27. $12 ¶ 6$ — $-6 + 3$ — **–3**

Define operation ∞ by $x ∞ y = -xy - 5$. For example, $2 ∞ 3 = -6 - 5 = -11$.

12. $8 ∞ -7$ — $56 - 5$ — **51**
13. $-4 ∞ -9$ — $-36 - 5$ — **–41**
14. $-3 ∞ 6$ — $18 - 5$ — **13**
15. $\frac{3}{4} ∞ \frac{4}{3}$ — $-1 - 5$ — **–6**
16. $\frac{12}{5} ∞ \frac{5}{6}$ — $-2 - 5$ — **–7**

Define operation • by $a • b = ab + a$. For example, $4 • 2 = 8 + 4 = 12$.

28. $3 • 4$ — $12 + 3$ — **15**
29. $\frac{7}{5} • \frac{5}{7}$ — $1 + \frac{7}{5}$ — $\frac{12}{5}$
30. $\frac{9}{8} • 16$ — $18 + \frac{9}{8}$ — $19\frac{1}{8}$
31. $-5 • 7$ — $-35 - 5$ — **–40**
32. $8 • 9$ — $72 + 8$ — **80**

Fictitious Operations 3

Evaluate the nested or grouped operations by following the definitions.

Define operation @ by @x = x² – 5.
For example, @5 = 20.

1. @(@1)   @(–4)   **11**

2. @(@2)   @(–1)   **–4**

3. @(@4)   @(11)   **116**

Define operation [ ] by [x] = 2x² + 1.
For example, [5] = 50 + 1 = 51.

18. [[2]]   [9]   **163**

19. [[–1]]   [3]   **19**

20. [[4]]   [33]   **2179**

Define operation ß by x ß y = 3xy – y.
For example, 5 ß 6 = 84.

4. (2 ß 3) ß 4   15 ß 4   **176**

5. 2 ß (3 ß 4)   2 ß 32   **160**

6. (.5 ß 6) ß 5   3 ß 5   **40**

7. –1 ß (7 ß 3)   –1 ß 60   **–240**

Define operation △ by a △ b = 10b ÷ a.
For example, 4 △ 2 = 5.

21. (2 △ 5) △ 5   25 △ 5   **2**

22. 2 △ (3 △ 6)   2 △ 20   **100**

23. (6 △ 3) △ 10   5 △ 10   **20**

24. (4 △ 8) △ 5   20 △ 5   **2.5**

Define operation ~ ~ by ~x~ = 4x + 3.
For example, ~4~ = 16 + 3 = 19.

8. ~(~5~)~   ~23~   **95**

9. ~(~.5~)~   ~5~   **23**

10. ~(~3~)~   ~15~   **63**

11. ~(~1.5~)~   ~9~   **39**

Define operation ‡ ‡ by ‡x‡ = –x² + 4.
For example, ‡7‡ = –45.

25. ‡(‡2‡)‡   ‡0‡   **4**

26. ‡(‡(–1)‡)‡   ‡3‡   **–5**

27. ‡(‡–3‡)‡   ‡(–5)‡   **–21**

28. ‡(‡4‡)‡   ‡(–12)‡   **–140**

Define operation ∞ by x ∞ y = –2xy + y.
For example, 2 ∞ 3 = –12 + 3 = –9.

12. (3 ∞ –4) ∞ 5   20 ∞ 5   **–195**

13. (–5 ∞ 2) ∞ 3   22 ∞ 3   **–129**

14. 6 ∞ (–2 ∞ 4)   6 ∞ 20   **–220**

15. –2 ∞ (4 ∞ 5)   –2 ∞ –35   **–175**

16. (–1 ∞ 9) ∞ 0   27 ∞ 0   **0**

17. –3 ∞ (–2 ∞ 7)   –3 ∞ 35   **245**

Define operation • by a • b = ab – |2a|.
For example, 7 • 8 = 56 – 14 = 42.

29. (–2 • 5) • 3   –14 • 3   **–70**

30. (8 • 2) • 4   0 • 4   **0**

31. –3 • (–2 • 3)   –3 • –10   **24**

32. –6 • (4 • 5)   –6 • 12   **–84**

33. (–1 • 6) • 5   –8 • 5   **–56**

34. –4 • (–3 • 5)   –4 • –21   **76**

Fictitious Operations 4

Evaluate the backwards operations by solving the equations.

Define operation @ by x@y = x³ − 3 + 2y.
For example, 2@12 = 8 − 3 + 24 = 29.

1. 3@a = 50   24 + 2a = 50   **13**

2. (−1)@z = 32   −4 + 2z = 32   **18**

3. w@5 = 71   w³ + 7 = 71   **4**

Define operation [ ] by [x, y] = y² − 3x + 4.
For example, [3, 2] = 4 − 9 + 4 = −1.

17. [w, 5] = 2   25 − 3w + 4 = 2   **9**

18. [a, −3] = 9   9 − 3a + 4 = 9   **4/3**

19. [2, c] = 34   c² − 2 = 34   **±6**

Define operation ß by x ß y = 3y + 4x − 2.
For example, 3 ß 2 = 6 + 12 − 2 = 16.

4. 4 ß w = 56   3w + 14 = 56   **14**

5. c ß 3 = 47   9 + 4c − 2 = 47   **10**

6. 5 ß e = 72   3e + 18 = 72   **18**

7. c ß 5 = 69   13 + 4c = 69   **14**

Define operation Δ by a Δ b = 30b ÷ a + 1.
For example, 10 Δ 2 = 6 + 1 = 7.

20. y Δ 4 = 13   120 ÷ y + 1 = 13   **10**

21. 5 Δ w = 9   30w ÷ 5 + 1 = 9   **4/3**

22. 6 Δ x = 16   30x ÷ 6 + 1 = 16   **3**

23. x Δ 8 = 49   240 ÷ x + 1 = 49   **5**

Define operation ~~ by x~~y = 5y − 7x.
For example, 4~~3 = 15 − 28 = −13.

8. c~~8 = 33   40 − 7c = 33   **1**

9. −3~~w = 46   5w + 21 = 46   **5**

10. b~~6.2 = 10   31 − 7b = 10   **3**

11. −9~~z = 98   5z + 63 = 98   **7**

Define operation ‡ by a‡b = 2a² − 1 + b.
For example, 3‡5 = 18 − 1 + 5 = 22.

24. 6‡z = 90   72 − 1 + z = 90   **19**

25. w‡(−6) = 43   2w² − 7 = 43   **±5**

26. −4‡x = 33   31 + x = 33   **2**

27. c‡11 = 42   2c² + 10 = 42   **±4**

Define operation μ by μx = 3x if x is even
and μx = 5x if x is odd. For example,
μ4 = 12 and μ9 = 45.

12. (μw)(μ10) = 450   (μw)(30) = 450   **3**

13. μ5 + μy = 43   25 + μy = 43   **6**

14. μ(μx) = 18   μx = 6   **2**

15. μc + μ8 = 60   μc + 24 = 60   **12**

16. (μ5)(μw) = 125   (25)(μw) = 125   **1**

Define operation £ by £x = 2x if x is prime
and £x = 4x if x is composite. For example,
£7 = 14 and £6 = 24. £1 is undefined.

28. £13 − £y = −14   26 − £y = −14   **10**

29. £w ÷ £5 = 8.8   £w ÷ 10 = 8.8   **22**

30. £(£x) = 16   £(x) = 4   **2**

31. (£3)(£w) = 96   6(£w) = 96   **4**

32. £6 + £y = 30   24 + £y = 30   **3**

Fractional Parts 1

Find the fractional part by forming the "is/of" fraction and simplying.

1.	15 seconds is what fractional part of 15 hours? $$\frac{15}{15 \times 60 \times 60} \quad \frac{1}{\textbf{3600}}$$	9.	5 pints is what fractional part of 10 gallons? $$\frac{5}{10 \times 8} \quad \frac{1}{\textbf{16}}$$
2.	2 cups is what fractional part of 3 gallons? $$\frac{2}{3 \times 16} \quad \frac{1}{\textbf{24}}$$	10.	What fraction of 20 hours is 10 seconds? $$\frac{10}{20 \times 60 \times 60} \quad \frac{1}{\textbf{7200}}$$
3.	440 is what fractional part of 1870? $$\frac{440}{1870} \quad \frac{44}{187} \quad \frac{4}{\textbf{17}}$$	11.	1020 is what fractional part of 8500? $$\frac{1020}{8500} \quad \frac{102}{850} \quad \frac{51}{425} \quad \frac{3}{\textbf{25}}$$
4.	What fractional part of one and one-third hours is two minutes? $$\frac{2}{80} \quad \frac{1}{\textbf{40}}$$	12.	What fraction of one and one-quarter hours is half a minute? $$\frac{.5}{75} \quad \frac{1}{\textbf{150}}$$
5.	Commercials aired for 16 minutes on an hour TV show. What fractional part of the entire show were the commercials? $$\frac{16}{60} \quad \frac{4}{\textbf{15}}$$	13.	Commercials played for 6 minutes on a half hour radio show. For what fractional part of the show were commercials not shown? $$\frac{24}{30} \quad \frac{4}{\textbf{5}}$$
6.	If 12 scripts were read and 18 were unread, what fraction of the scripts were unread? $$\frac{18}{30} \quad \frac{3}{\textbf{5}}$$	14.	A basketball team won 6 games and lost 30. What fraction of its games did the team win? $$\frac{6}{36} \quad \frac{1}{\textbf{6}}$$
7.	6 inches is what fractional part of 2 yards? $$\frac{6}{2 \times 36} \quad \frac{1}{\textbf{12}}$$	15.	In a class of 40 students, 24 are girls. What fraction of the students are boys? $$\frac{16}{40} \quad \frac{2}{\textbf{5}}$$
8.	The days of January, February, and March are what fractional part of a non-leap year? $$\frac{31 + 28 + 31}{365} \quad \frac{90}{365} \quad \frac{18}{\textbf{73}}$$	16.	In a jar of 12 red, 18 blue, 25 green and 50 purple marbles, what fraction of the marbles are not red? $$\frac{93}{105} \quad \frac{31}{\textbf{35}}$$

Fractional Parts 2

Answer as indicated, drawing vertical trees with discards to the left.

1. Twenty-four apples were on a tree. One third fell off and 1/4 of the remainder were picked. What fraction of the original apples were left?

$$\frac{12}{24} \quad \mathbf{\frac{1}{2}}$$

5. Eighteen apples are on a tree. One third fall off. Of those left, someone picks 1/3. What fraction of the original apples are still on the tree?

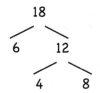

$$\frac{8}{18} \quad \mathbf{\frac{4}{9}}$$

2. A carpenter used 1/3 of his lumber for one project and 3/5 of what was left for another project. If he started with 30 units of lumber, what fraction of the original lumber remained?

$$\frac{8}{30} \quad \mathbf{\frac{4}{15}}$$

6. Abby baked 80 cookies and threw out 1/4 that crumbled. Of those whole, she gave 1/6 to her sisters. Of those then left, she wrapped and froze 3/5. What fraction of the original cookies remained?

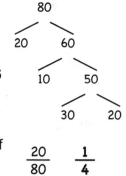

$$\frac{20}{80} \quad \mathbf{\frac{1}{4}}$$

3. Dan had 96 marbles. He lost 1/4. He gave 1/6 of those left to a friend and 2/5 of those then remaining to another friend. What fraction of the original marbles did Dan then have?

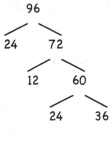

$$\frac{36}{96} \quad \mathbf{\frac{3}{8}}$$

7. Jon had 100 marbles. He lost 1/5. Of those left, he gave 1/5 to a friend, 1/4 to a brother, and kept the rest. What fraction of the marbles did Jon keep?

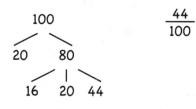

$$\frac{44}{100} \quad \mathbf{\frac{11}{25}}$$

4. Forty-eight lemons were on a tree. One third fell off and 1/8 of the remainder were picked. What fraction of the original apples were left?

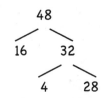

$$\frac{28}{48} \quad \mathbf{\frac{7}{12}}$$

8. If a tree has 36 bananas, 1/3 are picked, and then 1/8 of those left are picked, what fraction of the bananas were picked?

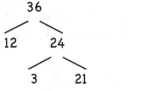

$$\frac{15}{36} \quad \mathbf{\frac{5}{12}}$$

Fractional Parts 3

Answer as indicated, drawing vertical trees with discards to the left.

1.

A baker gave 1/2 of his cookies to a charity. After he boxed 1/5 of those left, 40 cookies remained. Find the number of cookies at the start.

$$\dfrac{C}{5 \times 2} \quad \dfrac{4C}{5 \times 2} = 40$$

$$C = \textbf{100}$$

2. Jon gave 30% of his cards to his brother, 10% to his sister, and 20% of those left to a friend. If 24 cards remained, how many did Jon have to start?

$$\dfrac{C}{25} = 2$$
$$C = \textbf{50}$$

$$\dfrac{3C}{5 \times 5} \quad \dfrac{4 \times 3C}{5 \times 5} = 24$$

3.

One third of the students in a class went to the movies, 1/2 of those left went to the game, and the remaining 10 stayed home. How many students were in the class?

$$\dfrac{2S}{2 \times 3} \quad \dfrac{2S}{2 \times 3} = 10$$

$$S = \textbf{30}$$

4.

Hal boxed 1/6 of his books and gave away 1/4. Of those left, 1/7 went to charity. Of those left, he sold 1/2, leaving 30. What was his original number of books?

$$\dfrac{B}{4} = 30$$

$$B = \textbf{120}$$

5.

Ted saved 12% of his marbles and gave 24% to a friend. Of those left, he lost 25%. If 60 marbles remained, how many did Ted have originally?

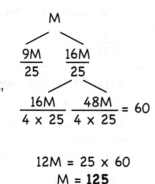

$$\dfrac{16M}{4 \times 25} \quad \dfrac{48M}{4 \times 25} = 60$$

$$12M = 25 \times 60$$
$$M = \textbf{125}$$

6.

Dana ate 1/4 of her candies and froze 1/3 of those left. She then packed 1/3 of those remaining, leaving 20. How many candies did Dana have originally?

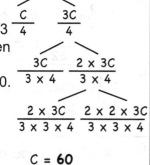

$$\dfrac{C}{3} = 20$$
$$C = \textbf{60}$$

7.

Of the apples on a tree, 20% fell off. Of those left, 15% were boxed. Of those left, 25% were discarded. How many were on the tree to start if 102 apples remained?

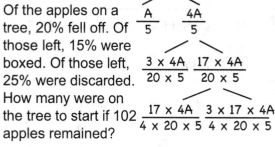

$$51A/100 = 102 \qquad A = \textbf{200}$$

8. One-sixth of a group ate pizza only, 1/4 ate sandwiches only, and 1/3 of the others ate soup only. If 24 people had no food, how many people were in the group?

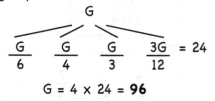

$$\dfrac{G}{6} \quad \dfrac{G}{4} \quad \dfrac{G}{3} \quad \dfrac{3G}{12} = 24$$

$$G = 4 \times 24 = \textbf{96}$$

MAVA Math: Enhanced Skills Solutions Copyright © 2015 Marla Weiss

Fractional Parts 4

Find the fractional part.

1. 1.04 is what fractional part of 4.16?

$$\frac{1.04}{4.16} \qquad \frac{1}{4}$$

2. 4.9 is what fractional part of 12.25?

$$\frac{4.9}{12.25} \qquad \frac{0.7}{1.75} \qquad \frac{70}{175} \qquad \frac{2}{5}$$

3. 5.2 is what fractional part of 21.6?

$$\frac{5.2}{21.6} \qquad \frac{52}{216} \qquad \frac{13}{54}$$

4. $10\frac{2}{5}$ is what fractional part of $12\frac{2}{5}$?

$$\frac{52}{62} \qquad \frac{26}{31}$$

5. $7\frac{1}{9}$ is what fractional part of $14\frac{2}{3}$?

$$\frac{64}{9} \times \frac{3}{44} \qquad \frac{16}{3} \times \frac{1}{11} \qquad \frac{16}{33}$$

6. 30% is what fractional part of 75%?

$$\frac{30}{75} \qquad \frac{2}{5}$$

7. 20% is what fractional part of 85%?

$$\frac{20}{85} \qquad \frac{4}{17}$$

8. $6\frac{1}{4}$ is what fractional part of $16\frac{2}{3}$?

$$\frac{25}{4} \times \frac{3}{50} \qquad \frac{1}{4} \times \frac{3}{2} \qquad \frac{3}{8}$$

9. $6\frac{5}{6}$ is what fractional part of $13\frac{2}{3}$?

$$\frac{41}{6} \times \frac{3}{41} \qquad \frac{1}{2}$$

10. 3.25 is what fractional part of 15.25?

$$\frac{3.25}{15.25} \qquad \frac{325}{1525} \qquad \frac{13}{61}$$

11. $1.\overline{5}$ is what fractional part of $1.\overline{7}$?

$$\frac{14}{9} \times \frac{9}{16} \qquad \frac{7}{8}$$

12. 15% is what fractional part of 35%?

$$\frac{15}{35} \qquad \frac{3}{7}$$

13. $11\frac{1}{9}$ is what fractional part of $15\frac{5}{6}$?

$$\frac{100}{9} \times \frac{6}{95} \qquad \frac{20}{3} \times \frac{2}{19} \qquad \frac{40}{57}$$

14. 10.5 is what fractional part of 9.1?

$$\frac{10.5}{9.1} \qquad \frac{105}{91} \qquad \frac{15}{13}$$

15. 8.6 is what fractional part of 60.2?

$$\frac{8.6}{60.2} \qquad \frac{43}{301} \qquad \frac{1}{7}$$

16. $8\frac{5}{9}$ is what fractional part of $17\frac{3}{5}$?

$$\frac{77}{9} \times \frac{5}{88} \qquad \frac{7}{9} \times \frac{5}{8} \qquad \frac{35}{72}$$

17. 12% is what fractional part of 66%?

$$\frac{12}{66} \qquad \frac{2}{11}$$

18. 20.1 is what fractional part of 21.3?

$$\frac{20.1}{21.3} \qquad \frac{201}{213} \qquad \frac{67}{71}$$

19. $6\frac{3}{4}$ is what fractional part of $12\frac{3}{8}$?

$$\frac{27}{4} \times \frac{8}{99} \qquad \frac{3}{1} \times \frac{2}{11} \qquad \frac{6}{11}$$

20. 3.12 is what fractional part of 5.16?

$$\frac{3.12}{5.16} \qquad \frac{312}{516} \qquad \frac{78}{129} \qquad \frac{26}{43}$$

21. $6\frac{3}{7}$ is what fractional part of $7\frac{13}{14}$?

$$\frac{45}{7} \times \frac{14}{111} \qquad \frac{15}{1} \times \frac{2}{37} \qquad \frac{30}{37}$$

22. $3.\overline{1}$ is what fractional part of $4.\overline{4}$?

$$\frac{28}{9} \times \frac{9}{40} \qquad \frac{7}{10}$$

Fractions 1

Simplify.

1. $\dfrac{150}{360}$ $\dfrac{15}{36}$ $\dfrac{5}{12}$	12. $\dfrac{112}{224}$ $\dfrac{1}{2}$	23. $\dfrac{147}{196}$ $\dfrac{21}{28}$ $\dfrac{3}{4}$
2. $\dfrac{630}{840}$ $\dfrac{63}{84}$ $\dfrac{3}{4}$	13. $\dfrac{135}{495}$ $\dfrac{15}{55}$ $\dfrac{3}{11}$	24. $\dfrac{510}{570}$ $\dfrac{51}{57}$ $\dfrac{17}{19}$
3. $\dfrac{105}{420}$ $\dfrac{21}{84}$ $\dfrac{1}{4}$	14. $\dfrac{315}{385}$ $\dfrac{63}{77}$ $\dfrac{9}{11}$	25. $\dfrac{280}{175}$ $\dfrac{56}{35}$ $\dfrac{8}{5}$
4. $\dfrac{280}{180}$ $\dfrac{28}{18}$ $\dfrac{14}{9}$	15. $\dfrac{168}{252}$ $\dfrac{42}{63}$ $\dfrac{2}{3}$	26. $\dfrac{352}{374}$ $\dfrac{32}{34}$ $\dfrac{16}{17}$
5. $\dfrac{130}{650}$ $\dfrac{13}{65}$ $\dfrac{1}{5}$	16. $\dfrac{720}{990}$ $\dfrac{72}{99}$ $\dfrac{8}{11}$	27. $\dfrac{126}{423}$ $\dfrac{14}{47}$
6. $\dfrac{112}{168}$ $\dfrac{28}{42}$ $\dfrac{2}{3}$	17. $\dfrac{105}{231}$ $\dfrac{35}{77}$ $\dfrac{5}{11}$	28. $\dfrac{504}{630}$ $\dfrac{56}{70}$ $\dfrac{8}{10}$ $\dfrac{4}{5}$
7. $\dfrac{770}{630}$ $\dfrac{77}{63}$ $\dfrac{11}{9}$	18. $\dfrac{462}{168}$ $\dfrac{231}{84}$ $\dfrac{33}{12}$ $\dfrac{11}{4}$	29. $\dfrac{128}{512}$ $\dfrac{1}{4}$
8. $\dfrac{225}{900}$ $\dfrac{25}{100}$ $\dfrac{1}{4}$	19. $\dfrac{560}{640}$ $\dfrac{56}{64}$ $\dfrac{7}{8}$	30. $\dfrac{325}{875}$ $\dfrac{13}{35}$
9. $\dfrac{810}{720}$ $\dfrac{81}{72}$ $\dfrac{9}{8}$	20. $\dfrac{126}{612}$ $\dfrac{14}{68}$ $\dfrac{7}{34}$	31. $\dfrac{594}{154}$ $\dfrac{54}{14}$ $\dfrac{27}{7}$
10. $\dfrac{225}{275}$ $\dfrac{9}{11}$	21. $\dfrac{484}{363}$ $\dfrac{44}{33}$ $\dfrac{4}{3}$	32. $\dfrac{300}{375}$ $\dfrac{12}{15}$ $\dfrac{4}{5}$
11. $\dfrac{385}{231}$ $\dfrac{35}{21}$ $\dfrac{5}{3}$	22. $\dfrac{147}{210}$ $\dfrac{21}{30}$ $\dfrac{7}{10}$	33. $\dfrac{198}{594}$ $\dfrac{18}{54}$ $\dfrac{1}{3}$

Fractions 2

Multiply, simplifying first.

1.

$$\frac{\overset{1}{\underset{2}{13}}}{14} \times \frac{\overset{3}{\underset{5}{21}}}{25} \times \frac{\overset{3}{\underset{\underset{1}{3}}{15}}}{39} \times \frac{\overset{1}{\underset{2}{5}}}{6} \qquad \frac{1}{4}$$

8.

$$\frac{\overset{1}{\underset{\underset{1}{6}}{11}}}{36} \times \frac{\overset{6}{\underset{2}{42}}}{22} \times \frac{\overset{1}{\underset{3}{13}}}{39} \times \frac{\overset{1}{\underset{1}{6}}}{7} \qquad \frac{1}{6}$$

2.

$$\frac{\overset{1}{\underset{14}{3}{45}}}{42} \times \frac{\overset{1}{\underset{1}{2}{10}}}{15} \times \frac{\overset{1}{\underset{\underset{1}{5}}{2}{20}}}{50} \times \frac{1}{4} \qquad \frac{1}{14}$$

9.

$$\frac{\overset{1}{\underset{\underset{1}{2}}{29}}}{78} \times \frac{\overset{2}{\underset{1}{34}}}{15} \times \frac{\overset{1}{39}}{17} \times \frac{\overset{8}{\underset{\underset{1}{2}}{16}}}{58} \qquad \frac{8}{15}$$

3.

$$\frac{\overset{1}{\underset{\underset{1}{4}}{7}{21}}}{12} \times \frac{\overset{1}{\underset{\underset{1}{7}}{4}{36}}}{35} \times \frac{3}{28}{}^{7} \times \frac{2}{9}{}_{1}^{8} \qquad \frac{6}{7}$$

4.

$$\frac{5}{14}{}^{1}_{1}{15} \times \frac{1}{20}{}^{11} \times \frac{1}{55}{}^{16}_{5} \times \frac{5}{80}{}^{70}_{5} \times \frac{1}{3}_{1} \qquad \frac{1}{20}$$

10.

$$\frac{1}{9}{}^{5} \times \frac{3}{26}{}^{27}_{1} \times \frac{3}{84}{}^{63}_{4} \times \frac{1}{15}{}^{26}_{3} \times \frac{4}{225}{}^{900}_{1} \qquad 3$$

11.

$$\frac{1}{5}{}^{2} \times \frac{5}{14}{}^{70}_{1} \times \frac{5}{21}{}^{10}_{3} \times \frac{5}{35}{}^{15}_{5} \times \frac{7}{24}{}^{49}_{12}{}_{6} \qquad \frac{5}{6}$$

5.

$$\frac{1}{15}{}^{16}_{3}{}_{1} \times \frac{5}{48}{}^{55}_{3}{}_{1} \times \frac{6}{25}{}^{30}_{5}{}_{1} \times \frac{6}{11}{}^{30}_{1} \times \frac{6}{5} \qquad \frac{24}{5}$$

12.

$$\frac{4}{55}{}^{16}_{5}{}_{1} \times \frac{3}{14}{}^{15}_{1} \times \frac{1}{20}{}^{11}_{4} \times \frac{1}{24}{}^{14}_{2}{}_{1} \times \frac{1}{2}{}^{5} \qquad \frac{1}{4}$$

6.

$$\frac{2}{12}{}^{22}_{1}{}_{1} \times \frac{1}{50}{}^{16}_{2}{}_{1} \times \frac{1}{48}{}^{25}_{3}{}_{1} \times \frac{3}{10}{}^{33}_{1} \times \frac{10}{121}{}^{120}_{11}{}_{1} \qquad 1$$

13.

$$\frac{3}{20}{}^{21}_{5}{}_{1} \times \frac{3}{35}{}^{15}_{7}{}_{1} \times \frac{4}{27}{}^{12}_{3}{}_{1} \times \frac{4}{10}{}^{16}_{2}{}_{1} \times \frac{1}{4}{}^{5}_{1} \qquad \frac{2}{5}$$

7.

$$\frac{5}{32}{}^{45}_{2}{}_{1} \times \frac{2}{14}{}^{56}_{1} \times \frac{3}{63}{}^{48}_{7}{}_{1} \times \frac{7}{50}{}^{70}_{5} \times \frac{2}{5}{}^{1} \qquad \frac{12}{5}$$

14.

$$\frac{2}{15}{}^{98}_{1}{}_{1} \times \frac{3}{56}{}^{45}_{8}{}_{1} \times \frac{4}{49}{}^{36}_{1} \times \frac{6}{54}{}^{42}_{9}{}_{1} \qquad 3$$

MAVA Math: Enhanced Skills Solutions Copyright © 2015 Marla Weiss

Fractions 3

Add or subtract as indicated.

1.

$$\overset{5}{\frac{7}{24}} + \overset{3}{\frac{3}{40}} \qquad \overset{11}{\frac{44}{8\cdot 3\cdot 5}} \qquad \frac{11}{30}$$

8·3   8·5

2

8·3·5

2.

$$\overset{7}{\frac{11}{50}} + \overset{5}{\frac{13}{70}} \qquad \overset{71}{\frac{142}{5\cdot 7\cdot 10}} \qquad \frac{71}{175}$$

5·10   7·10

5

5·7·10

3.

$$\overset{2}{\frac{7}{60}} - \overset{3}{\frac{3}{40}} \qquad \overset{1}{\frac{5}{2\cdot 3\cdot 20}} \qquad \frac{1}{24}$$

3·20   2·20

4

2·3·20

4.

$$\overset{5}{\frac{3}{76}} + \overset{19}{\frac{1}{20}} \qquad \overset{17}{\frac{34}{4\cdot 5\cdot 19}} \qquad \frac{17}{190}$$

4·19   4·5

2

4·5·19

5.

$$\overset{2}{\frac{2}{45}} + \overset{5}{\frac{7}{18}} \qquad \overset{13}{\frac{39}{2\cdot 5\cdot 9}} \qquad \frac{13}{30}$$

9·5   9·2

3

2·5·9

6.

$$\overset{7}{\frac{9}{40}} + \overset{5}{\frac{5}{56}} \qquad \overset{11}{\frac{88}{8\cdot 5\cdot 7}} \qquad \frac{11}{35}$$

8·5   8·7

1

8·5·7

7.

$$\overset{5}{\frac{5}{36}} + \overset{3}{\frac{7}{60}} \qquad \overset{23}{\frac{46}{3\cdot 5\cdot 12}} \qquad \frac{23}{90}$$

12·3   12·5

6

12·3·5

8.

$$\overset{6}{\frac{5}{21}} + \overset{9}{\frac{13}{14}} + \overset{14}{\frac{1}{9}} \qquad \overset{23}{\underset{}{\frac{161}{}}}\; \frac{30+117+14}{7\cdot 2\cdot 9} \qquad \frac{23}{18}$$

7·3   7·2   3·3

7·2·9   1

9.

$$\overset{4}{\frac{1}{15}} + \overset{5}{\frac{5}{12}} + \overset{15}{\frac{7}{4}} \qquad \frac{67}{134}\;\frac{4+25+105}{5\cdot 4\cdot 3} \qquad \frac{67}{30}$$

5·3   4·3   4

5·4·3   2

10.

$$\overset{8}{\frac{2}{21}} + \overset{7}{\frac{5}{24}} + \overset{6}{\frac{3}{28}} \qquad \frac{23}{69}\;\frac{16+35+18}{7\cdot 3\cdot 8} \qquad \frac{23}{56}$$

7·3   3·8   7·4

7·3·8   1

11.

$$\overset{3}{\frac{3}{20}} - \overset{4}{\frac{2}{15}} + \overset{15}{\frac{1}{4}} \qquad \frac{4}{16}\;\frac{9-8+15}{5\cdot 4\cdot 3} \qquad \frac{4}{15}$$

5·4   5·3   4

5·4·3   1

12.

$$\overset{15}{\frac{5}{22}} + \overset{10}{\frac{3}{33}} - \overset{6}{\frac{4}{55}} \qquad \frac{27}{81}\;\frac{75+30-24}{6\cdot 5\cdot 11} \qquad \frac{27}{110}$$

2·11   3·11   5·11

6·5·11   2

13.

$$\overset{4}{\frac{5}{18}} + \overset{3}{\frac{5}{24}} - \overset{2}{\frac{1}{36}} \qquad \frac{11}{33}\;\frac{20+15-2}{9\cdot 8} \qquad \frac{11}{24}$$

2·9   3·8   4·9

9·8   3

14.

$$\overset{5}{\frac{4}{21}} + \overset{7}{\frac{2}{15}} - \overset{3}{\frac{2}{35}} \qquad \frac{4}{28}\;\frac{20+14-6}{7\cdot 5\cdot 3} \qquad \frac{4}{15}$$

7·3   5·3   5·7

7·5·3   1

Fractions 4

Operate and simplify.

1.
$$\dfrac{\frac{1}{4} + \frac{3}{8} + \frac{5}{12}}{\frac{5}{12}} \qquad \dfrac{\frac{6+9+10}{24}}{\frac{5}{12}} \qquad \frac{25}{24} \times \frac{12}{5} \qquad \mathbf{\frac{5}{2}}$$

2.
$$\dfrac{\frac{3}{7} + \frac{1}{5}}{\frac{11}{40}} \qquad \dfrac{\frac{22}{35}}{\frac{11}{40}} \qquad \frac{\overset{2}{\cancel{22}}}{\underset{7}{\cancel{35}}} \times \frac{\overset{8}{\cancel{40}}}{\underset{1}{\cancel{11}}} \qquad \mathbf{\frac{16}{7}}$$

3.
$$\dfrac{\frac{5}{6}}{\frac{1}{2}} + \frac{1}{3} \qquad \frac{5}{6} \times \frac{2}{1} \qquad \frac{5}{3} + \frac{1}{3} \qquad \mathbf{2}$$

4.
$$\dfrac{\frac{5}{6} - \frac{1}{3}}{\frac{2}{9} + \frac{1}{6}} \qquad \dfrac{\frac{1}{2}}{\frac{7}{18}} \qquad \frac{1}{2} \times \frac{18}{7} \qquad \mathbf{\frac{9}{7}}$$

5.
$$\dfrac{\frac{3}{16} + 6}{9 - \frac{3}{8}} \qquad \dfrac{\frac{99}{16}}{\frac{69}{8}} \qquad \frac{99}{\underset{2}{\cancel{16}}} \times \frac{\overset{1}{\cancel{8}}}{\underset{23}{\cancel{69}}} \qquad \mathbf{\frac{33}{46}}$$

6.
$$\dfrac{\frac{3}{11} + 2}{2 - \frac{3}{4}} \qquad \dfrac{\frac{25}{11}}{\frac{5}{4}} \qquad \frac{25}{11} \times \frac{\overset{5}{\cancel{4}}}{\underset{1}{\cancel{5}}} \qquad \mathbf{\frac{20}{11}}$$

7.
$$\dfrac{\frac{2}{5} + \frac{5}{6} + \frac{7}{15}}{\frac{17}{60}} \qquad \dfrac{\frac{12+25+14}{30}}{\frac{17}{60}} \quad \frac{51}{30} \qquad \frac{17}{10} \times \frac{60}{17} \qquad \mathbf{6}$$

8.
$$\dfrac{\frac{4}{5} + \frac{3}{10}}{\frac{33}{20}} \qquad \dfrac{\frac{11}{10}}{\frac{33}{20}} \qquad \frac{11}{10} \times \frac{20}{33} \qquad \mathbf{\frac{2}{3}}$$

9.
$$\dfrac{\frac{9}{10} + \frac{5}{6}}{\frac{9}{10} - \frac{5}{6}} \qquad \dfrac{\frac{27}{30} + \frac{25}{30}}{\frac{27}{30} - \frac{25}{30}} \qquad \frac{52}{2} \qquad \mathbf{26}$$

10.
$$\dfrac{\frac{5}{6} + \frac{3}{4}}{\frac{5}{6} - \frac{3}{4}} \qquad \dfrac{\frac{10}{12} + \frac{9}{12}}{\frac{10}{12} - \frac{9}{12}} \qquad \frac{19}{1} \qquad \mathbf{19}$$

11.
$$\dfrac{\frac{5}{6} \times \frac{2}{3}}{\frac{2}{3} \div \frac{5}{6}} \qquad \dfrac{\frac{5}{9}}{\frac{4}{5}} \qquad \frac{5}{9} \times \frac{5}{4} \qquad \mathbf{\frac{25}{36}}$$

12.
$$\dfrac{\frac{5}{6} + \frac{2}{3}}{\frac{2}{3} \times \frac{5}{6}} \qquad \dfrac{\frac{9}{6}}{\frac{5}{9}} \qquad \frac{9}{6} \times \frac{9}{5} \qquad \mathbf{\frac{27}{10}}$$

13.
$$\dfrac{\frac{3}{4} \times \frac{2}{3}}{\frac{2}{3} + \frac{3}{4}} \qquad \dfrac{\frac{1}{2}}{\frac{17}{12}} \qquad \frac{1}{2} \times \frac{12}{17} \qquad \mathbf{\frac{6}{17}}$$

14.
$$\dfrac{\frac{15}{16} \times \frac{4}{5}}{\frac{5}{12} \div \frac{5}{6}} \qquad \dfrac{\frac{3}{4}}{\frac{1}{2}} \qquad \frac{3}{4} \times \frac{2}{1} \qquad \mathbf{\frac{3}{2}}$$

Fractions 5

Operate and simplify. Answer as a mixed number.

1. $5 + \cfrac{1}{3 + \cfrac{1}{2}}$ $\cfrac{7}{2}$ $\cfrac{2}{7}$ $\quad 5\dfrac{2}{7}$

9. $6 - \cfrac{3}{5 - \cfrac{7}{9}}$ $\cfrac{38}{9}$ $\cfrac{27}{38}$ $\quad 5\dfrac{11}{38}$

2. $2 + \cfrac{1}{1 + \cfrac{1}{6}}$ $\cfrac{7}{6}$ $\cfrac{6}{7}$ $\quad 2\dfrac{6}{7}$

10. $3 - \cfrac{2}{4 - \cfrac{1}{8}}$ $\cfrac{31}{8}$ $\cfrac{16}{31}$ $\quad 2\dfrac{15}{31}$

3. $1 + \cfrac{1}{2 + \cfrac{1}{4}}$ $\cfrac{9}{4}$ $\cfrac{4}{9}$ $\quad 1\dfrac{4}{9}$

11. $5 + \cfrac{4}{3 - \cfrac{1}{5}}$ $\cfrac{14}{5}$ $\cfrac{10}{7}$ $\quad 6\dfrac{3}{7}$

4. $8 + \cfrac{1}{3 + \cfrac{1}{3}}$ $\cfrac{10}{3}$ $\cfrac{3}{10}$ $\quad 8\dfrac{3}{10}$

12. $4 - \cfrac{2}{4 - \cfrac{7}{11}}$ $\cfrac{37}{11}$ $\cfrac{22}{37}$ $\quad 3\dfrac{15}{37}$

5. $6 + \cfrac{1}{3 + \cfrac{1}{5}}$ $\cfrac{16}{5}$ $\cfrac{5}{16}$ $\quad 6\dfrac{5}{16}$

13. $5 - \cfrac{2}{3 - \cfrac{4}{5}}$ $\cfrac{11}{5}$ $\cfrac{10}{11}$ $\quad 4\dfrac{1}{11}$

6. $7 - \cfrac{1}{1 + \cfrac{1}{1 + \cfrac{1}{7}}}$ $\cfrac{8}{7}$ $\cfrac{7}{8}$ $\cfrac{15}{8}$ $\cfrac{8}{15}$ $\quad 6\dfrac{7}{15}$

14. $3 + \cfrac{1}{5 - \cfrac{1}{4 - \cfrac{1}{5}}}$ $\cfrac{19}{5}$ $\cfrac{5}{19}$ $\cfrac{90}{19}$ $\cfrac{19}{90}$ $\quad 3\dfrac{19}{90}$

7. $2 + \cfrac{1}{4 - \cfrac{1}{2 + \cfrac{1}{3}}}$ $\cfrac{7}{3}$ $\cfrac{3}{7}$ $\cfrac{25}{7}$ $\cfrac{7}{25}$ $\quad 2\dfrac{7}{25}$

15. $4 - \cfrac{1}{2 + \cfrac{1}{3 - \cfrac{1}{4}}}$ $\cfrac{11}{4}$ $\cfrac{4}{11}$ $\cfrac{26}{11}$ $\cfrac{11}{26}$ $\quad 3\dfrac{15}{26}$

8. $5 - \cfrac{1}{1 + \cfrac{2}{3 + \cfrac{1}{4}}}$ $\cfrac{13}{4}$ $\cfrac{8}{13}$ $\cfrac{21}{13}$ $\cfrac{13}{21}$ $\quad 4\dfrac{8}{21}$

16. $1 + \cfrac{1}{3 + \cfrac{1}{3 + \cfrac{1}{3}}}$ $\cfrac{10}{3}$ $\cfrac{3}{10}$ $\cfrac{33}{10}$ $\cfrac{10}{33}$ $\quad 1\dfrac{10}{33}$

Fractions 6

Rewrite in ascending order using multiple techniques: mental math, decimal conversion, common denominator, versus 1/2, missing part (same N or D), and pairwise comparison.

1. $\dfrac{4}{7}$ $\dfrac{5}{9}$ $\dfrac{3}{5}$ $\dfrac{1}{2}$

$\boxed{\dfrac{1}{2}}$ $\boxed{\dfrac{5}{9}}$ $\boxed{\dfrac{4}{7}}$ $\boxed{\dfrac{3}{5}}$

7. $\dfrac{8}{13}$ $\dfrac{11}{15}$ $\dfrac{8}{11}$ $\dfrac{2}{3}$

$\boxed{\dfrac{8}{13}}$ $\boxed{\dfrac{2}{3}}$ $\boxed{\dfrac{8}{11}}$ $\boxed{\dfrac{11}{15}}$

2. $\dfrac{13}{20}$ $\dfrac{20}{33}$ $\dfrac{11}{23}$ $\dfrac{1}{2}$

$\boxed{\dfrac{11}{23}}$ $\boxed{\dfrac{1}{2}}$ $\boxed{\dfrac{20}{33}}$ $\boxed{\dfrac{13}{20}}$

8. $\dfrac{3}{20}$ $\dfrac{1}{9}$ $\dfrac{4}{25}$ $\dfrac{7}{40}$

$\boxed{\dfrac{1}{9}}$ $\boxed{\dfrac{3}{20}}$ $\boxed{\dfrac{4}{25}}$ $\boxed{\dfrac{7}{40}}$

3. $\dfrac{12}{25}$ $\dfrac{9}{17}$ $\dfrac{9}{19}$ $\dfrac{11}{21}$

$\boxed{\dfrac{9}{19}}$ $\boxed{\dfrac{12}{25}}$ $\boxed{\dfrac{11}{21}}$ $\boxed{\dfrac{9}{17}}$

9. $\dfrac{5}{6}$ $\dfrac{23}{30}$ $\dfrac{4}{5}$ $\dfrac{11}{15}$

$\boxed{\dfrac{11}{15}}$ $\boxed{\dfrac{23}{30}}$ $\boxed{\dfrac{4}{5}}$ $\boxed{\dfrac{5}{6}}$

4. $\dfrac{7}{10}$ $\dfrac{11}{20}$ $\dfrac{14}{25}$ $\dfrac{23}{50}$

$\boxed{\dfrac{23}{50}}$ $\boxed{\dfrac{11}{20}}$ $\boxed{\dfrac{14}{25}}$ $\boxed{\dfrac{7}{10}}$

10. $\dfrac{11}{17}$ $\dfrac{11}{18}$ $\dfrac{6}{11}$ $\dfrac{15}{31}$

$\boxed{\dfrac{15}{31}}$ $\boxed{\dfrac{6}{11}}$ $\boxed{\dfrac{11}{18}}$ $\boxed{\dfrac{11}{17}}$

5. $\dfrac{4}{5}$ $\dfrac{2}{3}$ $\dfrac{6}{7}$ $\dfrac{7}{9}$

$\boxed{\dfrac{2}{3}}$ $\boxed{\dfrac{7}{9}}$ $\boxed{\dfrac{4}{5}}$ $\boxed{\dfrac{6}{7}}$

11. $\dfrac{32}{33}$ $\dfrac{30}{31}$ $\dfrac{97}{99}$ $\dfrac{59}{62}$

$\boxed{\dfrac{59}{62}}$ $\boxed{\dfrac{30}{31}}$ $\boxed{\dfrac{32}{33}}$ $\boxed{\dfrac{97}{99}}$

6. $\dfrac{7}{12}$ $\dfrac{27}{48}$ $\dfrac{13}{24}$ $\dfrac{5}{8}$

$\boxed{\dfrac{13}{24}}$ $\boxed{\dfrac{27}{48}}$ $\boxed{\dfrac{7}{12}}$ $\boxed{\dfrac{5}{8}}$

12. $\dfrac{32}{95}$ $\dfrac{31}{99}$ $\dfrac{30}{91}$ $\dfrac{31}{94}$

$\boxed{\dfrac{31}{99}}$ $\boxed{\dfrac{30}{91}}$ $\boxed{\dfrac{31}{94}}$ $\boxed{\dfrac{32}{95}}$

Fractions 7

Operate using fractions. Answer as a simplified fraction.

1. 10% of 40% of $\frac{3}{4}$ of 1.25

$\frac{1}{10} \cdot \frac{2}{5} \cdot \frac{3}{4} \cdot \frac{5}{4}$ $\frac{3}{80}$

9. 52% of 25% of 22 + $\frac{2}{7}$ of 63%

$\frac{13}{25} \cdot \frac{1}{4} \cdot \frac{22}{1} + \frac{2}{7} \cdot \frac{63}{100}$ $\frac{152}{50}$

2. $\frac{2}{3}$ of 25% of 5% of 5.4

$\frac{2}{3} \cdot \frac{1}{4} \cdot \frac{1}{20} \cdot \frac{27}{5}$ $\frac{9}{200}$

10. 18.75% of 11.2 − $\frac{4}{9}$ of 12% of 22.5

$\frac{3}{16} \cdot \frac{56}{5} - \frac{4}{9} \cdot \frac{3}{25} \cdot \frac{45}{2}$ $\frac{9}{10}$

3. $\frac{1}{3}$ of 80% of 2.4 + $\frac{1}{2}$ of 6% of $1.\overline{3}$

$\frac{1}{3} \cdot \frac{4}{5} \cdot \frac{12}{5} + \frac{1}{2} \cdot \frac{3}{50} \cdot \frac{4}{3}$ $\frac{17}{25}$

11. $\frac{5}{9}$ of 50% of 6.3 + $\frac{5}{6}$ of 24%

$\frac{5}{9} \cdot \frac{1}{2} \cdot \frac{63}{10} + \frac{5}{6} \cdot \frac{6}{25}$ $\frac{39}{20}$

4. $\frac{4}{5}$ of 60% of 2.25 + $\frac{5}{3}$ of 8% of 5.1

$\frac{4}{5} \cdot \frac{3}{5} \cdot \frac{9}{4} + \frac{5}{3} \cdot \frac{2}{25} \cdot \frac{51}{10}$ $\frac{44}{25}$

12. $6.\overline{6}$% of 62.5% of $0.1\overline{6}$ of 0.48

$\frac{1}{15} \cdot \frac{5}{8} \cdot \frac{1}{6} \cdot \frac{12}{25}$ $\frac{1}{300}$

5. $\frac{5}{9}$ of 5% of 45% of $\frac{3}{4}$ of 20.8

$\frac{5}{9} \cdot \frac{1}{20} \cdot \frac{9}{20} \cdot \frac{3}{4} \cdot \frac{104}{5}$ $\frac{39}{200}$

13. 0.25 of 12% of $\frac{16}{27}$ of 8% of 4.5

$\frac{1}{4} \cdot \frac{3}{25} \cdot \frac{16}{27} \cdot \frac{2}{25} \cdot \frac{9}{2}$ $\frac{4}{625}$

6. .375 of 40% of $\frac{2}{3}$ of 62.5% of 3.2

$\frac{3}{8} \cdot \frac{2}{5} \cdot \frac{2}{3} \cdot \frac{5}{8} \cdot \frac{16}{5}$ $\frac{1}{5}$

14. 2.2 of 15% of 6.4 of $\frac{2}{11}$ of 15.625%

$\frac{11}{5} \cdot \frac{3}{20} \cdot \frac{32}{5} \cdot \frac{2}{11} \cdot \frac{5}{32}$ $\frac{3}{50}$

7. $\frac{4}{5}$ of 65% + $\frac{1}{3}$ of 5.4 of 80%

$\frac{4}{5} \cdot \frac{13}{20} + \frac{1}{3} \cdot \frac{27}{5} \cdot \frac{4}{5}$ $\frac{49}{25}$

15. $\frac{3}{5}$ of 4.5% of $\frac{2}{9}$ of .125 of $4.\overline{4}$

$\frac{3}{5} \cdot \frac{9}{200} \cdot \frac{2}{9} \cdot \frac{1}{8} \cdot \frac{40}{9}$ $\frac{1}{300}$

8. 30% of 62.5% of 4.8 + .125 of 8.8

$\frac{3}{10} \cdot \frac{5}{8} \cdot \frac{24}{5} + \frac{1}{8} \cdot \frac{44}{5}$ 2

16. $5.8\overline{3}$ of 87.5% of $\frac{20}{49}$ of 10.5%

$\frac{35}{6} \cdot \frac{7}{8} \cdot \frac{20}{49} \cdot \frac{21}{200}$ $\frac{7}{32}$

Fractions 8

Operate on natural numbers. Simplify before multiplying. Answer as a simplified fraction.

---

1. Divide the product of the 6 least composite numbers by the product of the next 6 composites.

$$\frac{4 \cdot 6 \cdot 8 \cdot 9 \cdot 10 \cdot 12}{14 \cdot 15 \cdot 16 \cdot 18 \cdot 20 \cdot 21}$$
$$\frac{2}{245}$$

7. Divide the product of the 6 least 2-digit composite numbers by the product of the next 5 composites.

$$\frac{10 \cdot 12 \cdot 14 \cdot 15 \cdot 16 \cdot 18}{20 \cdot 21 \cdot 22 \cdot 24 \cdot 25}$$
$$\frac{72}{55}$$

---

2. Divide the product of the first 5 2-digit composites greater than 19 by the product of the first 5 primes.

$$\frac{20 \cdot 21 \cdot 22 \cdot 24 \cdot 25}{2 \cdot 3 \cdot 5 \cdot 7 \cdot 11}$$
$20 \times 24 \times 5 =$ **2400**

8. Divide the product of the 2-digit composites with tens digit 4 by the product of the first 7 squares.

$$\frac{40 \cdot 42 \cdot 44 \cdot 45 \cdot 46 \cdot 48 \cdot 49}{1 \cdot 4 \cdot 9 \cdot 16 \cdot 25 \cdot 36 \cdot 49}$$
$8 \times 7 \times 11 \times 23 =$ **14,168**

---

3. Divide the product of the 2-digit composites with ones digit 9 by the product of the 1st 4 numbers with ones digit 3.

$$\frac{39 \cdot 49 \cdot 69 \cdot 99}{3 \cdot 13 \cdot 23 \cdot 33}$$
$3 \times 49 \times 3 =$ **441**

9. Divide the product of the 2-digit composites with ones digit 3 by the product of the 1st three 2-digit numbers with ones digit 1.

$$\frac{33 \cdot 63 \cdot 93}{11 \cdot 21 \cdot 31}$$
$3 \times 3 \times 3 =$ **27**

---

4. Divide the product of the 2-digit composites with ones digit 1 by the product of the first 9 odd numbers.

$$\frac{21 \cdot 51 \cdot 81 \cdot 91}{1 \cdot 3 \cdot 5 \cdot 7 \cdot 9 \cdot 11 \cdot 13 \cdot 15 \cdot 17}$$
$$\frac{63}{275}$$

10. Divide the product of the 2-digit composites with ones digit 1 by the product of the 1st 3 numbers with ones digit 7.

$$\frac{21 \cdot 51 \cdot 81 \cdot 91}{7 \cdot 17 \cdot 27}$$
$3 \times 9 \times 91 =$ **2457**

---

5. Divide the product of the first 6 composites greater than 29 by the product of the first 7 squares.

$$\frac{30 \cdot 32 \cdot 33 \cdot 34 \cdot 35 \cdot 36}{1 \cdot 4 \cdot 9 \cdot 16 \cdot 25 \cdot 36 \cdot 49}$$
$$\frac{374}{7}$$

11. Divide the product of the first 6 composites greater than 39 by the product of the first 6 cubes.

$$\frac{40 \cdot 42 \cdot 44 \cdot 45 \cdot 46 \cdot 48}{1 \cdot 8 \cdot 27 \cdot 64 \cdot 125 \cdot 216}$$
$$\frac{1771}{90}$$

---

6. Divide the product of the first 7 multiples of 4 by the product of the first 7 multiples of 6. Take the 7th root.

$$\frac{4 \cdot 8 \cdot 12 \cdot 16 \cdot 20 \cdot 24 \cdot 28}{6 \cdot 12 \cdot 18 \cdot 24 \cdot 30 \cdot 36 \cdot 42}$$
$$\frac{2}{3}$$

12. Divide the product of the first 6 multiples of 6 by the product of the first 6 multiples of 15. Take the 6th root.

$$\frac{6 \cdot 12 \cdot 18 \cdot 24 \cdot 30 \cdot 36}{15 \cdot 30 \cdot 45 \cdot 60 \cdot 75 \cdot 90}$$
$$\frac{2}{5}$$

Fractions 9

Answer by making a chart.

1. A flower garden is 1/8 pansies, 1/14 lilies, 1/2 roses, and 1/6 gardenias, while the remaining 46 flowers are irises. How many of the flowers are lilies?

	Fr	Fr	#
P	$\frac{1}{8}$	$\frac{21}{168}$	
L	$\frac{1}{14}$	$\frac{12}{168}$	**24**
R	$\frac{1}{2}$	$\frac{84}{168}$	
G	$\frac{1}{6}$	$\frac{28}{168}$	
I		$\frac{23}{168}$	46
tot		$\frac{168}{168}$	

3 x 7 x 8 = 168

4. A flower garden is 1/5 pansies, 2/15 lilies, 1/4 roses, and 3/20 gardenias, while the remaining 48 flowers are irises. How many of the flowers are pansies?

	Fr	Fr	#
P	$\frac{1}{5}$	$\frac{12}{60}$	**36**
L	$\frac{2}{15}$	$\frac{8}{60}$	
R	$\frac{1}{4}$	$\frac{15}{60}$	
G	$\frac{3}{20}$	$\frac{9}{60}$	
I		$\frac{16}{60}$	48
tot		$\frac{60}{60}$	

3 x 4 x 5 = 60

2. A vegetable garden is 1/3 onions, 1/10 carrots, 2/7 radishes, and 2/15 peppers. The other 93 plants are lettuces. How many of each type of vegetable are in the garden?

	Fr	Fr	#
O	$\frac{1}{3}$	$\frac{70}{210}$	**210**
C	$\frac{1}{10}$	$\frac{21}{210}$	**63**
R	$\frac{2}{7}$	$\frac{60}{210}$	**180**
P	$\frac{2}{15}$	$\frac{28}{210}$	**84**
L		$\frac{31}{210}$	**93**
tot		$\frac{210}{210}$	

2 x 7 x 15 = 210

5. A vegetable garden is 1/8 onions, 1/12 carrots, 2/5 radishes, and 4/15 peppers. The other 45 plants are lettuces. How many of each type of vegetable are in the garden?

	Fr	Fr	#
O	$\frac{1}{8}$	$\frac{15}{120}$	**45**
C	$\frac{1}{12}$	$\frac{10}{120}$	**30**
R	$\frac{2}{5}$	$\frac{48}{120}$	**144**
P	$\frac{4}{15}$	$\frac{32}{120}$	**96**
L		$\frac{15}{120}$	**45**
tot		$\frac{120}{120}$	

3 x 5 x 8 = 120

3. An herb garden is 1/12 parsley, 3/14 sage, 4/21 rosemary, and 5/18 thyme. The remaining 177 herb plants are dill. How many of the herbs are thyme?

	Fr	Fr	#
P	$\frac{1}{12}$	$\frac{21}{252}$	
S	$\frac{3}{14}$	$\frac{54}{252}$	
R	$\frac{4}{21}$	$\frac{48}{252}$	
T	$\frac{5}{18}$	$\frac{70}{252}$	**210**
D		$\frac{59}{252}$	177
tot		$\frac{252}{252}$	

4 x 7 x 9 = 252

6. An herb garden is 1/15 parsley, 1/3 sage, 3/10 rosemary, and 7/50 thyme. The remaining 96 herb plants are dill. Find the total number of herb plants in the garden.

	Fr	Fr	#
P	$\frac{1}{15}$	$\frac{10}{150}$	
S	$\frac{1}{3}$	$\frac{50}{150}$	
R	$\frac{3}{10}$	$\frac{45}{150}$	
T	$\frac{7}{50}$	$\frac{21}{150}$	
D		$\frac{24}{150}$	96
tot		$\frac{150}{150}$	**600**

50 x 3 = 150

Fractions 10

Find the number of unique values of a/b, where a and b are distinct members of the given set.	*Answer as indicated.*

1. {2, 4, 8}

$$\frac{2}{4} \quad \frac{2}{8} \quad \frac{\cancel{4}}{8} \quad \frac{4}{2} \quad \frac{8}{2} \quad \frac{\cancel{8}}{4} \qquad 4$$

2. {2, 6, 9}

$$\frac{2}{6} \quad \frac{2}{9} \quad \frac{6}{9} \quad \frac{6}{2} \quad \frac{9}{2} \quad \frac{9}{6} \qquad 6$$

3. {3, 9, 27}

$$\frac{3}{9} \quad \frac{3}{27} \quad \frac{\cancel{9}}{\cancel{27}} \quad \frac{9}{3} \quad \frac{27}{3} \quad \frac{\cancel{27}}{\cancel{9}} \qquad 4$$

4. {2, 8, 32}

$$\frac{2}{8} \quad \frac{2}{32} \quad \frac{\cancel{8}}{\cancel{32}} \quad \frac{8}{2} \quad \frac{32}{2} \quad \frac{\cancel{32}}{\cancel{8}} \qquad 4$$

5. {2, 4, 6}

$$\frac{2}{4} \quad \frac{2}{6} \quad \frac{4}{6} \quad \frac{4}{2} \quad \frac{6}{2} \quad \frac{6}{4} \qquad 6$$

6. {2, 5, 6, 15} 4x2 with reciprocals

$$\frac{2}{5} \quad \frac{2}{6} \quad \frac{2}{15} \quad \frac{5}{6} \quad \frac{\cancel{5}}{\cancel{15}} \quad \frac{\cancel{6}}{\cancel{15}} \qquad 8$$

7. {2, 4, 8, 16} 3x2 with reciprocals

$$\frac{2}{4} \quad \frac{2}{8} \quad \frac{2}{16} \quad \frac{\cancel{4}}{\cancel{8}} \quad \frac{\cancel{4}}{\cancel{16}} \quad \frac{\cancel{8}}{\cancel{16}} \qquad 6$$

8. {3, 6, 12, 24} 3x2 with reciprocals

$$\frac{3}{6} \quad \frac{3}{12} \quad \frac{3}{24} \quad \frac{\cancel{6}}{\cancel{12}} \quad \frac{\cancel{6}}{\cancel{24}} \quad \frac{\cancel{12}}{\cancel{24}} \qquad 6$$

9. {2, 6, 18, 54} 3x2 with reciprocals

$$\frac{2}{6} \quad \frac{2}{18} \quad \frac{2}{54} \quad \frac{\cancel{6}}{\cancel{18}} \quad \frac{\cancel{6}}{\cancel{54}} \quad \frac{\cancel{18}}{\cancel{54}} \qquad 6$$

10. {2, 3, 8, 12} 4x2 with reciprocals

$$\frac{2}{3} \quad \frac{2}{8} \quad \frac{2}{12} \quad \frac{3}{8} \quad \frac{\cancel{3}}{\cancel{12}} \quad \frac{\cancel{8}}{\cancel{12}} \qquad 8$$

11. Determine the greatest fraction using mental math.

$$\boxed{\frac{1790}{1791}} \quad \frac{1789}{1790} \quad \frac{1789}{1792} \quad \frac{1788}{1789}$$

missing only 1 of the smallest size piece

12. The sum of 2 positive consecutive integers is x. In terms of x, what fraction is the value of the lesser of these two integers?

$$y + (y + 1) = x$$
$$2y + 1 = x$$
$$2y = x - 1 \qquad \frac{x - 1}{2}$$

13. If t orders of tea, each bought at c cents, are purchased per hour at a cafe that is open h hours a day, find the amount of money in dollars paid in 1 day for the tea.

$$t \cdot \frac{c}{100} \text{ in 1 hr} \qquad \frac{tch}{100} \text{ in h hrs}$$

14. A pig weighs 3/8 of a horse's weight. A cow weighs 6/5 of the horse's weight. The pig weighs 75 pounds. Find the weight of the cow in pounds.

$$\frac{P}{H} = \frac{3}{8} = \frac{15}{40} \qquad \frac{P}{C} = \frac{15}{48} = \frac{75}{240}$$
$$\frac{C}{H} = \frac{6}{5} = \frac{48}{40} \qquad\qquad \mathbf{240}$$

15. In City C, 7/10 of the farmers grow only corn; the rest grow only wheat. Three-fifths of the corn farmers and 2/3 of the wheat farmers own their own land. Of the farmers in City C, what fraction own their own land?

$$\frac{3}{5} \times \frac{7}{10} + \frac{2}{3} \times \frac{3}{10} = \frac{31}{50}$$

Functions 1

Complete the function charts.

1. $f(x) = -2x^2 + 2x - 1$

x	0	1	−1	2	−2	3	−3	4	−4	5	−5	10	−10
$f(x)$	−1	−1	−5	−5	−13	−13	−25	−25	−41	−41	−61	−181	−221

2. $f(x) = 3x^2 - x + 4$

x	0	1	−1	2	−2	3	−3	4	−4	5	−5	10	−10
$f(x)$	4	6	8	14	18	28	34	48	56	74	84	294	314

3. $f(x) = -x^2 + 3x - 2$

x	0	1	−1	2	−2	3	−3	4	−4	5	−5	10	−10
$f(x)$	−2	0	−6	0	−12	−2	−20	−6	−30	−12	−42	−72	−132

4. $f(x) = 2x^2 - 4x + 5$

x	0	1	−1	2	−2	3	−3	4	−4	5	−5	10	−10
$f(x)$	5	3	11	5	21	11	35	21	53	35	75	165	245

5. $f(x) = -3x^2 + x - 3$

x	0	1	−1	2	−2	3	−3	4	−4	5	−5	10	−10
$f(x)$	−3	−5	−7	−13	−17	−27	−33	−47	−55	−73	−83	−293	−313

6. $f(x) = 4x^2 - 3x + 2$

x	0	1	−1	2	−2	3	−3	4	−4	5	−5	10	−10
$f(x)$	2	3	9	12	24	29	47	54	78	87	117	372	432

7. $f(x) = 5x^2 - 5x + 6$

x	0	1	−1	2	−2	3	−3	4	−4	5	−5	10	−10
$f(x)$	6	6	16	16	36	36	66	66	106	106	156	456	556

The 2 ordered pairs may vary for #4, 9, 12, 15, 18, 20.	Functions 2

Answer YES or NO as to whether the relation is a function. If NO, state 2 ordered pairs that violate the function definition.

1. $y = 5x$ **YES**	12. $y^2 = x^2$ (1, 1) (1, −1) **NO**
2. { (2, 3), (3, 3), (4, 3), (5, 3) } **YES**	13. **YES**
3. $y = x^2$ **YES**	14. { (1, 2), (2, 1), (4, 3), (3, 4) } **YES**
4. $x = y^2$ (4, 2) (4, −2) **NO**	15. $x = 2y^2 + 4y − 2$ (−2, −2) (−2, 0) **NO**
5. $2x − 7y = 1$ **YES**	16. (7, 2) (7, −4) **NO**
6. $y = 5x^2 + 3x − 2$ **YES**	17. $y = −5$ **YES**
7. $y = x^3$ **YES**	18. $x = 9$ (9, 0) (9, 1) **NO**
8. { (8, 6), (4, 2), (8, 3), (5, 1) } (8, 6) (8, 3) **NO**	19. $x = −4y + 12$ **YES**
9. $x = 3y^2 − 1$ (2, 1) (2, −1) **NO**	20. $x^2 + y^2 = 25$ (0, 5) (0, −5) **NO**
10. $x = y^3$ **YES**	21. { (9, 0), (3, 2), (9, 2), (2, 0) } (9, 0) (9, 2) **NO**
11. { (x, y)	x is a natural number, y is half x} **YES**	22. { (x, y)	x is an integer, y is twice x squared} **YES**

Functions 3

Evaluate each function based on the given definitions.

ABS(x) = x if x > 0, –x if x < 0, and 0 if x = 0.
SGN(x) = 1 if x > 0, –1 if x < 0, and 0 if x = 0.
INT(x) = the greatest integer contained in x.
TRUNC(x) = x with its fractional or decimal part truncated.
SIGMA(x) = the number of divisors of x where x is a natural number.
SQR(x) = x squared.
SQRT(x) = the square root of x if x ≥ 0, undefined if x < 0.

1. ABS(–90) **90**	14. TRUNC (7.75) **7**	27. SQRT(0) **0**
2. SQR(25) **625**	15. SIGMA(8) **4**	28. SQR(0) **0**
3. TRUNC(8.97) **8**	16. SQRT(–8) **undef**	29. ABS(0) **0**
4. SIGMA(10) **4**	17. INT(–9.2) **–10**	30. SIGMA(0) **undef**
5. SIGMA(1) **1**	18. SGN(–32) **–1**	31. INT(0) **0**
6. TRUNC(–53.5) **–53**	19. SQR(–11) **121**	32. SGN(0) **0**
7. INT(–53.5) **–54**	20. SQR(1.2) **1.44**	33. TRUNC(–1.32) **–1**
8. SGN(987) **1**	21. INT(–13) **–13**	34. ABS(–431) **431**
9. SGN(3 – 17) **–1**	22. SIGMA(12) **6**	35. SQR(0.5) **0.25**
10. SIGMA(25) **3**	23. TRUNC(–76.25) **–76**	36. TRUNC(9.97) **9**
11. SQRT(81) **9**	24. SGN(–3⁵) **–1**	37. INT(–6.91) **–7**
12. ABS(5 – 9) **4**	25. ABS(–5³) **125**	38. SGN(5 ÷ 3) **1**
13. INT(21.5) **21**	26. SQR(–6) **36**	39. SQRT(1,000,000) **1000**

MAVA Math: Enhanced Skills Solutions Copyright © 2015 Marla Weiss

Functions 4

Write the linear function equation, and complete the function chart.

1. $f(x) =$ **4x − 5**

x	−2	−1	0	1	2	3	4
y	**−13**	**−9**	−5	−1	**3**	**7**	11

7. $f(x) =$ **−4x + 7**

x	−2	−1	0	1	2	3	4
y	15	**11**	7	**3**	−1	**−5**	**−9**

2. $f(x) =$ **3x + 2**

x	−2	−1	0	1	2	3	4
y	−4	**−1**	2	**5**	**8**	11	**14**

8. $f(x) =$ **−3x − 6**

x	−2	−1	0	1	2	3	4
y	**0**	−3	−6	**−9**	**−12**	−15	**−18**

3. $f(x) =$ **6x − 4**

x	−2	−1	0	1	2	3	4
y	**−16**	**−10**	−4	**2**	8	**14**	20

9. $f(x) =$ **−2x + 9**

x	−2	−1	0	1	2	3	4
y	**13**	**11**	9	**7**	5	3	**1**

4. $f(x) =$ **2x + 3**

x	−2	−1	0	1	2	3	4
y	−1	**1**	3	**5**	7	**9**	**11**

10. $f(x) =$ **−5x + 1**

x	−2	−1	0	1	2	3	4
y	11	**6**	1	−4	**−9**	**−14**	**−19**

5. $f(x) =$ **5x − 1**

x	−2	−1	0	1	2	3	4
y	−11	**−6**	−1	**4**	9	14	**19**

11. $f(x) =$ **8x − 3**

x	−2	−1	0	1	2	3	4
y	**−19**	−11	−3	**5**	**13**	**21**	29

6. $f(x) =$ **7x − 8**

x	−2	−1	0	1	2	3	4
y	**−22**	**−15**	−8	**−1**	6	**13**	20

12. $f(x) =$ **9x + 5**

x	−2	−1	0	1	2	3	4
y	**−13**	−4	5	14	**23**	**32**	**41**

Functions 5

Answer YES or NO as to whether the graph depicts a function.

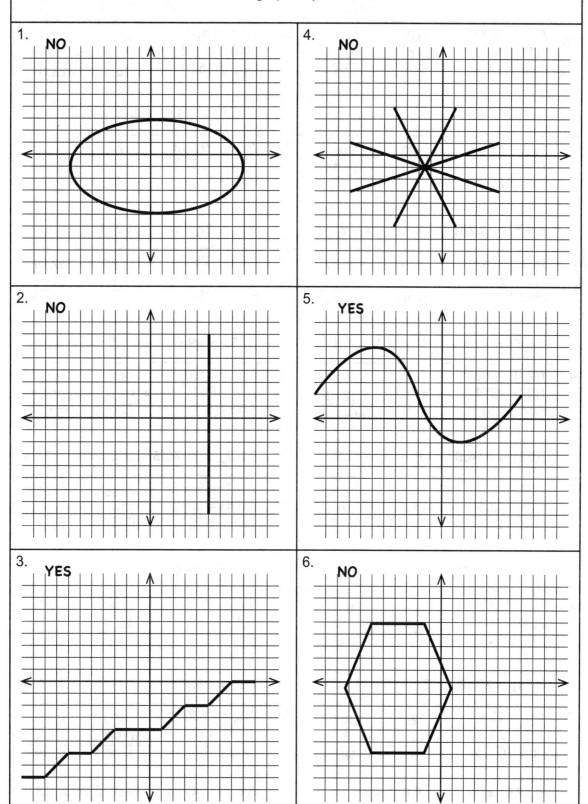

1. NO

2. NO

3. YES

4. NO

5. YES

6. NO

Functions 6

Evaluate using linear interpolation.

1. $f(10) = 500$
   $f(13) = 260$
   Find $f(12.5)$

500	260
− 260	+ 40
240	300

   f(12.5) = 300   240 / 6 = 40

2. $f(6) = 3.087$
   $f(8) = 1.871$
   Find $f(6.5)$

3.087	3.087
− 1.871	− 0.304
1.216	2.783

   f(6.5) = 2.783   1.216 / 4 = 0.304

3. $f(2) = 17.09$
   $f(6) = 33.97$
   Find $f(4.5)$

33.97	33.97
− 17.09	− 6.33
16.88	27.64

   f(4.5) = 27.64   16.88 / 8 = 2.11

4. $f(20) = 769$
   $f(30) = 354$
   Find $f(23)$

769	769.0
− 354	− 124.5
415	644.5

   f(23) = 644.5   415 / 10 = 41.5

5. $f(2) = 32.4$
   $f(3) = 53.5$
   Find $f(2.9)$

53.5	53.50
− 32.4	− 2.11
21.1	51.39

   f(2.9) = 51.39   21.1 / 10 = 2.11

6. $f(31) = 3084$
   $f(36) = 2269$
   Find $f(33)$

3084	3084
− 2269	− 326
815	2758

   f(33) = 2758   815 / 5 = 163

7. $f(6) = 72.9$
   $f(8) = 24.1$
   Find $f(7.25)$

72.9	24.1
− 24.1	+ 18.3
48.8	42.4

   f(7.25) = 42.4   48.8 / 8 = 6.1

8. $f(14) = 48.6$
   $f(16) = 22.2$
   Find $f(15.5)$

48.6	22.2
− 22.2	+ 6.6
26.4	28.8

   f(15.5) = 28.8   26.4 / 4 = 6.6

9. $f(6) = 39.043$
   $f(12) = 2.677$
   Find $f(7)$

39.043	39.043
− 2.677	− 6.061
36.366	32.982

   f(7) = 32.982   36.366 / 6 = 6.061

10. $f(22) = 22.74$
    $f(27) = 67.89$
    Find $f(25)$

67.89	67.89
− 22.74	− 18.06
45.15	49.83

    f(25) = 49.83   45.15 / 5 = 9.03

11. $f(6) = 5306$
    $f(7) = 8541$
    Find $f(6.7)$

8541	8541.0
− 5306	− 970.5
3235	7570.5

    f(6.7) = 7570.5   3235 / 10 = 323.5

12. $f(5.1) = 9465$
    $f(5.4) = 3762$
    Find $f(5.2)$

9465	9465
− 3762	− 1901
5703	7564

    f(5.2) = 7564   5703 / 3 = 1901

13. $f(7) = 36.22$
    $f(9) = 22.08$
    Find $f(8)$

36.22	22.08
− 22.08	+ 7.07
14.14	29.15

    f(8) = 29.15   14.14 / 2 = 7.07

14. $f(5) = 6850$
    $f(7) = 4390$
    Find $f(6.7)$

6850	4390
− 4390	+ 369
2460	4759

    f(6.7) = 4759   2460 / 20 = 123

Functions 7

Evaluate.

Find k, the unknown constant term, given one value of the quadratic function.

1. $f(x) = 3x + 2$
$g(x) = 2x - 1$
Find $g(f(4)) + f(g(2))$.
$f(4) = 14$
$g(14) = 27$
$g(2) = 3$
$f(3) = 11$
38

12. $f(x) = 3x^2 + 2x + k$
$f(2) = 6$
$12 + 4 + k = 6$
$k = $ **−10**

2. $f(x) = 4x - 5$
$g(x) = 5x + 6$
Find $g(f(2)) + f(g(1))$.
$f(2) = 3$
$g(3) = 21$
$g(1) = 11$
$f(11) = 39$
60

13. $f(x) = 5x^2 + 3x - k$
$f(3) = 50$
$45 + 9 - k = 50$
$k = $ **4**

3. $f(x) = 3x^2 - 1$
$g(x) = 2x + 3$
Find $g(f(2)) - f(g(-2))$.
$f(2) = 11$
$g(11) = 25$
$g(-2) = -1$
$f(-1) = 2$
23

14. $f(x) = -x^2 + 4x - k$
$f(4) = 8$
$-16 + 16 - k = 8$
$k = $ **−8**

4. $f(x) = x^x$
$g(x) = 3x + 1$
Find $g(f(2)) + f(g(1))$.
$f(2) = 4$
$g(4) = 13$
$g(1) = 4$
$f(4) = 256$
269

15. $f(x) = 2x^2 + 3x - k$
$f(-2) = 10$
$8 - 6 - k = 10$
$k = $ **−8**

5. $f(x) = 2x^2 + 3$
$g(x) = 5x + 6$
Find $g(f(3)) - f(g(-3))$.
$f(3) = 21$
$g(21) = 111$
$g(-3) = -9$
$f(-9) = 165$
−54

16. $f(x) = -4x^2 - 3x - k$
$f(-2) = 5$
$-16 + 6 - k = 5$
$k = $ **−15**

6. $f(x) = (x - 3) \div 2$
$g(x) = 4x + 5$
Find $g(f(7)) - f(g(1))$.
$f(7) = 2$
$g(2) = 13$
$g(1) = 9$
$f(9) = 3$
10

17. $f(x) = 6x^2 + 5x + k$
$f(-5) = 100$
$150 - 25 + k = 100$
$k = $ **−25**

7. $f(x) = 3x^2 + 2x - 3$
$g(x) = -3x + 8$
Find $g(f(-2)) + f(g(1))$.
$f(-2) = 5$
$g(5) = -7$
$g(1) = 5$
$f(5) = 82$
75

18. $f(x) = 4x^2 + 2x - k$
$f(-3) = 14$
$36 - 6 - k = 14$
$k = $ **16**

8. $f(x) = 2x - 3$
$g(x) = 4x + 3$
Find $g(f(2)) + f(g(2))$.
$f(2) = 1$
$g(1) = 7$
$g(2) = 11$
$f(11) = 19$
26

19. $f(x) = -x^2 + 7x + k$
$f(-4) = 4$
$-16 - 28 + k = 4$
$k = $ **48**

9. $f(x) = (3x + 1) \div 2$
$g(x) = (2x - 1) \div 3$
Find $g(f(3)) + f(g(2))$.
$f(3) = 5$
$g(5) = 3$
$g(2) = 1$
$f(1) = 2$
5

20. $f(x) = -4x^2 - 2x + k$
$f(-1) = 7$
$-4 + 2 + k = 7$
$k = $ **9**

10. $f(x) = x^{-x}$
$g(x) = 8x + 6$
Find $g(f(2)) \div f(g(-1))$.
$f(2) = 1/4$
$g(1/4) = 8$
$g(-1) = -2$
$f(-2) = 4$
2

21. $f(x) = 5x^2 + 6x - k$
$f(1) = 4$
$5 + 6 - k = 4$
$k = $ **7**

11. $f(x) = (-x)^3$
$g(x) = -2x + 6$
Find $g(f(3)) + f(g(1))$.
$f(3) = -27$
$g(-27) = 60$
$g(1) = 4$
$f(4) = -64$
−4

22. $f(x) = -3x^2 - 5x + k$
$f(-2) = 6$
$-12 + 10 + k = 6$
$k = $ **8**

Functions 8

Evaluate the recursive function, given the domain is the set of natural numbers.

1. $f(n) = 3f(n-1) - n$
$f(1) = 4$
Find $f(4)$.

$f(2) = 3 \times 4 - 2 = 10$
$f(3) = 3 \times 10 - 3$
$\quad = 27$
$f(4) = 3 \times 27 - 4$
$\quad = \mathbf{77}$

9. $f(n) = f(n + 2) + 1$
$f(7) = 10$ or
Find $f(1)$.

$f(1) = f(3) + 1$
$f(3) = f(5) + 1$
$f(5) = 11$ $f(5) = 10 + 1 = 11$
$f(3) = 12$ $f(3) = 12$
$f(1) = 13$ $f(1) = \mathbf{13}$

2. $f(mn) = f(m) + f(n)$
$f(8) = 20$
Find $f(64)$.

$f(64) = f(8 \times 8)$
$\quad = f(8) + f(8)$
$\quad = 20 + 20$
$\quad = \mathbf{40}$

10. $f(n) = 4f(n-1) + 2n$
$f(1) = 4$
Find $f(4)$.

$f(4) = 4f(3) + 8$
$f(3) = 4f(2) + 6$
$f(2) = 4f(1) + 4$
$f(2) = 20$
$f(4) = 344 + 8 = \mathbf{352}$ $f(3) = 86$

3. $f(n + 2) = 5f(n) - 5$
$f(1) = 5$
Find $f(7)$.

$f(3) = 5 \times 5 - 5 = 20$
$f(5) = 5 \times 20 - 5$
$\quad = 95$
$f(7) = 5 \times 95 - 5$
$\quad = \mathbf{470}$

11. $f(n) = 4f(n-2)$
$f(0) = 0, f(1) = 2$
Find $f(5)$.

$f(2) = 4f(0) = 0$
$f(3) = 4f(1) = 8$
$f(4) = 4f(2) = 0$
$f(5) = 4f(3) = \mathbf{32}$

4. $f(mn) = f(m) - f(n)$
Find $f(81)$.

regardless of the value of f(9)

$f(81) = f(9 \times 9)$
$\quad = f(9) - f(9)$
$\quad = \mathbf{0}$

12. $f(n) = 1 + f(3n - 1)$ if n odd and n>1
$f(n) = 2n$ if n even
$f(1) = 1$
Find $f(3)$. $f(3) = 1 + 16 = \mathbf{17}$

5. $f(n) = (f(n - 1))^2 - 2$
$f(3) = 10$
Find $f(0)$.

$f(0) = \sqrt{2}$

$10 = f(2)^2 - 2$
$12 = f(1)^4 - 2$
$14 = f(0)^8 - 2$
$16 = f(0)^8$

13. $f(n) = 2 + f(5n - 1)$ if n odd and n>1
$f(n) = n \div 2$ if n even
$f(1) = 1$
Find $f(5)$.
$f(5) = 2 + 12 = \mathbf{14}$

6. $f(n) = (f(n + 1))^2$
$f(4) = 10$
Find $f(1)$.
or f(3) = 100
$f(2) = 10{,}000$
$f(1) = 100{,}000{,}000$

$f(1) = f(2)^2$
$\quad = f(3)^4$
$\quad = f(4)^8$
$\quad = \mathbf{100{,}000{,}000}$

14. $f(n) = f(n + 1) - 2$ if n odd and n>1
$f(n) = 3n$ if n even
$f(1) = 5$
Find $f(7)$.
$f(7) = 24 - 2 = \mathbf{22}$

7. $f(n) = 2f(n + 1) + 3$
$f(10) = 30$
Find $f(7)$.

$f(9) = 60 + 3 = 63$
$f(8) = 126 + 3 = 129$
$f(7) = 258 + 3$
$\quad = \mathbf{261}$

15. $f(n) = 2f(n - 1) - f(n - 2)$
$f(1) = 4, f(2) = 9$
Find $f(5)$.
$f(3) = 18 - 4 = 14$
$f(4) = 28 - 9 = 19$
$f(5) = 38 - 14 = \mathbf{24}$

8. $f(n) = nf(n + 1)$
$f(4) = 5$
Find $f(1)$.
or
$f(3) = 15$
$f(2) = 30$
$f(1) = 30$

$f(1) = 1f(2)$
$\quad = 2f(3)$
$\quad = 6f(4)$
$\quad = \mathbf{30}$

16. $f(n) = -3f(n - 1) - 3f(n - 2)$
$f(1) = 2, f(2) = 5$
Find $f(5)$. $f(3) = -15 - 6 = -21$
$f(4) = 63 - 15 = 48$
$f(5) = -144 + 63 = \mathbf{-81}$

Geometric Structures 1

Name with proper notation, using the diagrams.	*Name with proper notation, using the diagram. (Answers may vary.)*

1. Name 6 different rays.

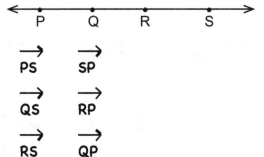

$\overrightarrow{PS}$ $\overrightarrow{SP}$

$\overrightarrow{QS}$ $\overrightarrow{RP}$

$\overrightarrow{RS}$ $\overrightarrow{QP}$

2. Name 6 different angles.

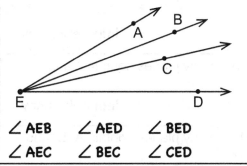

∠AEB   ∠AED   ∠BED
∠AEC   ∠BEC   ∠CED

3. Name 9 different line segments.

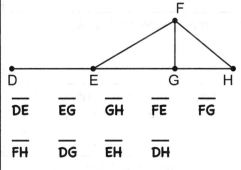

$\overline{DE}$ $\overline{EG}$ $\overline{GH}$ $\overline{FE}$ $\overline{FG}$

$\overline{FH}$ $\overline{DG}$ $\overline{EH}$ $\overline{DH}$

4. Name 4 different line segments that all intersect with each other.

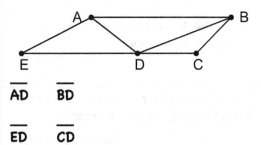

$\overline{AD}$ $\overline{BD}$

$\overline{ED}$ $\overline{CD}$

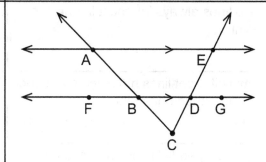

5. 3 noncollinear points — A, B, G

6. 4 collinear points — F, B, D, G

7. 1 scalene trapezoid — ABDE

8. 2 parallel rays — $\overrightarrow{FG}$ $\overrightarrow{AE}$

9. 2 parallel line segments — $\overline{AE}$ $\overline{FD}$

10. 4 line segments with endpoint C — $\overline{CB}$ $\overline{CA}$ $\overline{CD}$ $\overline{CE}$

11. 2 vertical angles — ∠CDG ∠BDE

12. 2 supplementary angles — ∠FBA ∠DBA

13. 2 alternate interior angles — ∠AED ∠EDG

14. 1 obtuse angle — ∠BDE

15. 1 acute angle — ∠GDE

16. a ray containing B, not as an endpoint — $\overrightarrow{FG}$

17. 2 rays intersecting in infinitely many points — $\overrightarrow{GF}$ $\overrightarrow{FG}$

18. 2 similar triangles — △CBD △CAE

19. 2 congruent line segments — none

20. 3 rays with endpoint B — $\overrightarrow{BF}$ $\overrightarrow{BA}$ $\overrightarrow{BD}$

MAVA Math: Enhanced Skills Solutions Copyright © 2015 Marla Weiss

Geometric Structures 2

Answer TRUE (T) or FALSE (F).	*Answer for 4 distinct, coplanar points A, B, C, and D, only pairwise collinear.*
1. Two lines always intersect in one point. **F**	12. How many lines contain A and B? **1**
2. Two different lines may never intersect in more than one point. **T**	13. How many lines contain A, B, and C? **0**
3. Two non-parallel lines can have an empty intersection. skew **T**	14. How many lines contain A, B, C, and D? **0**
4. Two line segments may never intersect in more than one point. **F**	15. How many planes contain A? **infinitely many**
5. Two non-parallel line segments in a plane can have an empty intersection. **T**	16. How many planes contain A and D? **infinitely many**
6. Two planes may intersect in one point. **F**	17. How many planes contain A, B, and D? **1**
7. If two planes do not intersect, they are parallel. **T**	18. How many planes contain A, B, C, and D? **1**
8. The intersection of a line and a plane is a point. **F**	19. If two lines intersect at D, at how many other points do they intersect? **0**
9. Three parallel lines must be coplanar. **F**	20. The line through A and B and the line through C and D intersect in how many points? **1**
10. If two points of a ray are contained in a line, the entire ray is contained in the line. **T**	21. Angle BAD and angle BCD intersect in how many points? **2 or 3**
11. If two points of a line segment are contained in a ray, the entire segment is contained in the ray. **F**	22. The intersection of 2 planes contains B and how many other points? **infinitely many**

Graphs 1

Graph on the number line.

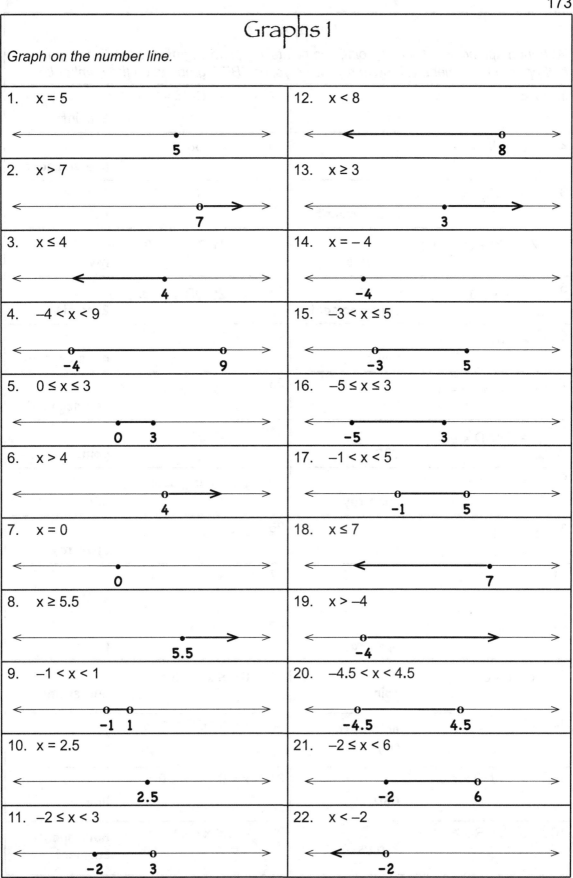

1. x = 5

2. x > 7

3. x ≤ 4

4. −4 < x < 9

5. 0 ≤ x ≤ 3

6. x > 4

7. x = 0

8. x ≥ 5.5

9. −1 < x < 1

10. x = 2.5

11. −2 ≤ x < 3

12. x < 8

13. x ≥ 3

14. x = − 4

15. −3 < x ≤ 5

16. −5 ≤ x ≤ 3

17. −1 < x < 5

18. x ≤ 7

19. x > −4

20. −4.5 < x < 4.5

21. −2 ≤ x < 6

22. x < −2

Graphs 2

Without graphing, identify as a point, two points, line, line segment, open line segment, half-open line segment, ray, open ray, two rays, or CBD if graphed on the number line.

1. $x \geq 8$	ray	17. $x = -7$ OR $x = -8$	2 points
2. $x = -9$	point	18. $0 \leq x \leq 99$	line segment
3. $-12 \leq x \leq 15$	line segment	19. $x \geq 34$	ray
4. $x = 4.3$ OR $x = -5.2$	2 points	20. $x \geq 11$ OR $x \geq 13$	ray
5. $-10 < x < 11$	open segment	21. $x = 25$ OR $x = -32$	2 points
6. $2x = 10$	point	22. $-7 < x < 14$	open segment
7. $x + 2 = 5 + x - 3$	line	23. $35 \geq x \geq 22$	line segment
8. $x = 2$ AND $x = 8$	CBD	24. $x = -109.6$	point
9. $x > 19$	open ray	25. $x < 5$ OR $x > -1$	line
10. $x \leq -13$	ray	26. $x > -23$	open ray
11. $x < 2$ OR $x > 0$	line	27. $x \geq 0$ OR $x \leq -6$	2 rays
12. $x < 12$	open ray	28. $3x + 1 = 8$	point
13. $x + 1 = 6$	point	29. $0.9 \leq x \leq 1.3$	line segment
14. $-4 < x \leq 24$	half-open segment	30. $x \leq 59$ OR $x \leq 46$	ray
15. $x > 1$ AND $x < 0$	CBD	31. $x < 0$ OR $x \geq 0$	line
16. $x \leq -5$ OR $x \geq 17$	2 rays	32. $-11 \leq x < -1$	half-open segment

Graphs 3

Graph on the number line.

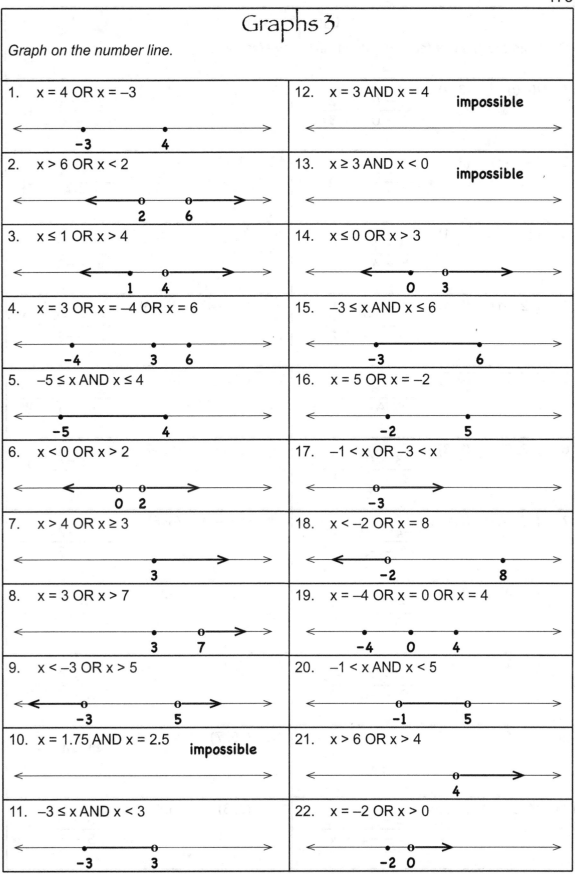

1. x = 4 OR x = –3

2. x > 6 OR x < 2

3. x ≤ 1 OR x > 4

4. x = 3 OR x = –4 OR x = 6

5. –5 ≤ x AND x ≤ 4

6. x < 0 OR x > 2

7. x > 4 OR x ≥ 3

8. x = 3 OR x > 7

9. x < –3 OR x > 5

10. x = 1.75 AND x = 2.5 **impossible**

11. –3 ≤ x AND x < 3

12. x = 3 AND x = 4 **impossible**

13. x ≥ 3 AND x < 0 **impossible**

14. x ≤ 0 OR x > 3

15. –3 ≤ x AND x ≤ 6

16. x = 5 OR x = –2

17. –1 < x OR –3 < x

18. x < –2 OR x = 8

19. x = –4 OR x = 0 OR x = 4

20. –1 < x AND x < 5

21. x > 6 OR x > 4

22. x = –2 OR x > 0

Graphs 4

Compute the slope of the line determined by the two points.

1.	(0, 0)	(6, 4)	$\dfrac{4-0}{6-0}$	$\dfrac{2}{3}$		12.	(3, 3)	(2, −2)	$\dfrac{3--2}{3-2}$	**5**
2.	(−1, −6)	(3, 2)	$\dfrac{2--6}{3--1}$	**2**		13.	(1, 4)	(6, −1)	$\dfrac{4--1}{1-6}$	**−1**
3.	(5, 3)	(4, 1)	$\dfrac{3-1}{5-4}$	**2**		14.	(6, 7)	(−3, −2)	$\dfrac{7--2}{6--3}$	**1**
4.	(−2, −1)	(0, 4)	$\dfrac{4--1}{0--2}$	$\dfrac{5}{2}$		15.	(−3, −4)	(5, −8)	$\dfrac{-8--4}{5--3}$	$\dfrac{-1}{2}$
5.	(−7, 6)	(6, 6)	$\dfrac{6-6}{DNA}$	**0**		16.	(−5, 2)	(3, −3)	$\dfrac{2--3}{-5-3}$	$\dfrac{-5}{8}$
6.	(−3, 0)	(2, 3)	$\dfrac{3-0}{2--3}$	$\dfrac{3}{5}$		17.	(8, 0)	(1, −7)	$\dfrac{0--7}{8-1}$	**1**
7.	(−5, −6)	(2, −4)	$\dfrac{-4--6}{2--5}$	$\dfrac{2}{7}$		18.	(−2, −2)	(8, 3)	$\dfrac{3--2}{8--2}$	$\dfrac{1}{2}$
8.	(0, −4)	(5, 1)	$\dfrac{-4-1}{0-5}$	**1**		19.	(−3, 2)	(−2, −5)	$\dfrac{2--5}{-3--2}$	**−7**
9.	(0, 0)	(−6, 4)	$\dfrac{0-4}{0--6}$	$\dfrac{-2}{3}$		20.	(8, 4)	(2, 0)	$\dfrac{4-0}{8-2}$	$\dfrac{2}{3}$
10.	(0, −8)	(11, 3)	$\dfrac{-8-3}{0-11}$	**1**		21.	(−5, 7)	(−5, −8)		**undefined**
11.	(4, 3)	(−4, 5)	$\dfrac{3-5}{4--4}$	$\dfrac{-1}{4}$		22.	(3, 3)	(5, 5)	$\dfrac{5-3}{5-3}$	**1**

Graphs 5

Complete the chart. Graph by plotting 3 points. Draw the line using a straightedge, and cross both axes.

1.   y = 4x − 1

x	y
−3	**−13**
−2	**−9**
−1	**−5**
0	−1
1	**3**
2	7
3	**11**

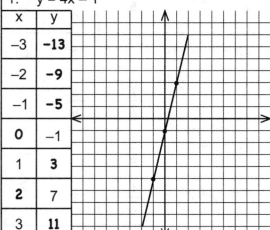

4.   y = −2x + 1

x	y
−3	**7**
−2	5
−1	**3**
0	**1**
1	**−1**
2	**−3**
3	**−5**

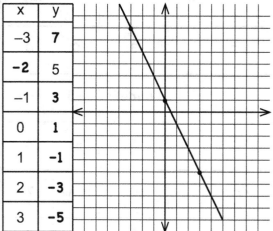

2.   y = 2x + 3

x	y
−2	**−1**
−1	**1**
0	**3**
1	5
2	7
3	**9**
4	**11**

5.   y = −3x − 2

x	y
−2	**4**
−1	**1**
0	**−2**
1	**−5**
2	−8
3	**−11**
4	**−14**

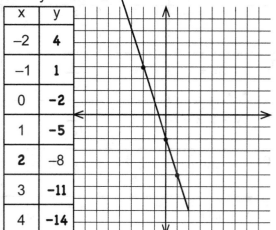

3.   y = 3x − 3

x	y
−2	**−9**
−1	**−6**
0	**−3**
1	0
2	3
3	**6**
4	**9**

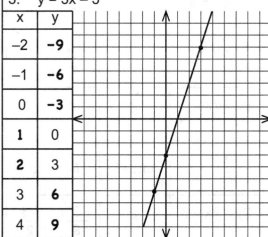

6.   y = −x + 4

x	y
−3	**7**
−2	6
−1	**5**
0	**4**
1	**3**
2	**2**
5	**−1**

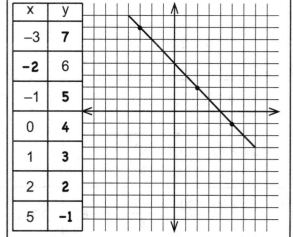

Graphs 6

Complete the chart for the linear equation.

	EQUATION	SLOPE	X-INT	Y-INT		EQUATION	SLOPE	X-INT	Y-INT
1.	$y = 2x - 2$	2	1	-2	17.	$y = 6x + 5$	6	$\frac{-5}{6}$	5
2.	$y = 3x$	3	0	0	18.	$3x + 7y = 21$	$\frac{-3}{7}$	7	3
3.	$x = -5$	undef	-5	none	19.	$8y = 4x - 3$	$\frac{1}{2}$	$\frac{3}{4}$	$\frac{-3}{8}$
4.	$y = 4x + 4$	4	-1	4	20.	$5y = x + 3$	$\frac{1}{5}$	-3	$\frac{3}{5}$
5.	$2y = 14$	0	none	7	21.	$x = 3y + 7$	$\frac{1}{3}$	7	$\frac{-7}{3}$
6.	$3y = 2x - 6$	$\frac{2}{3}$	3	-2	22.	$2x = y - 1$	2	$\frac{-1}{2}$	1
7.	$y = 7x - 3$	7	$\frac{3}{7}$	-3	23.	$y = 9$	0	none	9
8.	$y = 5x + 1$	5	$\frac{-1}{5}$	1	24.	$y - x = 4$	1	-4	4
9.	$3y = 12x - 7$	4	$\frac{7}{12}$	$\frac{-7}{3}$	25.	$x = 4y$	$\frac{1}{4}$	0	0
10.	$3x + 4y = 12$	$\frac{-3}{4}$	4	3	26.	$x + y = 10$	-1	10	10
11.	$y = 5x + 9$	5	$\frac{-9}{5}$	9	27.	$2y = 6x - 1$	3	$\frac{1}{6}$	$\frac{-1}{2}$
12.	$2x - 2y = 7$	1	$\frac{7}{2}$	$\frac{-7}{2}$	28.	$x + 5y = 4$	$\frac{-1}{5}$	4	$\frac{4}{5}$
13.	$2y = -9x - 2$	$\frac{-9}{2}$	$\frac{-2}{9}$	-1	29.	$4x + y = -5$	-4	$\frac{-5}{4}$	-5
14.	$x - y = 6$	1	6	-6	30.	$y = -7x$	-7	0	0
15.	$2x + 5y = 20$	$\frac{-2}{5}$	10	4	31.	$x = 6$	undef	6	none
16.	$5x + 3y = 30$	$\frac{-5}{3}$	6	10	32.	$2y = 8x - 3$	4	$\frac{3}{8}$	$\frac{-3}{2}$

Graphs 7

Write the equation of the line in slope-intercept form.

1. y = $\frac{1}{3}$ x − 1

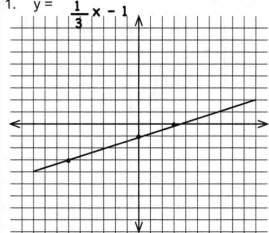

4. y = **2x + 6**

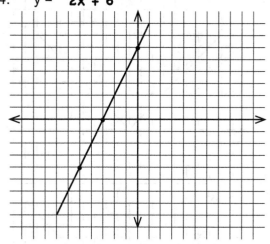

2. y = $\frac{-1}{4}$ x + 1

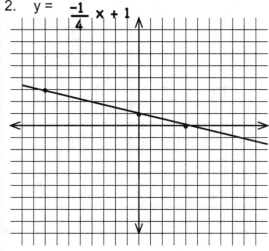

5. y = $\frac{2}{3}$ x − 2

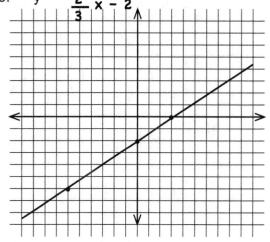

3. y = $\frac{-2}{5}$ x + 4

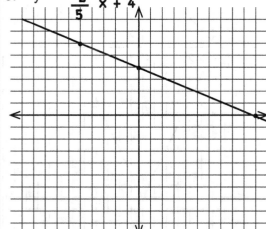

6. y = **−3x − 5**

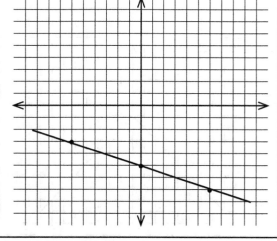

MAVA Math: Enhanced Skills Solutions Copyright © 2015 Marla Weiss

Graphs 8

Graph by plotting points.

1. y = |x|

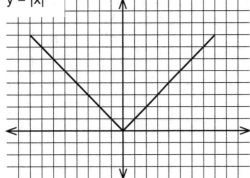

2. y = |x| + 1

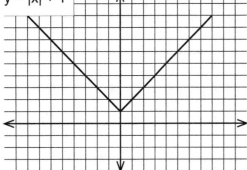

3. y = |x| − 2

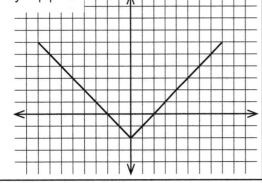

4. y = − |x|

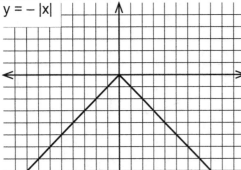

5. y = |3x|

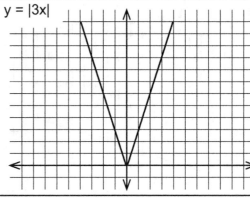

6. x = |y|

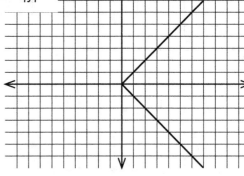

7. x = − |y|

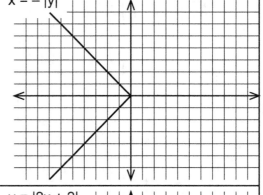

8. y = |2x + 2|

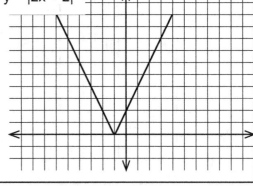

Graphs 9

Graph each line with a straightedge. Use the slope and y-intercept.

1. $y = 3x + 5$

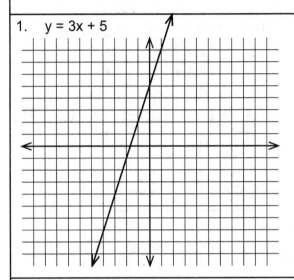

4. $y = \dfrac{-2}{3}x + 6$

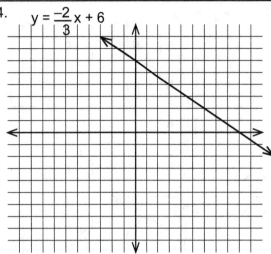

2. $y = -4x - 4$

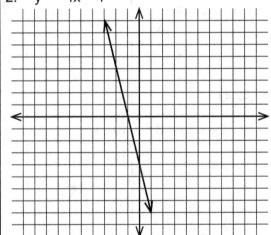

5. $y = \dfrac{3}{4}x - 3$

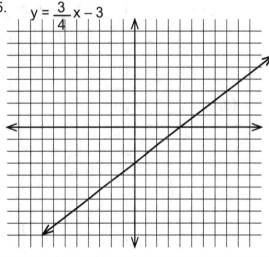

3. $y = 2x + 3$

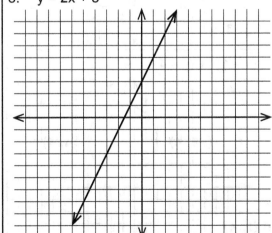

6. $y = \dfrac{-2}{5}x - 2$

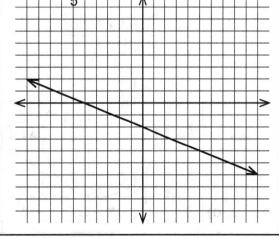

Graphs 10

Graph the circle. Identify the center (C). Label four lattice points on the circle.

1. $x^2 + y^2 = 16$

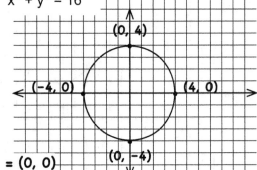

(0, 4)

(−4, 0) (4, 0)

(0, −4)

C = (0, 0)

5. $(x − 5)^2 + y^2 = 36$

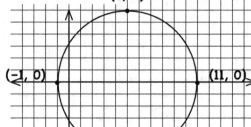

(5, 6)

(−1, 0) (11, 0)

(5, −6)

C = (5, 0)

2. $x^2 + y^2 = 36$

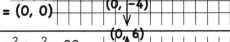

(0, 6)

(−6, 0) (6, 0)

(0, −6)

C = (0, 0)

6. $x^2 + (y − 2)^2 = 9$

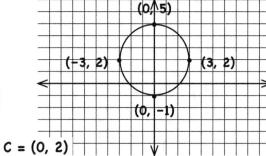

(0, 5)

(−3, 2) (3, 2)

(0, −1)

C = (0, 2)

3. $x^2 + (y − 1)^2 = 49$

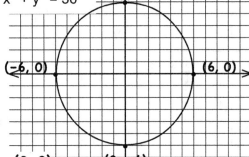

(0, 8)

(−7, 1) (7, 1)

C = (0, 1)

7. $(x − 2)^2 + (y − 3)^2 = 16$

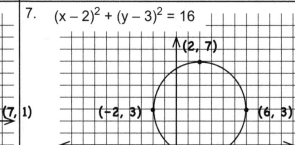

(2, 7)

(−2, 3) (6, 3)

(2, −1)

C = (2, 3)

4. $(x − 3)^2 + y^2 = 25$

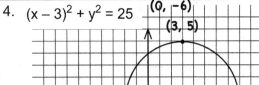

(0, −6)

(3, 5)

(−2, 0) (8, 0)

C = (3, 0) (3, −5)

8. $(x − 1)^2 + (y − 4)^2 = 9$

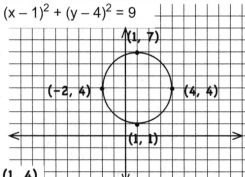

(1, 7)

(−2, 4) (4, 4)

(1, 1)

C = (1, 4)

Or calculate slope by delta y over delta x. Write a linear equation given two intercepts.	Graphs 11 Write a linear equation given slope and one intercept.
1. (0, 3) and (6, 0) $y = mx + 3$ $0 = 6m + 3$ $m = -1/2$ $y = \frac{-1}{2}x + 3$	9. slope = 5; y-intercept = (0, 4) $y = 5x + 4$
2. (0, 7) and (2, 0) $y = mx + 7$ $0 = 2m + 7$ $m = -7/2$ $y = \frac{-7}{2}x + 7$	10. slope = 4; x-intercept = (2, 0) $y = 4x + b$ $0 = 8 + b$ $b = -8$ $y = 4x - 8$
3. (0, -4) and (5, 0) $y = mx - 4$ $0 = 5m - 4$ $m = 4/5$ $y = \frac{4}{5}x - 4$	11. slope = 2/3; y-intercept = (0, -5) $y = \frac{2}{3}x - 5$
4. (0, 2) and (-8, 0) $y = mx + 2$ $0 = -8m + 2$ $m = 1/4$ $y = \frac{1}{4}x + 2$	12. slope = 2; x-intercept = (-3, 0) $y = 2x + b$ $0 = -6 + b$ $b = 6$ $y = 2x + 6$
5. (0, 5) and (6, 0) $y = mx + 5$ $0 = 6m + 5$ $m = -5/6$ $y = \frac{-5}{6}x + 5$	13. slope = -3/5; y-intercept = (0, 1) $y = \frac{-3}{5}x + 1$
6. (0, -2) and (7, 0) $y = mx - 2$ $0 = 7m - 2$ $m = 2/7$ $y = \frac{2}{7}x - 2$	14. slope = -3; x-intercept = (1, 0) $y = -3x + b$ $0 = -3 + b$ $b = 3$ $y = -3x + 3$
7. (0, 9) and (9, 0) $y = mx + 9$ $0 = 9m + 9$ $m = -1$ $y = -x + 9$	15. slope = 3/4; y-intercept = (0, -2) $y = \frac{3}{4}x - 2$
8. (0, 1) and (-7, 0) $y = mx + 1$ $0 = -7m + 1$ $m = 1/7$ $y = \frac{1}{7}x + 1$	16. slope = 6; x-intercept = (0, 0) $y = 6x$

Graphs 12

Write a linear equation given two points.	Write a linear equation given slope and one point.

1. (3, 6) and (4, 7)

m = 1	b = 3
y = x + b	
6 = 3 + b	**y = x + 3**

9. slope = 3; point = (1, 5)

y = 3x + b
5 = 3 + b
b = 2   **y = 3x + 2**

2. (4, 5) and (6, 9)

m = 2	b = −3
y = 2x + b	
5 = 8 + b	**y = 2x − 3**

10. slope = −4; point = (2, −7)

y = −4x + b
−7 = −8 + b
b = 1   **y = −4x + 1**

3. (3, −4) and (5, −6)

m = −1	b = −1
y = −x + b	
−4 = −3 + b	**y = −x − 1**

11. slope = 2/3; point = (6, 9)

y = 2/3x + b
9 = 4 + b
b = 5 $y = \dfrac{2}{3}x + 5$

4. (2, 4) and (4, 7)

m = 3/2	b = 1
4 = (3/2)(2) + b	
4 = 3 + b	$y = \dfrac{3}{2}x + 1$

12. slope = 3/4; point = (8, −3)

y = 3/4x + b
−3 = 6 + b
b = −9 $y = \dfrac{3}{4}x - 9$

5. (3, 7) and (6, 9)

m = 2/3	b = 5
7 = (2/3)(3) + b	
7 = 2 + b	$y = \dfrac{2}{3}x + 5$

13. slope = −1/5; point = (−10, 4)

y = −1/5x + b
4 = 2 + b
b = 2 $y = \dfrac{-1}{5}x + 2$

6. (6, −3) and (8, −8)

m = −5/2	b = 12
−3 = (−5/2)(6) + b	
−3 = −15 + b	$y = \dfrac{-5}{2}x + 12$

14. slope = 5/6; point = (12, 7)

y = 5/6x + b
7 = 10 + b
b = −3 $y = \dfrac{5}{6}x - 3$

7. (6, 3) and (9, 7)

m = 4/3	b = −5
3 = (4/3)(6) + b	
3 = 8 + b	$y = \dfrac{4}{3}x - 5$

15. slope = −3/7; point = (−14, 9)

y = −3/7x + b
9 = 6 + b
b = 3 $y = \dfrac{-3}{7}x + 3$

8. (1, −10) and (−1, −2)

m = −4	b = −6
y = −4x + b	
−2 = 4 + b	**y = −4x − 6**

16. slope = 7/8; point = (8, 6)

y = 7/8x + b
6 = 7 + b
b = −1 $y = \dfrac{7}{8}x - 1$

Greatest Common Factor 1

Find the GCF mentally.

1. 45 and 81 **9**	17. 11, 13, and 31 **1**	33. 36 and 70 **2**	49. 49 and 50 **1**
2. 12 and 72 **12**	18. 550 and 600 **50**	34. 8, 18, and 28 **2**	50. 21 and 91 **7**
3. 17 and 64 **1**	19. 230 and 590 **10**	35. 6, 18, and 42 **6**	51. 82 and 86 **2**
4. 28 and 70 **14**	20. 24, 30, and 39 **3**	36. 120 and 330 **30**	52. 624 and 625 **1**
5. 18 and 32 **2**	21. 22 and 451 **11**	37. 50, 90, and 360 **10**	53. 21 and 1001 **7**
6. 18, 45, and 90 **9**	22. 100 and 150 **50**	38. 8, 27, and 64 **1**	54. 91 and 1001 **91**
7. 5, 7, and 20 **1**	23. 56 and 140 **28**	39. 121 and 385 **11**	55. 84 and 88 **4**
8. 6, 9, and 15 **3**	24. 12, 15, and 16 **1**	40. 18, 30, and 66 **6**	56. 60 and 75 **15**
9. 24 and 42 **6**	25. 56 and 112 **56**	41. 31 and 62 **31**	57. 220 and 242 **22**
10. 26 and 56 **2**	26. 38, 57, and 190 **19**	42. 8, 10, and 12 **2**	58. 150 and 360 **30**
11. 34 and 51 **17**	27. 22, 33, and 55 **11**	43. 32 and 160 **32**	59. 200 and 260 **20**
12. 36, 54, and 90 **18**	28. 21 and 56 **7**	44. 5, 6, and 21 **1**	60. 33 and 363 **33**
13. 38 and 56 **2**	29. 5, 10, and 15 **5**	45. 4, 11, and 16 **1**	61. 120 and 330 **30**
14. 72 and 90 **18**	30. 27 and 32 **1**	46. 5, 6, and 10 **1**	62. 27 and 51 **3**
15. 6, 7, and 55 **1**	31. 12, 30, and 42 **6**	47. 18 and 45 **9**	63. 198 and 374 **22**
16. 30, 42, and 48 **6**	32. 9, 15, and 33 **3**	48. 44, 66, and 88 **22**	64. 35, 42, and 77 **7**

Greatest Common Factor 2

Find the greatest common factor.

1. $GCF(a^3b^4c^5 , a^2b^6c^3)$

    $a^2b^4c^3$

2. $GCF(16x^4y^5z^3 , 48x^3y^6z^3)$

    $16x^3y^5z^3$

3. $GCF(5w^7x^4y^5z^3 , 7w^2x^5y^4z)$

    $w^2x^4y^4z$

4. $GCF(5a^3s^4 , 6a^4s^3 , 12a^2s^3)$

    a^2s^3

5. $GCF(35a^2b^2c^2 , 42a^3b^5)$

    $7a^2b^2$

6. $GCF(6f^7g^3h^6 , 10f^4g^5h^8)$

    $2f^4g^3h^6$

7. $GCF(36d^5e^2f^6 , 144d^3e^6f^8)$

    $36d^3e^2f^6$

8. $GCF(5x^4y^5z^6 , 51x^3y^2z^8)$

    $x^3y^2z^6$

9. $GCF(12rst , 16r^2s^5t^4 , 8r^3s^3t^6)$

    $4rst$

10. $GCF(6b^3d^5g^6 , 9b^4d^5h^2)$

    $3b^3d^5$

11. $GCF(4d^2e^3 , 5d^4e^4 , 6d^6e^5)$

    d^2e^3

12. $GCF(15x^9y^6z^5 , 70x^5y^6z^9)$

    $5x^5y^6z^5$

13. $GCF(11a^4b^4c^3 , 44a^9b^3c^6)$

    $11a^4b^3c^3$

14. $GCF(5g^7s^6 , 10g^6s^7 , 15g^8s^3)$

    $5g^6s^3$

15. $GCF(36p^4q^2r^8 , 90p^5q^2r^3)$

    $18p^4q^2r^3$

16. $GCF(14a^5bc^7 , 63a^3b^9)$

    $7a^3b$

17. $GCF(14f^7g^6h^4 , 16f^3g^4h^2)$

    $2f^3g^4h^2$

18. $GCF(16r^4st^2 , 28r^6s^3t^5 , 4r^2s^3t)$

    $4r^2st$

19. $GCF(15x^2y^3z^4 , 20w^2y^4z^3)$

    $5y^3z^3$

20. $GCF(14p^9q^2r^4 , 70p^5q^5r^6)$

    $14p^5q^2r^4$

21. $GCF(6m^3n^6p^7 , 15m^2n^4p^8)$

    $3m^2n^4p^7$

22. $GCF(6e^3f^5g^5 , 8e^6f^5g^7)$

    $2e^3f^5g^5$

Greatest Common Factor 3

Find the greatest common factor by prime factorization.

---

1.   130   286

13 x 10   26 x 11
2 x 5 x 13   2 x 11 x 13
GCF = 2 x 13 = **26**

9.   780   910

2 x 39 x 10   91 x 10
2^2 x 3 x 5 x 13   2 x 5 x 7 x 13
GCF = 10 x 13 = **130**

---

2.   405   495

5 x 81   45 x 11
3^4 x 5 3^2 x 5 x 11
GCF = 9 x 5 = **45**

10.   9000   2772

9 x 10 x 10 x 10   9 x 308
2^3 x 3^2 x 5^3 9 x 4 x 77
GCF = 4 x 9 = **36** 2^2 x 3^2 x 7 x 11

---

3.   819   1260

9 x 91   9 x 14 x 10
3^2 x 7 x 13 2^2 x 3^2 x 5 x 7
GCF = 9 x 7 = **63**

11.   3850   2205

35 x 11 x 10   5 x 441
2 x 5^2 x 7 x 11   5 x 9 x 49
GCF = 5 x 7 = **35** 3^2 x 5 x 7^2

---

4.   1078   2450

11 x 98   5 x 49 x 10
2 x 7^2 x 11   2 x 5^2 x 7^2
GCF = 2 x 49 = **98**

12.   80   96   144

8 x 10   6 x 16   12 x 12
2^4 x 5 2^5 x 3 2^4 x 3^2
GCF = **16**

---

5.   4200   5000

6 x 7 x 10 x 10   5 x 10 x 10 x 10
2^3 x 3 x 5^2 x 7 2^3 x 5^4
GCF = 8 x 25 = **200**

13.   78   102   726

6 x 13   2 x 51   3 x 242
2 x 3 x 13   2 x 3 x 17   3 x 11 x 22
GCF = 2 x 3 = **6**   2 x 3 x 11^2

---

6.   341   1001

11 x 31   7 x 11 x 13

GCF = **11**

14.   24   54   72

8 x 3   2 x 27   8 x 9
2^3 x 3   2 x 3^3 2^3 x 3^2
GCF = 2 x 3 = **6**

---

7.   396   2070

36 x 11   207 x 10
2^2 x 3^2 x 11   9 x 23 x 2 x 5
GCF = 2 x 9 = **18**   2 x 3^2 x 5 x 23

15.   132   165   297

12 x 11   15 x 11   27 x 11
2^2 x 3 x 11   3 x 5 x 11 3^3 x 11
GCF = 3 x 11 = **33**

---

8.   2700   1080

27 x 10 x 10   108 x 10
2^2 x 3^3 x 5^2 9 x 12 x 2 x 5
GCF = 4 x 5 x 27 = **540** 2^3 x 3^3 x 5

16.   84   168   210

4 x 21   8 x 21   21 x 10
2^2 x 3 x 7 2^3 x 3 x 7   2 x 3 x 5 x 7
GCF = 6 x 7 = **42**

GCF(m,n) x LCM(m,n) = mn **Greatest Common Factor 4**	
Given one of two numbers m and their GCF and LCM, find the second number n.	*Evaluate as indicated.*

1. m = 20
   GCF(m, n) = 10
   LCM(m, n) = 60
10 x 60 = 20n
n = **30**
	9. LCM(GCF(45, 63), GCF(36, 42))
	
	   LCM(9, 6) = **18**

2. m = 60
   GCF(m, n) = 30
   LCM(m, n) = 4620
30 x 4620 = 60n
4620 = 2n
n = **2310**
	10. LCM(GCF(28, 42), GCF(45, 60))
	
	   LCM(14, 15) = **210**

3. m = 154
   GCF(m, n) = 7
   LCM(m, n) = 8008
7 x 8008 = 154n
8008 = 22n
11n = 4004
n = **364**
	11. GCF(LCM(30, 75), LCM(70, 45))
	
	   GCF(150, 630) = **30**

4. m = 48
   GCF(m, n) = 12
   LCM(m, n) = 432
12 x 432 = 48n
432 = 4n
n = **108**
	12. LCM(GCF(120, 168), GCF(28, 92))
	
	   LCM(24, 4) = **24**

5. m = 288
   GCF(m, n) = 18
   LCM(m, n) = 1440
18 x 1440 = 288n
1440 = 16n
4n = 360
n = **90**
	13. GCF(LCM(91, 49), LCM(222, 66))
	
	   GCF(91 x 7, 222 x 11) = **1**

6. m = 72
   GCF(m, n) = 24
   LCM(m, n) = 3960
24 x 3960 = 72n
3960 = 3n
n = **1320**
	14. LCM(GCF(147, 210), GCF(56, 64))
	
	   LCM(21, 8) = **168**

7. m = 210
   GCF(m, n) =15
   LCM(m, n) = 7350
15 x 7350 = 210n
7350 = 14n
2n = 1050
n = **525**
	15. GCF(LCM(30, 99), LCM(55, 60))
	
	   GCF(99 x 10, 60 x 11) = **330**

8. m = 198
   GCF(m, n) = 18
   LCM(m, n) = 4158
18 x 4158 = 198n
4158 = 11n
n = **378**
	16. LCM(GCF(225, 825), GCF(70, 84))
	
	   LCM(75, 14) = **1050**

Historical Math 1

Use the Sieve of Eratosthenes to determine all prime numbers listed. Circle a prime; then cross off all of its multiples. Proceed to the next prime, and repeat.

1	2	3	4	5	6	7	8	9	10	11	12
13	14	15	16	17	18	19	20	21	22	23	24
25	26	27	28	29	30	31	32	33	34	35	36
37	38	39	40	41	42	43	44	45	46	47	48
49	50	51	52	53	54	55	56	57	58	59	60
61	62	63	64	65	66	67	68	69	70	71	72
73	74	75	76	77	78	79	80	81	82	83	84
85	86	87	88	89	90	91	92	93	94	95	96
97	98	99	100	101	102	103	104	105	106	107	108
109	110	111	112	113	114	115	116	117	118	119	120
121	122	123	124	125	126	127	128	129	130	131	132
133	134	135	136	137	138	139	140	141	142	143	144
145	146	147	148	149	150	151	152	153	154	155	156
157	158	159	160	161	162	163	164	165	166	167	168
169	170	171	172	173	174	175	176	177	178	179	180
181	182	183	184	185	186	187	188	189	190	191	192
193	194	195	196	197	198	199	200	201	202	203	204
205	206	207	208	209	210	211	212	213	214	215	216
217	218	219	220	221	222	223	224	225	226	227	228
229	230	231	232	233	234	235	236	237	238	239	240
241	242	243	244	245	246	247	248	249	250	251	252
253	254	255	256	257	258	259	260	261	262	263	264

If YES, has exactly 0 or 2 odd vertices.

Historical Math 2

Answer YES or NO as to whether the figure is an Euler graph (may be traced drawing each line exactly once without lifting the pencil). Mark the start and stop.

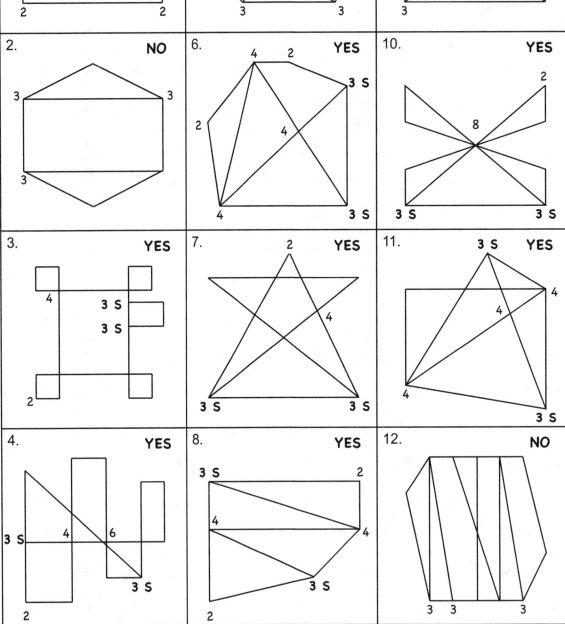

Historical Math 3

Verify Goldbach's conjecture (through 130) that every even integer greater than 2 is the sum of two primes. Answers may vary.

1. 4	2 + 2	17. 36	5 + 31	33. 68	7 + 61	49. 100	3 + 97
2. 6	3 + 3	18. 38	7 + 31	34. 70	3 + 67	50. 102	5 + 97
3. 8	3 + 5	19. 40	3 + 37	35. 72	5 + 67	51. 104	7 + 97
4. 10	5 + 5	20. 42	5 + 37	36. 74	3 + 71	52. 106	3 + 103
5. 12	5 + 7	21. 44	3 + 41	37. 76	5 + 71	53. 108	5 + 103
6. 14	7 + 7	22. 46	5 + 41	38. 78	7 + 71	54. 110	3 + 107
7. 16	5 + 11	23. 48	5 + 43	39. 80	7 + 73	55. 112	5 + 107
8. 18	7 + 11	24. 50	7 + 43	40. 82	3 + 79	56. 114	7 + 107
9. 20	7 + 13	25. 52	5 + 47	41. 84	5 + 79	57. 116	7 + 109
10. 22	11 + 11	26. 54	7 + 47	42. 86	3 + 83	58. 118	5 + 113
11. 24	5 + 19	27. 56	3 + 53	43. 88	5 + 83	59. 120	7 + 113
12. 26	13 + 13	28. 58	5 + 53	44. 90	17 + 73	60. 122	13 + 109
13. 28	11 + 17	29. 60	7 + 53	45. 92	19 + 73	61. 124	17 + 107
14. 30	11 + 19	30. 62	3 + 59	46. 94	5 + 89	62. 126	19 + 107
15. 32	3 + 29	31. 64	3 + 61	47. 96	7 + 89	63. 128	19 + 109
16. 34	17 + 17	32. 66	5 + 61	48. 98	19 + 79	64. 130	29 + 101

Historical Math 4

Verify the first 4 pairs of amicable numbers by showing the sum of the proper divisors of each equals the other.

Verify the first 4 perfect numbers by summing proper divisors.

1. 220 1 + 2 + 4 + 5 + 10 + 11 + 20 + 22 + 44 + 55 + 110 = 284 2 and 11 are divisors implies 22 is divisor, etc.	284 1 + 2 + 4 + 71 + 142 = 220	5. 6 1 + 2 + 3 = 6
2. 1184 1 + 2 + 4 + 8 + 16 + 32 + 37 + 74 + 148 + 296 + 592 = 1210	1210 1 + 2 + 5 + 10 + 11 + 22 + 55 + 110 + 121 + 242 + 605 = 1184	6. 28 1 + 2 + 4 + 7 + 14 = 28
3. 2620 1 + 2 + 4 + 5 + 10 + 20 + 131 + 262 + 524 + 655 + 1310 = 2924	2924 1 + 2 + 4 + 17 + 34 + 43 + 68 + 86 + 172 + 731 + 1462 = 2620	7. 496 1 + 2 + 4 + 8 + 16 + 31 + 62 + 124 + 248 = 496
4. 5020 1 + 2 + 4 + 5 + 10 + 20 + 251 + 502 + 1004 + 1255 + 2510 = 5564	5564 1 + 2 + 4 + 13 + 26 + 52 + 107 + 214 + 428 + 1391 + 2782 = 5020	8. 8128 1 + 2 + 4 + 8 + 16 + 32 + 64 + 127 + 254 + 508 + 1016 + 2032 + 4064 = 8128

Inequalities 1

Solve.

1. $5x < 35$ **$x < 7$**	12. $9 - x \geq -11$ $-x \geq -20$ **$x \leq 20$**	23. $3 - 10x \geq 36 + x$ $-11x \geq 33$ **$x \leq -3$**
2. $3x \leq 33$ **$x \leq 11$**	13. $10 - 3x < 40$ $-3x < 30$ **$x > -10$**	24. $-7x \leq x + 88$ $-8x \leq 88$ **$x \geq -11$**
3. $-7x > -63$ **$x < 9$**	14. $3x - 1 \leq -40$ $3x \leq -39$ **$x \leq -13$**	25. $9x - 2 \geq -12 - x$ $10x \geq -10$ **$x \geq -1$**
4. $13x \geq -52$ **$x \geq -4$**	15. $11x + 4 \leq -28 + 3x$ $8x \leq -32$ **$x \leq -4$**	26. $100 > 37 - 7x$ $63 > -7x$ **$x > -9$**
5. $2x + 3 < -55$ $2x < -58$ **$x < -29$**	16. $11 - x > 44$ $-x > 33$ **$x < -33$**	27. $5x + 5 < x - 43$ $4x < -48$ **$x < -12$**
6. $-x - 2 \leq 24$ $-x \leq 26$ **$x \geq -26$**	17. $8x + 5 \leq -27$ $8x \leq -32$ **$x \leq -4$**	28. $3x - 8 \geq 32 - 5x$ $8x \geq 40$ **$x \geq 5$**
7. $4x + 6 > 7x$ $6 > 3x$ **$x < 2$**	18. $-6x + 1 \geq -71$ $-6x \geq -72$ **$x \leq 12$**	29. $12 - 5x < -48 + x$ $-6x < -60$ **$x > 10$**
8. $6x + 11 < 3x - 1$ $3x < -12$ **$x < -4$**	19. $4x < -x + 55$ $5x < 55$ **$x < 11$**	30. $8 - 7x > -48$ $-7x > -56$ **$x < 8$**
9. $-2x + 7 \leq 39 - x$ $-x \leq 32$ **$x \geq -32$**	20. $9x - 4 \leq -28 + x$ $8x \leq -24$ **$x \leq -3$**	31. $3 - 2x \leq 45 + x$ $-3x \leq 42$ **$x \geq -14$**
10. $8x + 5 < -61$ $8x < -66$ **$x < \dfrac{-33}{4}$**	21. $4 - x \geq x - 22$ $-2x \geq -26$ **$x \leq 13$**	32. $9x + 5 > 65$ $9x > 60$ **$x > \dfrac{20}{3}$**
11. $-10x - 8 \leq -73$ $-10x \leq -65$ **$x \geq \dfrac{13}{2}$**	22. $3x - 7 \leq 35 + 9x$ $-6x \leq 42$ **$x \geq -7$**	33. $-5x - 4 < 64 + 3x$ $-8x < 68$ **$x > \dfrac{-17}{2}$**

194

Inequalities 2

Solve.

1. $\frac{1}{6}x \geq -12$

   $x \geq -72$

2. $\frac{2}{5}x + 1 < 31$

   $\frac{2}{5}x < 30$

   $x < 75$

3. $\frac{3}{8}x - 6 \leq -30$

   $\frac{3}{8}x \leq -24$

   $x \leq -64$

4. $2 - \frac{5}{7}x > -43$

   $\frac{-5}{7}x > -45$

   $x < 63$

5. $10 - \frac{5}{9}x < 25$

   $\frac{-5}{9}x < 15$

   $x > -27$

6. $\frac{1}{7}x - 2 \geq 9$

   $\frac{1}{7}x \geq 11$

   $x \geq 77$

7. $\frac{2}{3}x + 5 < -13$

   $\frac{2}{3}x < -18$

   $x < -27$

8. $\frac{11}{20}x - 2 \geq 86$

   $\frac{11}{20}x \geq 88$

   $x \geq 160$

9. $\frac{3}{4}(8x - 4) \geq 21$

   $6x - 3 \geq 21$

   $6x \geq 24$

   $x \geq 4$

10. $\frac{2}{7}(14x + 21) < -5$

    $4x + 6 < -5$

    $4x < -11 \quad x < \frac{-11}{4}$

11. $\frac{-3}{14}(42x - 14) > 11$

    $-9x + 3 > 11$

    $-9x > 8 \quad x > \frac{-8}{9}$

12. $\frac{11}{15}(45 + 30x) \leq 44$

    $33 + 22x \leq 44$

    $22x \leq 11 \quad x \leq \frac{1}{2}$

13. $\frac{-4}{21}(63x) \geq 22 - x$

    $-12x \geq 22 - x$

    $-11x \geq 22$

    $x \leq -2$

14. $\frac{5}{9}(18x - 27) \leq 5x$

    $10x - 15 \leq 5x$

    $5x \leq 15$

    $x \leq 3$

15. $\frac{3}{5}(45x + 20) \geq 3x$

    $27x + 12 \geq 3x$

    $24x \geq -12 \quad x \geq \frac{-1}{2}$

16. $\frac{-5}{13}(13 + 26x) > -25$

    $-5 - 10x > -25$

    $-10x > -20$

    $x < 2$

17. $\frac{2}{3}(5x - 1) \geq 11$

    $10x - 2 \geq 33$

    $10x \geq 35$

    $x \geq 3.5$

18. $\frac{4}{5}(8x + 3) < -4$

    $32x + 12 < -20$

    $32x < -32$

    $x < -1$

19. $\frac{-3}{7}(5x - 8) > 12$

    $-15x + 24 > 84$

    $-15x > 60$

    $x < -4$

20. $\frac{5}{8}(9 - 5x) \leq 15$

    $45 - 25x \leq 120$

    $-25x \leq 75$

    $x \geq -3$

21. $\frac{-3}{4}(10x) \geq -14 - 3x$

    $-30x \geq -56 - 12x$

    $-18x \geq -56 \quad x \leq \frac{28}{9}$

22. $\frac{5}{6}(9x - 18) \leq 15x$

    $45x - 90 \leq 90x$

    $-45x \leq 90$

    $x \geq -2$

23. $\frac{1}{2}(17x + 18) \geq 4x$

    $17x + 18 \geq 8x$

    $9x \geq -18$

    $x \geq -2$

24. $\frac{-4}{9}(6 + 20x) > -16$

    $-24 - 80x > -144$

    $-80x > -120 \quad x < \frac{3}{2}$

MAVA Math: Enhanced Skills Solutions Copyright © 2015 Marla Weiss

Integers 1

Add or subtract as indicated. Use mental math.

1. 15 − 30 **−15**	17. 11 − 20 **−9**	33. 50 − 90 **−40**	49. −34 + 87 **53**
2. −20 + (−11) **−31**	18. −35 + −55 **−90**	34. −15 + −46 **−61**	50. −45 + −54 **−99**
3. −45 + 45 **0**	19. 29 − 50 **−21**	35. 25 − 66 **−41**	51. 20 − 72 **−52**
4. 13 − 18 **−5**	20. 0 − 99 **−99**	36. −23 + −67 **−90**	52. 33 + −79 **−46**
5. −30 − (−12) **−18**	21. −23 − (−54) **31**	37. 20 − 30 **−10**	53. 33 − 79 **−46**
6. −16 − 16 **−32**	22. 53 − 70 **−17**	38. −53 + −46 **−99**	54. −33 + −46 **−79**
7. 14 − 21 **−7**	23. 88 + −77 **11**	39. −12 + −79 **−91**	55. 46 + −79 **−33**
8. −89 + 13 **−76**	24. −76 + 78 **2**	40. 11 − 16 **−5**	56. 79 − 46 **33**
9. 89 − 13 **76**	25. −17 + −18 **−35**	41. −19 − 15 **−34**	57. −47 − 27 **−74**
10. −13 + 89 **76**	26. 52 − 98 **−46**	42. 40 − 48 **−8**	58. 30 − 88 **−58**
11. 13 − 89 **−76**	27. 11 − 13 **−2**	43. −18 + −19 **−37**	59. −26 + −36 **−62**
12. 41 − (−47) **88**	28. 19 − 16 **3**	44. −14 + −53 **−67**	60. −28 + −28 **−56**
13. −34 − 36 **−70**	29. 16 − 19 **−3**	45. 25 − 35 **−10**	61. 37 − 70 **−33**
14. 84 + (−52) **32**	30. −19 + −16 **−35**	46. −55 + 59 **4**	62. −19 + 89 **70**
15. 45 − 50 **−5**	31. −16 + 19 **3**	47. 18 − 36 **−18**	63. 30 − 55 **−25**
16. −51 + 51 **0**	32. −19 + 16 **−3**	48. −15 + 75 **60**	64. 42 + −90 **−48**

Integers 2

Multiply or divide as indicated. Use mental math.

1. 19 x (– 2) **-38**	17. (–101)(–9) **909**	33. –121 ÷ 11 **-11**
2. –17 x (–4) **68**	18. (–20)(–30)(–40) **-24,000**	34. 64 ÷ (–16) **-4**
3. –20 x 60 **-1200**	19. –4907 ÷ 7 **-701**	35. (–11)(–16) **176**
4. –51 ÷ 3 **-17**	20. –21 x 5 **-105**	36. –45 ÷ (–3) **15**
5. (–4)(–7)(–5) **-140**	21. (–35)(–3) **105**	37. (–99) ÷ (–9) **11**
6. (–16)(–3) **48**	22. (–5)(6)(–25) **750**	38. –169 ÷ 13 **-13**
7. (–12)(–7)(–5) **-420**	23. –17 x 3 **-51**	39. –810 ÷ (–9) **90**
8. (–8)(6)(–5) **240**	24. –600 ÷ 15 **-40**	40. –743 x –10 **7430**
9. –93 ÷ 3 **-31**	25. –560 ÷ (–70) **8**	41. –369 ÷ 3 **-123**
10. –45 x 11 **-495**	26. –6 x 13 **-78**	42. –816 ÷ (–4) **204**
11. (–880) ÷ (–11) **80**	27. –824 ÷ (–4) **206**	43. –11 x –27 **297**
12. –9036 ÷ 9 **-1004**	28. (–53)(–11) **583**	44. (–100)(–10) **1000**
13. –150 ÷ (–30) **5**	29. 1000 ÷ (–5) **-200**	45. (–12)(–12) **144**
14. 102 ÷ (–3) **-34**	30. (–15)(–7) **105**	46. 444 ÷ (–4) **-111**
15. –105 ÷ 5 **-21**	31. 25 x (–8) **-200**	47. –36 ÷ (–2) **18**
16. –(33 x 5) **-165**	32. –864 ÷ 8 **-108**	48. 43 x (–11) **-473**

Integers 3

Find two consecutive positive integers with the given description.	*Find consecutive integers with the given description.*
1. sum is 37 **18 and 19**	20. 2 consecutive even, sum is 54 **26, 28**
2. half their product is 45 **9 and 10**	21. 5 consecutive odd, sum is 15 **−1, 1, 3, 5, 7**
3. triple their greater quotient is 4 **3 and 4**	22. 5 consecutive even, sum is 0 **−4, −2, 0, 2, 4**
4. twice their sum is 22 **5 and 6**	23. 3 consecutive odd, sum is 69 **21, 23, 25**
5. sum is 53 **26 and 27**	24. 4 consecutive even, sum is 36 **6, 8, 10, 12**
6. half their product is 66 **11 and 12**	25. 6 consecutive, sum is −9 **−4, −3, −2, −1, 0, 1**
7. average is 31.5 **31 and 32**	26. 6 consecutive odd, sum is 0 **−5, −3, −1, 1, 3, 5**
8. triple their sum is 105 **17 and 18**	27. 9 consecutive, sum is −9 **−5, −4, −3, −2, −1, 0, 1, 2, 3**
9. product is 2550 **50 and 51**	28. 6 consecutive even, sum is −6 **−6, −4, −2, 0, 2, 4**
10. double their lesser quotient is 1 **1 and 2**	29. 8 consecutive, sum is 12 **−2, −1, 0, 1, 2, 3, 4, 5**
11. half their sum is 15.5 **15 and 16**	30. 8 consecutive odd, sum is −16 **−9, −7, −5, −3, −1, 1, 3, 5**
12. average is 89.5 **89 and 90**	31. 5 consecutive odd, sum is 75 **11, 13, 15, 17, 19**
13. product is 650 **25 and 26**	32. 7 consecutive odd, sum is 21 **−3, −1, 1, 3, 5, 7, 9**
14. sum is 151 **75 and 76**	33. 4 consecutive even, sum is 140 **32, 34 36, 38**
15. double their average is 27 **13 and 14**	34. 8 consecutive, sum is −4 **−4, −3, −2, −1, 0, 1, 2, 3**
16. product is 600 **24 and 25**	35. 5 consecutive odd, sum is 465 **89, 91, 93, 95, 97**
17. half their product is 55 **10 and 11**	36. 9 consecutive, sum is −18 **−6, −5, −4, −3, −2, −1, 0, 1, 2**
18. product is 210 **14 and 15**	37. 3 consecutive odd, sum is 153 **49, 51, 53**
19. triple their sum is 135 **22 and 23**	38. 7 consecutive even, sum is −14 **−8, −6, −4, −2, 0, 2, 4**

Integers 4

Answer as indicated.

1. Find 4 consecutive odd integers such that the sum of the two greatest subtracted from twice the sum of the two least is 76.
$x, x + 2, x + 4, x + 6$
$2(2x + 2) - (2x + 10) = 76$
$2x - 6 = 76$
$x = 41$
41, 43, 45, 47

2. Find 5 consecutive odd integers such that the 4th is the sum of the 1st and twice the 3rd.
$x, x + 2, x + 4, x + 6, x + 8$
$x + 6 = x + 2(x + 4)$
$6 = 2x + 8$
$x = -1$
−1, 1, 3, 5, 7

3. Find 3 consecutive multiples of 4 such that triple the 3rd minus twice the 2nd is 4 less than twice the 1st.
$x, x + 4, x + 8$
$(3x + 24) - (2x + 8) = 2x - 4$
$x + 16 = 2x - 4$
$x = 20$
20, 24, 28

4. Find 4 consecutive integers such that 4 times the 3rd decreased by triple the 2nd is 39.
$x, x + 1, x + 2, x + 3$
$4x + 8 - (3x + 3) = 39$
$x + 5 = 39$
$x = 34$
34, 35, 36, 37

5. Find 3 consecutive even integers such that the sum of the least and greatest is 108.
$x, x + 2, x + 4$
$x + x + 4 = 108$
$2x = 104$
$x = 52$
52, 54, 56

6. Find 5 consecutive multiples of 3 such that twice the sum of the 2nd and 4th is 30 more than the sum of the greatest and least.
$x, x + 3, x + 6, x + 9, x + 12$
$2(2x + 12) = 30 + (2x + 12)$
$2x = 18$
$x = 9$
9, 12, 15, 18, 21

7. Find 6 consecutive integers such that twice the sum of the 2 least minus the sum of the 2 greatest is 33.
$x, x + 1, x + 2, x + 3, x + 4, x + 5$
$2(2x + 1) - (2x + 9) = 33$
$2x - 7 = 33$
$x = 20$
20, 21, 22, 23, 24, 25

8. Find 4 consecutive even integers such that the sum of the 3 greatest is 90 more than the sum of the 2 least.
$x, x + 2, x + 4, x + 6$
$(3x + 12) = 90 + 2x + 2$
$x = 80$
80, 82, 84, 86

9. Find 4 consecutive multiples of 5 such that the sum of the 3 least is 20 more than twice the 4th.
$x, x + 5, x + 10, x + 15$
$3x + 15 = 20 + 2(x + 15)$
$3x = 5 + 2x + 30$
$x = 35$
35, 40, 45, 50

10. Find two consecutive even integers such that twice the greater is 22 more than triple the lesser.
$x, x + 2$
$2(x + 2) = 22 + 3x$
$-18 = x$
−18, −16

Interest 1

Complete the chart of annual simple interest by mental math.

	PRINCIPAL	RATE	INTEREST		PRINCIPAL	RATE	INTEREST
1.	$100	6%	**$6**	20.	$1000	18%	**$180**
2.	**$200**	7%	$14	21.	**$450**	10%	$45
3.	$250	10%	**$25**	22.	$350	20%	**$70**
4.	**$450**	6%	$27	23.	$275	**8%**	$22
5.	**$300**	8%	$24	24.	**$9000**	5%	$450
6.	$300	**11%**	$33	25.	$6000	**3%**	$180
7.	$500	**15%**	$75	26.	$5200	**2%**	$104
8.	$400	17%	**$68**	27.	$6100	11%	**$671**
9.	$5000	**7%**	$350	28.	$3100	**3%**	$93
10.	$1000	5%	**$50**	29.	$4860	25%	**$1215**
11.	**$660**	10%	$66	30.	**$4700**	1%	$47
12.	$850	**12%**	$102	31.	**$4400**	5%	$220
13.	**$820**	15%	$123	32.	**$2100**	6%	$126
14.	$600	20%	**$120**	33.	$5060	15%	**$759**
15.	$900	4%	**$36**	34.	**$5100**	8%	$408
16.	$150	3%	**$4.50**	35.	$4100	7%	**$287**
17.	$3000	**2%**	$60	36.	$7500	**11%**	$825
18.	**$8100**	8%	$648	37.	**$2040**	15%	$306
19.	$4000	**9%**	$360	38.	$8400	**4%**	$336

Interest 2

Find the simple interest by writing and simplifying an I=PRT equation.

1. on $1200 at 8% for 7 months

$$I = \frac{1200 \cdot 8 \cdot 7}{100 \cdot 12} \qquad \textbf{\$56}$$

2. on $6000 at 6% for 3 months

$$I = \frac{6000 \cdot 6 \cdot 3}{100 \cdot 12} \qquad \textbf{\$90}$$

3. on $2000 at 8% for 9 months

$$I = \frac{2000 \cdot 8 \cdot 9}{100 \cdot 12} \qquad \textbf{\$120}$$

4. on $24,000 at 5% for 15 months

$$I = \frac{24,000 \cdot 5 \cdot 15}{100 \cdot 12} \qquad \textbf{\$1500}$$

5. on $6000 at 6% for 3 months

$$I = \frac{6000 \cdot 6 \cdot 3}{100 \cdot 12} \qquad \textbf{\$90}$$

6. on $4800 at 4% for 9 months

$$I = \frac{4800 \cdot 4 \cdot 9}{100 \cdot 12} \qquad \textbf{\$144}$$

7. on $5500 at 9% for 8 months

$$I = \frac{5500 \cdot 9 \cdot 8}{100 \cdot 12} \qquad \textbf{\$330}$$

8. on $3200 at 10% for 9 months

$$I = \frac{3200 \cdot 10 \cdot 9}{100 \cdot 12} \qquad \textbf{\$240}$$

9. on $2600 at 9% for 6 months

$$I = \frac{2600 \cdot 9 \cdot 6}{100 \cdot 12} \qquad \textbf{\$117}$$

10. on $70,000 at 5% for 3 months

$$I = \frac{70,000 \cdot 5 \cdot 3}{100 \cdot 12} \qquad \textbf{\$875}$$

11. on $3600 at 8% for 8 months

$$I = \frac{3600 \cdot 8 \cdot 8}{100 \cdot 12} \qquad \textbf{\$192}$$

12. on $2400 at 6% for 18 months

$$I = \frac{2400 \cdot 6 \cdot 18}{100 \cdot 12} \qquad \textbf{\$216}$$

13. on $60,000 at 11% for 2 months

$$I = \frac{60,000 \cdot 11 \cdot 2}{100 \cdot 12} \qquad \textbf{\$1100}$$

14. on $59,400 at 12% for 1 month

$$I = \frac{59,400 \cdot 12 \cdot 1}{100 \cdot 12} \qquad \textbf{\$594}$$

15. on $7200 at 7% for 15 months

$$I = \frac{7200 \cdot 7 \cdot 15}{100 \cdot 12} \qquad \textbf{\$630}$$

16. on $6400 at 10% for 9 months

$$I = \frac{6400 \cdot 10 \cdot 9}{100 \cdot 12} \qquad \textbf{\$480}$$

Interest 3

Complete the chart of compound interest using a calculator given the annual rate.

	PRINCIPAL	RATE	COMPOUND	YEARS	FORMULA	VALUE
1.	$2000	6%	annually	5	$(2000)(1.06)^5$	$2676.45
2.	$350	8%	quarterly	2	$(350)(1.02)^8$	$410.08
3.	$5000	4.8%	monthly	1	$(5000)(1.004)^{12}$	$5245.35
4.	$3000	2%	quarterly	2.5	$(3000)(1.005)^{10}$	$3153.42
5.	$6000	10%	semi-annually	6	$(6000)(1.05)^{12}$	$10,775.14
6.	$500	3.6%	semi-monthly	1.5	$(500)(1.0015)^{36}$	$527.72
7.	$2500	5%	annually	2	$(2500)(1.05)^2$	$2756.25
8.	$5500	2.4%	monthly	2	$(5500)(1.002)^{24}$	$5770.16
9.	$600	12%	monthly	1.5	$(600)(1.01)^{18}$	$717.69
10.	$1750	9%	quarterly	4	$(1750)(1.0225)^{16}$	$2498.34
11.	$1000	12%	semi-annually	7	$(1000)(1.06)^{14}$	$2260.90
12.	$900	6.25%	semi-annually	3	$(900)(1.03125)^6$	$1082.50
13.	$8000	4%	quarterly	2	$(8000)(1.01)^8$	$8662.85
14.	$250	6%	quarterly	3	$(250)(1.015)^{12}$	$298.90
15.	$2600	10%	annually	3	$(2600)(1.1)^3$	$3460.60
16.	$1100	15%	semi-monthly	1/2	$(1100)(1.00625)^{12}$	$1185.40
17.	$1000	12%	semi-annually	1	$(1000)(1.06)^2$	$1123.60
18.	$35,000	8%	annually	3	$(35,000)(1.08)^3$	$44,089.92
19.	$2000	10%	quarterly	1	$(2000)(1.025)^4$	$2207.63

MAVA Math: Enhanced Skills Solutions Copyright © 2015 Marla Weiss

Interest 4

Answer by writing and solving an I=PRT equation.

---

1. Find the principal that yields $198 simple interest at 5.5% for 16 months.

$$198 = \frac{P \cdot 5.5 \cdot 4}{100 \cdot 3}$$

$198 \cdot 300 = 22 \cdot P$

$P = 2700$

$2700

---

2. Simple annual interest for 4 years on $9000 was $1980. Find the rate.

$$1980 = \frac{9000 \cdot R \cdot 4}{100}$$

$1980 = 90 \cdot 4 \cdot R$

$R = 5.5$

5.5%

---

3. Find the principal that yields $27 simple interest at 4.5% for 9 months.

$$27 = \frac{P \cdot 4.5 \cdot 3}{100 \cdot 4}$$

$10,800 = 13.5 \cdot P$

$P = 800$

$800

---

4. Find the rate that yields $3250 simple interest on $6500 after 5 years.

$$3250 = \frac{6500 \cdot R \cdot 5}{100}$$

$3250 = 65 \cdot 5 \cdot R$

$R = 10$

10%

---

5. Find the principal that yields $195 simple interest at 6.5% for 10 months.

$$195 = \frac{P \cdot 6.5 \cdot 5}{100 \cdot 6}$$

$117,000 = 32.5 \cdot P$

$P = 3600$

$3600

---

6. Simple annual interest paying 6% was $840 after 2 years. Find the principal.

$$840 = \frac{P \cdot 6 \cdot 2}{100}$$

$84,000 = 12 \cdot P$

$P = 7000$

$7000

---

7. Simple annual interest for 42 months on $1300 was $182. Find the rate.

$$182 = \frac{1300 \cdot R \cdot 3.5}{100}$$

$182 = 13 \cdot 3.5 \cdot R$

$R = 4$

4%

---

8. Find the rate that yields $100 simple interest on $500 after 8 years.

$$100 = \frac{500 \cdot R \cdot 8}{100}$$

$100 = 5 \cdot 8 \cdot R$

$R = 2.5$

2.5%

---

9. Simple annual interest for 3.5 years on $3500 was $735. Find the rate.

$$735 = \frac{3500 \cdot R \cdot 7}{100 \cdot 2}$$

$1470 = 35 \cdot 7 \cdot R$

$R = 6$

6%

---

10. Find the principal that yields $22 simple interest at 5.5% for 8 months.

$$22 = \frac{P \cdot 5.5 \cdot 2}{100 \cdot 3}$$

$6600 = 11 \cdot P$

$P = 600$

$600

---

11. Simple annual interest for 6 years on $6000 was $2880. Find the rate.

$$2880 = \frac{6000 \cdot R \cdot 6}{100}$$

$2880 = 60 \cdot 6 \cdot R$

$R = 8$

8%

---

12. Simple annual interest paying 11% was $3080 after 7 years. Find the principal.

$$3080 = \frac{P \cdot 11 \cdot 7}{100}$$

$308,000 = 11 \cdot 7 \cdot P$

$P = 4000$

$4000

Least Common Multiple 1

Find the LCM by mental math.

1. 12 and 60 **60**	17. 26 and 39 **78**	33. 2, 9, and 20 **180**
2. 19 and 20 **180**	18. 6, 11, and 12 **132**	34. 18 and 24 **72**
3. 9 and 12 **36**	19. 15 and 80 **240**	35. 6, 22, and 33 **66**
4. 16 and 24 **48**	20. 10 and 18 **90**	36. 9 and 22 **198**
5. 14 and 77 **154**	21. 25 and 40 **200**	37. 4, 5, and 10 **20**
6. 8, 10, and 12 **120**	22. 3, 8, and 16 **48**	38. 44 and 121 **484**
7. 20 and 21 **420**	23. 40 and 56 **280**	39. 10, 15, and 25 **150**
8. 50 and 60 **300**	24. 10 and 21 **210**	40. 5, 9, and 10 **90**
9. 3, 11, and 66 **66**	25. 2, 3, and 4 **12**	41. 6, 9, and 15 **90**
10. 16 and 20 **80**	26. 9, 10, and 11 **990**	42. 15 and 40 **120**
11. 14 and 21 **42**	27. 2, 5, and 13 **130**	43. 2, 5, and 6 **30**
12. 6, 7, and 10 **210**	28. 25 and 35 **175**	44. 5, 6, and 21 **210**
13. 50 and 80 **400**	29. 8, 10, and 20 **40**	45. 4, 11, and 16 **176**
14. 20 and 36 **180**	30. 8, 9, and 10 **360**	46. 5, 6, and 10 **30**
15. 6, 12, and 18 **36**	31. 20 and 39 **780**	47. 25 and 325 **325**
16. 2, 9, and 11 **198**	32. 5, 7, and 20 **140**	48. 18, 45, and 90 **90**

MAVA Math: Enhanced Skills Solutions Copyright © 2015 Marla Weiss

Least Common Multiple 2

Find the least common multiple by prime factorization.

1.	45	75	9.	168	224

1. 45   75

  9 x 5   3 x 25

  3^2 x 5   3 x 5^2

LCM = 3^2 x 5^2 = 3 x 75 = **225**

9. 168   224

  8 x 21   4 x 56

  2^3 x 3 x 7 2^5 x 7

LCM = 2^5 x 3 x 7 = 3 x 224 = **672**

2. 36   80

  4 x 9   8 x 10

  2^2 x 3^2 2^4 x 5

LCM = 2^4 x 3^2 x 5 = 8 x 9 x 10 = **720**

10. 135   225

  27 x 5   9 x 25

  3^3 x 5 3^2 x 5^2

LCM = 3^3 x 5^2 = 3 x 225 = **675**

3. 63   77

  9 x 7   7 x 11

  3^2 x 7

LCM = 3^2 x 7 x 11 = 63 x 11 = **693**

11. 108   405

  4 x 27   81 x 5

  2^2 x 3^3 3^4 x 5

LCM = 2^2 x 3^4 x 5 = 4 x 405 = **1620**

4. 48   72

  16 x 3   8 x 9

  2^4 x 3 2^3 x 3^2

LCM = 2^4 x 3^2 = 2 x 72 = **144**

12. 126   297

  9 x 14   27 x 11

  2 x 3^2 x 7 3^3 x 11

LCM = 2 x 3^3 x 7 x 11 = 126 x 33 = **4158**

5. 54   60

  2 x 27   4 x 15

  2 x 3^3 2^2 x 3 x 5

LCM = 2^2 x 3^3 x 5 = 54 x 10 = **540**

13. 21   24   84

  3 x 7 2^3 x 3   4 x 21

      2^2 x 3 x 7

LCM = 2^3 x 3 x 7 = 2 x 84 = **168**

6. 27   90

  3^3 9 x 10

      2 x 3^2 x 5

LCM = 2 x 3^3 x 5 = 3 x 90 = **270**

14. 42   52   70

  2 x 3 x 7 2^2 x 13   2 x 5 x 7

LCM = 2^2 x 3 x 5 x 7 x 13 =
42 x 130 = **5460**

7. 144   216

  12 x 12   9 x 24

  2^4 x 3^2 2^3 x 3^3

LCM = 2^4 x 3^3 = 2 x 216 = **432**

15. 16   56   98

  2^4 2^3 x 7   2 x 7^2

LCM = 2^4 x 7^2 = 16 x 49 = **784**

8. 675   750

  25 x 27   3 x 25 x 10

  3^3 x 5^2 2 x 3 x 5^3

LCM = 2 x 3^3 x 5^3 = 9 x 750 = **6750**

16. 40   88   96

  8 x 5   8 x 11   16 x 6

  2^3 x 5 2^3 x 11 2^5 x 3

LCM = 2^5 x 3 x 5 x 11 = 480 x 11 = **5280**

Least Common Multiple 3

Find the LCM.

1. $LCM(a^3b^4c^5, a^2b^6c^3)$ $a^3b^6c^5$	14. $LCM(11a^4b^4c, 14a^9b^3c^6)$ $154a^9b^4c^6$
2. $LCM(16x^4y^5z^3, 48x^3y^6z^3)$ $48x^4y^6z^3$	15. $LCM(15a^5c^7, 9a^3b^9)$ $45a^5b^9c^7$
3. $LCM(5a^3s^4, 35a^4s^3, 10a^2s^3)$ $70a^4s^4$	16. $LCM(14f^7g^6h^4, 16f^3g^4h^2)$ $112f^7g^6h^4$
4. $LCM(35a^2b^2c^2, 42a^3b^5)$ $210a^3b^5c^2$	17. $LCM(5w^7x^4y^5z^3, 7w^2x^5y^4z)$ $35w^7x^5y^5z^3$
5. $LCM(6f^7g^3h^6, 10f^4g^5h^8)$ $30f^7g^5h^8$	18. $LCM(6b^3d^5g^6, 18b^4d^5h^2)$ $18b^4d^5g^6h^2$
6. $LCM(36d^5e^2h^6, 144d^3e^6h^8)$ $144d^5e^6h^8$	19. $LCM(10p^9q^2r^4, 14p^5q^5r^6)$ $70p^9q^5r^6$
7. $LCM(3x^4y^5z^6, 51x^3y^2z^8)$ $51x^4y^5z^8$	20. $LCM(6m^3n^6p^7, 15m^2n^4p^8)$ $30m^3n^6p^8$
8. $LCM(15x^2y^3z^4, 20w^2y^4z^3)$ $60w^2x^2y^4z^4$	21. $LCM(6e^3f^5g^5, 8e^6f^5g^7)$ $24e^6f^5g^7$
9. $LCM(4d^2e^3, 5d^4e^4, 6d^6e^5)$ $60d^6e^5$	22. $LCM(5g^7s^6, 10g^6s^7, 15g^8s^3)$ $30g^8s^7$
10. $LCM(36p^4q^2r^8, 90p^5q^2r^3)$ $180p^5q^2r^8$	23. $LCM(14x^3z^7, 21y^2z^6, 35x^4y^3)$ $210x^4y^3z^7$
11. $LCM(15x^9y^6z^5, 70x^5y^6z^9)$ $210x^9y^6z^9$	24. $LCM(8m^8n^2, 12p^9n^5, 20m^3p^4)$ $120m^8n^5p^9$
12. $LCM(16r^4st, 56r^6s^3t^5, 4r^2s^3t)$ $112r^6s^3t^5$	25. $LCM(6a^5b^5, 33a^6b^6c, 22a^5b^6)$ $66a^6b^6c$
13. $LCM(12rst, 16r^2s^5t^4, 8r^3s^3t^6)$ $48r^3s^5t^6$	26. $LCM(9r^3s^6, 8pr^6s^3, 12r^5s^5)$ $72pr^6s^6$

Least Common Multiple 4

$$GCF(m,n) \times LCM(m,n) = mn$$

Answer as indicated.

Given one of two numbers m and their GCF and LCM, find the second number n.

1. A collection of \$100 bills may be put in stacks of 2, 5, or 7 with none left over. If the number of bills is between 100 and 200, find the value of the money.
 	
LCM (2, 5, 7) = 70
70 bills: no   **\$14,000**
140 bills: yes
	7. m = 550
	    GCF(m, n) = 2 x 5 x 11
	    LCM(m, n) = 2^3 x 5^2 x 11^2
	
	  m = 2 x 5^2 x 11
	  n = 2^3 x 5 x 11^2 = 40 x 121 = **4840**

2. Two runners start a race at the same time from the same line. One completes a lap every 8 minutes, the other a lap every 10 minutes. How long after they start do they first cross the line together?   LCM (8, 10) = 40
40 minutes
	8. m = 630
	    GCF(m, n) = 2 x 5 x 7
	    LCM(m, n) = 2^2 x 3^2 x 5^4 x 7
	
	  m = 2 x 3^2 x 5 x 7
	  n = 2^2 x 5^4 x 7 = 25 x 700 = **17,500**

3. Find the least number of \$50 bills that may be put in stacks of 3, 7, or 11 with none left over. What is the value of the money?
LCM (3, 7, 11) = 231
231 bills
231 x 50 = **\$11,550**
	9. m = 520
	    GCF(m, n) = 2 x 5 x 13
	    LCM(m, n) = 2^2 x 3 x 5^2 x 7
	
	  m = 2^3 x 5 x 13
	  n = 3 x 5^2 x 7 = 21 x 25 = **525**

4. A school that is in session Monday through Friday has an 8-day-cycle schedule of classes that starts on a Tuesday. After how many days is the day of the week again Tuesday and the order of classes matches the first day's?
LCM (5, 8) = 40   **40 days**
	10. m = 1485
	    GCF(m, n) = 3 x 5 x 11
	    LCM(m, n) = 3^3 x 5^2 x 7 x 11
	
	  m = 3^3 x 5 x 11 n = 75 x 7 x 11
	  n = 3 x 5^2 x 7 x 11 n = 525 x 11
	                            n = **5775**

5. A set of marbles in groups of 2s, 3s, or 4s has 1 marble left but 0 left in groups of 5s. Find the number of marbles greater than 25 and less than 100.
    2: end in 5
    3: not 45, 75
    35 = 3x11 + 2
    55 = 4x13 + 3
    65 = 3x21 + 2
    95 = 3x31 + 2
    85
	11. m = 1540
	    GCF(m, n) = 2 x 5 x 7
	    LCM(m, n) = 2^2 x 5^2 x 7^2 x 11
	
	  m = 2^2 x 5 x 7 x 11
	  n = 2 x 5^2 x 7^2 = 50 x 49 = **2450**

6. A mother is bringing cookies for 1 class, a teacher for another. The mother cannot remember whether the class of 32 or 24 students is hers. Find the least number of cookies to ensure all cookies are given out and each child gets the same number. LCM (24, 32) = **96**	12. m = 5390
	    GCF(m, n) = 2 x 5 x 11
	    LCM(m, n) = 2^2 x 5^2 x 7^2 x 11^2
	
	  m = 2 x 5 x 7^2 x 11
	  n = 2^2 x 5^2 x 11^2 = 100 x 121 = **12,100**

Line Segments 1

Find the midpoint or endpoint of the vertical or horizontal segment.

	ENDPOINT	ENDPOINT	MIDPOINT		ENDPOINT	ENDPOINT	MIDPOINT
1.	(6, 6)	(18, 6)	**(12, 6)**	20.	(5, 3)	**(5, 19)**	(5, 11)
2.	(−10, 10)	(1, 10)	**(−4.5, 10)**	21.	(4, −3)	**(17, −3)**	(10.5, −3)
3.	(15, −3)	(18, −3)	**(16.5, −3)**	22.	**(14, −7)**	(14, 1)	(14, −3)
4.	(7, 13)	(7, −11)	**(7, 1)**	23.	**(18, 0)**	(18, 9)	(18, 4.5)
5.	(−8, 4)	(7, 4)	**(−.5, 4)**	24.	(12, 3)	**(12, −4)**	(12, −.5)
6.	(0, 9)	(0, 12)	**(0, 10.5)**	25.	(3, 5)	**(8, 5)**	(5.5, 5)
7.	(9, 5)	(16, 5)	**(12.5, 5)**	26.	**(−2, 13)**	(−2, 1)	(−2, 7)
8.	(11, −2)	(11, 4)	**(11, 1)**	27.	**(10, 2)**	(−5, 2)	(2.5, 2)
9.	(−3, 1)	(19, 1)	**(8, 1)**	28.	(2.9, 2)	**(9.9, 2)**	(6.4, 2)
10.	(−20, 6)	(13, 6)	**(−3.5, 6)**	29.	**(8.1, 9)**	(8.9, 9)	(8.5, 9)
11.	(11, 4)	(18, 4)	**(14.5, 4)**	30.	(35, 36)	**(39, 36)**	(37, 36)
12.	(8, 0)	(−7, 0)	**(.5, 0)**	31.	**(24, 7)**	(24, 17)	(24, 12)
13.	(1, 3)	(1, −9)	**(1, −3)**	32.	(−2, 15)	**(6, 15)**	(2, 15)
14.	(0, 8)	(7, 8)	**(3.5, 8)**	33.	(30, 35)	**(40, 35)**	(35, 35)
15.	(8, 18)	(11, 18)	**(9.5, 18)**	34.	**(15.5, 1)**	(9.5, 1)	(12.5, 1)
16.	(20, 16)	(29, 16)	**(24.5, 16)**	35.	(11, −12)	**(11, 15)**	(11, 1.5)
17.	(6.3, 8)	(1.3, 8)	**(3.8, 8)**	36.	(15, 1)	**(26, 1)**	(20.5, 1)
18.	(9, 13)	(16, 13)	**(12.5, 13)**	37.	**(3, −3.5)**	(3, 8.5)	(3, 2.5)
19.	(6, −10)	(6, 10)	**(6, 0)**	38.	**(11, −25)**	(11, 15)	(11, −5)

Line Segments 2

Find the endpoint or consecutive points of trisection of the vertical or horizontal segment.

	ENDPOINT	TRI-POINT	TRI-POINT	ENDPOINT
1.	(6, 6)	**(13, 6)**	**(20, 6)**	(27, 6)
2.	(−3, 10)	**(−3, 22)**	**(−3, 34)**	(−3, 46)
3.	(8, 7)	**(22, 7)**	(36, 7)	**(50, 7)**
4.	**(3, 17)**	(3, 25)	**(3, 33)**	(3, 41)
5.	(4, 9)	**(19, 9)**	**(34, 9)**	(49, 9)
6.	(8, −3)	**(8, 6)**	**(8, 15)**	(8, 24)
7.	**(14, −7)**	(14, 3)	**(14, 13)**	(14, 23)
8.	(2, 13)	**(19, 13)**	(36, 13)	**(53, 13)**
9.	(−4, 1.5)	**(7, 1.5)**	**(18, 1.5)**	(29, 1.5)
10.	(−7, 1)	**(−7, 14)**	**(−7, 27)**	(−7, 40)
11.	(12, −6)	**(4, −6)**	(−4, −6)	**(−12, −6)**
12.	**(5, 48)**	(5, 38)	**(5, 28)**	(5, 18)
13.	(0, −1)	**(0, −6)**	**(0, −11)**	(0, −16)
14.	(3, 11)	**(22, 11)**	**(41, 11)**	(60, 11)
15.	**(−8, −8)**	(−8, −1)	**(−8, 6)**	(−8, 13)
16.	(6, 10)	**(22, 10)**	**(38, 10)**	(54, 10)
17.	(−5, 2)	**(15, 2)**	**(35, 2)**	(55, 2)
18.	(−1, 40)	**(−1, 30)**	**(−1, 20)**	(−1, 10)
19.	(42, 12)	**(31, 12)**	(20, 12)	**(9, 12)**

Line Segments 3

	Find the midpoint of the diagonal segment.			Find the endpoint of the diagonal segment.	
ENDPOINT	**ENDPOINT**	**MIDPOINT**	**ENDPOINT**	**ENDPOINT**	**MIDPOINT**
1. (6, −6)	(10, 12)	**(8, 3)**	20. (13, 30)	**(16, 42)**	(14.5, 36)
2. (−1, 3)	(7, 6)	**(3, 4.5)**	21. (−5, −9)	**(2, −1)**	(−1.5, −5)
3. (2, 4)	(14, 7)	**(8, 5.5)**	22. (12, −6)	**(10, −4)**	(11, −5)
4. (4, −8)	(−4, 7)	**(0, −.5)**	23. (−20, 7)	**(−10, −1)**	(−15, 3)
5. (−3, 5)	(1, −7)	**(−1, −1)**	24. (15, 9)	**(14, −6)**	(14.5, 1.5)
6. (9, 5)	(−2, 7)	**(3.5, 6)**	25. (5, 16)	**(0, 4)**	(2.5, 10)
7. (13, −6)	(6, 6)	**(9.5, 0)**	26. (8, −10)	**(9, −9)**	(8.5, −9.5)
8. (6, 2)	(12, −7)	**(9, −2.5)**	27. (14, 5)	**(3, −1)**	(8.5, 2)
9. (8, 7)	(−4, 10)	**(2, 8.5)**	28. (4, 20)	**(2, 16)**	(3, 18)
10. (1, 8)	(11, −6)	**(6, 1)**	29. (5, 5)	**(−11, 13)**	(−3, 9)
11. (2, 18)	(7, −8)	**(4.5, 5)**	30. (0, 9)	**(5, 0)**	(2.5, 4.5)
12. (3, 5)	(−3, −5)	**(0, 0)**	31. (8, 7)	**(3, −5)**	(5.5, 1)
13. (−3, −4)	(−5, −6)	**(−4, −5)**	32. (20, 0)	**(13, 8)**	(16.5, 4)
14. (11, −3)	(17, −8)	**(14, −5.5)**	33. (5, 13)	**(18, 5)**	(11.5, 9)
15. (3, 12)	(−7, 15)	**(−2, 13.5)**	34. (30, 2)	**(20, 4)**	(25, 3)
16. (−4, 11)	(−9, 15)	**(−6.5, 13)**	35. (5, 5)	**(−1, −3)**	(2, 1)
17. (12, 12)	(22, −4)	**(17, 4)**	36. (58, 8)	**(50, 2)**	(54, 5)
18. (−3, −8)	(9, −1)	**(3, −4.5)**	37. (−2, 10)	**(18, −32)**	(8, −11)
19. (14, −12)	(11, 15)	**(12.5, 1.5)**	38. (22, 9)	**(6, 17)**	(14, 13)

Line Segments 4

Find the endpoint or consecutive points of trisection of the diagonal segment.

	ENDPOINT	TRI-POINT	TRI-POINT	ENDPOINT
1.	(0, 0)	**(7, 7)**	**(14, 14)**	(21, 21)
2.	(2, 4)	**(7, 7)**	**(12, 10)**	(17, 13)
3.	(1, 18)	**(7, 11)**	(13, 4)	**(19, −3)**
4.	**(−5, −5)**	(−1, 3)	**(3, 11)**	(7, 19)
5.	(9, 3)	**(4, 0)**	**(−1, −3)**	(−6, −6)
6.	(7, 2)	**(12.5, 11)**	**(18, 20)**	(23.5, 29)
7.	**(11, −3)**	(7.6, 3.1)	**(4.2, 9.2)**	(0.8, 15.3)
8.	(8, 13)	**(1, 9)**	(−6, 5)	**(−13, 1)**
9.	(−4, 3.5)	**(2, 7)**	**(8, 10.5)**	(14, 14)
10.	(−7, 6)	**(−2, 0)**	**(3, −6)**	(8, −12)
11.	(14, −4)	**(11.4, −1.5)**	(8.8, 1)	**(6.2, 3.5)**
12.	**(5, 10)**	(18, 3)	**(31, −4)**	(44, −11)
13.	(4, −1)	**(7.7, −7)**	**(11.4, −13)**	(15.1, −19)
14.	(−1, 9)	**(−10, 6.5)**	**(−19, 4)**	(−28, 1.5)
15.	**(−6, −6)**	(−7.5, −1)	**(−9, 4)**	(−10.5, 9)
16.	(3, 7)	**(4.4, 9.6)**	**(5.8, 12.2)**	(7.2, 14.8)
17.	(−8, 0)	**(0, 1.1)**	**(8, 2.2)**	(16, 3.3)
18.	(−1, 20)	**(−2, 13.5)**	**(−3, 7)**	(−4, 0.5)
19.	(21, 12)	**(23.7, 3)**	(26.4, −6)	**(29.1, −15)**

MAVA Math: Enhanced Skills Solutions   Copyright © 2015 Marla Weiss

Line Segments 5

Find the coordinate of the point on the number line.

1.　3/5 of the way from –7 to 13 $$\frac{3}{5}\ (20) = 12 \quad -7 + 12 = \mathbf{5}$$	10.　1/3 of the way from –4 to –2.5 $$\frac{1}{3}\ (1.5) = .5 \quad -4 + .5 = \mathbf{-3.5}$$
2.　1/4 of the way from 10 to 0 $$\frac{1}{4}\ (10) = 2.5 \quad 10 - 2.5 = \mathbf{7.5}$$	11.　1/6 of the way from –1.2 to 1.2 $$\frac{1}{6}\ (2.4) = 0.4 \quad -1.2 + .4 = \mathbf{-0.8}$$
3.　.6 of the way from –1 to 3 $$\frac{3}{5}\ (4) = 2.4 \quad -1 + 2.4 = \mathbf{1.4}$$	12.　2/3 of the way from –4 to 5 $$\frac{2}{3}\ (9) = 6 \quad -4 + 6 = \mathbf{2}$$
4.　3/4 of the way from 4 to –2 $$\frac{3}{4}\ (6) = 4.5 \quad 4 - 4.5 = \mathbf{-.5}$$	13.　.75 of the way from 4.2 to –2.2 $$\frac{3}{4}\ (6.4) = 4.8 \quad 4.2 - 4.8 = \mathbf{-0.6}$$
5.　2/3 of the way from –2 to 4 $$\frac{2}{3}\ (6) = 4 \quad -2 + 4 = \mathbf{2}$$	14.　2/3 of the way from .15 to .9 $$\frac{2}{3}\ (.75) = .5 \quad .15 + .5 = \mathbf{.65}$$
6.　2/5 of the way from 2 to –.5 $$\frac{2}{5}\ (2.5) = 1 \quad 2 - 1 = \mathbf{1}$$	15.　2/5 of the way from –3 to 2 $$\frac{2}{5}\ (5) = 2 \quad -3 + 2 = \mathbf{-1}$$
7.　.4 of the way from 2 to –3 $$\frac{2}{5}\ (5) = 2 \quad 2 - 2 = \mathbf{0}$$	16.　3/4 of the way from –2 to 10 $$\frac{3}{4}\ (12) = 9 \quad -2 + 9 = \mathbf{7}$$
8.　5/6 of the way from –1 to 11 $$\frac{5}{6}\ (12) = 10 \quad -1 + 10 = \mathbf{9}$$	17.　.6 of the way from 9 to –6 $$\frac{3}{5}\ (15) = 9 \quad 9 - 9 = \mathbf{0}$$
9.　5/7 of the way from 7 to –7 $$\frac{5}{7}\ (14) = 10 \quad 7 - 10 = \mathbf{-3}$$	18.　7/9 of the way from –9 to 18 $$\frac{7}{9}\ (27) = 21 \quad -9 + 21 = \mathbf{12}$$

Line Segments 6

Answer as indicated.   Be clear when marking a value for a segment split by a point.

---

1. On $\overline{PR}$, PQ = 12 and QR = 8. What is the length of the segment joining the midpoints of $\overline{PQ}$ and $\overline{QR}$?

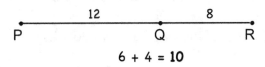

P   12   Q   8   R

6 + 4 = **10**

---

7. A, B, C, and D lie on a segment in that order. AB:BC = 4:5, and BC:CD = 7:2. AB = 14. Find AD.

AB:BC:CD = 28:35:10
AB:BC:CD = 14:17.5:5
AD = 14 + 17.5 + 5 = **36.5**

---

2. A is the midpoint of $\overline{TH}$, and TA = 12. Points E and C are trisection points of $\overline{TH}$. Find AC.

T   E   A   C   H

TH = 24, TE = 8, TC = 16, TA = 12
16 − 12 = **4**

---

8. B and C are on $\overline{AD}$ with AD = 36. BD is twice AB, and CD is one third BC. Find CD.   AD = 36 in 3 parts
BD = 24 in 4 parts

A   12   B   18   C 6 D

---

3. A, B, C, and D are points on a line with D the midpoint of BC. AB = 16. AC = 4. Find all possible values of AD.

10

B 6 D 6 C 4 A

6

B 10 D 6 A 4 C

---

9. E, F, G, and H are on a segment. H is the midpoint of $\overline{FG}$. EF = 60. EG = 12. FG = 72. Find EH.

F 36 H **24** E 12 G

---

4. E divides $\overline{DF}$ into $\overline{DE}$ and $\overline{EF}$ in the ratio 5:2. DF = 21. Find EF.

D   E   F

DE:EF = 5:2
DE:EF = 15:6
EF = **6**

---

10. GT = 70. R is the midpoint of $\overline{GE}$. E is the midpoint of $\overline{RA}$. A is the midpoint of $\overline{RT}$. Find RA.   GR = RE = EA
AT = 2RE
70/5 = 14
RA = **28**

G 14 R 14 E 14 A 28 T

---

5. AC = DF. AB = EF. Find BC − DE.

A   B   C

D   E   F

BC − DE = **0**

---

11. A and B are each placed to the right of C on a segment such that 3BC = 5AC. Find AB:AC.

Let BC = 10
AC = 6
AB = 4

C   A   B AB:AC = **2:3**

---

6. R and Q are on $\overline{PS}$ with PS = 36. QS = 3PQ. QR = 2RS. Find RS.

Cut 36 into 4 parts. Cut 27 into 3 parts.

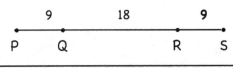

P 9 Q 18 R 9 S

---

12. P, Q, R, S, and T lie on a segment in that order. R is a midpoint of $\overline{QS}$. QS = 9. RT = 12. PT = 22.5. Find PQ.

P 6 Q 4.5 R 4.5 S 7.5 T

Logic 1

Prove the statement by evaluating the logical equivalency of both sides of the equation.

1. A AND (B OR C) = (A AND B) OR (A AND C)   Distributive Property

A	B	C	X=B OR C	A AND X	Y=A AND B	Z=A AND C	Y OR Z
TRUE	TRUE	TRUE	T	**T**	T	T	**T**
TRUE	TRUE	FALSE	T	**T**	T	F	**T**
TRUE	FALSE	TRUE	T	**T**	F	T	**T**
TRUE	FALSE	FALSE	F	**F**	F	F	**F**
FALSE	TRUE	TRUE	T	**F**	F	F	**F**
FALSE	TRUE	FALSE	T	**F**	F	F	**F**
FALSE	FALSE	TRUE	T	**F**	F	F	**F**
FALSE	FALSE	FALSE	F	**F**	F	F	**F**

2. A OR (B AND C) = (A OR B) AND (A OR C)   Distributive Property

A	B	C	X=B AND C	A OR X	Y=A OR B	Z=A OR C	Y AND Z
TRUE	TRUE	TRUE	T	**T**	T	T	**T**
TRUE	TRUE	FALSE	F	**T**	T	T	**T**
TRUE	FALSE	TRUE	F	**T**	T	T	**T**
TRUE	FALSE	FALSE	F	**T**	T	T	**T**
FALSE	TRUE	TRUE	T	**T**	T	T	**T**
FALSE	TRUE	FALSE	F	**F**	T	F	**F**
FALSE	FALSE	TRUE	F	**F**	F	T	**F**
FALSE	FALSE	FALSE	F	**F**	F	F	**F**

3. NOT (A OR B) = (NOT A) AND (NOT B)   DeMorgan's Law

A	B		X=A OR B	NOT X	Y = NOT A	Z = NOT B	Y AND Z
TRUE	TRUE		T	**F**	F	F	**F**
TRUE	FALSE		T	**F**	F	T	**F**
FALSE	TRUE		T	**F**	T	F	**F**
FALSE	FALSE		F	**T**	T	T	**T**

4. NOT (A AND B) = (NOT A) OR (NOT B)   DeMorgan's Law

A	B		X=A AND B	NOT X	Y=NOT A	Z=NOT B	Y OR Z
TRUE	TRUE		T	**F**	F	F	**F**
TRUE	FALSE		F	**T**	F	T	**T**
FALSE	TRUE		F	**T**	T	F	**T**
FALSE	FALSE		F	**T**	T	T	**T**

Logic 2

Write the converse, inverse, and contrapositive of the true implication. Label TRUE or FALSE.

1. If I live in Reno, then I live in Nevada.

Converse: **If I live in Nevada, then I live in Reno. FALSE**

Inverse: **If I do not live in Reno, then I do not live in Nevada. FALSE**

Contrapositive: **If I do not live in Nevada, then I do not live in Reno. TRUE**

2. If a triangle has two congruent sides, then the angles opposite the sides are congruent.

Converse: **If the angles opposite the sides of a triangle are congruent, then the sides are congruent. TRUE**

Inverse: **If a triangle does not have two congruent sides, then the angles opposite the sides are not congruent. TRUE**

Contrapositive: **If the angles opposite the sides of a triangle are not congruent, then the sides are not congruent. TRUE**

3. If two lines in a plane are parallel, then the alternate interior angles are congruent.

Converse: **If alternate interior angles in a plane are congruent, then the two lines forming them are parallel. TRUE**

Inverse: **If two lines in a plane are not parallel, then the alternate interior angles are not congruent. TRUE**

Contrapositive: **If alternate interior angles in a plane are not congruent, then the two lines forming them are not parallel. TRUE**

4. If two lines in a plane are perpendicular, then the lines have negative reciprocal slopes.

Converse: **If two lines in a plane have negative reciprocal slopes, then the lines are perpendicular. TRUE**

Inverse: **If two lines in a plane are not perpendicular, then the lines do not have negative reciprocal slopes. TRUE**

Contrapositive: **If two lines in a plane do not have negative reciprocal slopes, then the lines are not perpendicular. TRUE**

5. If a median, altitude, and angle bisector of a triangle are all the same line segment, then the triangle is isosceles.

Converse: **If a triangle is isosceles, then a median, altitude, and angle bisector are all the same line segment. TRUE**

Inverse: **If a median, altitude, and angle bisector of a triangle are not all the same line segment, then the triangle is not isosceles. TRUE**

Contrapositive: **If a triangle is not isosceles, then a median, altitude, and angle bisector are not all the same line segment. TRUE**

Matrices 1

Add or subtract as indicated.

1. $\begin{bmatrix} 17 & -5 & 33 \\ 16 & 17 & -42 \end{bmatrix} + \begin{bmatrix} -3 & 11 & 44 \\ -4 & -9 & 57 \end{bmatrix} + \begin{bmatrix} 21 & -6 & -9 \\ 24 & 53 & 42 \end{bmatrix}$ $\begin{bmatrix} \mathbf{35} & \mathbf{0} & \mathbf{68} \\ \mathbf{36} & \mathbf{61} & \mathbf{57} \end{bmatrix}$

2. $\begin{bmatrix} -5 & 65 \\ 14 & -23 \end{bmatrix} + \begin{bmatrix} 17 & 15 \\ 16 & -12 \end{bmatrix} - \begin{bmatrix} -4 & 24 \\ 13 & -10 \end{bmatrix} + \begin{bmatrix} 12 & 11 \\ 10 & -13 \end{bmatrix}$ $\begin{bmatrix} \mathbf{28} & \mathbf{67} \\ \mathbf{27} & \mathbf{-38} \end{bmatrix}$

3. $\begin{bmatrix} 10 & 45 \\ 13 & -22 \\ 56 & -30 \end{bmatrix} - \begin{bmatrix} 23 & 25 \\ -7 & -40 \\ 16 & -60 \end{bmatrix} - \begin{bmatrix} 12 & 24 \\ 16 & -18 \\ 28 & -50 \end{bmatrix} + \begin{bmatrix} 50 & 12 \\ 24 & -10 \\ 46 & -20 \end{bmatrix}$ $\begin{bmatrix} \mathbf{25} & \mathbf{8} \\ \mathbf{28} & \mathbf{26} \\ \mathbf{58} & \mathbf{60} \end{bmatrix}$

4. $\begin{bmatrix} -7 & -1 \\ 17 & 20 \\ 16 & -42 \\ 29 & 14 \end{bmatrix} - \begin{bmatrix} -11 & 45 \\ 17 & -25 \\ -13 & -42 \\ 18 & -15 \end{bmatrix} + \begin{bmatrix} -13 & 22 \\ 17 & -35 \\ 13 & 42 \\ 21 & -11 \end{bmatrix} + \begin{bmatrix} 9 & 24 \\ 17 & -31 \\ 13 & -42 \\ -5 & 38 \end{bmatrix}$ $\begin{bmatrix} \mathbf{0} & \mathbf{0} \\ \mathbf{34} & \mathbf{-21} \\ \mathbf{55} & \mathbf{0} \\ \mathbf{27} & \mathbf{56} \end{bmatrix}$

5. $\begin{bmatrix} -8 & 17 & 35 \\ 39 & 21 & -22 \end{bmatrix} + \begin{bmatrix} -9 & 37 & 35 \\ 16 & -5 & -22 \end{bmatrix} + \begin{bmatrix} -11 & 33 & 25 \\ 44 & 16 & 40 \end{bmatrix}$ $\begin{bmatrix} \mathbf{-28} & \mathbf{87} & \mathbf{95} \\ \mathbf{99} & \mathbf{32} & \mathbf{-4} \end{bmatrix}$

6. $\begin{bmatrix} -5 & -4 & 18 \\ 35 & 16 & -29 \end{bmatrix} + \begin{bmatrix} 25 & 19 & 21 \\ -8 & 16 & -31 \end{bmatrix} + \begin{bmatrix} -6 & -3 & 19 \\ 55 & 16 & -11 \end{bmatrix}$ $\begin{bmatrix} \mathbf{14} & \mathbf{12} & \mathbf{58} \\ \mathbf{82} & \mathbf{48} & \mathbf{-71} \end{bmatrix}$

7. $\begin{bmatrix} 29 & 9 \\ 13 & -34 \\ 14 & -42 \end{bmatrix} - \begin{bmatrix} -5 & 20 \\ 20 & -40 \\ 16 & -42 \end{bmatrix} + \begin{bmatrix} -11 & 52 \\ 27 & -12 \\ 12 & -43 \end{bmatrix} + \begin{bmatrix} -11 & 11 \\ 14 & -4 \\ 38 & 50 \end{bmatrix}$ $\begin{bmatrix} \mathbf{12} & \mathbf{52} \\ \mathbf{34} & \mathbf{-10} \\ \mathbf{48} & \mathbf{7} \end{bmatrix}$

8. $\begin{bmatrix} 50 & 15 \\ 36 & -21 \\ 52 & -33 \end{bmatrix} + \begin{bmatrix} -9 & 27 \\ 34 & -12 \\ 28 & -17 \end{bmatrix} - \begin{bmatrix} -11 & -5 \\ 20 & -11 \\ -8 & -22 \end{bmatrix} - \begin{bmatrix} -5 & 17 \\ 13 & -10 \\ 12 & -18 \end{bmatrix}$ $\begin{bmatrix} \mathbf{57} & \mathbf{30} \\ \mathbf{37} & \mathbf{-12} \\ \mathbf{76} & \mathbf{-10} \end{bmatrix}$

Matrices 2

Multiply by the scalar using mental math.

1.

$5 \begin{bmatrix} 12 & -3 & 10 \\ 15 & 13 & 11 \end{bmatrix}$ $\begin{bmatrix} 60 & -15 & 50 \\ 75 & 65 & 55 \end{bmatrix}$

9.

$2 \begin{bmatrix} 2.9 & 3.6 & 4.6 \\ 1.7 & 3.9 & 3.8 \end{bmatrix}$ $\begin{bmatrix} 5.8 & 7.2 & 9.2 \\ 3.4 & 7.8 & 7.6 \end{bmatrix}$

2.

$4 \begin{bmatrix} -6 & 14 \\ 40 & -17 \end{bmatrix}$ $\begin{bmatrix} -24 & 56 \\ 160 & -68 \end{bmatrix}$

10.

$6 \begin{bmatrix} -7 & 19 \\ 45 & -25 \end{bmatrix}$ $\begin{bmatrix} -42 & 114 \\ 270 & -150 \end{bmatrix}$

3.

$11 \begin{bmatrix} 17 & 31 \\ 27 & 87 \\ 76 & 54 \end{bmatrix}$ $\begin{bmatrix} 187 & 341 \\ 297 & 957 \\ 836 & 594 \end{bmatrix}$

11.

$25 \begin{bmatrix} 11 & -30 \\ 25 & -15 \\ 19 & -12 \end{bmatrix}$ $\begin{bmatrix} 275 & -750 \\ 625 & -375 \\ 475 & -300 \end{bmatrix}$

4.

$12 \begin{bmatrix} 20 & 11 \\ 8.5 & -5 \\ 12 & 15 \end{bmatrix}$ $\begin{bmatrix} 240 & 132 \\ 102 & -60 \\ 144 & 180 \end{bmatrix}$

12.

$15 \begin{bmatrix} 60 & 15 \\ 2.5 & -7 \\ 11 & 16 \end{bmatrix}$ $\begin{bmatrix} 900 & 225 \\ 37.5 & -105 \\ 165 & 240 \end{bmatrix}$

5.

$1.5 \begin{bmatrix} -6 & 32 & 60 \\ 50 & 24 & -8 \end{bmatrix}$ $\begin{bmatrix} -9 & 48 & 90 \\ 75 & 36 & -12 \end{bmatrix}$

13.

$1.1 \begin{bmatrix} -33 & 45 \\ -1.3 & 32 \end{bmatrix}$ $\begin{bmatrix} -36.3 & 49.5 \\ -1.43 & 35.2 \end{bmatrix}$

6.

$-6 \begin{bmatrix} -8 & 12 & -9 \\ 1 & -11 & 0 \end{bmatrix}$ $\begin{bmatrix} 48 & -72 & 54 \\ -6 & 66 & 0 \end{bmatrix}$

14.

$-9 \begin{bmatrix} 31 & -25 \\ 14 & -72 \end{bmatrix}$ $\begin{bmatrix} -279 & 225 \\ -126 & 648 \end{bmatrix}$

7.

$20 \begin{bmatrix} -9 & 35 \\ 13 & -7 \\ 22 & 46 \end{bmatrix}$ $\begin{bmatrix} -180 & 700 \\ 260 & -140 \\ 440 & 920 \end{bmatrix}$

15.

$30 \begin{bmatrix} -9 & 111 \\ 99 & -75 \\ 80 & 151 \end{bmatrix}$ $\begin{bmatrix} -270 & 3330 \\ 2970 & -2250 \\ 2400 & 4530 \end{bmatrix}$

8.

$10 \begin{bmatrix} 6.5 & 2.36 \\ .92 & .345 \\ 5.1 & 89.2 \end{bmatrix}$ $\begin{bmatrix} 65 & 23.6 \\ 9.2 & 3.45 \\ 51 & 892 \end{bmatrix}$

16.

$50 \begin{bmatrix} 8.2 & .12 \\ .5 & 2.4 \\ 3.5 & 10.1 \end{bmatrix}$ $\begin{bmatrix} 410 & 6 \\ 25 & 120 \\ 175 & 505 \end{bmatrix}$

Matrices 3

Multiply the matrices.

(r by c) X (R by C)
need c = R to multiply
result is r by C

1.

$$\begin{bmatrix} 6 & 9 \\ 3 & 7 \end{bmatrix} \begin{bmatrix} 5 & 8 \\ 4 & 2 \end{bmatrix} \begin{bmatrix} \mathbf{66} & \mathbf{66} \\ \mathbf{43} & \mathbf{38} \end{bmatrix}$$

6x5 + 9x4 = 30 + 36 = 66
6x8 + 9x2 = 48 + 18 = 66
3x5 + 7x4 = 15 + 28 = 43
3x8 + 7x2 = 24 + 14 = 38

5.

$$\begin{bmatrix} 5 & -9 & 11 \\ 6 & -5 & 7 \end{bmatrix} \begin{bmatrix} 6 & 11 \\ 5 & 10 \\ 7 & 12 \end{bmatrix} \begin{bmatrix} \mathbf{62} & \mathbf{97} \\ \mathbf{60} & \mathbf{100} \end{bmatrix}$$

5x6 + −9x5 + 11x7 = 30 − 45 + 77 = 62
5x11 + −9x10 + 11x12 = 55 − 90 + 132 = 97
6x6 + −5x5 + 7x7 = 36 − 25 + 49 = 60
6x11 + −5x10 + 7x12 = 66 − 50 + 84 = 100

2.

$$\begin{bmatrix} 9 & -8 \\ -5 & 6 \\ -4 & 3 \\ -3 & 2 \end{bmatrix} \begin{bmatrix} 11 & -2 \\ 10 & -3 \end{bmatrix} \begin{bmatrix} \mathbf{19} & \mathbf{6} \\ \mathbf{5} & \mathbf{-8} \\ \mathbf{-14} & \mathbf{-1} \\ \mathbf{-13} & \mathbf{0} \end{bmatrix}$$

99 − 80 = 19	−44 + 30 = −14
−18 + 24 = 6	8 − 9 = −1
−55 + 60 = 5	−33 + 20 = −13
10 − 18 = −8	6 − 6 = 0

6.

$$\begin{bmatrix} -7 & -6 & -8 \\ -4 & -2 & -3 \end{bmatrix} \begin{bmatrix} -5 & -6 \\ 10 & -1 \\ -9 & 11 \end{bmatrix} \begin{bmatrix} \mathbf{47} & \mathbf{-40} \\ \mathbf{27} & \mathbf{-7} \end{bmatrix}$$

35 − 60 + 72 = 47
42 + 6 − 88 = −40
20 − 20 + 27 = 27
24 + 2 − 33 = −7

3.

$$\begin{bmatrix} 100 & -25 & 125 \\ 200 & -15 & 300 \end{bmatrix} \begin{bmatrix} -1 & 2 \\ 2 & -4 \\ 3 & 1 \end{bmatrix}$$

−100 − 50 + 375 = 225
200 + 100 + 125 = 425
−200 − 30 + 900 = 670
400 + 60 + 300 = 760

$$\begin{bmatrix} \mathbf{225} & \mathbf{425} \\ \mathbf{670} & \mathbf{760} \end{bmatrix}$$

7.

$$\begin{bmatrix} 1 & -2 & 3 \\ 4 & 0 & -5 \\ 6 & -7 & 8 \end{bmatrix} \begin{bmatrix} 11 & 10 & 9 \\ 8 & 7 & -6 \\ 5 & -4 & 3 \end{bmatrix}$$

$$\begin{bmatrix} \mathbf{10} & \mathbf{-16} & \mathbf{30} \\ \mathbf{-19} & \mathbf{60} & \mathbf{21} \\ \mathbf{50} & \mathbf{-21} & \mathbf{120} \end{bmatrix}$$

4.

$$\begin{bmatrix} 80 & 20 \\ 40 & 10 \\ 50 & 70 \\ 60 & 90 \end{bmatrix} \begin{bmatrix} 1.5 & 3.5 & 5.5 \\ 2.5 & 4.5 & 6.5 \end{bmatrix}$$

$$\begin{bmatrix} \mathbf{170} & \mathbf{370} & \mathbf{570} \\ \mathbf{85} & \mathbf{185} & \mathbf{285} \\ \mathbf{250} & \mathbf{490} & \mathbf{730} \\ \mathbf{315} & \mathbf{615} & \mathbf{915} \end{bmatrix}$$

8.

$$\begin{bmatrix} 10 & -5 & 11 \\ 30 & 12 & 25 \\ 15 & -2 & 20 \end{bmatrix} \begin{bmatrix} 16 & 14 & 20 \\ 15 & 10 & -3 \\ 11 & 16 & 12 \end{bmatrix}$$

$$\begin{bmatrix} \mathbf{206} & \mathbf{266} & \mathbf{347} \\ \mathbf{935} & \mathbf{940} & \mathbf{864} \\ \mathbf{430} & \mathbf{510} & \mathbf{546} \end{bmatrix}$$

Matrices 4

Solve for the variables.

1.
$$\begin{bmatrix} 3x+1 & 33 \\ 5y-4 & 42 \end{bmatrix} = \begin{bmatrix} 52 & 3w+3 \\ -64 & 4z-6 \end{bmatrix}$$

3x = 51   3w = 30   5y = -60   4z = 48
x = 17   **w = 10**   **y = -12**   **z = 12**

8.
$$\begin{bmatrix} 2x+4 & 70 \\ 3y-6 & 31 \end{bmatrix} = \begin{bmatrix} 47 & 9w+7 \\ 42 & 3z-8 \end{bmatrix}$$

2x+4=47   9w+7=70   3y-6=42   3z-8=31
x = 21.5   **x = 7**   **y = 16**   **z = 13**

2.
$$\begin{bmatrix} 7w-2 & 65 \\ 5z-5 & 35 \end{bmatrix} = \begin{bmatrix} 12 & 3x+8 \\ 15 & 6y-7 \end{bmatrix}$$

7w-2=12   3x+8=65   5z-5=15   6y-7=35
w = 2   **x = 19**   **z = 4**   **y = 7**

9.
$$\begin{bmatrix} 75 & 4x+1 \\ 9y-6 & 42 \end{bmatrix} = \begin{bmatrix} 7w+5 & 89 \\ 66 & 4z-6 \end{bmatrix}$$

7w+5=75   4x+1=89   9y-6=66   4z-6=42
w = 10   **x = 22**   **y = 8**   **z = 12**

3.
$$\begin{bmatrix} 24 & 5x+7 \\ 4y-3 & 41 \end{bmatrix} = \begin{bmatrix} 3w+6 & 77 \\ 45 & 7z+6 \end{bmatrix}$$

3w+6=24   5x+7=77   4y-3=45   7z+6=41
w = 6   **x = 14**   **y = 12**   **z = 5**

10.
$$\begin{bmatrix} 8a+2 & 50 \\ 7c+6 & 68 \end{bmatrix} = \begin{bmatrix} 90 & 9b+5 \\ -8 & 7d-9 \end{bmatrix}$$

8a+2=90   9b+5=50   7c+6=-8   7d-9=68
a = 11   **b = 5**   **c = -2**   **d = 11**

4.
$$\begin{bmatrix} 9x+2 & 28 \\ 5y+1 & 38 \end{bmatrix} = \begin{bmatrix} 56 & 3w-2 \\ -4 & 3z+5 \end{bmatrix}$$

9x+2=56   3w-2=28   5y+1=-4   3z+5=38
x = 6   **w = 10**   **y = -1**   **z = 11**

11.
$$\begin{bmatrix} 3x+5 & 38 \\ 9y-9 & 80 \end{bmatrix} = \begin{bmatrix} 80 & 9w+2 \\ 90 & 4z-4 \end{bmatrix}$$

3x+5=80   9w+2=38   9y-9=90   4z-4=80
x = 25   **w = 4**   **y = 11**   **z = 21**

5.
$$\begin{bmatrix} 7x+1 & 43 \\ 4y+9 & 53 \end{bmatrix} = \begin{bmatrix} -34 & 8w+3 \\ 49 & 5z-7 \end{bmatrix}$$

7x+1=-34   8w+3=43   4y+9=49   5z-7=53
x = -5   **w = 5**   **y = 10**   **z = 12**

12.
$$\begin{bmatrix} 6x+9 & 33 \\ 5z-4 & 50 \end{bmatrix} = \begin{bmatrix} -21 & 6w+3 \\ 66 & 9y-4 \end{bmatrix}$$

6x+9=-21   6w+3=33   5z-4=66   9y-4=50
x = -5   **w = 5**   **z = 14**   **y = 6**

6.
$$\begin{bmatrix} 2a+9 & 32 \\ 7c+6 & 55 \end{bmatrix} = \begin{bmatrix} -11 & 5b+7 \\ 69 & 8d-9 \end{bmatrix}$$

2a+9=-11   5b+7=32   7c+6=69   8d-9=55
a = -10   **b = 5**   **c = 9**   **d = 8**

13.
$$\begin{bmatrix} 4a+1 & 57 \\ 3c+4 & 41 \end{bmatrix} = \begin{bmatrix} -51 & 6b-3 \\ 52 & 7d-8 \end{bmatrix}$$

4a+1=-51   6b-3=57   3c+4=52   7d-8=41
a = -13   **b = 10**   **c = 16**   **d = 7**

7.
$$\begin{bmatrix} 69 & 7x+5 \\ 7y+8 & 45 \end{bmatrix} = \begin{bmatrix} 6w+3 & 40 \\ -6 & 3z-6 \end{bmatrix}$$

6w+3=69   7x+5=40   7y+8=-6   3z-6=45
w = 11   **x = 5**   **y = -2**   **z = 17**

14.
$$\begin{bmatrix} 43 & 5x+7 \\ 8y-3 & 66 \end{bmatrix} = \begin{bmatrix} 6w+1 & 47 \\ 61 & 5z-9 \end{bmatrix}$$

6w+1=43   5x+7=47   8y-3=61   5z-9=66
w = 7   **x = 8**   **y = 8**   **z = 15**

Measurement 1

Convert the measurements in the standard system.

1.	2.5 gallon =	**10**	quarts	20.	3.5 pints =	**7** cups
2.	5.5 quarts =	**11**	pints	21.	4/3 yards =	**48** inches
3.	6.5 dozen =	**78**		22.	2.75 ton =	**5500** pounds
4.	3/4 quart =	**3**	cups	23.	1.4 ton =	**2800** pounds
5.	5/4 yard =	**45**	inches	24.	1/4 ton =	**8000** ounces
6.	1/4 gallon =	**2**	pints	25.	5/4 pound =	**20** ounces
7.	7/4 feet =	**21**	inches	26.	4/3 dozen =	**16**
8.	3.25 yard =	**9.75**	feet	27.	3/4 gallon =	**3** quarts
9.	2.75 pound =	**44**	ounces	28.	2/5 ton =	**800** pounds
10.	3.1 tons =	**6200**	pounds	29.	.375 quart =	**1.5** cups
11.	.125 gallon =	**2**	cups	30.	.3 mile =	**1584** feet
12.	.6 ton =	**1200**	pounds	31.	1.125 quart =	**4.5** cups
13.	1.75 dozen =	**21**		32.	.001 ton =	**2** pounds
14.	8/3 yard =	**96**	inches	33.	5/9 yard =	**20** inches
15.	2.6 ton =	**5200**	pounds	34.	1.5 pints =	**3** cups
16.	1/4 mile =	**1320**	feet	35.	.875 gallon =	**3.5** quarts
17.	7/2 pound =	**56**	ounces	36.	.05 ton =	**100** pounds
18.	19.5 feet =	**6.5**	yards	37.	54 inches =	**1.5** yards
19.	7/6 dozen =	**14**		38.	40 ounces =	**2.5** pounds

Measurement 2

Operate and regroup.

1. 3 mi 101 yd 2 ft 5 in + 1 mi 200 yd 9 ft 8 in 4 mi 301 yd 11 ft 13 in 4 mi 301 yd 12 ft 1 in **4 mi 305 yd 0 ft 1 in**	**9.** 6 3 14 7 yd 1 ft 2 in − 3 yd 2 ft 7 in **3 yd 1 ft 7 in**
2. $6 \cdot (3\text{ lb}\ \ 7\text{ oz})$ 18 lb 42 oz **20 lb 10 oz**	**10.** $(8\text{ yd}\ \ 2\text{ ft}\ \ 8\text{ in}) \div 5$ 5 yd 11 ft 8 in 5 yd 10 ft 20 in **1 yd 2 ft 4 in**
3. 5 3 13 6 yd 1 ft 1 in − 2 yd 2 ft 8 in **3 yd 1 ft 5 in**	**11.** $(8\text{ yd}\ \ 2\text{ ft}\ \ 6\text{ in}) \div 3$ 6 yd 8 ft 6 in 6 yd 6 ft 30 in **2 yd 2 ft 10 in**
4. 14 29 67 15 days 6 hr 7 min − 5 days 17 hr 57 min **9 days 12 hr 10 min**	**12.** $4 \cdot (2\text{ yd}\ \ 2\text{ ft}\ \ 7\text{ in})$ 8 yd 8 ft 28 in 8 yd 10 ft 4 in **11 yd 1 ft 4 in**
5. 5 tons 560 lb 17 oz + 2 tons 1950 lb 15 oz 7 tons 2510 lb 32 oz **8 tons 512 lb**	**13.** 2 mi 750 yd 74 ft 7 in + 3 mi 1100 yd 70 ft 9 in 5 mi 1850 yd 144 ft 16 in 6 mi 138 yd 16 in **6 mi 138 yd 1 ft 4 in**
6. $4 \cdot (5\text{ hr}\ \ 20\text{ min}\ \ 40\text{ sec})$ 20 hr 80 min 160 sec 20 hr 82 min 40 sec **21 hr 22 min 40 sec**	**14.** 7 28 62 65 2 65 8 days 5 hr 3 min 5 sec − 4 days 5 hr 9 min 17 sec **3 days 23 hr 53 min 48 sec**
7. $\frac{1}{3}$ (7 days 8 hr 9 min) 6 days 32 hr 9 min 6 days 30 hr 129 min **2 days 10 hr 43 min**	**15.** 9 2 12 10 yd − 3 yd 1 ft 8 in **6 yd 1 ft 4 in**
8. $7 \cdot (3\text{ yd}\ \ 3\text{ ft}\ \ 8\text{ in})$ 21 yd 21 ft 56 in 21 yd 25 ft 8 in **29 yd 1 ft 8 in**	**16.** $\frac{1}{4}$ (5 hr 19 min 4 sec) 4 hr 79 min 4 sec 4 hr 76 min 184 sec **1 hr 19 min 46 sec**

221

Measurement 3

Answer as indicated.

1.	1 cm + 1 m =	**1.01** m	12.	1.5 L – 150 mL =	**1350** mL
2.	1 cm + 1 m =	**101** cm	13.	6 cm + 7 dm =	**7.6** dm
3.	1 mm + 1 m =	**1001** mm	14.	20 dm + 8 cm =	**208** cm
4.	1 dm + 1 km =	**1000.1** m	15.	6.1 km + 4.1 m =	**6104.1** m
5.	1 m + 1 dm =	**110** cm	16.	3 cm + 2 m =	**2030** mm
6.	1 cm + 1 dm =	**0.11** m	17.	2 kg + 2 g =	**2002** g
7.	1 cm + 1 mm =	**0.011** m	18.	5 g – 200 mg =	**4800** mg
8.	1 cm + 1 mm =	**0.11** dm	19.	3.5 L + 1.5 mL =	**3501.5** mL
9.	1 cm + 1 mm =	**1.1** cm	20.	0.6 L + 100 mL =	**0.7** L
10.	1 cm + 1 km =	**1000.01** m	21.	1.5 km + 0.5 m =	**1500.5** m
11.	1 dm + 1 mm =	**10.1** cm	22.	0.5 km + 2 m =	**0.502** km

MAVA Math: Enhanced Skills Solutions Copyright © 2015 Marla Weiss

Measurement 4

Answer as indicated.

1. 10 sq yd = **90** sq ft 10 x (3 x 3)	12. 321 ft by 30 ft = **1070** sq yd 107 x 10
2. 10 sq yd = **12,960** sq in 10 x (36 x 36)	13. 144 in by 120 in = **120** sq ft 12 x 10
3. 120 sq ft = **17,280** sq in 120 x (12 x 12)	14. 108 in by 180 in = **15** sq yd 3 x 5
4. 10 cu ft = **17,280** cu in 10 x (12 x 12 x 12)	15. 11 yd by 9 yd = **891** sq ft 33 x 27 = 11 x 81
5. 2 sq yd = **2592** sq in 2 x (36 x 36)	16. 5 ft by 2 ft = **1440** sq in 60 x 24
6. 3 sq ft = **432** sq in 3 x (12 x 12)	17. 111 ft by 33 ft by 21 ft = **2849** cu yd 37 x 11 x 7
7. 3 cu yd = **81** cu ft 3 x (3 x 3 x 3)	18. 36 in by 72 in by 96 in = **144** cu ft 3 x 6 x 8
8. 5 sq ft = **720** sq in 5 x (12 x 12)	19. 54 in by 36 in by 72 in = **3** cu yd 1.5 x 1 x 2
9. 2 sq yd = **18** sq ft 2 x (3 x 3)	20. 5 ft by 10 ft by 2 ft = **172,800** cu in 12 x 12 x 12 x 10 x 10
10. 10 cu yd = **270** cu ft 10 x (3 x 3 x 3)	21. 15 ft by 12 yd by 48 in = **80** cu yd 5 x 12 x (4/3) = 5 x 16
11. 5 cu ft = **8640** cu in 5 x (12 x 12 x 12)	22. 90 in by 18 yd by 66 ft = **990** cu yd 2.5 x 18 x 22 = 5 x 198

Mental Math 1

Multiply mentally using the rule for 11.

1. 81 x 11 **891**	17. 54 x 11 **594**	33. 77 x 11 **847**	49. 38 x 11 **418**
2. 33 x 11 **363**	18. 14 x 11 **154**	34. 11 x 55 **605**	50. 46 x 11 **506**
3. 22 x 11 **242**	19. 90 x 11 **990**	35. 48 x 11 **528**	51. 56 x 11 **616**
4. 63 x 11 **693**	20. 11 x 43 **473**	36. 69 x 11 **759**	52. 88 x 11 **968**
5. 16 x 11 **176**	21. 11 x 61 **671**	37. 89 x 11 **979**	53. 75 x 11 **825**
6. 51 x 11 **561**	22. 11 x 23 **253**	38. 95 x 11 **1045**	54. 19 x 11 **209**
7. 40 x 11 **440**	23. 72 x 11 **792**	39. 47 x 11 **517**	55. 37 x 11 **407**
8. 11 x 52 **572**	24. 17 x 11 **187**	40. 39 x 11 **429**	56. 76 x 11 **836**
9. 26 x 11 **286**	25. 32 x 11 **352**	41. 29 x 11 **319**	57. 68 x 11 **748**
10. 44 x 11 **484**	26. 53 x 11 **583**	42. 11 x 57 **627**	58. 11 x 49 **539**
11. 11 x 15 **165**	27. 34 x 11 **374**	43. 74 x 11 **814**	59. 99 x 11 **1089**
12. 53 x 11 **583**	28. 11 x 21 **231**	44. 67 x 11 **737**	60. 11 x 78 **858**
13. 11 x 24 **264**	29. 25 x 11 **275**	45. 59 x 11 **649**	61. 79 x 11 **869**
14. 11 x 71 **781**	30. 13 x 11 **143**	46. 11 x 64 **704**	62. 66 x 11 **726**
15. 62 x 11 **682**	31. 36 x 11 **396**	47. 58 x 11 **638**	63. 28 x 11 **308**
16. 35 x 11 **385**	32. 42 x 11 **462**	48. 65 x 11 **715**	64. 11 x 73 **803**

Mental Math 2

Multiply mentally, pulling out factors of 10.

1. 40 x 30 x 700 **840,000**	17. 770 x 50 x 20 **770,000**	33. 120 x 60 x 50 **360,000**
2. 1600 x 300 x 200 **96,000,000**	18. 20 x 450 x 100 **900,000**	34. 100 x 190 x 300 **5,700,000**
3. 20 x 90 x 20 **36,000**	19. 2150 x 200 x 100 **43,000,000**	35. 2900 x 20 x 10 **580,000**
4. 500 x 70 x 50 **1,750,000**	20. 50 x 60 x 50 x 60 **9,000,000**	36. 50 x 300 x 200 **3,000,000**
5. 140 x 4000 **560,000**	21. 20 x 30 x 40 x 50 **1,200,000**	37. 590 x 100 x 20 **1,180,000**
6. 110 x 40 x 30 **132,000**	22. 250 x 20 x 300 **1,500,000**	38. 270 x 200 x 30 **1,620,000**
7. 160 x 20 x 20 **64,000**	23. 260 x 30 x 100 **780,000**	39. 200 x 130 x 30 **780,000**
8. 80 x 200 x 80 **1,280,000**	24. 20 x 60 x 400 **480,000**	40. 60 x 600 x 20 **720,000**
9. 150 x 60 x 20 **180,000**	25. 10 x 220 x 400 **880,000**	41. 70 x 700 x 20 **980,000**
10. 170 x 400 x 10 **680,000**	26. 140 x 30 x 20 **84,000**	42. 80 x 10 x 5000 **4,000,000**
11. 10 x 1450 x 20 **290,000**	27. 130 x 40 x 20 **104,000**	43. 5500 x 200 x 30 **33,000,000**
12. 330 x 100 x 20 **660,000**	28. 40 x 60 x 2000 **4,800,000**	44. 10 x 3520 x 20 **704,000**
13. 20 x 40 x 90 **72,000**	29. 70 x 20 x 50 x 10 **700,000**	45. 30 x 800 x 20 **480,000**
14. 60 x 30 x 50 **90,000**	30. 250 x 900 x 20 **4,500,000**	46. 150 x 3000 x 40 **18,000,000**
15. 80 x 400 x 30 **960,000**	31. 350 x 30 x 20 **210,000**	47. 200 x 9000 x 30 **54,000,000**
16. 150 x 40 x 200 **1,200,000**	32. 370 x 20 x 200 **1,480,000**	48. 90 x 70 x 200 **1,260,000**

Mental Math 3

Operate by mental math.

1. $6012 \div 3$ **2004**	17. $872 + 227$ **1099**	33. $46{,}036 \div 2$ **23,018**
2. $1775 + 9125$ **10,900**	18. $1253 + 8625$ **9878**	34. $7000 - 1620$ **5380**
3. 1719×3 **5157**	19. $2842 \div 14$ **203**	35. $6000 - 2370$ **3630**
4. $641{,}638 \div 2$ **320,819**	20. $2613 \div 13$ **201**	36. 1312×7 **9184**
5. 25×19 **475**	21. $9999 \div 11$ **909**	37. $4842 \div 6$ **807**
6. $8000 - 1650$ **6,350**	22. $5105 \div 5$ **1021**	38. 1415×5 **7075**
7. $9775 - 1825$ **7950**	23. $6789 - 4321$ **2468**	39. 1315×6 **7890**
8. $35{,}028 \div 7$ **5004**	24. $6045 \div 5$ **1209**	40. $36{,}168 \div 4$ **9042**
9. $3624 \div 12$ **302**	25. $9250 - 8725$ **525**	41. $701{,}015 \div 5$ **140,203**
10. $4530 \div 15$ **302**	26. 815×4 **3260**	42. $8725 + 1270$ **9995**
11. 1418×3 **4254**	27. $241{,}554 \div 3$ **80,518**	43. $9065 - 7255$ **1810**
12. $42{,}018 \div 6$ **7003**	28. $1765 + 1815$ **3580**	44. $6090 \div 15$ **406**
13. $4000 - 2380$ **1620**	29. $9876 - 1234$ **8642**	45. $8430 + 1570$ **10,000**
14. $7590 \div 15$ **506**	30. $4880 \div 16$ **305**	46. $777{,}777 \div 11$ **70,707**
15. 1312×6 **7872**	31. 1213×7 **8491**	47. 1211×9 **10,899**
16. $324{,}072 \div 8$ **40,509**	32. $543{,}624 \div 6$ **90,604**	48. $526{,}036 \div 4$ **131,509**

Mental Math 4

Multiply mentally using DPMA.

1.	18 x 5	50 + 40 **90**	17.	24 x 7	140 + 28 **168**	33.	59 x 5	250 + 45 **295**
2.	16 x 7	70 + 42 **112**	18.	39 x 8	240 + 72 **312**	34.	64 x 3	180 + 12 **192**
3.	17 x 4	40 + 28 **68**	19.	23 x 6	120 + 18 **138**	35.	33 x 6	180 + 18 **198**
4.	15 x 5	50 + 25 **75**	20.	35 x 6	180 + 30 **210**	36.	62 x 9	540 + 18 **558**
5.	58 x 8	400 + 64 **464**	21.	29 x 7	140 + 63 **203**	37.	37 x 4	120 + 28 **148**
6.	42 x 9	360 + 18 **378**	22.	19 x 8	80 + 72 **152**	38.	54 x 7	350 + 28 **378**
7.	57 x 6	300 + 42 **342**	23.	33 x 7	210 + 21 **231**	39.	16 x 6	60 + 36 **96**
8.	65 x 3	180 + 15 **195**	24.	27 x 5	100 + 35 **135**	40.	29 x 4	80 + 36 **116**
9.	14 x 8	80 + 32 **112**	25.	13 x 4	40 + 12 **52**	41.	55 x 8	400 + 40 **440**
10.	39 x 4	120 + 36 **156**	26.	17 x 7	70 + 49 **119**	42.	56 x 5	250 + 30 **280**
11.	15 x 8	80 + 40 **120**	27.	34 x 4	120 + 16 **136**	43.	39 x 3	90 + 27 **117**
12.	19 x 5	50 + 45 **95**	28.	19 x 9	90 + 81 **171**	44.	22 x 8	160 + 16 **176**
13.	17 x 9	90 + 63 **153**	29.	56 x 4	200 + 24 **224**	45.	26 x 3	60 + 18 **78**
14.	13 x 8	80 + 24 **104**	30.	14 x 3	30 + 12 **42**	46.	45 x 7	280 + 35 **315**
15.	19 x 6	60 + 54 **114**	31.	48 x 8	320 + 64 **384**	47.	39 x 9	270 + 81 **351**
16.	23 x 9	180 + 27 **207**	32.	46 x 6	240 + 36 **276**	48.	79 x 6	420 + 54 **474**

Mental Math 5

Multiply or divide mentally by the power of 10.

1. 3.143 x 10 **31.43**	17. 5.632 ÷ 100 **.05632**	33. 5.8 ÷ 1000 **.0058**
2. 3.145 ÷ 100 **.03145**	18. 23.32 x 1000 **23,320**	34. 235.8 ÷ 100 **2.358**
3. 0.2976 x 1000 **297.6**	19. 0.78 x 100 **78**	35. 7.442 x 100 **744.2**
4. 4.751 x 100 **475.1**	20. 8.9 ÷ 1000 **.0089**	36. 8.176 ÷ 10 **0.8176**
5. 2.31 ÷ 100 **.0231**	21. 0.3276 ÷ 10 **.03276**	37. 5.187 x 1000 **5187**
6. 35.78 x 1000 **35,780**	22. 4.21 x 100,000 **421,000**	38. 78.99 ÷ 100 **0.7899**
7. 643.89 ÷ 100 **6.4389**	23. 0.9345 x 100 **93.45**	39. 532.77 x 1000 **532,770**
8. 5.33 ÷ 1000 **.00533**	24. 236.7 x 1000 **236,700**	40. 56.333 x 100 **5633.3**
9. 9.544 x 100 **954.4**	25. 0.543 ÷ 10 **.0543**	41. 78.8 x 10,000 **788,000**
10. 82.3 ÷ 1000 **.0823**	26. 643.1 ÷ 100 **6.431**	42. 5.1 x 100,000 **510,000**
11. 306.7 ÷ 10,000 **.03067**	27. 0.22 ÷ 10 **0.022**	43. 2356.77 ÷ 10,000 **0.235677**
12. 876.9 ÷ 100 **8.769**	28. 9.11 x 1000 **9110**	44. 7.324 x 10,000 **73,240**
13. 123.45 x 10 **1234.5**	29. 25.75 x 10 **257.5**	45. 21.88 ÷ 1000 **.02188**
14. 56.72 x 100 **5672**	30. 68.7 x 100 **6870**	46. 0.832 x 100 **83.2**
15. 2.15 x 10,000 **21,500**	31. 0.5 ÷ 100 **.005**	47. 431.9 ÷ 10 **43.19**
16. 43.99 ÷ 1000 **.04399**	32. 3.1 ÷ 10 **0.31**	48. 0.7 x 1000 **700**

Mental Math 6

Operate mentally using compensation.

1. 998 x 4 1000 x 4 – 8 4000 – 8 **3992**	12. 92 x 8 100 x 8 – 64 800 – 64 **736**	23. 3355 + 997 3355 + 1000 – 3 4355 – 3 **4352**
2. 9375 – 97 9375 – 100 + 3 9275 + 3 **9278**	13. 398 + 97 400 + 100 – 5 500 – 5 **495**	24. 996 x 5 1000 x 5 – 20 5000 – 20 **4980**
3. 999 x 48 1000 x 48 – 48 48,000 – 48 **47,952**	14. 12 x 999 1000 x 12 – 12 12,000 – 12 **11,988**	25. 52 x 999 52 x 1000 – 52 52,000 – 52 **51,948**
4. 996 + 8117 1000 + 8117 – 4 9117 – 4 **9113**	15. 94 x 3 100 x 3 – 18 300 – 18 **282**	26. 6545 – 999 6545 – 1000 + 1 5545 + 1 **5546**
5. 99 x 17 100 x 17 – 17 1700 – 17 **1683**	16. 3929 – 996 3930 – 1000 – 1 + 4 2930 + 3 **2933**	27. 99 x 13 100 x 13 – 13 1300 – 13 **1287**
6. 2985 – 98 2985 – 100 + 2 2885 + 2 **2887**	17. 96 x 9 100 x 9 – 36 900 – 36 **864**	28. 4356 – 98 4356 – 100 + 2 4256 + 2 **4258**
7. 97 x 3 100 x 3 – 9 300 – 9 **291**	18. 998 + 281 1000 + 280 – 2 + 1 1280 – 1 **1279**	29. 102 x 41 100 x 41 + 82 4100 + 82 **4182**
8. 8463 + 997 8463 + 1000 – 3 9463 – 3 **9460**	19. 95 x 7 100 x 7 – 35 700 – 35 **665**	30. 5495 + 999 5495 + 1000 – 1 6495 – 1 **6494**
9. 95 x 9 100 x 9 – 45 900 – 45 **855**	20. 70 x 99 100 x 70 – 70 7000 – 70 **6930**	31. 4 x 9999 10,000 x 4 – 4 40,000 – 4 **39,996**
10. 7469 + 999 7469 + 1000 – 1 8469 – 1 **8468**	21. 8265 – 99 8265 – 100 + 1 8165 + 1 **8166**	32. 6109 – 997 6109 – 1000 + 3 5109 + 3 **5112**
11. 8 x 999 8 x 1000 – 8 8000 – 8 **7992**	22. 97 x 8 100 x 8 – 24 800 – 24 **776**	33. 997 x 3 1000 x 3 – 9 3000 – 9 **2991**

Mental Math 7

Multiply by 101, 1001, or 10001 mentally using DPMA.

1. 34 x 101 **3434**	17. 89 x 101 **8989**	33. 101 x 49 **4949**
2. 56 x 1001 **56,056**	18. 35 x 10,001 **350,035**	34. 31 x 1001 **31,031**
3. 101 x 23 **2323**	19. 76 x 1001 **76,076**	35. 60 x 101 **6060**
4. 547 x 1001 **547,547**	20. 192 x 10,001 **1,920,192**	36. 996 x 1001 **996,996**
5. 756 x 10,001 **7,560,756**	21. 1224 x 10,001 **12,241,224**	37. 246 x 10,001 **2,460,246**
6. 27 x 101 **2727**	22. 94 x 101 **9494**	38. 874 x 1001 **874,874**
7. 99 x 10,001 **990,099**	23. 513 x 1001 **513,513**	39. 635 x 1001 **635,635**
8. 1001 x 79 **79,079**	24. 57 x 10,001 **570,057**	40. 97 x 101 **9797**
9. 52 x 101 **5252**	25. 78 x 101 **7878**	41. 99 x 101 **9999**
10. 295 x 1001 **295,295**	26. 926 x 1001 **926,926**	42. 1001 x 141 **141,141**
11. 25 x 101 **2525**	27. 72 x 101 **7272**	43. 10,001 x 478 **4,780,478**
12. 125 x 1001 **125,125**	28. 1001 x 379 **379,379**	44. 723 x 1001 **723,723**
13. 1234 x 10,001 **12,341,234**	29. 70 x 1001 **70,070**	45. 28 x 101 **2828**
14. 41 x 101 **4141**	30. 22 x 101 **2222**	46. 75 x 101 **7575**
15. 82 x 1001 **82,082**	31. 137 x 1001 **137,137**	47. 54 x 1001 **54,054**
16. 65 x 10,001 **650,065**	32. 624 x 10,001 **6,240,624**	48. 1923 x 10,001 **19,231,923**

Mental Math 8

Multiply using DPMA.

1. 27 x 12 (27)(10) + (27)(2) 270 + 54 **324**	12. 41 x 23 (41)(20) + (41)(3) 820 + 123 **943**	23. 25 x 47 (25)(40) + (25)(7) 1000 + 175 **1175**
2. 52 x 13 (52)(10) + (52)(3) 520 + 156 **676**	13. 17 x 32 (17)(30) + (17)(2) 510 + 34 **544**	24. 80 x 14 (80)(10) + (80)(4) 800 + 320 **1120**
3. 41 x 15 (41)(10) + (41)(5) 410 + 205 **615**	14. 65 x 23 (65)(20) + (65)(3) 1300 + 195 **1495**	25. 85 x 21 (85)(20) + (85)(1) 1700 + 85 **1785**
4. 90 x 15 (90)(10) + (90)(5) 900 + 450 **1350**	15. 37 x 21 (37)(20) + (37)(1) 740 + 37 **777**	26. 33 x 15 (33)(10) + (33)(5) 330 + 165 **495**
5. 75 x 22 (75)(20) + (75)(2) 1500 + 150 **1650**	16. 55 x 31 (55)(30) + (55)(1) 1650 + 55 **1705**	27. 45 x 32 (45)(30) + (45)(2) 1350 + 90 **1440**
6. 18 x 12 (18)(10) + (18)(2) 180 + 36 **216**	17. 31 x 26 (31)(20) + (31)(6) 620 + 186 **806**	28. 25 x 38 (25)(30) + (25)(8) 750 + 200 **950**
7. 51 x 23 (51)(20) + (51)(3) 1020 + 153 **1173**	18. 35 x 16 (35)(10) + (35)(6) 350 + 210 **560**	29. 45 x 26 (45)(20) + (45)(6) 900 + 270 **1170**
8. 31 x 14 (31)(10) + (31)(4) 310 + 124 **434**	19. 23 x 32 (23)(30) + (23)(2) 690 + 46 **736**	30. 18 x 14 (18)(10) + (18)(4) 180 + 72 **252**
9. 42 x 41 (42)(40) + (42)(1) 1680 + 42 **1722**	20. 85 x 42 (85)(40) + (85)(2) 3400 + 170 **3570**	31. 95 x 24 (95)(20) + (95)(4) 1900 + 380 **2280**
10. 75 x 43 (75)(40) + (75)(3) 3000 + 225 **3225**	21. 39 x 21 (39)(20) + (39)(1) 780 + 39 **819**	32. 97 x 12 (97)(10) + (97)(2) 970 + 194 **1164**
11. 35 x 42 (35)(40) + (35)(2) 1400 + 70 **1470**	22. 55 x 42 (55)(40) + (55)(2) 2200 + 110 **2310**	33. 75 x 26 (75)(20) + (75)(6) 1500 + 450 **1950**

Mental Math 9

Operate efficiently using APA and CPA.

1. 32 + 7 + 59 + 8 + 6 + 25 + 83 + 44 + 1 32 + 8 + 83 + 7 + 59 + 1 + 44 + 6 + 25 40 + 90 + 60 + 50 + 25 = 240 + 25 = **265**	12. 53 + 66 + 95 + 21 + 19 + 15 − 43 + 24 53 − 43 + 66 + 24 + 95 + 15 + 21 + 19 10 + 90 + 110 + 40 = 100 + 150 = **250**
2. 9 + 23 + 6 + 51 + 28 + 15 + 7 + 34 + 2 9 + 51 + 23 + 7 + 6 + 34 + 28 + 2 + 15 60 + 30 + 40 + 30 + 15 = 160 + 15 = **175**	13. 89 + 63 + 72 + 80 + 38 − 44 + 77 + 25 89 − 44 + 25 + 63 + 77 + 80 + 72 + 38 70 + 140 + 80 + 110 = 150 + 250 = **400**
3. 69 + 53 + 66 + 48 + 67 + 71 + 32 + 44 69 + 71 + 53 + 67 + 66 + 44 + 48 + 32 140 + 120 + 110 + 80 = 250 + 200 = **450**	14. 119 + 346 + 153 + 427 + 154 + 331 119 + 331 + 427 + 153 + 346 + 154 450 + 580 + 500 = 450 + 1080 = **1530**
4. 21 + 76 + 52 + 79 + 87 + 54 + 33 + 18 21 + 79 + 76 + 54 + 52 + 18 + 87 + 33 100 + 130 + 70 + 120 = 200 + 220 = **420**	15. 51 + 65 + 142 + 95 + 9 + 27 + 28 − 87 51 + 9 + 65 + 95 + 142 + 28 + 27 − 87 60 + 160 + 170 − 60 = 220 + 110 = **330**
5. 17 + 81 + 74 + 92 + 83 + 88 + 96 + 99 17 + 83 + 81 + 99 + 74 + 96 + 92 + 88 100 + 180 + 170 + 180 = 270 + 360 = **630**	16. 532 − 24 + 257 + 74 + 20 + 133 + 138 532 + 138 + 74 − 24 + 257 + 133 + 20 670 + 50 + 390 + 20 = 700 + 410 = **1130**
6. 63 + 58 + 86 + 29 + 24 + 91 + 37 + 72 63 + 37 + 58 + 72 + 29 + 91 + 86 + 24 100 + 130 + 120 + 110 = 230 + 230 = **460**	17. 406 + 52 + 173 + 54 + 218 + 87 + 75 406 + 54 + 52 + 218 + 173 + 87 + 75 460 + 270 + 260 + 75 = 720 + 345 = **1065**
7. 77 + 82 + 26 + 41 + 23 + 64 + 39 + 28 77 + 23 + 82 + 28 + 26 + 64 + 41 + 39 100 + 110 + 90 + 80 = 180 + 200 = **380**	18. 64 − 77 + 49 − 53 + 27 + 41 − 84 + 93 49 + 41 + 64 − 84 + 27 − 77 + 93 − 53 90 − 20 − 50 + 40 = 70 − 10 = **60**
8. 84 + 93 + 57 + 89 + 22 + 36 + 98 + 31 84 + 36 + 93 + 57 + 89 + 31 + 22 + 98 120 + 150 + 120 + 120 = 150 + 360 = **510**	19. 105 + 501 + 13 + 149 + 327 + 90 − 55 105 − 55 + 501 + 149 + 327 + 13 + 90 50 + 650 + 340 + 90 = 790 + 340 = **1130**
9. 62 + 94 + 49 + 38 + 73 + 16 + 61 + 97 62 + 38 + 94 + 16 + 49 + 61 + 73 + 97 100 + 110 + 110 + 170 = 220 + 270 = **490**	20. 215 + 342 + 107 + 425 + 333 + 118 215 + 425 + 342 + 118 + 107 + 333 640 + 460 + 440 = 640 + 900 = **1540**
10. 78 + 59 + 47 + 25 + 51 + 85 + 12 + 13 78 + 12 + 59 + 51 + 47 + 13 + 25 + 85 90 + 110 + 60 + 110 = 150 + 220 = **370**	21. 95 + 26 + 32 + 77 + 88 + 54 + 43 − 15 95 − 15 + 26 + 54 + 32 + 88 + 77 + 43 80 + 80 + 120 + 120 = 160 + 240 = **400**
11. 27 + 65 + 19 + 34 + 56 + 75 + 43 + 11 27 + 43 + 65 + 75 + 19 + 11 + 34 + 56 70 + 140 + 30 + 90 = 190 + 140 = **330**	22. 72 + 89 + 17 − 85 + 55 − 22 − 29 + 73 72 − 22 + 89 − 29 + 17 + 73 + 55 − 85 50 + 60 + 90 − 30 = 110 + 60 = **170**

Mental Math 10

Operate by mental math.   Multiply by 10; divide by 2.

1.	567 x 5	5670 **2835**	17.	7461 x 5	74,610 **37,305**	33.	4429 x 5	44,290 **22,145**
2.	7288 x 5	72,880 **36,440**	18.	3665 x 5	36,650 **18,325**	34.	7656 x 5	76,560 **38,280**
3.	6615 x 5	66,150 **33,075**	19.	4656 x 5	46,560 **23,280**	35.	3447 x 5	34,470 **17,235**
4.	4813 x 5	48,130 **24,065**	20.	5025 x 5	50,250 **25,125**	36.	983 x 5	9830 **4915**
5.	389 x 5	3890 **1945**	21.	5423 x 5	54,230 **27,115**	37.	2654 x 5	26,540 **13,270**
6.	3662 x 5	36,620 **18,310**	22.	1688 x 5	16,880 **8440**	38.	8894 x 5	88,940 **44,470**
7.	8692 x 5	86,920 **43,460**	23.	487 x 5	4870 **2435**	39.	9065 x 5	90,650 **45,325**
8.	16,854 x 5	168,540 **84,270**	24.	1846 x 5	18,460 **9230**	40.	964 x 5	9640 **4820**
9.	925 x 5	9250 **4625**	25.	5436 x 5	54,360 **27,180**	41.	7292 x 5	72,920 **36,460**
10.	6672 x 5	66,720 **33,360**	26.	926 x 5	9260 **4630**	42.	3872 x 5	38,720 **19,360**
11.	983 x 5	9830 **4915**	27.	18,662 x 5	186,620 **93,310**	43.	3065 x 5	30,650 **15,325**
12.	745 x 5	7450 **3725**	28.	7845 x 5	78,450 **39,225**	44.	5678 x 5	56,780 **28,390**
13.	2847 x 5	28,470 **14,235**	29.	526 x 5	5260 **2630**	45.	18,463 x 5	184,630 **92,315**
14.	589 x 5	5890 **2945**	30.	6855 x 5	68,550 **34,275**	46.	9256 x 5	92,560 **46,280**
15.	9078 x 5	90,780 **45,390**	31.	9445 x 5	94,450 **47,225**	47.	3229 x 5	32,290 **16,145**
16.	643 x 5	6430 **3215**	32.	2889 x 5	28,890 **14,445**	48.	245 x 5	2450 **1225**

Mental Math 11

Operate as follows: for #1–8 think of a circle; for #9–16 think of quarters; for #17–24 think of a deck of cards or a clock; for #25–32 think of a calendar; for #33–40 think of polygons; for #41–48 think of palindromes.

1.	360 ÷ 8	**45**	17.	13 x 4	**52**	33.	108 x 5	**540**
2.	72 x 5	**360**	18.	3 x 4 + 10 x 4	**52**	34.	45 + 45 + 90	**180**
3.	360 ÷ 15	**24**	19.	.25 x 52	**13**	35.	180 x 5	**900**
4.	.25 x 360	**90**	20.	(60)(2/3)	**40**	36.	60 x 3	**180**
5.	360 ÷ 40	**9**	21.	12 x 5	**60**	37.	90 x 4	**360**
6.	360 ÷ 18	**20**	22.	60 x 60	**3600**	38.	12 x 9 / 2	**54**
7.	60 x 6	**360**	23.	52 ÷ 13	**4**	39.	1260 ÷ 7	**180**
8.	2 x 180	**360**	24.	60 ÷ 4	**15**	40.	180 x 6	**1080**
9.	25 x 16	**400**	25.	365 ÷ 7	$52\frac{1}{7}$	41.	3669663 ÷ 3	**1223221**
10.	7 x 25	**175**	26.	7 x 31 + 4 x 30 + 29	**366**	42.	98689 – 12321	**86368**
11.	25 x 32	**800**	27.	.25 x 12	**3**	43.	212 + 424	**636**
12.	23 x 25	**575**	28.	365 x 4 + 1	**1461**	44.	8558 – 4114	**4444**
13.	17 x 25	**425**	29.	52 x 7	**364**	45.	3443 + 1221	**4664**
14.	9 x 25	**225**	30.	365 ÷ 12	$30\frac{5}{12}$	46.	989 – 454	**535**
15.	25 x 26	**650**	31.	31 + 28 + 31	**90**	47.	6776 – 2332	**4444**
16.	12 x 25	**300**	32.	365 ÷ 4	**91.25**	48.	5335 + 2552	**7887**

Mental Math 12

Operate by factoring (DPMA backwards).

1.

$$\frac{258^2 - 258}{258} \qquad \frac{258\,(258 - 1)}{258}$$

257

11.

$$\frac{3^4 - 3^2 + 3}{3} \qquad 3^3 - 3 + 1$$
$$27 - 3 + 1$$

25

2.

12.5% of 52 − 12.5% of 4

12.5% of (52 − 4)
(1/8) (48)

6

12.

497 x 784 + 503 x 784

(497 + 503) (784)
(1000) (784)

784,000

3.

1378 x 56 + 1378 x 44

(1378) (56 + 44)
(1378) (100)

137,800

13.

173 x 85 − 85 x 52 + 79 x 85

(173 − 52 + 79) (85)
(200) (85)

17,000

4.

689 x 499 + 311 x 499

(689 + 311) (499)
(1000) (499)

499,000

14.

5% of 77 − 5% of 32 + 5% of 45

5% of (77 − 32 + 45)
(1/20) (90)

4.5

5.

0.5 of 71 + 0.5 of 39

(0.5) (71 + 39)
(0.5) (110)

55

15.

87 x 7145 + 7145 x 13

(87 + 13) (7145)
(100) (7145)

714,500

6.

$$\frac{175^2 + 350}{175} \qquad \frac{175\,(175 + 2)}{175}$$

177

16.

$$\frac{343^2 + 343}{343} \qquad \frac{343\,(343 + 1)}{343}$$

344

7.

1536 x 78 + 1536 x 22

(1536) (78 + 22)
(1536) (100)

153,600

17.

393 x 65 + 65 x 47 − 40 x 65

(393 + 47 − 40) (65)
(400) (65)

26,000

8.

0.75 x 99 − 0.75 x 15

(0.75) (99 − 15)
(3/4) (84)

63

18.

62.5% of 908 − 62.5% of 108

62.5% of (908 − 108)
(5/8) (800)

500

9.

$$\frac{25^3 - 25^2 + 75}{25} \qquad 25^2 - 25 + 3$$
$$625 - 25 + 3$$

603

19.

$$\frac{15^3 - 15^2 + 30}{15} \qquad 15^2 - 15 + 2$$
$$225 - 15 + 2$$

212

10.

0.25 x 157 − 0.25 x 33

(0.25) (157 − 33)
(1/4) (124)

31

20.

half of 357 + half of 143

(1/2) (357 + 143)
(1/2) (500)

250

Mental Math 13

Multiply mentally using the rule for 11.

1. 235 x 11 **2585**	17. 11 x 7152 **78,672**	33. 11 x 23,625 **259,875**
2. 363 x 11 **3993**	18. 6324 x 11 **69,564**	34. 40,817 x 11 **448,987**
3. 451 x 11 **4961**	19. 11 x 1253 **13,783**	35. 57,391 x 11 **631,301**
4. 526 x 11 **5786**	20. 1632 x 11 **17,952**	36. 43,625 x 11 **479,875**
5. 716 x 11 **7876**	21. 5436 x 11 **59,796**	37. 85,157 x 11 **936,727**
6. 11 x 815 **8965**	22. 6145 x 11 **67,595**	38. 57,248 x 11 **629,728**
7. 352 x 11 **3872**	23. 4445 x 11 **48,895**	39. 35,451 x 11 **389,961**
8. 11 x 534 **5874**	24. 5674 x 11 **62,414**	40. 11 x 29,415 **323,565**
9. 262 x 11 **2882**	25. 2953 x 11 **32,483**	41. 11,632 x 11 **127,952**
10. 436 x 11 **4796**	26. 4652 x 11 **51,172**	42. 66,743 x 11 **734,173**
11. 11 x 624 **6864**	27. 11 x 2396 **26,356**	43. 72,343 x 11 **795,773**
12. 814 x 11 **8954**	28. 8546 x 11 **94,006**	44. 11 x 35,814 **393,954**
13. 11 x 636 **6996**	29. 4539 x 11 **49,929**	45. 48,365 x 11 **532,015**
14. 11 x 271 **2981**	30. 7683 x 11 **84,513**	46. 66,327 x 11 **729,597**
15. 345 x 11 **3795**	31. 3939 x 11 **43,329**	47. 27,248 x 11 **299,728**
16. 11 x 535 **5885**	32. 6758 x 11 **74,338**	48. 17,962 x 11 **197,582**

Mental Math 14

Multiply using the difference of two squares method.

1. 53 x 47	(50 + 3) (50 − 3) 2500 − 9 **2491**	12. 31 x 29	(30 + 1) (30 − 1) 900 − 1 **899**
2. 603 x 597	(600 + 3) (600 − 3) 360,000 − 9 **359,991**	13. 137 x 123	(130 + 7) (130 − 7) 16,900 − 49 **16,851**
3. 41 x 39	(40 + 1) (40 − 1) 1600 − 1 **1599**	14. 76 x 84	(80 − 4) (80 + 4) 6400 − 16 **6384**
4. 66 x 74	(70 − 4) (70 + 4) 4900 − 16 **4884**	15. 109 x 91	(100 + 9) (100 − 9) 10,000 − 81 **9919**
5. 67 x 53	(60 + 7) (60 − 7) 3600 − 49 **3551**	16. 153 x 147	(150 + 3) (150 − 3) 22,500 − 9 **22,491**
6. 95 x 105	(100 − 5) (100 + 5) 10,000 − 25 **9975**	17. 37 x 23	(30 + 7) (30 − 7) 900 − 49 **851**
7. 36 x 44	(40 − 4) (40 + 4) 1600 − 16 **1584**	18. 85 x 95	(90 − 5) (90 + 5) 8100 − 25 **8075**
8. 409 x 391	(400 + 9) (400 − 9) 160,000 − 81 **159,919**	19. 204 x 196	(200 + 4) (200 − 4) 40,000 − 16 **39,984**
9. 91 x 89	(90 + 1) (90 − 1) 8100 − 1 **8099**	20. 55 x 65	(60 − 5) (60 + 5) 3600 − 25 **3575**
10. 32 x 28	(30 + 2) (30 − 2) 900 − 4 **896**	21. 98 x 102	(100 − 2) (100 + 2) 10,000 − 4 **9996**
11. 122 x 118	(120 + 2) (120 − 2) 14,400 − 4 **14,396**	22. 294 x 306	(300 − 6) (300 + 6) 90,000 − 36 **89,964**

Mental Math 15

Multiply or divide by 4 as indicated.

Multiply by 4 by doubling twice.
Divide by 4 by halving twice.

1. 65 x 4 130 **260**	17. 960 ÷ 4 480 **240**	33. 135 x 4 270 **540**	49. 674 ÷ 4 337 **168.5**
2. 46 x 4 92 **184**	18. 780 ÷ 4 390 **195**	34. 240 x 4 480 **960**	50. 838 ÷ 4 419 **209.5**
3. 53 x 4 106 **212**	19. 624 ÷ 4 312 **156**	35. 345 x 4 690 **1380**	51. 562 ÷ 4 281 **140.5**
4. 75 x 4 150 **300**	20. 740 ÷ 4 370 **185**	36. 275 x 4 550 **1100**	52. 786 ÷ 4 393 **196.5**
5. 4 x 19 38 **76**	21. 340 ÷ 4 170 **85**	37. 4 x 185 370 **740**	53. 382 ÷ 4 191 **95.5**
6. 4 x 38 76 **152**	22. 568 ÷ 4 284 **142**	38. 4 x 460 920 **1840**	54. 926 ÷ 4 463 **231.5**
7. 27 x 4 54 **108**	23. 520 ÷ 4 260 **130**	39. 280 x 4 560 **1120**	55. 462 ÷ 4 231 **115.5**
8. 95 x 4 190 **380**	24. 920 ÷ 4 460 **230**	40. 475 x 4 950 **1900**	56. 470 ÷ 4 235 **117.5**
9. 4 x 34 68 **136**	25. 680 ÷ 4 340 **170**	41. 4 x 630 1260 **2520**	57. 286 ÷ 4 143 **71.5**
10. 56 x 4 112 **224**	26. 660 ÷ 4 330 **165**	42. 370 x 4 740 **1480**	58. 654 ÷ 4 327 **163.5**
11. 4 x 44 88 **176**	27. 344 ÷ 4 172 **86**	43. 4 x 235 470 **940**	59. 814 ÷ 4 407 **203.5**
12. 29 x 4 58 **116**	28. 760 ÷ 4 380 **190**	44. 531 x 4 1062 **2124**	60. 542 ÷ 4 271 **135.5**
13. 17 x 4 34 **68**	29. 588 ÷ 4 294 **147**	45. 175 x 4 350 **700**	61. 746 ÷ 4 373 **186.5**
14. 4 x 85 170 **340**	30. 984 ÷ 4 492 **246**	46. 4 x 914 1828 **3656**	62. 490 ÷ 4 245 **122.5**
15. 47 x 4 94 **188**	31. 388 ÷ 4 194 **97**	47. 724 x 4 1448 **2896**	63. 366 ÷ 4 183 **91.5**
16. 39 x 4 78 **156**	32. 676 ÷ 4 338 **169**	48. 617 x 4 1234 **2468**	64. 722 ÷ 4 361 **180.5**

MAVA Math: Enhanced Skills Solutions Copyright © 2015 Marla Weiss

Mental Math 16

Square using the ones-digit-5 method.

1.	35^2	3 x 4 = 12	**1225**	17.	125^2	12 x 13 = 144 + 12 = 156	**15,625**	
2.	75^2	7 x 8 = 56	**5625**	18.	495^2	49 x 50 = 2500 − 50	**245,025**	
3.	45^2	4 x 5 = 20	**2025**	19.	905^2	90 x 91 = 8100 + 90	**819,025**	
4.	85^2	8 x 9 = 72	**7225**	20.	395^2	39 x 40 = 1600 − 40	**156,025**	
5.	95^2	9 x 10 = 90	**9025**	21.	505^2	50 x 51 = 2500 + 50	**255,025**	
6.	25^2	2 x 3 = 6	**625**	22.	805^2	80 x 81 = 6400 + 80	**648,025**	
7.	55^2	5 x 6 = 30	**3025**	23.	595^2	59 x 60 = 3600 − 60	**354,025**	
8.	15^2	1 x 2 = 2	**225**	24.	165^2	16 x 17 = 256 + 16	**27,225**	
9.	65^2	6 x 7 = 42	**4225**	25.	135^2	13 x 14 = 169 + 13	**18,225**	
10.	115^2	11 x 12 = 132	**13,225**	26.	895^2	89 x 90 = 8100 − 90	**801,025**	
11.	195^2	19 x 20 = 380	**38,025**	27.	145^2	14 x 15 = 196 + 14	**21,025**	
12.	205^2	20 x 21 = 420	**42,025**	28.	255^2	25 x 26 = 625 + 25	**65,025**	
13.	105^2	10 x 11 = 110	**11,025**	29.	155^2	15 x 16 = 225 + 15	**24,025**	
14.	305^2	30 x 31 = 930	**93,025**	30.	1005^2	100 x 101 = 10,100	**1,010,025**	
15.	405^2	40 x 41 = 1640	**164,025**	31.	995^2	99 x 100 = 9900	**990,025**	
16.	605^2	60 x 61 = 3660	**366,025**	32.	705^2	70 x 71 = 4900 + 70	**497,025**	

Mixed Numbers 1

Convert from mixed number to fraction by mental math.

Convert from fraction to mixed number by mental math.

1.	$12\frac{13}{15}$	$\frac{193}{15}$	17.	$49\frac{6}{11}$	$\frac{545}{11}$	33.	$\frac{94}{15}$	$6\frac{4}{15}$	49.	$\frac{97}{4}$	$24\frac{1}{4}$
2.	$11\frac{12}{55}$	$\frac{617}{55}$	18.	$5\frac{23}{32}$	$\frac{183}{32}$	34.	$\frac{101}{12}$	$8\frac{5}{12}$	50.	$\frac{186}{19}$	$9\frac{15}{19}$
3.	$10\frac{16}{45}$	$\frac{466}{45}$	19.	$25\frac{13}{25}$	$\frac{638}{25}$	35.	$\frac{173}{13}$	$13\frac{4}{13}$	51.	$\frac{437}{25}$	$17\frac{12}{25}$
4.	$13\frac{7}{13}$	$\frac{176}{13}$	20.	$17\frac{12}{17}$	$\frac{301}{17}$	36.	$\frac{152}{3}$	$50\frac{2}{3}$	52.	$\frac{199}{18}$	$11\frac{1}{18}$
5.	$6\frac{10}{19}$	$\frac{124}{19}$	21.	$30\frac{21}{23}$	$\frac{711}{23}$	37.	$\frac{245}{6}$	$40\frac{5}{6}$	53.	$\frac{283}{35}$	$8\frac{3}{35}$
6.	$19\frac{17}{20}$	$\frac{397}{20}$	22.	$12\frac{11}{12}$	$\frac{155}{12}$	38.	$\frac{97}{24}$	$4\frac{1}{24}$	54.	$\frac{89}{22}$	$4\frac{1}{22}$
7.	$14\frac{9}{14}$	$\frac{205}{14}$	23.	$26\frac{4}{11}$	$\frac{290}{11}$	39.	$\frac{168}{11}$	$15\frac{3}{11}$	55.	$\frac{195}{19}$	$10\frac{5}{19}$
8.	$11\frac{21}{34}$	$\frac{395}{34}$	24.	$11\frac{11}{53}$	$\frac{594}{53}$	40.	$\frac{599}{9}$	$66\frac{5}{9}$	56.	$\frac{169}{11}$	$15\frac{4}{11}$
9.	$18\frac{17}{30}$	$\frac{557}{30}$	25.	$43\frac{7}{11}$	$\frac{480}{11}$	41.	$\frac{635}{7}$	$90\frac{5}{7}$	57.	$\frac{319}{20}$	$15\frac{19}{20}$
10.	$15\frac{8}{15}$	$\frac{233}{15}$	26.	$40\frac{19}{20}$	$\frac{819}{20}$	42.	$\frac{107}{26}$	$4\frac{3}{26}$	58.	$\frac{151}{35}$	$4\frac{11}{35}$
11.	$38\frac{8}{11}$	$\frac{426}{11}$	27.	$20\frac{13}{37}$	$\frac{753}{37}$	43.	$\frac{113}{14}$	$8\frac{1}{14}$	59.	$\frac{207}{40}$	$5\frac{7}{40}$
12.	$13\frac{37}{50}$	$\frac{687}{50}$	28.	$25\frac{15}{31}$	$\frac{790}{31}$	44.	$\frac{107}{21}$	$5\frac{2}{21}$	60.	$\frac{333}{31}$	$10\frac{23}{31}$
13.	$16\frac{15}{16}$	$\frac{271}{16}$	29.	$22\frac{8}{25}$	$\frac{558}{25}$	45.	$\frac{137}{17}$	$8\frac{1}{17}$	61.	$\frac{60}{19}$	$3\frac{3}{19}$
14.	$12\frac{5}{21}$	$\frac{257}{21}$	30.	$14\frac{9}{10}$	$\frac{149}{10}$	46.	$\frac{373}{75}$	$4\frac{73}{75}$	62.	$\frac{229}{55}$	$4\frac{9}{55}$
15.	$19\frac{9}{19}$	$\frac{370}{19}$	31.	$10\frac{42}{43}$	$\frac{472}{43}$	47.	$\frac{308}{51}$	$6\frac{2}{51}$	63.	$\frac{779}{10}$	$77\frac{9}{10}$
16.	$32\frac{15}{32}$	$\frac{1039}{32}$	32.	$15\frac{7}{12}$	$\frac{187}{12}$	48.	$\frac{55}{16}$	$3\frac{7}{16}$	64.	$\frac{817}{9}$	$90\frac{7}{9}$

Mixed Numbers 2

Multiply by converting to fractions.

Divide by converting to fractions.

1. $3\frac{1}{2}$ x $5\frac{1}{7}$ $\frac{7}{2}$ x $\frac{36}{7}$ **18**

17. $7\frac{2}{5} \div 4\frac{5}{8}$ $\frac{37}{5}$ x $\frac{8}{37}$ $\frac{8}{5}$

2. $6\frac{2}{5}$ x $4\frac{3}{8}$ $\frac{32}{5}$ x $\frac{35}{8}$ **28**

18. $3\frac{7}{9} \div 5\frac{2}{3}$ $\frac{34}{9}$ x $\frac{3}{17}$ $\frac{2}{3}$

3. $2\frac{7}{9}$ x $4\frac{1}{5}$ $\frac{25}{9}$ x $\frac{21}{5}$ $\frac{35}{3}$

19. $4\frac{5}{6} \div 6\frac{4}{9}$ $\frac{29}{6}$ x $\frac{9}{58}$ $\frac{3}{4}$

4. $7\frac{1}{7}$ x $4\frac{1}{5}$ $\frac{50}{7}$ x $\frac{21}{5}$ **30**

20. $6\frac{1}{9} \div 3\frac{2}{3}$ $\frac{55}{9}$ x $\frac{3}{11}$ $\frac{5}{3}$

5. $6\frac{3}{4}$ x $7\frac{1}{9}$ $\frac{27}{4}$ x $\frac{64}{9}$ **48**

21. $5\frac{3}{7} \div 4\frac{3}{4}$ $\frac{38}{7}$ x $\frac{4}{19}$ $\frac{8}{7}$

6. $3\frac{4}{7}$ x $8\frac{2}{5}$ $\frac{25}{7}$ x $\frac{42}{5}$ **30**

22. $5\frac{4}{7} \div 6\frac{1}{2}$ $\frac{39}{7}$ x $\frac{2}{13}$ $\frac{6}{7}$

7. $4\frac{3}{8}$ x $2\frac{2}{5}$ $\frac{35}{8}$ x $\frac{12}{5}$ $\frac{21}{2}$

23. $8\frac{2}{5} \div 1\frac{2}{5}$ $\frac{42}{5}$ x $\frac{5}{7}$ **6**

8. $4\frac{2}{7}$ x $5\frac{4}{9}$ $\frac{30}{7}$ x $\frac{49}{9}$ $\frac{70}{3}$

24. $5\frac{2}{3} \div 9\frac{4}{9}$ $\frac{17}{3}$ x $\frac{9}{85}$ $\frac{3}{5}$

9. $2\frac{4}{5}$ x $8\frac{3}{4}$ $\frac{14}{5}$ x $\frac{35}{4}$ $\frac{49}{2}$

25. $11\frac{3}{5} \div 7\frac{1}{4}$ $\frac{58}{5}$ x $\frac{4}{29}$ $\frac{8}{5}$

10. $9\frac{4}{5}$ x $3\frac{4}{7}$ $\frac{49}{5}$ x $\frac{25}{7}$ **35**

26. $12\frac{1}{2} \div 5\frac{5}{6}$ $\frac{25}{2}$ x $\frac{6}{35}$ $\frac{15}{7}$

11. $11\frac{2}{3}$ x $4\frac{5}{7}$ $\frac{35}{3}$ x $\frac{33}{7}$ **55**

27. $10\frac{5}{8} \div 3\frac{7}{9}$ $\frac{85}{8}$ x $\frac{9}{34}$ $\frac{45}{16}$

12. $12\frac{1}{2}$ x $3\frac{3}{5}$ $\frac{25}{2}$ x $\frac{18}{5}$ **45**

28. $9\frac{3}{4} \div 3\frac{1}{4}$ $\frac{39}{4}$ x $\frac{4}{13}$ **3**

13. $13\frac{1}{3}$ x $9\frac{3}{4}$ $\frac{40}{3}$ x $\frac{39}{4}$ **130**

29. $2\frac{3}{5} \div 3\frac{9}{10}$ $\frac{13}{5}$ x $\frac{10}{39}$ $\frac{2}{3}$

14. $6\frac{2}{5}$ x $3\frac{1}{8}$ $\frac{32}{5}$ x $\frac{25}{8}$ **20**

30. $15\frac{5}{6} \div 5\frac{3}{7}$ $\frac{95}{6}$ x $\frac{7}{38}$ $\frac{35}{12}$

15. $2\frac{2}{5}$ x $4\frac{4}{9}$ $\frac{12}{5}$ x $\frac{40}{9}$ $\frac{32}{3}$

31. $8\frac{2}{7} \div 2\frac{1}{14}$ $\frac{58}{7}$ x $\frac{14}{29}$ **4**

16. $5\frac{1}{2}$ x $5\frac{1}{11}$ $\frac{11}{2}$ x $\frac{56}{11}$ **28**

32. $8\frac{2}{3} \div 5\frac{1}{5}$ $\frac{26}{3}$ x $\frac{5}{26}$ $\frac{5}{3}$

MAVA Math: Enhanced Skills Solutions   Copyright © 2015 Marla Weiss

Mixed Numbers 3

Add and subtract as indicated.

1.
$$\begin{array}{ccc} 36 & 45 & 10 \\ 3\dfrac{18}{25} + 2\dfrac{9}{10} - 1\dfrac{1}{5} \\ 50 & 50 & 50 \end{array}$$
$4\dfrac{71}{50}$

$\mathbf{5\dfrac{21}{50}}$

2.
$$\begin{array}{ccc} 32 & 26 & 15 \\ 6\dfrac{4}{5} + 3\dfrac{13}{20} - 4\dfrac{3}{8} \\ 40 & 40 & 40 \end{array}$$
$5\dfrac{43}{40}$

$\mathbf{6\dfrac{3}{40}}$

3.
$$\begin{array}{ccc} 52 & 35 & 24 \\ 5\dfrac{26}{45} + 8\dfrac{7}{18} - 2\dfrac{4}{15} \\ 90 & 90 & 90 \end{array}$$
$11\dfrac{63}{90}$

$\mathbf{11\dfrac{7}{10}}$

4.
$$\begin{array}{ccc} 88 & 25 & 51 \\ 3\dfrac{11}{15} + 9\dfrac{5}{24} - 2\dfrac{17}{40} \\ 120 & 120 & 120 \end{array}$$
$10\dfrac{62}{120}$

$\mathbf{10\dfrac{31}{60}}$

5.
$$\begin{array}{cc} 34 & 65 \\ 9\dfrac{17}{35} + 2\dfrac{13}{14} - 3\dfrac{31}{70} \\ 70 & 70 \end{array}$$
$8\dfrac{68}{70}$

$\mathbf{8\dfrac{34}{35}}$

6.
$$\begin{array}{ccc} 49 & 33 & 20 \\ 3\dfrac{7}{12} + 2\dfrac{11}{28} + 4\dfrac{5}{21} \\ 84 & 84 & 84 \end{array}$$
$9\dfrac{102}{84}$
$\dfrac{18}{84}$

$\mathbf{10\dfrac{3}{14}}$

7.
$$\begin{array}{ccc} 95 & 50 & 63 \\ 5\dfrac{19}{36} + 1\dfrac{5}{18} + 2\dfrac{7}{20} \\ 180 & 180 & 180 \end{array}$$
$8\dfrac{208}{180}$
$\dfrac{28}{180}$

$\mathbf{9\dfrac{7}{45}}$

8.
$$\begin{array}{ccc} 70 & 27 & 15 \\ 9\dfrac{7}{9} - 1\dfrac{3}{10} - 2\dfrac{1}{6} \\ 90 & 90 & 90 \end{array}$$
$6\dfrac{28}{90}$

$\mathbf{6\dfrac{14}{45}}$

9.
$$\begin{array}{ccc} 64 & 24 & 15 \\ 9\dfrac{4}{5} - 3\dfrac{3}{10} + 4\dfrac{3}{16} \\ 80 & 80 & 80 \end{array}$$
$10\dfrac{55}{80}$

$\mathbf{10\dfrac{11}{16}}$

10.
$$\begin{array}{ccc} 44 & 25 & 14 \\ 4\dfrac{11}{15} + 2\dfrac{5}{12} - 1\dfrac{7}{30} \\ 60 & 60 & 60 \end{array}$$
$5\dfrac{55}{60}$

$\mathbf{5\dfrac{11}{12}}$

11.
$$\begin{array}{ccc} 27 & 12 & 49 \\ 1\dfrac{3}{7} - 4\dfrac{4}{21} + 8\dfrac{7}{9} \\ 63 & 63 & 63 \end{array}$$
$5\dfrac{64}{63}$

$\mathbf{6\dfrac{1}{63}}$

12.
$$\begin{array}{ccc} 30 & 51 & 42 \\ 8\dfrac{15}{33} + 5\dfrac{17}{22} - 4\dfrac{7}{11} \\ 66 & 66 & 66 \end{array}$$
$9\dfrac{39}{66}$

$\mathbf{9\dfrac{13}{22}}$

13.
$$\begin{array}{ccc} 27 & 35 & 15 \\ 5\dfrac{9}{25} + 1\dfrac{7}{15} + 2\dfrac{1}{5} \\ 75 & 75 & 75 \end{array}$$
$8\dfrac{77}{75}$

$\mathbf{9\dfrac{2}{75}}$

14.
$$\begin{array}{ccc} 33 & 90 & 22 \\ 5\dfrac{11}{33} + 1\dfrac{10}{11} - 2\dfrac{2}{9} \\ 99 & 99 & 99 \end{array}$$
$4\dfrac{101}{99}$

$\mathbf{5\dfrac{2}{99}}$

242

Mixed Numbers 4

Operate using DPMA and simplify.

1. $9\frac{1}{2}$ x $6\frac{2}{3}$

$54 + 6 + 3 + \frac{1}{3}$ $\quad 63\frac{1}{3}$

2. $5\frac{3}{4}$ x $8\frac{3}{5}$

$40 + 3 + 6 + \frac{9}{20}$ $\quad 49\frac{9}{20}$

3. $14\frac{1}{2}$ x $2\frac{3}{7}$

$28 + 6 + 1 + \frac{3}{14}$ $\quad 35\frac{3}{14}$

4. $16\frac{3}{4}$ x $4\frac{3}{8}$

$64 + 6 + 3 + \frac{9}{32}$ $\quad 73\frac{9}{32}$

5. $12\frac{1}{3}$ x $6\frac{1}{4}$

$72 + 3 + 2 + \frac{1}{12}$ $\quad 77\frac{1}{12}$

6. $10\frac{6}{7}$ x $7\frac{4}{5}$

$70 + 8 + 6 + \frac{24}{35}$ $\quad 84\frac{24}{35}$

7. $11\frac{5}{6}$ x $6\frac{2}{11}$

$66 + 2 + 5 + \frac{5}{33}$ $\quad 73\frac{5}{33}$

8. $12\frac{1}{4}$ x $4\frac{5}{12}$

$48 + 5 + 1 + \frac{5}{48}$ $\quad 54\frac{5}{48}$

9. $15\frac{2}{3}$ x $6\frac{2}{5}$

$90 + 6 + 4 + \frac{4}{15}$ $\quad 100\frac{4}{15}$

10. $45\frac{5}{6}$ x $2\frac{8}{15}$

$90 + 24 + \frac{5}{3} + \frac{4}{9}$ $\quad 116\frac{1}{9}$

11. $36\frac{3}{4}$ x $2\frac{7}{12}$

$72 + 21 + \frac{3}{2} + \frac{7}{16}$ $\quad 94\frac{15}{16}$

12. $22\frac{1}{6}$ x $3\frac{6}{11}$

$66 + 12 + \frac{1}{2} + \frac{1}{11}$ $\quad 78\frac{13}{22}$

13. $39\frac{3}{5}$ x $2\frac{5}{13}$

$78 + 15 + \frac{6}{5} + \frac{3}{13}$ $\quad 94\frac{28}{65}$

14. $18\frac{4}{7}$ x $4\frac{7}{18}$

$72 + 7 + \frac{16}{7} + \frac{2}{9}$ $\quad 81\frac{32}{63}$

15. $15\frac{1}{4}$ x $5\frac{2}{3}$

$75 + 10 + \frac{5}{4} + \frac{1}{6}$ $\quad 86\frac{5}{12}$

16. $35\frac{1}{3}$ x $2\frac{6}{7}$

$70 + 30 + \frac{2}{3} + \frac{2}{7}$ $\quad 100\frac{20}{21}$

17. $10\frac{5}{6}$ x $9\frac{3}{5}$

$90 + 6 + \frac{15}{2} + \frac{1}{2}$ $\quad 104$

18. $16\frac{4}{5}$ x $5\frac{5}{8}$

$80 + 10 + 4 + \frac{1}{2}$ $\quad 94\frac{1}{2}$

19. $18\frac{1}{4}$ x $4\frac{8}{9}$

$72 + 16 + 1 + \frac{2}{9}$ $\quad 89\frac{2}{9}$

20. $21\frac{7}{8}$ x $4\frac{4}{7}$

$84 + 12 + \frac{7}{2} + \frac{1}{2}$ $\quad 100$

21. $44\frac{2}{5}$ x $2\frac{5}{22}$

$88 + 10 + \frac{4}{5} + \frac{1}{11}$ $\quad 98\frac{49}{55}$

22. $24\frac{2}{7}$ x $3\frac{7}{12}$

$72 + 14 + \frac{6}{7} + \frac{1}{6}$ $\quad 87\frac{1}{42}$

23. $12\frac{4}{5}$ x $6\frac{3}{4}$

$72 + 9 + \frac{24}{5} + \frac{3}{5}$ $\quad 86\frac{2}{5}$

24. $30\frac{3}{4}$ x $8\frac{4}{15}$

$240 + 8 + 6 + \frac{1}{5}$ $\quad 254\frac{1}{5}$

Modulo Arithmetic 1

Complete each congruence.

1. 15 ≡ __7__ (MOD 8)	17. 36 ≡ __3__ (MOD 11)	33. 47 ≡ __3__ (MOD 4)
2. 21 ≡ __1__ (MOD 4)	18. 54 ≡ __2__ (MOD 13)	34. 50 ≡ __2__ (MOD 8)
3. 12 ≡ __5__ (MOD 7)	19. 35 ≡ __2__ (MOD 3)	35. 52 ≡ __0__ (MOD 13)
4. 63 ≡ __3__ (MOD 5)	20. 39 ≡ __4__ (MOD 7)	36. 57 ≡ __0__ (MOD 3)
5. 29 ≡ __9__ (MOD 10)	21. 86 ≡ __6__ (MOD 8)	37. 66 ≡ __3__ (MOD 7)
6. 22 ≡ __4__ (MOD 9)	22. 73 ≡ __1__ (MOD 9)	38. 35 ≡ __3__ (MOD 8)
7. 19 ≡ __3__ (MOD 8)	23. 81 ≡ __1__ (MOD 5)	39. 47 ≡ __2__ (MOD 15)
8. 22 ≡ __0__ (MOD 11)	24. 63 ≡ __3__ (MOD 4)	40. 73 ≡ __1__ (MOD 3)
9. 42 ≡ __2__ (MOD 4)	25. 70 ≡ __10__ (MOD 20)	41. 80 ≡ __8__ (MOD 9)
10. 81 ≡ __0__ (MOD 9)	26. 43 ≡ __4__ (MOD 13)	42. 55 ≡ __1__ (MOD 6)
11. 52 ≡ __2__ (MOD 5)	27. 64 ≡ __4__ (MOD 6)	43. 50 ≡ __2__ (MOD 16)
12. 40 ≡ __4__ (MOD 9)	28. 27 ≡ __3__ (MOD 12)	44. 96 ≡ __1__ (MOD 5)
13. 23 ≡ __2__ (MOD 7)	29. 29 ≡ __14__ (MOD 15)	45. 73 ≡ __3__ (MOD 10)
14. 24 ≡ __0__ (MOD 6)	30. 92 ≡ __2__ (MOD 10)	46. 30 ≡ __2__ (MOD 14)
15. 62 ≡ __8__ (MOD 9)	31. 68 ≡ __2__ (MOD 3)	47. 40 ≡ __4__ (MOD 12)
16. 40 ≡ __1__ (MOD 13)	32. 28 ≡ __1__ (MOD 9)	48. 79 ≡ __2__ (MOD 11)

Modulo Arithmetic 2

Find the opposite.	*Find the reciprocal.*	*Find the square root.*
1. $-5 \equiv \underline{3}$ (MOD 8)	17. of 3 MOD 7 $\underline{5}$	33. $\sqrt{5} \equiv \underline{5}$ (MOD 10)
2. $-4 \equiv \underline{1}$ (MOD 5)	18. of 2 MOD 5 $\underline{3}$	34. $\sqrt{1} \equiv \underline{1,3,5}$ (MOD 8)
3. $-7 \equiv \underline{5}$ (MOD 12)	19. of 4 MOD 11 $\underline{3}$	35. $\sqrt{5} \equiv \underline{none}$ (MOD 6)
4. $-5 \equiv \underline{2}$ (MOD 7)	20. of 5 MOD 13 $\underline{8}$	36. $\sqrt{1} \equiv \underline{1, 3}$ (MOD 4)
5. $-1 \equiv \underline{7}$ (MOD 8)	21. of 4 MOD 19 $\underline{5}$	37. $\sqrt{5} \equiv \underline{4, 7}$ (MOD 11)
6. $-7 \equiv \underline{7}$ (MOD 14)	22. of 5 MOD 8 $\underline{5}$	38. $\sqrt{4} \equiv \underline{2, 4}$ (MOD 6)
7. $-3 \equiv \underline{6}$ (MOD 9)	23. of 3 MOD 6 $\underline{none}$	39. $\sqrt{9} \equiv \underline{3, 7}$ (MOD 10)
8. $-8 \equiv \underline{2}$ (MOD 10)	24. of 3 MOD 10 $\underline{7}$	40. $\sqrt{6} \equiv \underline{none}$ (MOD 8)
9. $-6 \equiv \underline{5}$ (MOD 11)	25. of 7 MOD 12 $\underline{7}$	41. $\sqrt{1} \equiv \underline{1, 8}$ (MOD 9)
10. $-5 \equiv \underline{1}$ (MOD 6)	26. of 6 MOD 7 $\underline{6}$	42. $\sqrt{5} \equiv \underline{none}$ (MOD 7)
11. $-9 \equiv \underline{3}$ (MOD 12)	27. of 9 MOD 11 $\underline{5}$	43. $\sqrt{0} \equiv \underline{0, 4}$ (MOD 8)
12. $-6 \equiv \underline{7}$ (MOD 13)	28. of 9 MOD 10 $\underline{9}$	44. $\sqrt{5} \equiv \underline{4, 7}$ (MOD 11)
13. $-4 \equiv \underline{3}$ (MOD 7)	29. of 3 MOD 13 $\underline{9}$	45. $\sqrt{6} \equiv \underline{4, 6}$ (MOD 10)
14. $-9 \equiv \underline{7}$ (MOD 16)	30. of 13 MOD 19 $\underline{3}$	46. $\sqrt{2} \equiv \underline{3, 4}$ (MOD 7)
15. $-2 \equiv \underline{13}$ (MOD 15)	31. of 4 MOD 5 $\underline{4}$	47. $\sqrt{1} \equiv \underline{1, 5}$ (MOD 6)
16. $-0 \equiv \underline{0}$ (MOD 8)	32. of 4 MOD 9 $\underline{7}$	48. $\sqrt{3} \equiv \underline{3}$ (MOD 6)

InPM in prime mods only.	Modulo Arithmetic 3										

Answer YES or NO as to whether the MOD has the specified property. See Abbreviations.

	MOD 2	MOD 3	MOD 4	MOD 5	MOD 6	MOD 7	MOD 8	MOD 9	MOD 10	MOD 11	MOD 12
1. CíPA	✔	✔	✔	✔	✔	✔	✔	✔	✔	✔	✔
2. APA	✔	✔	✔	✔	✔	✔	✔	✔	✔	✔	✔
3. IdPA	✔	✔	✔	✔	✔	✔	✔	✔	✔	✔	✔
4. InPA	✔	✔	✔	✔	✔	✔	✔	✔	✔	✔	✔
5. CPA	✔	✔	✔	✔	✔	✔	✔	✔	✔	✔	✔
6. CíPM	✔	✔	✔	✔	✔	✔	✔	✔	✔	✔	✔
7. APM	✔	✔	✔	✔	✔	✔	✔	✔	✔	✔	✔
8. IdPM	✔	✔	✔	✔	✔	✔	✔	✔	✔	✔	✔
9. InPM	✔	✔	−	✔	−	✔	−	−	−	✔	−
10. CPM	✔	✔	✔	✔	✔	✔	✔	✔	✔	✔	✔
11. DPMA	✔	✔	✔	✔	✔	✔	✔	✔	✔	✔	✔
12. ZPM	✔	✔	✔	✔	✔	✔	✔	✔	✔	✔	✔
13. RPC	✔	✔	✔	✔	✔	✔	✔	✔	✔	✔	✔
14. SPC	✔	✔	✔	✔	✔	✔	✔	✔	✔	✔	✔
15. TPC	✔	✔	✔	✔	✔	✔	✔	✔	✔	✔	✔
16. APC	✔	✔	✔	✔	✔	✔	✔	✔	✔	✔	✔
17. MPC	✔	✔	✔	✔	✔	✔	✔	✔	✔	✔	✔

Think of reciprocals.

Modulo Arithmetic 4   –6 mod 7 is 1, and so on.

Find the specified numbers in the mods.

1. $\frac{1}{2} \equiv$ __3__ (MOD 5)	17. $\frac{2}{3} \equiv$ __7__ (MOD 19)	33. $\sqrt{-2} \equiv$ __none__ (MOD 7)
2. $\frac{2}{3} \equiv$ __4__ (MOD 5)	18. $\frac{2}{3} \equiv$ __12__ (MOD 17)	34. $\sqrt{-4} \equiv$ __1, 4__ (MOD 5)
3. $\frac{1}{2} \equiv$ __4__ (MOD 7)	19. $\frac{4}{9} \equiv$ __9__ (MOD 11)	35. $\sqrt{-5} \equiv$ __3, 4__ (MOD 7)
4. $\frac{2}{3} \equiv$ __3__ (MOD 7)	20. $\frac{3}{7} \equiv$ __15__ (MOD 17)	36. $\sqrt{-3} \equiv$ __2, 5__ (MOD 7)
5. $\frac{1}{2} \equiv$ __6__ (MOD 11)	21. $\frac{5}{6} \equiv$ __4__ (MOD 19)	37. $\sqrt{-7} \equiv$ __2, 9__ (MOD 11)
6. $\frac{2}{3} \equiv$ __8__ (MOD 11)	22. $\frac{3}{8} \equiv$ __10__ (MOD 11)	38. $\sqrt{-5} \equiv$ __5__ (MOD 10)
7. $\frac{1}{2} \equiv$ __7__ (MOD 13)	23. $\frac{3}{8} \equiv$ __2__ (MOD 13)	39. $\sqrt{-9} \equiv$ __2, 11__ (MOD 13)
8. $\frac{2}{3} \equiv$ __5__ (MOD 13)	24. $\frac{4}{9} \equiv$ __12__ (MOD 13)	40. $\sqrt{-2} \equiv$ __2, 4__ (MOD 6)
9. $\frac{3}{4} \equiv$ __9__ (MOD 11)	25. $\sqrt{-1} \equiv$ __2, 3__ (MOD 5)	41. $\sqrt{-1} \equiv$ __none__ (MOD 11)
10. $\frac{3}{4} \equiv$ __6__ (MOD 7)	26. $\sqrt{-1} \equiv$ __none__ (MOD 7)	42. $\sqrt{-1} \equiv$ __5, 8__ (MOD 13)
11. $\frac{2}{5} \equiv$ __7__ (MOD 11)	27. $\sqrt{-1} \equiv$ __4, 13__ (MOD 17)	43. $\sqrt{-4} \equiv$ __2, 6__ (MOD 8)
12. $\frac{3}{5} \equiv$ __11__ (MOD 13)	28. $\sqrt{-4} \equiv$ __3, 10__ (MOD 13)	44. $\sqrt{-1} \equiv$ __3, 7__ (MOD 10)
13. $\frac{2}{5} \equiv$ __6__ (MOD 7)	29. $\sqrt{-6} \equiv$ __1, 6__ (MOD 7)	45. $\sqrt{-4} \equiv$ __none__ (MOD 7)
14. $\frac{5}{6} \equiv$ __10__ (MOD 11)	30. $\sqrt{-2} \equiv$ __7, 10__ (MOD 17)	46. $\sqrt{-2} \equiv$ __3, 8__ (MOD 11)
15. $\frac{5}{6} \equiv$ __3__ (MOD 13)	31. $\sqrt{-13} \equiv$ __2, 15__ (MOD 17)	47. $\sqrt{-4} \equiv$ __4, 6__ (MOD 10)
16. $\frac{4}{7} \equiv$ __10__ (MOD 11)	32. $\sqrt{-8} \equiv$ __5, 6__ (MOD 11)	48. $\sqrt{-8} \equiv$ __3, 14__ (MOD 17)

Money 1

Find the number of each coin using smart trial and error. | Answer as indicated.

1. 17 quarters and dimes have a value of $3.05.

1 Q, 16 D: .25 + 1.60 = 1.85
need 1.20 more
+ 2 Q, – 2 D = +.30
add 4 to original
9 Q, 8 D: 2.25 + .80

8. 7 coins = 58¢ How many quarters?

3 pennies
1 quarter
3 dimes

2. 16 quarters and nickels have a value of $2.40.

2 Q, 14 N: .50 + .70 = 1.20
need 1.20 more
+ 2 Q, – 2 N = +.40
add 3 to original
8 Q, 8 N: 2.00 + .40

9. 8 coins = 37¢ How many dimes?

2 pennies
5 nickels
1 dime

3. 29 nickels and dimes have a value of $2.05.

1 N, 28 D: .05 + 2.80 = 2.85
need 0.80 less
+ 1 N, – 1 D = –.05
add 16 to original
17 N, 12 D: .85 + 1.20

10. 10 coins = 69¢ How many nickels?

4 pennies
1 quarter
3 dimes
2 nickels

4. 22 quarters and dimes have a value of $3.70.

2 Q, 20 D: .50 + 2.00 = 2.50
need 1.20 more
+ 2 Q, – 2 D = +.30
add 4 to original
10 Q, 12 D: 2.50 + 1.20

11. 9 coins = 135¢ How many dimes?

4 quarters
2 dimes
3 nickels

5. 40 nickels and dimes have a value of $2.75.

1 N, 39 D: .05 + 3.90 = 3.95
need 1.20 less
+ 3 N, – 3 D = –.15
add 8 to original
25 N, 15 D: 1.25 + 1.50

12. 10 coins = 37¢ How many nickels?

7 pennies
3 dimes
0 nickels

6. 20 quarters and nickels have a value of $3.60.

19 Q, 1 N: 4.75 + .05 = 4.80
need 1.20 less
– 1 Q, + 1 N = –.20
add 6 to original
13 Q, 7 N: 3.25 + .35

13. 10 coins = 110¢ How many nickels?

2 quarters　　1 quarter
4 dimes　OR　8 dimes
4 nickels　　**1 nickel**

7. 28 quarters and dimes have a value of $4.30.

2 Q, 26 D: .50 + 2.60 = 3.10
need 1.20 more
+ 2 Q, – 2 D = +.30
add 4 to original
10 Q, 18 D: 2.50 + 1.80

14. 10 coins = 100¢ How many nickels?

2 quarters
2 dimes
6 nickels

248

Money 2

Find the number of coins (at least 1 of each) for the given value and conditions.

Find the number of different ways to make $1 using . . .

1.	pennies, dimes, and quarters to make $1.67; fewest number of coins	8.	quarters and dimes.	
	2 P		4 Q, 0 D	
	6 Q		2 Q, 5 D	
	1 D		0 Q, 10 D	
	1 N	**10**		**3**

2.	nickels and dimes to make $1.75; twice as many dimes as nickels	9.	quarters and nickels.	
	2 D + 1 N = .25		4 Q, 0 N 1 Q, 15 N	
	.25 x 7 = 1.75		3 Q, 5 N 0 Q, 20 N	
	14 D		2 Q, 10 N	
	7 N	**21**		**5**

3.	dimes and quarters to make $2.50; 3 more quarters than dimes	10.	dimes and nickels.	
	4 Q + 1 D = 1.10		10 D, 0 N	
	need 1.40; 1.40/.35 = 4		9 D, 2 N	
	8 Q		through	
	5 D	**13**	0 D, 20 N	**11**

4.	nickels and dimes to make $2.20; twice as many nickels as dimes	11.	dimes and pennies.	
	2 N + 1 D = .20		10 D, 0 P	
	.20 x 11 = 2.20		9 D, 10 P	
	22 N		through	
	11 D	**33**	0 D, 100 P	**11**

5.	dimes and quarters to make $1.95; 9 more dimes than quarters	12.	quarters, dimes, and nickels.	
	10 D + 1 Q = 1.25		4 Q, 0 D, 0 N 2 Q, 5 D through 0 D	
	need .70; .70/.35 = 2		3 Q, 1 D, 3 N 1 Q, 7 D through 0 D	
	12 D		3 Q, 2 D, 1 N 0 Q, 10 D through 0 D	
	3 Q	**15**		**28**

6.	pennies, nickels, dimes, and quarters to make $7.98; fewest number of coins	13.	half dollars, nickels, and pennies.	
	3 P		2 H, 0 N, 0 P	
	31 Q		1 H, 10 N through 0 N	
	1 D		0 H, 20 N through 0 N	
	2 N	**37**		**33**

7.	nickels and quarters to make $1.60; 3 times as many nickels as quarters	14.	half dollars, quarters, and dimes.	
	3 N + 1 Q = .40		2 H 4 Q	
	.40 x 4 = 1.60		1 H, 2 Q 2 Q, 5 D	
	12 N		1 H, 5 D 10 D	
	4 Q	**16**		**6**

MAVA Math: Enhanced Skills Solutions   Copyright © 2015 Marla Weiss

Money 3

Answer as indicated.	*Find the total number of coins given an equal number of each.*

1. 14 quarters, 15 dimes, and 40 nickels is what percent of $5?
 	
 14x.25 + 15x.10 + 40x.05 =   7/5 = 14/10
 3.50 + 1.50 + 2.00 =
 7.00   **140%**
	8. nickels, dimes, and quarters
	      value = $2.00 *Here, a variable represents the value of a coin.*
	
	 1 N + 1 D + 1 Q = .40
	 5 N + 5 D + 5 Q = 2.00
	
	                                        15

2. 19 quarters, 16 dimes, and 55 pennies is what percent of $10?
 	
 19x.25 + 16x.10 + 55x.01 =   6.9/10
 4.75 + 1.60 + 0.55 =
 6.90   **69%**
	9. pennies, nickels, and dimes
	      value = $4.00
	
	 1 P + 1 N + 1 D = .16
	 25 P + 25 N + 25 D = 4.00
	                                        75

3. 21 quarters, 21 dimes, and 18 nickels is what percent of $20?
 	
 21x.25 + 21x.10 + 18x.05 =   8.25/20
 5.25 + 2.10 + 0.90 =
 8.25   **41.25%**
	10. pennies, dimes, and quarters
	      value = $2.52
	
	 1 P + 1 D + 1 Q = .36
	 7 P + 7 D + 7 Q = 2.52
	                                        21

4. 22 quarters, 19 dimes, and 55 nickels is what percent of $50?
 	
 22x.25 + 19x.10 + 55x.05 =   10.15/50
 5.50 + 1.90 + 2.75 =
 10.15   **20.3%**
	11. nickels, dimes, and half-dollars
	      value = $7.15
	
	 1 N + 1 D + 1 H = .65
	 11 (N + D + Q) = 7.15
	                                        33

5. 11 quarters, 9 dimes, and 18 nickels is what percent of $5?
 	
 11x.25 + 9x.10 + 18x.05 =   45.5/50
 2.75 + .90 + .90 =
 4.55   **91%**
	12. nickels, quarters, and half-dollars
	      value = $10.40
	
	 1 N + 1 Q + 1 H = .80
	 13 (N + D + Q) = 10.40
	                                        39

6. 16 quarters, 19 dimes, and 15 pennies is what percent of $10?
 	
 16x.25 + 19x.10 + 15x.01 =   6.05/10
 4.00 + 1.90 + .15 =
 6.05   **60.5%**
	13. dimes, quarters, and half-dollars
	      value = $10.20
	
	 1 D + 1 Q + 1 H = .85
	 12 (N + D + Q) = 10.20
	                                        36

7. 26 quarters, 33 dimes, and 7 nickels is what percent of $20?
 	
 26x.25 + 33x.10 + 7x.05 =   10.15/20
 6.50 + 3.30 + .35 =
 10.15   **50.75%**
	14. pennies, nickels, and quarters
	      value = $4.34
	
	 1 P + 1 N + 1 Q = .31
	 14 (P + N + Q) = 4.34
	                                        42

Money 4

Find the number coins using algebra.

Find the sum of the values that may be made using one or more of the coins.

---

1. A group of 57 coins, only nickels and quarters, has value $7.05. Find the number of each coin.

$5N + 25Q = 705$
$N + 5Q = 141$
$N + Q = 57$
$57 - Q + 5Q = 141$
$4Q = 84$

Q = 21
N = 36

---

6. 1 quarter, 1 dime, 1 penny

0Q, 0D, 1P	0.01		1Q, 1D, 0P	0.35
0Q, 1D, 0P	0.10		1Q, 1D, 1P	0.36
0Q, 1D, 1P	0.11			
1Q, 0D, 0P	0.25			
1Q, 0D, 1P	0.26			

Note order of 0s and 1s forms counting in base two.

sum = **$1.44**

---

2. A group of dimes and quarters, with 4 more quarters than dimes, is worth $13.60. Find the number of each.

$10D + 25Q = 1360$
$2D + 5Q = 272$
$Q = D + 4$
$2D + 5D + 20 = 272$
$7D = 252$

D = 36
Q = 40

---

7. 1 dime, 1 nickel, 1 penny

0D, 0N, 1P	0.01		1D, 1N, 0P	0.15
0D, 1N, 0P	0.05		1D, 1N, 1P	0.16
0D, 1N, 1P	0.06			
1D, 0N, 0P	0.10			
1D, 0N, 1P	0.11		sum =	**$0.64**

---

3. A group of 46 coins, only nickels and quarters, has value $8.70. Find the number of each coin.

$5N + 25Q = 870$
$N + 5Q = 174$
$N + Q = 46$
$46 - Q + 5Q = 174$
$4Q = 128$

Q = 32
N = 14

---

8. 1 quarter, 1 nickel, 1 penny

0Q, 0N, 1P	0.01		1Q, 1N, 0P	0.30
0Q, 1N, 0P	0.05		1Q, 1N, 1P	0.31
0Q, 1N, 1P	0.06			
1Q, 0N, 0P	0.25			
1Q, 0N, 1P	0.26		sum =	**$1.24**

---

4. A group of dimes and quarters, with 9 more dimes than quarters, is worth $8.95. Find the number of each.

$10D + 25Q = 895$
$2D + 5Q = 179$
$D = Q + 9$
$2Q + 18 + 5Q = 179$
$7Q = 161$

Q = 23
D = 32

---

9. 1 dime, 2 nickels, 1 penny

0D, 2N, 0P	0.10	plus all from #7 above
0D, 2N, 1P	0.11	
1D, 2N, 0P	0.20	
1D, 2N, 1P	0.21	

sum = $0.64 + 0.62 = **$1.26**

---

5. A group of coins, with twice as many dimes as quarters and no others, has $10.80 value. Find how many of each.

$10D + 25Q = 1080$
$2D + 5Q = 216$
$D = 2Q$
$4Q + 5Q = 216$
$9Q = 216$

Q = 24
D = 48

---

10. 2 quarters, 1 dime, 1 penny

2Q, 0D, 0P	0.50	plus all from #6 above
2Q, 0D, 1P	0.51	
2Q, 1D, 0P	0.60	
2Q, 1D, 1P	0.61	

sum = $1.44 + 2.22 = **$3.66**

Order of Operations 1

Operate. Show one line of work before the answer.

1.
$2 + 2 \times 2 - 2 \div 2 \times 2 + 2 - 2 + 2^2$
$2 + 4 - 2 + 2 - 2 + 4$
8

2.
$18 \div 6 \times 3 - 2^3 + (2 \times 4) + 18 \div 2$
$9 - 8 + 8 + 9$
18

3.
$7 + 5(1 - 5) - 2 \times (3 - 8) + 6^2 \div 3$
$7 - 20 + 10 + 12$
9

4.
$3 \div 3 \times 3 \times 3 \div 3 - 3 \times 3 \times 3 \div 3^2$
$3 - 3$
0

5.
$9 + 8 \div 4 - 3 \times 4 \div 6 - 45 \div 3 - 1$
$11 - 2 - 15 - 1$
-7

6.
$3(12 \div 3) - 20 \div 5 + 3 - 5 \times 4 - 2$
$12 - 4 + 3 - 20 - 2$
-11

7.
$3 - 2(5 - 7)^3 - (3 - 12) + 40 \div 8$
$3 + 16 + 9 + 5$
33

8.
$2 \times 5^2 - 30 \div 3 \times 5 + 60 \div 12 \times 5$
$50 - 50 + 25$
25

9.
$4 - 4(3 - 8)^2 - 2 \times (7 - 8) + 5^2$
$4 - 100 + 2 + 25$
-69

10.
$3 \times (7 - 7) - 3^3 - (3 + 7) + 14 \div 7$
$0 - 27 - 10 + 2$
-35

11.
$3 \times 4^2 - 80 \div 5 \times 2 - 35 \div 7 \times 5$
$48 - 32 - 25$
-9

12.
$6 - 6 \div 3 - 6 \times 3 \div 6 - 36 \div 3 - 3$
$6 - 2 - 3 - 12 - 3$
-14

13.
$1 - (7 - 1)^2 - 1 + 7 \times (1 - 7) - 1^7$
$1 - 36 - 1 - 42 - 1$
-79

14.
$8 + 8 - 8 \times 8 \div 8 - 8 + 8 \div 8 - 8^2$
$8 + 8 - 8 - 8 + 1 - 64$
-63

15.
$11 + 9 \div 3 - 7 \times 4 \div 2 - 60 \div 5 - 2$
$11 + 3 - 14 - 12 - 2$
-14

16.
$-2(2 - 5)^3 - (7 - 12) + 52 \div 4 - 4$
$54 + 5 + 13 - 4$
68

Order of Operations 2

Operate. Show one line of work before the answer.

1. $10 \div 3 \times 6 - 8 \div 5 \times 10 - 2 \div 2^2$

   $20 - 16 - .5$

   3.5

9. $7 - 5(3 - 6)^2 - 2 \div (7 - 12) \times 10^2$

   $7 - 45 + 40$

   2

2. $13 \div 3 \times 9 \times 2 - (1 - 4)^3 - (4 - 8)$

   $78 + 27 + 4$

   109

10. $9 \times (7 - 8) - 4^3 - (3 + 8) \div 22 \times 2$

    $-9 - 64 - 1$

    −74

3. $-1 - 4 - 6 \times (3 - 8) \div 7 \times 21 - 5^2$

   $-5 + 90 - 25$

   60

11. $-4 \times 5^2 - 7 \div 3 \times 33 - 7 \div 10 \times 5$

    $-100 - 77 - 3.5$

    −180.5

4. $4 \div 3 \times 9 \div 5 \times 15 - 9 \times 9 \div 3^2 \times 2$

   $36 - 18$

   18

12. $16 - 13 \div 7 \times 21 \div 5 \times 10 - 2 \div 5$

    $16 - 78 - .4$

    −62.4

5. $(5 + 12 \div 5 \times 20 \times 2 - 6) \div 5 - 10$

   $(5 + 96 - 6) \div 5 - 10$

   9

13. $10 - (3 - 10)^2 - 1 \div 7 \times (1 - 15)$

    $10 - 49 + 2$

    −37

6. $2(11 \div 5 \times 20) \div 8 - 2(5 \times 4 - 2)$

   $88 \div 8 - 36$

   −25

14. $6 \div 5 \times 20 \div 7 \times 28 - 6 - (5 - 5^2)$

    $96 - 6 + 20$

    110

7. $-3(6 - 8)^3 - 11 \div 3 \times 15 + 1 \div 4$

   $24 - 55 + .25$

   −30.75

15. $5 \div 3 \times 12 - 4 \div 8 \times 7 - 9 \div 18 + 1$

    $20 - 3.5 - 0.5 + 1$

    17

8. $3 \times 4^2 - 21 \div 5 \times 15 - 7 \div 14 \times 60$

   $48 - 63 - 30$

   −45

16. $-2(2 - 7)^3 - (5 - 11) \div 4 \div 24 \times 16$

    $250 + 1$

    251

Parallelograms 1

Find the area in square units of parallelogram ABCD with the given vertices.

1. A(4, 4), B(–8, 4),
   C(–14, –4), D(–2, –4)

   B = 12
   H = 8
   A = **96**

11. A(–3.5, 8), B(6.5, 8),
    C(–1, –8), D(–11, –8)

    B = 10
    H = 16
    A = **160**

2. A(–3, 7), B(6, 7),
   C(–2, –5), D(–11, –5)

   B = 9
   H = 12
   A = **108**

12. A(18.4, –1), B(2.4, –1),
    C(7, 6.5), D(23, 6.5)

    B = 16
    H = 7.5
    A = **120**

3. A(20, –5), B(–2, –5),
   C(–7, 6), D(15, 6)

   B = 22
   H = 11
   A = **242**

13. A(2, –4), B(–8, –4),
    C(–14, –9), D(–4, –9)

    B = 10
    H = 5
    A = **50**

4. A(1, 10), B(13, 10),
   C(10, –2), D(–2, –2)

   B = 12
   H = 12
   A = **144**

14. A(13, 9), B(–7, 9),
    C(–16, –7), D(4, –7)

    B = 20
    H = 16
    A = **320**

5. A(0, 0), B(4, 7),
   C(13, 7), D(9, 0)

   B = 9
   H = 7
   A = **63**

15. A(0.5, –3.5), B(2.5, 9.5),
    C(7.5, 9.5), D(5.5, –3.5)

    B = 5
    H = 13
    A = **65**

6. A(8, 5), B(–5, 5),
   C(–11, –4), D(2, –4)

   B = 13
   H = 9
   A = **117**

16. A(1, 10.3), B(14, 10.3),
    C(11, –2.7), D(–2, –2.7)

    B = 13
    H = 13
    A = **169**

7. A(0, –3), B(3, 9),
   C(17, 9), D(14, –3)

   B = 14
   H = 12
   A = **168**

17. A(–8, 7), B(7, 7),
    C(–1, –8), D(–16, –8)

    B = 15
    H = 15
    A = **225**

8. A(0, 0), B(–4, 0),
   C(9, 7), D(13, 7)

   B = 4
   H = 7
   A = **28**

18. A(20, –3), B(2, –3),
    C(4, 8), D(22, 8)

    B = 18
    H = 11
    A = **198**

9. A(9, 6), B(–7, 6),
   C(–13, –7), D(3, –7)

   B = 16
   H = 13
   A = **208**

19. A(2, –2), B(–10, –2),
    C(–15, –9), D(–3, –9)

    B = 12
    H = 7
    A = **84**

10. A(0, –3), B(3, 9),
    C(17, 9), D(14, –3)

    B = 14
    H = 12
    A = **168**

20. A(2, 11), B(–7, 11),
    C(–14, –6), D(–5, –6)

    B = 9
    H = 17
    A = **153**

Parallelograms 2

Find the four angles of parallelogram ABCD.

1. m∠A = 8x − 5
   m∠B = 2x + 23

   10x + 18 = 180   129.6 − 5 = 124.6
   10x = 162   32.4 + 23 = 55.4
   x = 16.2   **55.4, 124.6, 55.4, 124.6**

8. The measure of one angle of the parallelogram is 5 more than 4 times another.

   A = 5 + 4(180 − A)
   A = 725 − 4A
   5A = 725
   A = 145
   35, 145, 35, 145

2. m∠A = 2x + 3
   m∠C = −x + 21

   2x + 3 = −x + 21
   3x = 18
   x = 6   **15, 165, 15, 165**

9. The sum of 3 of the angles of the parallelogram is 40 more than 3 times the fourth.

   360 − A = 40 + 3A
   320 = 4A
   A = 80
   80, 100, 80, 100

3. m∠A = 7x − 4
   m∠B = 5x + 10

   12x + 6 = 180   101.5 − 4 = 97.5
   12x = 174   72.5 + 10 = 82.5
   x = 14.5   **97.5, 82.5, 97.5, 82.5**

10. The measure of one angle of the parallelogram is 4 less than 7 times another.

    A = 7(180 − A) − 4
    A = 1260 −7A − 4
    8A = 1256
    A = 157
    23, 157, 23, 157

4. m∠A = 6x + 11
   m∠C = −2x + 35

   6x + 11 = −2x + 35
   8x = 24
   x = 3   **29, 151, 29, 151**

11. The sum of 3 of the angles of the parallelogram is 60 less than 3 times the fourth.

    360 − A = 3A − 60
    420 = 4A
    A = 105
    105, 75, 105, 75

5. m∠A = 6x − 5
   m∠B = 9x + 20

   15x + 15 = 180   66 − 5 = 61
   15x = 165   99 + 20 = 119
   x = 11   **61, 119, 61, 119**

12. The measure of one angle of the parallelogram is 16 more than 3 times another.

    A = 16 + 3(180 − A)
    A = 16 + 540 − 3A
    4A = 556
    A = 139
    41, 139, 41, 139

6. m∠A = 2x + 3y
   m∠B = −x + 4y
   m∠C = 6x + y

   x + 7y = 180
   x + 14x = 180
   15x = 180
   x = 12, y = 24
   84, 96, 84, 96

   2x + 3y = 6x + y
   y = 2x

13. The measure of one angle of the parallelogram is 2 less than 6 times another.

    A = 6(180 − A) − 2
    A = 1080 −6A − 2
    7A = 1078
    A = 154
    26, 154, 26, 154

7. m∠A = 4x + y
   m∠B = −2x + 7y
   m∠C = 2x + 3y

   5x + 5x = 180
   10x = 180
   x = 18, y = 18
   90, 90, 90, 90

   4x + y = 2x + 3y
   x = y   A rectangle is a parallelogram.

14. The sum of 3 of the angles of the parallelogram is 78 more than twice the fourth.

    360 − A = 78 + 2A
    282 = 3A
    A = 94
    94, 86, 94, 86

Parallelograms 3

Diagonals bisect each other.

Find the perimeter of parallelogram ABCD in units. Angles given are for parallelograms. NTS

1.	AB = 9x − 4 BC = 11 AD = 10x + 1	10x + 1 = 11 x = 1 AB = 5 BC = 11	**32**	10. one angle 30° BE = 6 BC = 12 △ABE is 30-60-90 AB = 12 12 × 4 = **48**
2.	AB = 2x + 3 BC = 10x + 1 CD = 18x − 5	2x + 3 = 18x − 5 8 = 16x x = 1/2 AB = 4, BC = 6	**20**	11. one angle 45° BE = 10 AD = 30 △ABE is 45-45-90 AB = 10√2 **60 + 20√2**
3.	BA = 5x − 2 CD = 9x − 6 AD = 4x + 1	5x − 2 = 9x − 6 4 = 4x x = 1 AB = 3, AD = 5	**16**	12. BE = 12 AE = 9 DE = 11 △ABE is 9, 12, 15 AB = 15, AD = 20 15 + 20 = 35 35 × 2 = **70**
4.	BA = 5x + 3 BC = 17 AD = 10x − 3	10x − 3 = 17 x = 2 AB = 13 BC = 17	**60**	13. one angle 45° BE = 14 AD = 27 △ABE is 45-45-90 AB = 14√2 **54 + 28√2**
5.	BA = 6x − 5 AD = 3x + 5 DC = 25	6x − 5 = 25 x = 5 AB = 25 AD = 20	**90**	14. one angle 60° BE = 12 BC = 15 △ABE is 30-60-90 AB = 8√3 **30 + 16√3**
6.	BC = 6x + 4 DC = 4x + 1 DA = 8x − 3	6x + 4 = 8x − 3 7 = 2x x = 3.5 BC = 25, DC = 15	**80**	15. BE = 24 AE = 10 DE = 13 △ABE is 10, 24, 26 AB = 26, AD = 23 26 × 2 + 13 × 2 = 52 + 26 = **78**

7.	AB = 8x + 4
	CD = 10x − 1
	BC = 6x + 7
		8x + 4 = 10x − 1
		5 = 2x
		x = 2.5
		AB = 24, BC = 22
			92	16. perimeter of △ABD = 33
				BF = 5 AB+BF+FD+AD=33
				     BF = FD; AB+AD = 33−10 = 23
				     23 × 2 = **46**

8.	BA = 2x + 18
	BC = 26
	DA = 7x + 5
		7x + 5 = 26
		x = 3
		AB = 24
		BC = 26
			100	17. perimeter of △CAD = 44
				AF = 9 AF+FC+CD+DA=44
				     AF = FC; CD+DA = 44−18 = 26
				     26 × 2 = **52**

9.	BC = 11x − 8
	DC = 4x + 1
	DA = 5x + 16
		11x − 8 = 5x + 16
		6x = 24
		x = 4
		BC = 36, DC = 17
			106	18. altitude from B to CD = 16
				BC = 28 area = area
				BE = 20 AD × BE = CD × alt
				     28 × 20 = CD × 16 | CD = 35
				                       56 + 70
				                       = **126**

Parallelograms 4

Answer as indicated. Linear measurements are in units.

---

1. Randomly select two angles of a parallelogram. Find the probability that they are congruent.

   A, A, 180−A, 180−A
   C(4, 2) = 6

   $\dfrac{2}{6} = \dfrac{1}{3}$

---

6. The perimeter of parallelogram ABCD is 320. AB:BC = 11:5. Find AB.

   P = 320
   SP = 320/2 = 160
   11 + 5 = 16
   blow-up = 160/16 = 10
   AB = 11 × 10 = **110**

---

2. Randomly select two angles of a parallelogram. Find the probability that they are congruent or supplementary.

   1

---

7. The perimeter of parallelogram ABCD is 540. BC:CD = 8:7. Find CD.

   P = 540
   SP = 540/2 = 270
   8 + 7 = 15
   blow-up = 270/15 = 18
   CD = 7 × 18 = **126**

---

3. $\overline{AE}$ and $\overline{DE}$ are angle bisectors of parallelogram ABCD. Find m ∠ AED.

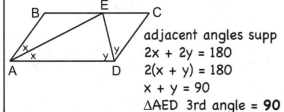

   adjacent angles supp
   2x + 2y = 180
   2(x + y) = 180
   x + y = 90
   △AED 3rd angle = **90**

---

8. The perimeter of parallelogram ABCD = 70. The altitude from B to $\overline{CD}$ = 16. Find the altitude from B to $\overline{AD}$.

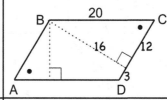

   70 − 40 = 30
   30/2 = 15
   ~ △s
   12, 16, 20 ~
   9, **12**, 15

---

4. The perimeter of parallelogram ABCD is 88. The perimeter of FECD is 56. EC = AF. Find EF.

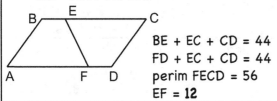

   BE + EC + CD = 44
   FD + EC + CD = 44
   perim FECD = 56
   EF = **12**

---

9. The perimeter of parallelogram ABCD is 64. The perimeter of ABEF is 47. BE = FD. Find EF.

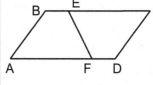

   AB + BE + EC = 32
   AB + BE + AF = 32
   perim ABEF = 47
   EF = **15**

---

5. A parallelogram is inscribed in a circle. What must be true about the parallelogram?

   It is a **rectangle**.

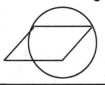

---

10. The perimeter of parallelogram ABCD is 160. AB:BC = 3:2. Find BC.

   P = 160
   SP = 160/2 = 80
   3 + 2 = 5
   blow-up = 80/5 = 16
   BC = 2 × 16 = **32**

Percents 1

Find the percent by mental math.

1. 10% of 50	**5**	17. 3% of 800	**24**	33. 12.5% of 16	**2**
2. 10% of 660	**66**	18. 27% of 1000	**270**	34. 75% of 84	**63**
3. 20% of 200	**40**	19. 25% of 500	**125**	35. 11% of 900	**99**
4. 30% of 20	**6**	20. 75% of 60	**45**	36. 4% of 700	**28**
5. 5% of 300	**15**	21. 40% of 25	**10**	37. 30% of 160	**48**
6. 15% of 500	**75**	22. 30% of 5000	**1500**	38. 66 2/3% of 33	**22**
7. 25% of 48	**12**	23. 37.5% of 80	**30**	39. 80% of 170	**136**
8. 50% of 102	**51**	24. 100% of 561	**561**	40. 15% of 80	**12**
9. 90% of 450	**405**	25. 15% of 2000	**300**	41. 50% of 846	**423**
10. 12.5% of 48	**6**	26. 20% of 570	**114**	42. 25% of 440	**110**
11. 33 1/3% of 99	**33**	27. 70% of 500	**350**	43. 12% of 600	**72**
12. 13% of 100	**13**	28. 50% of 64	**32**	44. 37.5% of 888	**333**
13. 1% of 300	**3**	29. 15% of 880	**132**	45. 75% of 32	**24**
14. 2% of 150	**3**	30. 10% of 1210	**121**	46. 12.5% of 800	**100**
15. 90% of 900	**810**	31. 200% of 45	**90**	47. 30% of 120	**36**
16. 50% of 350	**175**	32. 33 1/3% of 990	**330**	48. 25% of 1000	**250**

Percents 2

Complete the chart of equivalent values. Simplify fractions.

	PERCENT	DECIMAL	FRACTION		PERCENT	DECIMAL	FRACTION
1.	12%	.12	$\frac{3}{25}$	17.	52%	.52	$\frac{13}{25}$
2.	34%	.34	$\frac{17}{50}$	18.	72%	.72	$\frac{18}{25}$
3.	64%	.64	$\frac{16}{25}$	19.	39%	.39	$\frac{39}{100}$
4.	32%	.32	$\frac{8}{25}$	20.	65%	.65	$\frac{13}{20}$
5.	98%	.98	$\frac{49}{50}$	21.	28%	.28	$\frac{7}{25}$
6.	240%	2.4	$\frac{12}{5}$	22.	15%	.15	$\frac{3}{20}$
7.	55%	.55	$\frac{11}{20}$	23.	435%	4.35	$\frac{87}{20}$
8.	16%	.16	$\frac{4}{25}$	24.	54%	.54	$\frac{27}{50}$
9.	45%	.45	$\frac{9}{20}$	25.	106%	1.06	$\frac{53}{50}$
10.	3%	.03	$\frac{3}{100}$	26.	74%	.74	$\frac{37}{50}$
11.	66%	.66	$\frac{33}{50}$	27.	8%	.08	$\frac{2}{25}$
12.	78%	.78	$\frac{39}{50}$	28.	84%	.84	$\frac{21}{25}$
13.	30%	.3	$\frac{3}{10}$	29.	24%	.24	$\frac{6}{25}$
14.	85%	.85	$\frac{17}{20}$	30.	105%	1.05	$\frac{21}{20}$
15.	245%	2.45	$\frac{49}{20}$	31.	95%	.95	$\frac{19}{20}$
16.	56%	.56	$\frac{14}{25}$	32.	144%	1.44	$\frac{36}{25}$

MAVA Math: Enhanced Skills Solutions   Copyright © 2015 Marla Weiss

Percents 3

Answer each "forward" percent word problem using written mental math.

1. Find the price of a $640 keyboard sold at a 25% discount with 5% sales tax.	640 −160 480 + 24 **$504**	9. Find the total cost if each of 20 people has a bill of $12.50 plus 5% tax and 15% tip. No tax on tip; no tip on tax.	250 + 12.50 + 37.50 **$300**
2. Find the percent discount on the total purchase of a $30 book at a 20% discount and a $20 book at a 30% discount.	30−6 = 24 20−6 = 14 38/50 76/100 **24%**	10. Tim earns $175 a week base plus 15% commission on sales. Find his weekly earnings when his sales for 4 weeks are $2640.	2640 ÷ 4 660 + 99 **$274** +175
3. Find the price of a $350 suit sold at a 30% discount with 5% sales tax.	350 −105 245 +12.25 **$257.25**	11. Find the price of a $900 item sold at a 20% discount with 5% sales tax.	900 −180 720 + 36 **$756**
4. With 4% sales tax, find the total cost of 8 items at $13.25 each.	80+24+2 106 + 4.24 **$110.24**	12. Lily earns $220 a week base salary plus 12% commission. Find her total earnings in a week when her sales are $850.	850 102 +220 **$322**
5. With 3% sales tax, find the total cost of a $9.75 item and a $13.25 item.	23.00 +0.69 **$23.69**	13. If a population of 64,000 decreases by 25% each year, find the population after 2 years.	64,000 −16,000 48,000 −12,000 **36,000**
6. With 6% sales tax, find the total cost of six items at $11.25 each.	66 +1.50 67.50 +4.05 **$71.55**	14. Find the percent discount when an item that retails for $4000 is on sale for $3480.	4000 −520 3480 13 x 4 = 52 **13%**
7. Find the price of an $800 item sold at a 37.5% discount with 4% sales tax.	880 −330 (3/8) 550 + 22 **$572**	15. The value of a multi-media system depreciates by 20% a year. If it was purchased 2 years ago for $8000, find its current value. **$5120**	8000 −1600 6400 −1280 5120
8. Find the price of a $450 chair sold at a 20% discount with 5% sales tax.	450 −90 360 + 18 **$378**	16. Find the price of a $240 item sold at a 35% discount with 10% sales tax.	240 −72 −12 156 +15.60 **$171.60**

Percents 4

Complete the chart of equivalent values. Simplify fractions.

	PERCENT	DECIMAL	FRACTION		PERCENT	DECIMAL	FRACTION
1.	12.5%	.125	$\frac{1}{8}$	17.	.75%	.0075	$\frac{3}{400}$
2.	0.8%	.008	$\frac{1}{125}$	18.	.25%	.0025	$\frac{1}{400}$
3.	37.5%	.375	$\frac{3}{8}$	19.	3.6%	.036	$\frac{9}{250}$
4.	3.25%	.0325	$\frac{13}{400}$	20.	87.5%	.875	$\frac{7}{8}$
5.	0.5%	.005	$\frac{1}{200}$	21.	7.5%	.075	$\frac{3}{40}$
6.	650%	6.5	$\frac{13}{2}$	22.	.4%	.004	$\frac{1}{250}$
7.	0.65%	.0065	$\frac{13}{2000}$	23.	1.75%	.0175	$\frac{7}{400}$
8.	.6%	.006	$\frac{3}{500}$	24.	5.4%	.054	$\frac{27}{500}$
9.	4.5%	.045	$\frac{9}{200}$	25.	.35%	.0035	$\frac{7}{2000}$
10.	8.75%	.0875	$\frac{7}{80}$	26.	31.25%	.3125	$\frac{5}{16}$
11.	8.25%	.0825	$\frac{33}{400}$	27.	3.75%	.0375	$\frac{3}{80}$
12.	2.5%	.025	$\frac{1}{40}$	28.	2.75%	.0275	$\frac{11}{400}$
13.	.1%	.001	$\frac{1}{1000}$	29.	5.5%	.055	$\frac{11}{200}$
14.	8.5%	.085	$\frac{17}{200}$	30.	6.25%	.0625	$\frac{1}{16}$
15.	62.5%	.625	$\frac{5}{8}$	31.	1.5%	.015	$\frac{3}{200}$
16.	0.2%	.002	$\frac{1}{500}$	32.	2.2%	.022	$\frac{11}{500}$

Solutions 1–3, 7 show the methods. **Percents 5**

Answer as indicated by forming "is/of," following the English into an equation, writing a proportion, or forming a 2-column list.

1. 16 is what percent of 20? $\dfrac{16}{20}$ $\dfrac{80}{100}$ **80%**	10. 75% of what number is 120? $\dfrac{3}{4}(x) = 120$ $\quad 3x = 480$ $x = \textbf{160}$
2. 20 is what percent of 16? $20 = \dfrac{x}{100}(16)$ $\quad 2000 = 16x$ $x = 125$ $\quad$ **125%**	11. 75 is what percent of 120? $\dfrac{75}{120}$ $\dfrac{5}{8}$ **62.5%**
3. 16 is 20% of what number? $\dfrac{16}{x} = \dfrac{1}{5}$ $\quad$ **80**	12. What percent of 40 is 84? $\dfrac{84}{40}$ $\dfrac{21}{10}$ **210%**
4. 20 is 16% of what number? $\dfrac{20}{x} = \dfrac{16}{100} = \dfrac{4}{25}$ $\quad$ **125**	13. 84 is 40% of what number? $84 = \dfrac{40}{100}(x)$ $\quad 420x = 2x$ $x = \textbf{210}$
5. 2.4% of what number is 12? $\dfrac{12}{x} = \dfrac{2.4}{100} = \dfrac{24}{1000}$ $\quad 2x = 1000$ $x = 500$ $\quad$ **500**	14. 36 is 30% of what number? $\dfrac{36}{x} = \dfrac{3}{10}$ $\quad$ **120**
6. 2.4 is what percent of 12? $\dfrac{2.4}{12} = \dfrac{24}{120} = \dfrac{4}{20}$ $\quad$ **20%**	15. What percent of 150 is 48? $\dfrac{48}{150}$ $\dfrac{16}{50}$ **32%**
7. 2.4 is 12% of what number? 2.4 12% 24 120% 4 20% **20** 20 100%	16. 52 is 4% of what number? 52 4% 13 1% 1300 100% **1300**
8. 35% of what number is 56? $56 = \dfrac{35}{100}(x)$ $\quad x = 56 \cdot 20/7$ $x = 160$ $\quad$ **160**	17. 12 is 150% of what number? 12 150% 4 50% **8** 8 100%
9. 35 is what percent of 56? $\dfrac{35}{56}$ $\dfrac{5}{8}$ **62.5%**	18. What percent of 225 is 90? $\dfrac{90}{225}$ $\dfrac{10}{25}$ **40%**

Percents 6

Find the percent increase (I) or decrease (D) using change/original.

1. from 25 to 28 $\frac{3}{25}$ $\frac{12}{100}$ **12% I**	12. from 25 to 80 $\frac{55}{25}$ $\frac{220}{100}$ **220% I**	23. from 40 to 35 $\frac{5}{40}$ $\frac{1}{8}$ **12.5% D**
2. from 480 to 720 $\frac{240}{480}$ $\frac{1}{2}$ **50% I**	13. from 9 to 17 $\frac{8}{9}$ **88.$\overline{8}$ % I**	24. from 10 to 50 $\frac{40}{10}$ $\frac{4}{1}$ **400% I**
3. from 20 to 11 $\frac{9}{20}$ $\frac{45}{100}$ **45% D**	14. from 82 to 205 $\frac{123}{82}$ $\frac{3}{2}$ **150% I**	25. from 50 to 42 $\frac{8}{50}$ $\frac{16}{100}$ **16% D**
4. from 45 to 50 $\frac{5}{45}$ $\frac{1}{9}$ **11.$\overline{1}$ % I**	15. from 55 to 75 $\frac{20}{55}$ $\frac{4}{11}$ **36.$\overline{36}$ % I**	26. from 32 to 40 $\frac{8}{32}$ $\frac{1}{4}$ **25% I**
5. from 1000 to 20 $\frac{980}{1000}$ $\frac{98}{100}$ **98% D**	16. from 50 to 80 $\frac{30}{50}$ $\frac{60}{100}$ **60% I**	27. from 11 to 13 $\frac{2}{11}$ $\frac{18}{99}$ **18.$\overline{18}$% I**
6. from 560 to 420 $\frac{140}{560}$ $\frac{1}{4}$ **25% D**	17. from 18 to 12 $\frac{6}{18}$ $\frac{1}{3}$ **33.$\overline{3}$ % D**	28. from 19 to 57 $\frac{38}{19}$ $\frac{2}{1}$ **200% I**
7. from 360 to 480 $\frac{120}{360}$ $\frac{1}{3}$ **33.$\overline{3}$ % I**	18. from 60 to 160 $\frac{100}{60}$ $\frac{5}{3}$ **100.$\overline{6}$% I**	29. from 20 to 50 $\frac{30}{20}$ $\frac{3}{2}$ **150% I**
8. from 99 to 66 $\frac{33}{99}$ $\frac{1}{3}$ **33.$\overline{3}$ % D**	19. from 75 to 100 $\frac{25}{75}$ $\frac{1}{3}$ **33.$\overline{3}$ % I**	30. from 60 to 54 $\frac{6}{60}$ $\frac{1}{10}$ **10% D**
9. from 140 to 560 $\frac{420}{140}$ $\frac{3}{1}$ **300% I**	20. from 320 to 200 $\frac{120}{320}$ $\frac{3}{8}$ **37.5% D**	31. from 12 to 20 $\frac{8}{12}$ $\frac{2}{3}$ **66.$\overline{6}$% I**
10. from 40 to 15 $\frac{25}{40}$ $\frac{5}{8}$ **62.5% D**	21. from 480 to 72 $\frac{408}{480}$ $\frac{51}{60}$ $\frac{17}{20}$ **85% D**	32. from 20 to 12 $\frac{8}{20}$ $\frac{40}{100}$ **40% D**
11. from 72 to 36 $\frac{36}{72}$ $\frac{1}{2}$ **50% D**	22. from 500 to 50 $\frac{450}{500}$ $\frac{90}{100}$ **90% D**	33. from 55 to 60 $\frac{5}{55}$ $\frac{1}{11}$ $\frac{9}{99}$ **9.$\overline{09}$% I**

MAVA Math: Enhanced Skills Solutions Copyright © 2015 Marla Weiss

Percents 7

Represent the value as a fraction.

1.	5% increase	$\frac{21}{20}$	20.	40% decrease	$\frac{3}{5}$
2.	5% decrease	$\frac{19}{20}$	21.	45% increase	$\frac{29}{20}$
3.	10% increase	$\frac{11}{10}$	22.	45% decrease	$\frac{11}{20}$
4.	10% decrease	$\frac{9}{10}$	23.	50% increase	$\frac{3}{2}$
5.	12.5% increase	$\frac{9}{8}$	24.	50% decrease	$\frac{1}{2}$
6.	12.5% decrease	$\frac{7}{8}$	25.	60% increase	$\frac{8}{5}$
7.	15% increase	$\frac{23}{20}$	26.	60% decrease	$\frac{2}{5}$
8.	15% decrease	$\frac{17}{20}$	27.	62.5% increase	$\frac{13}{8}$
9.	20% increase	$\frac{6}{5}$	28.	62.5% decrease	$\frac{3}{8}$
10.	20% decrease	$\frac{4}{5}$	29.	66 2/3% increase	$\frac{5}{3}$
11.	25% increase	$\frac{5}{4}$	30.	66 2/3% decrease	$\frac{1}{3}$
12.	25% decrease	$\frac{3}{4}$	31.	70% increase	$\frac{17}{10}$
13.	30% increase	$\frac{13}{10}$	32.	70% decrease	$\frac{3}{10}$
14.	30% decrease	$\frac{7}{10}$	33.	75% increase	$\frac{7}{4}$
15.	33 1/3% increase	$\frac{4}{3}$	34.	75% decrease	$\frac{1}{4}$
16.	33 1/3% decrease	$\frac{2}{3}$	35.	80% increase	$\frac{9}{5}$
17.	37.5% increase	$\frac{11}{8}$	36.	80% decrease	$\frac{1}{5}$
18.	37.5% decrease	$\frac{5}{8}$	37.	90% increase	$\frac{19}{10}$
19.	40% increase	$\frac{7}{5}$	38.	90% decrease	$\frac{1}{10}$

Percents 8

Answer the "backward" percent word problem. Use a calculator after writing the set-up.

1. Find the original price of an item that costs $12.60 after a 16% discount.

$$.84P = 12.60$$
$$P = 12.60 \div .84 = \mathbf{\$15}$$

2. A book was bought on sale for $40.95 at 9% off. What was the regular price?

$$.91P = 40.95$$
$$P = 40.95 \div .91 = \mathbf{\$45}$$

3. A game was purchased for $32.19 at 13% off. What was the original price?

$$.87P = 32.19$$
$$P = 32.19 \div .87 = \mathbf{\$37}$$

4. A shirt cost $18 after a 20% discount. What was the original price?

$$.80P = 18.00$$
$$P = 18.00 \div .80 = \mathbf{\$22.50}$$

5. A coat cost $132 after a 12% discount. What was the original price?

$$.88P = 132$$
$$P = 132 \div .88 = \mathbf{\$150}$$

6. The price of a sofa with 5% tax was $924. What was the price without tax?

$$1.05P = 924.00$$
$$P = 924 \div 1.05 = \mathbf{\$880.00}$$

7. A television was bought for $507 at 22% off the regular price. Find the regular price.

$$.78P = 507$$
$$P = 507 \div .78 = \mathbf{\$650}$$

8. An item cost $48.15 including the 7% sales tax. What was the price without the tax?

$$1.07P = 48.15$$
$$P = 48.15 \div 1.07 = \mathbf{\$45}$$

9. The price of a chair with a 15% discount was $187. What was the original price?

$$.85P = 187$$
$$P = 187 \div .85 = \mathbf{\$220}$$

10. $270.40 is in a bank paying 4% simple annual interest. Find the value of the account 1 year ago assuming no other activity.

$$1.04P = 270.40$$
$$P = 270.4 \div 1.04 = \mathbf{\$260}$$

11. A shirt was bought for $15.90 including the 6% tax. What was the price without the tax?

$$1.06P = 15.90$$
$$P = 15.90 \div 1.06 = \mathbf{\$15}$$

12. Find the original price of an item that costs $63 after a 16% discount.

$$.84P = 63$$
$$P = 63 \div .84 = \mathbf{\$75}$$

13. The price of a stereo set with a 20% discount was $924. Find the original price.

$$.80P = 924.00$$
$$P = 924.00 \div .80 = \mathbf{\$1155.00}$$

14. A boat costs $13,250 with the 6% sales tax. What was the price prior to the tax?

$$1.06P = 13,250$$
$$P = 13,250 \div 1.06 = \mathbf{\$12,500}$$

15. Jake spent $380 for an item at 24% off the regular price. What was the regular price?

$$.76P = 380$$
$$P = 380 \div .76 = \mathbf{\$500}$$

16. $705.25 is invested paying 8.5% simple annual interest. Find the value of the account 1 year ago assuming no other activity.

$$1.085P = 705.25$$
$$P = 705.25 \div 1.085 = \mathbf{\$650}$$

Percents 9

Find the final percent increase (I) or decrease (D) after successive changes.

	FRACTION METHOD	100 METHOD	ANSWER
1. 50% I, 50% D	$\dfrac{3}{2} \cdot \dfrac{1}{2} = \dfrac{3}{4}$	100, 150, 75	**25% D**
2. 30% I, 30% D	$\dfrac{13}{10} \cdot \dfrac{7}{10} = \dfrac{91}{100}$	100, 130, 91	**9% D**
3. 25% I, 20% I	$\dfrac{5}{4} \cdot \dfrac{6}{5} = \dfrac{3}{2}$	100, 125, 150	**50% I**
4. 20% D, 15% D	$\dfrac{4}{5} \cdot \dfrac{17}{20} = \dfrac{68}{100}$	100, 80, 68	**32% D**
5. 25% D, 20% D	$\dfrac{3}{4} \cdot \dfrac{4}{5} = \dfrac{3}{5}$	100, 75, 60	**40% D**
6. 50% I, 33 1/3% D	$\dfrac{3}{2} \cdot \dfrac{2}{3} = 1$	100, 150, 100	**no change**
7. 50% I, 33 1/3% I	$\dfrac{3}{2} \cdot \dfrac{4}{3} = \dfrac{2}{1}$	100, 150, 200	**100% I**
8. 70% D, 33 1/3% I	$\dfrac{3}{10} \cdot \dfrac{4}{3} = \dfrac{2}{5}$	100, 30, 40	**60% D**
9. 75% D, 20% D	$\dfrac{1}{4} \cdot \dfrac{4}{5} = \dfrac{1}{5}$	100, 25, 20	**80% D**
10. 50% D, 50% D	$\dfrac{1}{2} \cdot \dfrac{1}{2} = \dfrac{1}{4}$	100, 50, 25	**75% D**
11. 90% D, 40% I	$\dfrac{1}{10} \cdot \dfrac{7}{5} = \dfrac{7}{50}$	100, 10, 14	**86% D**
12. 100% I, 50% I	$\dfrac{2}{1} \cdot \dfrac{3}{2} = \dfrac{3}{1}$	100, 200, 300	**200% I**
13. 20% I, 20% D	$\dfrac{6}{5} \cdot \dfrac{4}{5} = \dfrac{24}{25}$	100, 120, 96	**4% D**
14. 40% D, 33 1/3% I	$\dfrac{3}{5} \cdot \dfrac{4}{3} = \dfrac{4}{5}$	100, 60, 80	**20% D**
15. 20% I, 25% D, 30% D	$\dfrac{6}{5} \cdot \dfrac{3}{4} \cdot \dfrac{7}{10} = \dfrac{63}{100}$	100, 120, 90, 63	**37% D**
16. 25% D, 66 2/3% I, 20% D	$\dfrac{3}{4} \cdot \dfrac{5}{3} \cdot \dfrac{4}{5} = 1$	100, 75, 125, 100	**no change**
17. 60% I, 80% D, 50% I	$\dfrac{8}{5} \cdot \dfrac{1}{5} \cdot \dfrac{3}{2} = \dfrac{12}{25}$	100, 160, 32, 48	**52% D**
18. 20% D, 37.5% I, 20% I	$\dfrac{4}{5} \cdot \dfrac{11}{8} \cdot \dfrac{6}{5} = \dfrac{66}{50}$	100, 80, 110, 132	**32% I**
19. 20% D, 20% I, 50% D	$\dfrac{4}{5} \cdot \dfrac{6}{5} \cdot \dfrac{1}{2} = \dfrac{12}{25}$	100, 80, 96, 48	**52% D**

Percents 10

Answer as indicated.

1. A price increases by 10% and then again by 10% to $6.05. What was the original price?

$$\frac{11}{10} \cdot \frac{11}{10} \cdot P = 6.05$$
$$11 \cdot 11\, P = 605$$
$$P = \mathbf{\$5}$$

7. A price decreases by 30% and then increases by 40%. Find the ratio of the last price to the original.

Use 100 WLOG
30% D = 70
40% I = 98

$$\frac{98}{100} = \frac{49}{50}$$

2. A salary is reduced by 20%. Find the percent increase to return it to the original amount.   2nd method

Use 100 WLOG
− 20% = 80
+ **25% = 100**

$$\frac{4}{5} \cdot \frac{5}{4}\, S = S$$

times 5/4 is **25% I**

8. A price increases by 30% and then by 40% to $662.48. What was the original price?

$$\frac{13}{10} \cdot \frac{14}{10} \cdot P = 662.48$$

13 · 14 P = 66248
13 · 7 P = 33124
13P = 4732
P = **$364**

3. A price increases by 25% and then decreases by 40%. Find the ratio of the last price to the original.

Use 100 WLOG
25% I = 125
40% D = 75

$$\frac{75}{100} = \frac{3}{4}$$

9. A price increases by 20% and then again by 20% to $100.80. What was the original price?

$$\frac{12}{10} \cdot \frac{12}{10} \cdot P = 100.80$$

12 · 12 P = 10080
36P = 2520
18P = 1260
9P = 630
P = **$70**

4. A price increases by 40% and then decreases by 25%. Find the ratio of the last price to the original.

Use 100 WLOG
40% I = 140
25% D = 105

$$\frac{105}{100} = \frac{21}{20}$$

10. A salary is reduced by 25%. Find the percent increase to return it to the original amount.

Use 100 WLOG
− 25% = 75
+ **33.3̄% = 100**

5. A price increases by 20% and then again by 20% to $86.40. What was the original price?

$$\frac{12}{10} \cdot \frac{12}{10}\, P = 86.4$$

12 · 12 P = 8640
18P = 1080
9P = 540
P = **60**

11. A number is reduced by 60%. Find the percent increase to return it to the original value.

Use 100 WLOG
− 60% = 40
+ **150% = 100**

6. A price increases by 75% and then decreases by 60%. Find the ratio of the last price to the original.

Use 100 WLOG
75% I = 175
60% D = 70

$$\frac{70}{100} = \frac{7}{10}$$

12. A salary is reduced by 50%. Find the percent increase to return it to the original amount.

Use 100 WLOG
− 50% = 50
+ **100% = 100**

Percents 11

Both fraction parts are over 100, which simplify.

Find how much more must be added to the given amount to make 100%.

Find a percent of a percent.

1.	4800 is 60%.	1600 is 20% **3200** is 40%	14.	35% is what percent of 80%? $\dfrac{35}{80} = \dfrac{7}{16}$ **43.75%**
2.	135 is 30%.	45 is 10% **315** is 70%	15.	25% is what percent of 60%? $\dfrac{25}{60} = \dfrac{5}{12}$ **41.$\overline{6}$%**
3.	640 is 40%.	320 is 20% **960** is 60%	16.	30% is what percent of 80%? $\dfrac{30}{80} = \dfrac{3}{8}$ **37.5%**
4.	21 is 5%.	42 is 10% 420 100% **399** is 95%	17.	30% is what percent of 90%? $\dfrac{30}{90} = \dfrac{1}{3}$ **33.$\overline{3}$%**
5.	6300 is 70%.	900 is 10% **2700** is 30%	18.	50% is what percent of 75%? $\dfrac{50}{75} = \dfrac{2}{3}$ **66.$\overline{6}$%**
6.	3434 is 85%.	202 is 5% **606** is 15%	19.	60% is what percent of 72%? $\dfrac{60}{72} = \dfrac{5}{6}$ **83.$\overline{3}$%**
7.	609 is 15%.	203 is 5% **3451** is 85%	20.	3% is what percent of 48%? $\dfrac{3}{48} = \dfrac{1}{16}$ **6.25%**
8.	763 is 35%.	109 is 5% **1417** is 65%	21.	30% is what percent of 96%? $\dfrac{30}{96} = \dfrac{5}{16}$ **31.25%**
9.	459 is 45%.	51 is 5% **561** is 55%	22.	15% is what percent of 90%? $\dfrac{15}{90} = \dfrac{1}{6}$ **16.$\overline{6}$%**
10.	7742 is 70%.	1106 is 10% **3318** is 30%	23.	49% is what percent of 84%? $\dfrac{49}{84} = \dfrac{7}{12}$ **58.$\overline{3}$%**
11.	303 is 15%.	101 is 5% **1717** is 85%	24.	60% is what percent of 96%? $\dfrac{60}{96} = \dfrac{5}{8}$ **62.5%**
12.	2639 is 65%.	203 is 5% **1421** is 35%	25.	44% is what percent of 99%? $\dfrac{44}{99} = \dfrac{4}{9}$ **44.$\overline{4}$%**
13.	352 is 55%.	32 is 5% **288** is 45%	26.	12% is what percent of 64%? $\dfrac{12}{64} = \dfrac{3}{16}$ **18.75%**

Percents 12

Answer as indicated.

#8 and #9 show 2 different methods.

---

1. 7 out of 19 marbles are pink. How many pink marbles must be added to make the number of pinks 52%?

$$\frac{7 + x}{19 + x} = \frac{52}{100} = \frac{13}{25} \quad \textbf{6}$$

---

8. If x is 40% of y, what % of 8y is 7x?

$$\frac{x}{y} = \frac{40}{100} = \frac{2}{5} \quad \bigg| \quad \frac{7x}{8y} = \frac{7x}{20x} = \frac{7}{20}$$

$$5x = 2y$$
$$20x = 8y \qquad\qquad \textbf{35\%}$$

---

2. 20 out of 31 marbles are blue. How many blue marbles must be added to make the number of blues 78%?

$$\frac{20 + x}{31 + x} = \frac{78}{100} = \frac{39}{50} \quad \textbf{19}$$

---

9. If x is 30% of y, what % of 3y is 5x?

WLOG y = 10, x = 3 $\quad\bigg|\quad \dfrac{15}{30} = \dfrac{1}{2}$

3y = 30, 5x = 15

$$\textbf{50\%}$$

---

3. 41 out of 63 marbles are red. How many red marbles must be removed to make the number of reds 56%?

$$\frac{41 - x}{63 - x} = \frac{56}{100} = \frac{28}{50} \quad \textbf{13}$$

---

10. If x is 60% of y, what % of 3y is 4x?

WLOG y = 10, x = 6 $\quad\bigg|\quad \dfrac{24}{30} = \dfrac{4}{5}$

3y = 30, 4x = 24

$$\textbf{80\%}$$

---

4. 9 out of 18 buttons are white. How many white buttons must be added to make the number of whites 64%?

$$\frac{9 + x}{18 + x} = \frac{64}{100} = \frac{16}{25} \quad \textbf{7}$$

---

11. If x is 140% of y, 2x is what % of 7y?

WLOG y = 10, x = 14 $\quad\bigg|\quad \dfrac{28}{70} = \dfrac{4}{10}$

7y = 70, 2x = 28

$$\textbf{40\%}$$

---

5. 10 out of 17 peppers are green. How many green peppers must be added to make the number of greens 72%?

$$\frac{10 + x}{17 + x} = \frac{72}{100} = \frac{18}{25} \quad \textbf{8}$$

---

12. If x is 90% of y, what % of 9y is 2x?

WLOG y = 10, x = 9 $\quad\bigg|\quad \dfrac{18}{90} = \dfrac{2}{10}$

9y = 90, 2x = 18

$$\textbf{20\%}$$

---

6. 37 out of 64 marbles are red. How many red marbles must be removed to make the number of reds 46%?

$$\frac{37 - x}{64 - x} = \frac{46}{100} = \frac{23}{50} \quad \textbf{14}$$

---

13. If x is 80% of y, 4x is what % of 8y?

WLOG y = 10, x = 8 $\quad\bigg|\quad \dfrac{32}{80} = \dfrac{8}{20}$

8y = 80, 4x = 32

$$\textbf{40\%}$$

---

7. 20 out of 31 marbles are blue. How many blue marbles must be removed to make the number of blues 45%?

$$\frac{20 - x}{31 - x} = \frac{45}{100} = \frac{9}{20} \quad \textbf{11}$$

---

14. If x is 70% of y, 4x is what % of 8y?

WLOG y = 10, x = 7 $\quad\bigg|\quad \dfrac{28}{80} = \dfrac{7}{20}$

8y = 80, 4x = 28

$$\textbf{35\%}$$

Percents 13

#10 and 11 show 2 methods.

Evaluate using fraction simplification.	*Find the percent change in the area of a:*
1. 10% of 20% of 40% of 250 $\dfrac{1}{10} \cdot \dfrac{1}{5} \cdot \dfrac{2}{5} \cdot 250$ **2**	10. rectangle if each side is increased by 20%. $\dfrac{6}{5} \cdot \dfrac{6}{5} = \dfrac{36}{25}$ 11/25 = **44% I**
2. 25% of 20% of 75% of 64,000 $\dfrac{1}{4} \cdot \dfrac{1}{5} \cdot \dfrac{3}{4} \cdot 64{,}000$ **2400**	11. rectangle if 1 side is increased by 50% and 1 side is increased by 100%. 2 by 4: a = 8 $\dfrac{ch}{orig} = \dfrac{16}{8}$ **200% I** 3 by 8: A = 24
3. 10% of 20% of 35% of 50,000 $\dfrac{1}{10} \cdot \dfrac{1}{5} \cdot \dfrac{7}{20} \cdot 50{,}000$ **350**	12. square if each side is increased by 50%. 2 by 2: a = 4 $\dfrac{ch}{orig} = \dfrac{5}{4}$ **125% I** 3 by 3: A = 9
4. 30% of 40% of 80% of 7500 $\dfrac{3}{10} \cdot \dfrac{2}{5} \cdot \dfrac{4}{5} \cdot 7500$ **720**	13. rectangle if 1 side is decreased by 20% and 1 side is increased by 50%. 10 by 12: a = 120 $\dfrac{ch}{orig} = \dfrac{24}{120} = \dfrac{4}{20}$ **20% I** 8 by 18: A = 144
5. 25% of 10% of 30% of 60,000 $\dfrac{1}{4} \cdot \dfrac{1}{10} \cdot \dfrac{3}{10} \cdot 60{,}000$ **450**	14. rectangle if each side is decreased by 40%. 10 by 20: a = 200 $\dfrac{ch}{orig} = \dfrac{128}{200}$ **64% D** 6 by 12: A = 72
6. 15% of 60% of 12.5% of 84,000 $\dfrac{3}{20} \cdot \dfrac{3}{5} \cdot \dfrac{1}{8} \cdot 84{,}000$ **945**	15. rectangle if each side is increased by 40%. 10 by 20: a = 200 $\dfrac{ch}{orig} = \dfrac{192}{200}$ **96% I** 14 by 28: A = 392
7. 20% of 12% of 87.5% of 9000 $\dfrac{1}{5} \cdot \dfrac{3}{25} \cdot \dfrac{7}{8} \cdot 9000$ **189**	16. square if 1 side is increased by 40% and 1 side is decreased by 40%. 10 by 10: a = 100 $\dfrac{ch}{orig} = \dfrac{16}{100}$ **16% D** 14 by 6: A = 84
8. 45% of 28% of 37.5% of 80,000 $\dfrac{9}{20} \cdot \dfrac{7}{25} \cdot \dfrac{3}{8} \cdot 80{,}000$ **3780**	17. rectangle if each side is increased by 30%. 10 by 20: a = 200 $\dfrac{ch}{orig} = \dfrac{138}{200}$ **69% I** 13 by 26: A = 338
9. 16% of 65% of 62.5% of 100,000 $\dfrac{4}{25} \cdot \dfrac{13}{20} \cdot \dfrac{5}{8} \cdot 100{,}000$ **6500**	18. rectangle if each side is increased by 10%. 10 by 20: a = 200 $\dfrac{ch}{orig} = \dfrac{42}{200}$ **21% I** 11 by 22: A = 242

Percents 14

Answer as indicated. | *Answer as indicated for the solid figures.*

1. If 15% of a number is subtracted from the number, the result is 68. Find the number.

$$\frac{85}{100} x = 68$$

85x = 6800
17x = 1360
x = **80**

6. A square pyramid has its altitude tripled and each edge of its base halved. Find the percent change in its volume.   **25% D**

E = 20	e = 10	$\frac{ch}{orig}$ =
H = 12	h = 36	
V = 20x20x12/3	v = 10x10x36/3	$\frac{400}{1600}$
V = 1600	v = 1200	

2. If 45% of a number is added to the number, the result is 580. Find the number.

$$\frac{145}{100} x = 580$$

145x = 58,000
29x = 11,600
x = **400**

7. A square pyramid has its altitude halved and each edge of its base tripled. Find the percent change in its volume.   **350% I**

E = 10	e = 30	$\frac{ch}{orig}$ =
H = 12	h = 6	
V = 10x10x12/3	v = 30x30x6/3	$\frac{1400}{400}$
V = 400	v = 1800	

3. If 36% of a number is subtracted from the number, the result is 88. Find the number.

$$\frac{64}{100} x = 88$$

64x = 8800
8x = 1100
x = **137.5**

8. A rectangular pyramid with L=2W has its altitude halved and each edge of its base doubled. Find the percent change in its volume.   **100% I**

W = 10, L = 20	w = 20, l = 40	$\frac{ch}{orig}$ =
H = 12	h = 6	
V = 20x10x12/3	v = 40x20x6/3	$\frac{800}{800}$
V = 800	v = 1600	

4. If 400 critters are 13 1/3% of the critter population, how many are 100%?

400 is 13 1/3%
1200 is 40%
600 is 20%
3000 is 100%

9. A rectangular pyramid with L=4W has its altitude doubled and each edge of its base halved. Find the percent change in its volume. **50% D**

W = 10, L = 40	w = 5, l = 20	$\frac{ch}{orig}$ =
H = 12	h = 24	
V = 40x10x12/3	v = 20x5x24/3	$\frac{800}{1600}$
V = 1600	v = 800	

5. If 252 critters are 11 2/3% of the critter population, how many are 100%?

252 is 11 2/3%
756 is 35%
108 is 5%
2160 is 100%

10. A solid has a pentagonal base of area 72, total altitude 20, and pyramidal top with altitude 8. The volume of the top is what percent of the volume of the solid?

72x8/3=192 $\frac{192}{1056}$ $\frac{2}{11}$
72x12=864
192+864=1056
18.1̄8̄%

Perfect Squares & Cubes 1

Find the least natural number n such that the product is a perfect square.			*Find the least natural number n such that the product is a perfect cube.*		
1. 98n	49 x 2	**2**	14. 98n	7 x 7 x 2	**28**
2. 160n	16 x 10	**10**	15. 160n	8 x 20 8 x 4 x 5	**50**
3. 540n	9 x 6 x 10 9 x 4 x 15	**15**	16. 540n	27 x 2 x 10 27 x 4 x 5	**50**
4. 343n	7 x 7 x 7	**7**	17. 343n	7 x 7 x 7	**1**
5. 48n	16 x 3	**3**	18. 48n	8 x 2 x 3	**36**
6. 360n	36 x 10	**10**	19. 360n	8 x 45 8 x 9 x 5	**75**
7. 500n	25 x 4 x 5	**5**	20. 500n	125 x 4	**2**
8. 108n	9 x 12 9 x 4 x 3	**3**	21. 108n	9 x 12 27 x 4	**2**
9. 504n	9 x 56 9 x 4 x 14	**14**	22. 504n	9 x 56 9 x 8 x 7	**147**
10. 144n	12 x 12	**1**	23. 144n	12 x 12 4 x 4 x 3 x 3	**12**
11. 180n	9 x 20 9 x 4 x 5	**5**	24. 180n	9 x 20 9 x 4 x 5	**150**
12. 250n	25 x 10	**10**	25. 250n	125 x 2	**4**
13. 320n	16 x 2 x 10 16 x 4 x 5	**5**	26. 320n	8 x 4 x 10 8 x 8 x 5	**25**

Perfect Squares & Cubes 2

Answer as indicated.

1. The sum of the first 20 positive perfect squares is 2870. Find the sum of the first 19 positive perfect squares.

20th is 20^2 = 400
2870 − 400 = **2470**

2. Find the product of the 50th and 100th positive perfect squares.

50th is 50^2 = 2500
100th is 100^2
(2500)(100)(100) = **25,000,000**

3. Find the least positive integer n for which 5n is both an even integer and a perfect square.

x 2 even
x 5 and x 2 perfect square
2 x 5 x 2 = **20**

4. Find the least perfect square divisible by the first 4 prime numbers.

x 2, 3, 5, 7 twice
(210)(210) = **44,100**

5. Find the natural numbers less than 200 that have exactly 3 factors.

only squares of primes have 3 factors
4, 9, 25, 49, 121, and 169

6. The product mn is a perfect square. Both m and n are neither prime nor square. Find the least possible sum m + n.

Two variables can be different or assume the same value.

not 1, 4, 9, 16, 25
36 = 6 x 6
6 + 6 = **12**

7. The sum of the squares of three consecutive positive integers is 677. Find the sum of the integers.

100 + 121 + 144 too small
196 + 225 + 256 = 677
14 + 15 + 16 = **45**

8. How many perfect squares are between 400 and 3600?

20^2 = 400
60^2 = 3600
21 through 59: **39** numbers, each squared

9. A perfect cube was increased by 700%. By what percent was its cube root increased?

1x + 7x = 8x is 700% increase of x
(R)(R)(R)(8) = (2R)(2R)(2R)
cube root is doubled; increase of **100%**

10. The four-digit number 7AB1 is a perfect square. Find A + B.

The square root of 7AB1 ends in 1 or 9.
By estimating, 90 x 90 = 8100.
89 x 89 = 7921
A = 9, B = 2, 9 + 2 = **11**

11. Find the greatest 4-digit number that has exactly 3 factors.

only squares of primes have 3 factors
97^2 = **9409**

12. The sum of five consecutive integers is a perfect square. The sum of six consecutive integers is a perfect cube. Find the least product of the two sums.

3+4+5+6+7 = 25
2+3+4+5+6+7 = 27
(25)(27) = (25)(25+2) = 625 + 50 = **675**

MAVA Math: Enhanced Skills Solutions Copyright © 2015 Marla Weiss

Perimeter 1

Calculate perimeter in units. Assume right angles and semicircles.

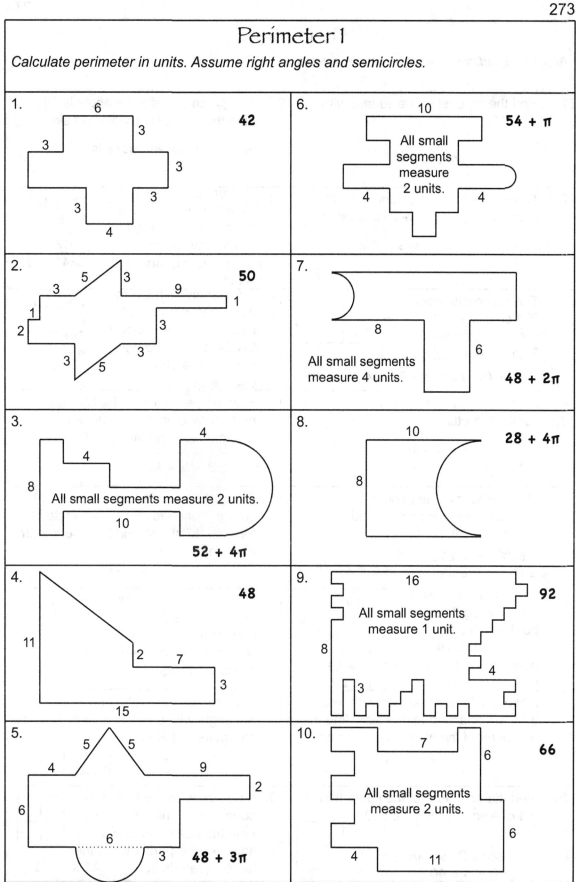

1. **42**

2. **50**

3. All small segments measure 2 units. **52 + 4π**

4. **48**

5. **48 + 3π**

6. All small segments measure 2 units. **54 + π**

7. All small segments measure 4 units. **48 + 2π**

8. **28 + 4π**

9. All small segments measure 1 unit. **92**

10. All small segments measure 2 units. **66**

Perimeter 2

Answer as indicated. All lengths are in units.

---

1. Find the perimeter of a square with side 0.25.

   .25 x 4 = **1**

---

2. The perimeter of a rectangle is 1/9. Find the sum of two adjacent sides.

   sum of 2 adjacent sides is 1/2 P

   $\dfrac{1}{18}$

---

3. The figure with area 294 consists of 6 congruent squares. Find the perimeter.

   294 ÷ 6 = 49   side sq = 7   14 x 7 = **98**

---

4. If a square has area 49, find its semi-perimeter.

   7 x 2 = **14**

---

5. Find the sum of the perimeters: a square with diagonal 12 and a square with diagonal 12√2.

   s = $6\sqrt{2}$ p = $24\sqrt{2}$
   S = 12   P = 48   **48 + 24√2**

---

6. A rectangle with W 4 times its L gets a uniform border of 3 units added to all 4 sides. The new perimeter is 174. Find the original perimeter.

   L by 4L
   L+6 by 4L+6
   5L+12=87
   L = 15
   W = 60
   P = **150**

---

7. If two equilateral triangles with side 3.25 are attached at one side, find the perimeter of the quadrilateral formed.

   3.25 x 4 = **13**

---

8. Find the perimeter of the figure that is 3/4 of a square with area 100.

   same P as square
   10 x 4 = **40**

---

9. The perimeter of a rectangle is 1/5. Find the sum of two adjacent sides.

   sum of 2 adjacent sides is 1/2 P $\dfrac{1}{10}$

---

10. A square has area 144. Triple the length of one side and halve the other. Find the ratio of the perimeters: square to rectangle.

    12 by 12
    36 by 6
    $\dfrac{4\times12}{2\times42}$ $\dfrac{4}{7}$

---

11. A rectangle with L 6 more than its W gets a uniform border of 5 units added to all 4 sides. The new perimeter is 104. Find the original perimeter.

    w by w+6
    w+10 by w+16
    2w+26=52
    w = 13
    L = 19
    P = **64**

---

12. Find the perimeter of the figure: an equilateral triangle is attached at one side of a square with area 81.

    9 x 5 = **45**

---

13. A 6 by 8 rectangle has its upper right corner "punched in" at right angles to form an "L" shape. Find the perimeter of the new figure.

    14 x 2 = **28**

---

14. A square has area 400. Double one side and quarter the other. Find the ratio of the perimeters: rectangle to square.

    20 by 20
    40 by 5
    $\dfrac{2\times45}{4\times20}$ $\dfrac{9}{8}$

---

15. Four squares are attached to form a rectangle with shortest side 5. Find the perimeter of the rectangle.

    10 x 5 = **50**

---

16. A square is divided into four congruent smaller squares. Find the perimeter of a small square if the perimeter of the large square is 2.

    .25 x 4 = **1**

Perimeter 3

Answer as indicated.

1. Find the perimeter of a regular octagon if a side has length 5.125.

   5.125 × 8 = **41**

8. Find the perimeter of an equilateral triangle if 5/6 the perimeter is 45.

   1/6 of P is 9
   9 × 6 = **54**

2. Find the perimeter of an equilateral triangle that is attached at one side of a square with area 81.

   9 × 3 = **27**

9. Find the perimeter of a regular pentagon if a side has length 1.2.

   1.2 × 5 = **6**

3. Find the perimeter of an equilateral triangle with area $16\sqrt{3}$.

   $\frac{s^2\sqrt{3}}{4} = 16\sqrt{3}$

   $s^2 = 64$
   $s = 8$
   P = **24**

10. Find the side of an equilateral triangle if a scalene triangle with sides 5, 6, and 10 has the same perimeter.

   21 ÷ 3 = **7**

4. Find the perimeter of an equilateral triangle with area $12\sqrt{3}$.

   $\frac{s^2\sqrt{3}}{4} = 12\sqrt{3}$

   $s^2 = 48$
   $s = 4\sqrt{3}$
   P = $\mathbf{12\sqrt{3}}$

11. Find the perimeter of an equilateral triangle if 2/3 the perimeter is 12.

   6 × 3 = **18**

5. The ratio of 2 adjacent sides of a parallelogram is 5:4. If the perimeter is 117, find the shorter side.

   5 + 4 = 9
   117/9 = 13
   13 × 4 = **52**

12. The ratio of 2 adjacent sides of a parallelogram is 7:3. If the perimeter is 170, find the longer side.

   7 + 3 = 10 | 85/10 = 8.5
   170/2 = 85 | 8.5 × 7 = **59.5**

6. Find the greatest possible perimeter of a right triangle with area 10 if the legs are whole numbers.

   bh = 20
   20 = 20 × 1 $21 + \sqrt{401}$

13. Find the least possible perimeter of a right triangle with area 1800 if the legs are whole numbers.

   bh = 3600
   3600 = 60 × 60 $120 + 60\sqrt{2}$

7. Find the perimeter of a regular hexagon with area $24\sqrt{3}$.

   $\frac{s^2\sqrt{3}}{4} = 4\sqrt{3}$

   $s^2 = 16$
   $s = 4$
   P = **24**

14. Find the perimeter of a regular hexagon with area $30\sqrt{3}$.

   $\frac{s^2\sqrt{3}}{4} = 5\sqrt{3}$

   $s^2 = 20$
   $s = 2\sqrt{5}$
   P = $\mathbf{12\sqrt{5}}$

Perimeter 4

Calculate the perimeter in units. Neighboring grid lines are one unit apart.

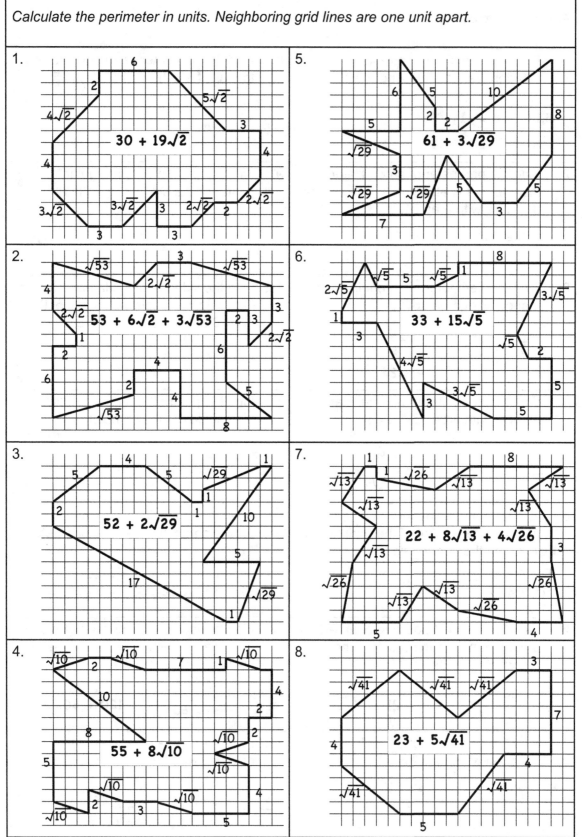

1. $30 + 19\sqrt{2}$

2. $53 + 6\sqrt{2} + 3\sqrt{53}$

3. $52 + 2\sqrt{29}$

4. $55 + 8\sqrt{10}$

5. $61 + 3\sqrt{29}$

6. $33 + 15\sqrt{5}$

7. $22 + 8\sqrt{13} + 4\sqrt{26}$

8. $23 + 5\sqrt{41}$

MAVA Math: Enhanced Skills Solutions Copyright © 2015 Marla Weiss

Permutations 1

Find the number of arrangements of the letters of the word.	Answer as indicated using the multiplication principle.
1. MATH 4 x 3 x 2 x 1 = **24**	11. Find the number of arrangements of 4 letters of the word EDUCATION. 9 x 8 x 7 x 6 = **3024**
2. FRACTIONS 9 x 8 x 7 x 6 x 5 x 4 x 3 x 2 x 1 = **362,880**	12. Find the number of ways to arrange 6 different books on a shelf. 6 x 5 x 4 x 3 x 2 x 1 = **720**
3. PRIME 5 x 4 x 3 x 2 x 1 = **120**	13. Find the number of arrangements of 3 letters of the word PRISM if letters may repeat. 5 x 5 x 5 = **125**
4. RHOMBUS 7 x 6 x 5 x 4 x 3 x 2 x 1 = **5040**	14. Find the number of 3-digit numbers containing only 3s and 7s. 2 x 2 x 2 = **8**
5. TWO 3 x 2 x 1 = **6**	15. Find the number of 4-letter arrangements of all vowels including Y where a letter may repeat. 6 x 6 x 6 x 6 = **1296**
6. EQUATION 8 x 7 x 6 x 5 x 4 x 3 x 2 x 1 = **40,320**	16. Find the number of 3-digit numbers with all digits odd. 5 x 5 x 5 = **125**
7. TRAPEZOIDS 10 x 9 x 8 x 7 x 6 x 5 x 4 x 3 x 2 x 1 = **3,628,800**	17. Find the number of 3-digit numbers with all digits even. 4 x 5 x 5 = **100**
8. NUMBER 6 x 5 x 4 x 3 x 2 x 1 = **720**	18. How many codes can be made with 2 letters followed by 3 digits if the letter "O" is not used? 25 x 25 x 10 x 10 x 10 = **625,000**
9. TWICE 5 x 4 x 3 x 2 x 1 = **120**	19. Find the number of 3-digit numbers without the digits 4 and 8. 7 x 8 x 8 = **448**
10. SQUARE 6 x 5 x 4 x 3 x 2 x 1 = **720**	20. Find the number of 4-digit even numbers without the digit 2. 8 x 9 x 9 x 9 = **5832**

Permutations 2

Find the number of arrangements of the letters of the word.

1. MISSISSIPPI		12. HELPLESS	
$\dfrac{11 \cdot 10 \cdot 9 \cdot 8 \cdot 7 \cdot 6 \cdot 5 \cdot 4 \cdot 3 \cdot 2 \cdot 1}{4 \cdot 3 \cdot 2 \cdot 1 \cdot 4 \cdot 3 \cdot 2 \cdot 1 \cdot 2 \cdot 1}$	**34,650**	$\dfrac{8 \cdot 7 \cdot 6 \cdot 5 \cdot 4 \cdot 3 \cdot 2 \cdot 1}{2 \cdot 1 \cdot 2 \cdot 1 \cdot 2 \cdot 1}$	**5040**
2. QUEEN		13. SASSY	
$\dfrac{5 \cdot 4 \cdot 3 \cdot 2 \cdot 1}{2 \cdot 1}$	**60**	$\dfrac{5 \cdot 4 \cdot 3 \cdot 2 \cdot 1}{3 \cdot 2 \cdot 1}$	**20**
3. SPINNER		14. SHELLFISH	
$\dfrac{7 \cdot 6 \cdot 5 \cdot 4 \cdot 3 \cdot 2 \cdot 1}{2 \cdot 1}$	**2520**	$\dfrac{9 \cdot 8 \cdot 7 \cdot 6 \cdot 5 \cdot 4 \cdot 3 \cdot 2 \cdot 1}{2 \cdot 1 \cdot 2 \cdot 1 \cdot 2 \cdot 1}$	**45,360**
4. LOLLYPOP		15. POSSESS	
$\dfrac{8 \cdot 7 \cdot 6 \cdot 5 \cdot 4 \cdot 3 \cdot 2 \cdot 1}{3 \cdot 2 \cdot 1 \cdot 2 \cdot 1 \cdot 2 \cdot 1}$	**1680**	$\dfrac{7 \cdot 6 \cdot 5 \cdot 4 \cdot 3 \cdot 2 \cdot 1}{4 \cdot 3 \cdot 2 \cdot 1}$	**210**
5. BALL		16. GUERRILLA	
$\dfrac{4 \cdot 3 \cdot 2 \cdot 1}{2 \cdot 1}$	**12**	$\dfrac{9 \cdot 8 \cdot 7 \cdot 6 \cdot 5 \cdot 4 \cdot 3 \cdot 2 \cdot 1}{2 \cdot 1 \cdot 2 \cdot 1}$	**90,720**
6. DEEDED		17. HIGHER	
$\dfrac{6 \cdot 5 \cdot 4 \cdot 3 \cdot 2 \cdot 1}{3 \cdot 2 \cdot 1 \cdot 3 \cdot 2 \cdot 1}$	**20**	$\dfrac{6 \cdot 5 \cdot 4 \cdot 3 \cdot 2 \cdot 1}{2 \cdot 1}$	**360**
7. GOOGOL		18. OVERWORK	
$\dfrac{6 \cdot 5 \cdot 4 \cdot 3 \cdot 2 \cdot 1}{2 \cdot 1 \cdot 3 \cdot 2 \cdot 1}$	**60**	$\dfrac{8 \cdot 7 \cdot 6 \cdot 5 \cdot 4 \cdot 3 \cdot 2 \cdot 1}{2 \cdot 1 \cdot 2 \cdot 1}$	**10,080**
8. TOMFOOL		19. PACIFIC	
$\dfrac{7 \cdot 6 \cdot 5 \cdot 4 \cdot 3 \cdot 2 \cdot 1}{3 \cdot 2 \cdot 1}$	**840**	$\dfrac{7 \cdot 6 \cdot 5 \cdot 4 \cdot 3 \cdot 2 \cdot 1}{2 \cdot 1 \cdot 2 \cdot 1}$	**1260**
9. BEEKEEPER		20. PAYDAY	
$\dfrac{9 \cdot 8 \cdot 7 \cdot 6 \cdot 5 \cdot 4 \cdot 3 \cdot 2 \cdot 1}{5 \cdot 4 \cdot 3 \cdot 2 \cdot 1}$	**3024**	$\dfrac{6 \cdot 5 \cdot 4 \cdot 3 \cdot 2 \cdot 1}{2 \cdot 1 \cdot 2 \cdot 1}$	**180**
10. PUPPET		21. PARABOLA	
$\dfrac{6 \cdot 5 \cdot 4 \cdot 3 \cdot 2 \cdot 1}{3 \cdot 2 \cdot 1}$	**120**	$\dfrac{8 \cdot 7 \cdot 6 \cdot 5 \cdot 4 \cdot 3 \cdot 2 \cdot 1}{3 \cdot 2 \cdot 1}$	**6720**
11. RASPBERRY		22. NONSENSE	
$\dfrac{9 \cdot 8 \cdot 7 \cdot 6 \cdot 5 \cdot 4 \cdot 3 \cdot 2 \cdot 1}{3 \cdot 2 \cdot 1}$	**60,480**	$\dfrac{8 \cdot 7 \cdot 6 \cdot 5 \cdot 4 \cdot 3 \cdot 2 \cdot 1}{3 \cdot 2 \cdot 1 \cdot 2 \cdot 1 \cdot 2 \cdot 1}$	**1680**

Permutations 3

Find the number of arrangements using multiplication principle, placing the restrictions first.

1. Find the number of ways to arrange 4 boys and 3 girls in a row with a boy at each end. 4 5 4 3 2 1 3 = **1440**	9. Find the number of ways to park 5 different colored cars with the red one in the middle. 4 3 1 2 1 = **24**
2. Find the number of ways to arrange 4 boys and 3 girls in a row, alternating boys and girls. 4 3 3 2 2 1 1 = **144**	10. Find the number of ways to park 5 different colored cars with the red one at either end. $\dfrac{1\ \ 4\ \ 3\ \ 2\ \ 1}{4\ \ 3\ \ 2\ \ 1\ \ 1} \dfrac{24}{+\ 24}$ = **48**
3. Find the number of ways to arrange 4 boys and 3 girls in a row with the girls completely centered. 4 3 3 2 1 2 1 = **144**	11. Find the number of ways to park 5 different colored cars with the red one next to the blue one. $\dfrac{4\ \ 3\ \ 2\ \ 1\ \ (RB)}{4\ \ 3\ \ 2\ \ 1\ \ (BR)} \dfrac{24}{+\ 24}$ = **48**
4. Find the number of 5-letter codes that have a vowel (not Y) in the middle with no letter repeating. 25 24 5 23 22 = **1,518,000**	12. Find the number of 5-digit numbers with the middle digit prime and no repeated digits. 8 8 4 7 6 = **10,752**
5. Find the number of arrangements of the letters PRIME if the first letter must be a vowel. 2 4 3 2 1 = **48**	13. Find the number of 5-digit odd numbers with the middle digit 7, first digit an odd prime, and no repeated digits. 2 7 1 6 3 = **252**
6. Find the number of arrangements of the letters PRIME if the vowels must be first and last. 2 3 2 1 1 = **12**	14. Find the number of ways to arrange the letters of SUPER with the first and last letters vowels. 2 3 2 1 1 = **12**
7. Find the number of arrangements of the 10 digits if the first one must be odd and the last one must be even. 5 8 7 6 5 4 3 2 1 5 = **1,008,000**	15. Find the number of 3-digit numbers that contain at least one 1. total # − no 1s (9×10×10) − (8×9×9) = 900 − 648 = **252**
8. Find the number of 3-digit even numbers selecting from 3, 4, 5, and 6 without repeating digits. 3 2 2 = **12**	16. Find the number of batting orders of 9 baseball players with the pitcher last and the 3 basemen starting the lineup. (3×2×1) × (5×4×3×2×1) × (1) 6 × 120 = **720**

Permutations 4

Find the number of different:	*Find the sum of:*
1. choral groups, selecting 6 of 8 boys and 6 of 8 girls. C(8,6) = 7x8/2 = 28 28 x 28 = **784**	9. all 2-digit permutations of 12. 12 + 21 = **33** Each digit appears once in each place.
2. school groups, selecting 5 of 7 teachers and 4 of 6 principals. C(7,5) = 6x7/2 = 21 C(6,4) = 5x6/2 = 15 21 x 15 = **315**	10. all 3-digit permutations of 123. 123 + 132 + 213 + 231 + 312 + 321 = **1332** No need to list. Each digit appears twice in each place. 1 + 2 + 3 = 6. 6 x 2 = 12. 1200 + 120 + 12 = 1332
3. bipartisan committees, selecting 3 of 5 Republicans and 3 of 5 Democrats. C(5,3) = 4x5/2 = 10 10 x 10 = **100**	11. all 4-digit permutations of 1234. (1+2+3+4) = 10 10 x 6 = 60 (each digit appears 6 times) 60,000 + 6000 + 600 + 60 = **66,660**
4. "words" formed by concatenating permutations of MAN to those of WOMAN. (3x2x1) x (5x4x3x2x1) 6 x 120 = **720**	12. all 3-digit permutations of 159. (1+5+9) = 15 15 x 2 = 30 (each digit appears 2 times) 3000 + 300 + 30 = **3330**
5. "words" formed by concatenating permutations of MOM to those of GRAMM. (6/2) x (120/2) 3 x 60 = **180**	13. all 4-digit permutations of 1256. (1+2+5+6) = 14 14 x 6 = 84 (each digit appears 6 times) 84,000 + 8400 + 840 + 84 = **93,324**
6. debate teams, selecting 2 of 9 boys and 2 of 10 girls. C(9,2) = 8x9/2 = 36 C(10,2) = 9x10/2 = 45 36 x 45 = **1620**	14. all 3-digit permutations of 234. (2+3+4) = 9 9 x 2 = 18 (each digit appears 2 times) 1800 + 180 + 18 = **1998**
7. "words" formed by concatenating permutations of GAME to those of BOARDS. (4x3x2x1) x (6x5x4x3x2x1) 24 x 720 = **17,280**	15. all 3-digit permutations of 178. (1+7+8) = 16 16 x 2 = 32 (each digit appears 2 times) 3200 + 320 + 32 = **3552**
8. "words" formed by concatenating permutations of SASS to those of GREIGG. (24/6) x (720/6) 4 x 120 = **480**	16. all 4-digit permutations of 1357. (1+3+5+7) = 16 16 x 6 = 96 (each digit appears 6 times) 96,000 + 9600 + 960 + 96 = **106,656**

Place Value 1

Write the numeral for the English expression.   "and" means the decimal point

1.	six hundred ten thousand **610,000**	12.	six and one hundred-thousandth **6.00001**
2.	six hundred and ten thousandths **600.010**	13.	sixty ten-thousandths **0.0060**
3.	six ten-thousandths **0.0006**	14.	sixty and six tenths **60.6**
4.	six hundred ten-thousandths **0.0600**	15.	six hundred and six hundredths **600.06**
5.	six hundred ten thousandths **0.610**	16.	six thousand six ten-thousandths **0.6006**
6.	six and ten thousandths **6.010**	17.	sixty and sixty hundredths **60.60**
7.	six hundred and one ten-thousandth **600.0001**	18.	six hundred and sixty hundredths **600.60**
8.	six hundred thousand **600,000**	19.	sixty thousand six **60,006**
9.	six hundred thousandths **0.600**	20.	sixty and sixty-one thousandths **60.061**
10.	six hundred-thousandths **0.00006**	21.	six and sixty-one ten-thousandths **6.0061**
11.	six and one hundred thousandths **6.100**	22.	six thousand six hundred six **6,606**

| Digits in sums may swap between numbers, keeping place value. | Place Value 2 | | | |

Arrange the digits as specified, using each digit once.

Round to the specified decimal place.

			10th	100th	1000th
1. digits 2, 3, 5, 6, 7, and 9 into two 3-digit numbers with least positive difference	623 −597 26	12. 4.1146	4.1	4.11	4.115
2. digits 1, 3, 6, 7, 8, and 9 into two 3-digit numbers with greatest positive difference	987 − 136 851	13. 3.5876	3.6	3.59	3.588
3. digits 1, 2, 4, 6, 7, and 8 into two 3-digit numbers with least positive sum	147 +268 415	14. 0.4392	0.4	0.44	0.439
4. digits 2, 3, 5, 6, 8, and 9 into two 3-digit numbers with greatest positive sum	963 +852 1815	15. 1.1515	1.2	1.15	1.152
5. digits 0, 1, 3, 4, 7, and 9 into two 3-digit numbers with least positive difference	401 −397 4	16. 7.0629	7.1	7.06	7.063
6. digits 0, 2, 4, 6, 8, and 9 into two 3-digit numbers with greatest positive difference	986 −204 782	17. 6.2656	6.3	6.27	6.266
7. digits 1, 3, 5, 7, 8, and 9 into two 3-digit numbers with least positive sum	159 +378 537	18. 2.7717	2.8	2.77	2.772
8. digits 1, 4, 5, 6, 7, and 8 into two 3-digit numbers with greatest positive sum	864 +751 1615	19. 4.3645	4.4	4.36	4.365
9. digits 1, 2, 3, 5, 6, 7, 8, and 9 into the two 4-digit numbers with the least positive difference	6123 −5987 136	20. 9.2288	9.2	9.23	9.229
10. digits 0, 1, 3, 4, 5, 6, 7, and 9 into the two 4-digit numbers with the least positive difference	5013 −4976 37	21. 4.5812	4.6	4.58	4.581
11. digits 1, 2, 3, 4, 5, 6, 8, and 9 into the two 4-digit numbers with the greatest positive difference	9865 −1234 8631	22. 7.5555	7.6	7.56	7.556

MAVA Math: Enhanced Skills Solutions   Copyright © 2015 Marla Weiss

Polygons 1

diag = $\dfrac{N(N - 3)}{2}$

Find the measure of one angle.	*Find the number of sides.*	*Find the number of sides.*
1. regular dodecagon ext angle = 360/12 = 30 int angle = 180 – 30 = **150**	12. each angle = 168° ext angle = 180 – 168 = 12 360/12 = 30 turns, **30** sides	23. # of diagonals = 90 N(N – 3) = 180 15 x 12 = 180 N = **15**
2. regular pentagon ext angle = 360/5 = 72 int angle = 180 – 72 = **108**	13. each angle = 165° ext angle = 180 – 165 = 15 360/15 = 24 turns, **24** sides	24. # of diagonals = 20 N(N – 3) = 40 8 x 5 = 40 N = **8**
3. regular 36-gon ext angle = 360/36 = 10 int angle = 180 – 10 = **170**	14. each angle = 172° ext angle = 180 – 172 = 8 360/8 = 45 turns, **45** sides	25. # of diagonals = 54 N(N – 3) = 108 12 x 9 = 108 N = **12**
4. regular decagon ext angle = 360/10 = 36 int angle = 180 – 36 = **144**	15. each angle = 120° ext angle = 180 – 120 = 60 360/60 = 6 turns, **6** sides	26. # of diagonals = 35 N(N – 3) = 70 10 x 7 = 70 N = **10**
5. regular pentadecagon ext angle = 360/15 = 24 int angle = 180 – 24 = **156**	16. each angle = 177° ext angle = 180 – 177 = 3 360/3 = 120 turns, **120** sides	27. # of diagonals = 209 N(N – 3) = 418 22 x 19 = 418 N = **22**
6. regular 18-gon ext angle = 360/18 = 20 int angle = 180 – 20 = **160**	17. each angle = 135° ext angle = 180 – 135 = 45 360/45 = 8 turns, **8** sides	28. # of diagonals = 104 N(N – 3) = 208 16 x 13 = 208 N = **16**
7. regular icosagon ext angle = 360/20 = 18 int angle = 180 – 18 = **162**	18. each angle = 176° ext angle = 180 – 176 = 4 360/4 = 90 turns, **90** sides	29. # of diagonals = 170 N(N – 3) = 340 20 x 17 = 340 N = **20**
8. regular 60-gon ext angle = 360/60 = 6 int angle = 180 – 6 = **174**	19. each angle = 171° ext angle = 180 – 171 = 9 360/9 = 40 turns, **40** sides	30. # of diagonals = 275 N(N – 3) = 550 25 x 22 = 550 N = **25**
9. regular 30-gon ext angle = 360/30 = 12 int angle = 180 – 12 = **168**	20. each angle = 60° ext angle = 180 – 60 = 120 360/120 = 3 turns, **3** sides	31. # of diagonals = 495 N(N – 3) = 990 33 x 30 = 990 N = **33**
10. regular octagon ext angle = 360/8 = 45 int angle = 180 – 45 = **135**	21. each angle = 140° ext angle = 180 – 140 = 40 360/40 = 9 turns, **9** sides	32. # of diagonals = 27 N(N – 3) = 54 9 x 6 = 54 N = **9**
11. regular 72-gon ext angle = 360/72 = 5 int angle = 180 – 5 = **175**	22. each angle = 162° ext angle = 180 – 162 = 18 360/18 = 20 turns, **20** sides	33. # of diagonals = 44 N(N – 3) = 88 11 x 8 = 88 N = **11**

Polygons 2

Find the sum of the angles in degrees.	Find the number of sides given the sum of the angles.
1. dodecagon 180 x 10 = **1800**	12. 2340° 2340/180 = 117/9 = 13 triangles **15** sides
2. 17-gon 180 x 15 = **2700**	13. 7200° 7200/180 = 360/9 = 40 triangles **42** sides
3. octagon 180 x 6 = **1080**	14. 6120° 6120/180 = 306/9 = 34 triangles **36** sides
4. hexagon 180 x 4 = **720**	15. 3960° 3960/180 = 132/6 = 22 triangles **24** sides
5. decagon 180 x 8 = **1440**	16. 1260° 1260/180 = 63/9 = 7 triangles **9** sides
6. 14-gon 180 x 12 = **2160**	17. 1980° 1980/180 = 99/9 = 11 triangles **13** sides
7. 32-gon 180 x 30 = **5400**	18. 8460° 8460/180 = 423/9 = 47 triangles **49** sides
8. heptagon 180 x 5 = **900**	19. 5040° 5040/180 = 252/9 = 28 triangles **30** sides
9. pentagon 180 x 3 = **540**	20. 9000° 9000/180 = 100/2 = 50 triangles **52** sides
10. 11-gon 180 x 9 = **1620**	21. 4500° 4500/180 = 50/2 = 25 triangles **27** sides
11. 20-gon 180 x 18 = **3240**	22. 2520° 2520/180 = 126/9 = 14 triangles **16** sides

Polygons 3

Find the number of diagonals for the convex polygon.

1. number of sides = 20 $\dfrac{20 \times 17}{2}$	**170**	12. each angle = 176° ext angle = 4 $\dfrac{90 \times 87}{2}$ 360/4 = 90 sides	**3915**
2. each angle = 108° ext angle = 72 $\dfrac{5 \times 2}{2}$ 360/72 = 5 sides	**5**	13. sum of angles = 2520° # Δ = 14 $\dfrac{16 \times 13}{2}$ # sides = 16	**104**
3. sum of angles = 3600° # Δ = 20 $\dfrac{22 \times 19}{2}$ # sides = 22	**209**	14. each angle = 135° ext angle = 45 $\dfrac{8 \times 5}{2}$ 360/45 = 8 sides	**20**
4. number of sides = 18 $\dfrac{18 \times 15}{2}$	**135**	15. each angle = 168° ext angle = 12 $\dfrac{30 \times 27}{2}$ 360/12 = 30 sides	**405**
5. sum of angles = 900° # Δ = 5 $\dfrac{7 \times 4}{2}$ # sides = 7	**14**	16. number of sides = 16 $\dfrac{16 \times 13}{2}$	**104**
6. shape is dodecagon $\dfrac{12 \times 9}{2}$	**54**	17. sum of angles = 7200° # Δ = 40 $\dfrac{42 \times 39}{2}$ # sides = 42	**819**
7. number of sides = 23 $\dfrac{23 \times 20}{2}$	**230**	18. each angle = 120° ext angle = 60 $\dfrac{6 \times 3}{2}$ 360/60 = 6 sides	**9**
8. sum of angles = 1980° # Δ = 11 $\dfrac{13 \times 10}{2}$ # sides = 13	**65**	19. shape is pentadecagon $\dfrac{15 \times 12}{2}$	**90**
9. number of sides = 40 $\dfrac{40 \times 37}{2}$	**740**	20. each angle = 165° ext angle = 15 $\dfrac{24 \times 21}{2}$ 360/15 = 24 sides	**252**
10. shape is nonagon $\dfrac{9 \times 6}{2}$	**27**	21. shape is decagon $\dfrac{10 \times 7}{2}$	**35**
11. sum of angles = 2700° # Δ = 15 $\dfrac{17 \times 14}{2}$ # sides = 17	**119**	22. shape is icosagon $\dfrac{20 \times 17}{2}$	**170**

Polygons 4

Answer as indicated. Angles are in degrees.

1. The measures of the interior angles of a quadrilateral are x, 3x, 2x + 30, and 2x − 10. Find the sum of the least and the greatest angles.

   8x + 20 = 360 | least = 42.5
   8x = 340 | greatest = 127.5
   x = 42.5 | sum = **170**

2. The measures of the interior angles of a triangle are 5x, 4x + 8, and 3x − 2. Find the median of the angles.

   12x + 6 = 180 | median = 4x + 8 = **66**
   12x = 174
   x = 14.5

3. The measures of the interior angles of a pentagon are x + 30, 4x − 10, 2x, 3x + 10, and 2x + 18. Find the angle measure that is a perfect square.

   12x + 48 = 540 | 2x + 18
   12x = 492 | = 82 + 18
   x = 41 | = **100**

4. The measures of the interior angles of a quadrilateral are 4x, 3x − 1, 2x + 21, and 5x − 10. Find the median of the angles.

   14x + 10 = 360 | 100, 74, 71, 115
   14x = 350 | 71, 74, 100, 115
   x = 25 | median = 174/2 = **87**

5. The measures of the interior angles of a hexagon are 3x, 4x − 7, 2x, 5x − 13, 6x − 15, and 7x − 55. Find the greatest angle.

   27x − 90 = 720 | greatest = 6x − 15
   27x = 810 | = **165**
   x = 30

6. One angle of a regular decagon is how much greater than one angle of a regular nonagon?

   360/10 = 36 | 144 − 140 = **4**
   180 − 36 = 144
   360/9 = 40
   180 − 40 = 140

7. Find the measure of the smaller angle formed by a side and the nearest short diagonal of a regular pentagon.

   ext ang of pent = 360/5 = 72
   int ang of pent = 108
   isos tri
   72/2 = **36**

8. Find the measure of the smaller angle formed by a side and the nearest short diagonal of a regular dodecagon.

   ext ang of dodec = 360/12 = 30
   int ang of dodec = 150
   isos tri: 150, x, x
   30/2 = **15**

9. The sum of the angles of a 22-gon is how much greater than the sum of the angles of a 14-gon?

   sum angles 22-gon = 20 x 180
   sum angles 14-gon = 12 x 180
   difference = 8 x 180 = **1440**

10. The sum of the angles of a 54-gon is how much greater than the sum of the angles of a 53-gon?

    sum angles 54-gon = 52 x 180
    sum angles 53-gon = 51 x 180
    difference = **180**

11. Find the measure of the smaller angle formed by a side and the nearest short diagonal of a regular octagon.

    ext ang of oct = 360/8 = 45
    int ang of oct = 135
    isos tri: 135, x, x
    45/2 = **22.5**

12. One angle of a regular 16-gon is how much greater than one angle of a regular 15-gon?

    360/16 = 22.5 | 157.5 − 156 =
    180 − 22.5 = 157.5 | **1.5**
    360/15 = 24
    180 − 24 = 156

Primes 1

Use divisibility rules to label the number prime, composite, or neither. If composite, state one factor other than 1 or the number. Answers may vary for the factors.

1. 1 **N**	17. 133 **C 7**	33. 411 **C 3**	49. 1001 **C 11**
2. 37 **P**	18. 134 **C 2**	34. 417 **C 3**	50. 1111 **C 11**
3. 51 **C 3**	19. 137 **P**	35. 451 **C 11**	51. 1221 **C 3**
4. 53 **P**	20. 143 **C 11**	36. 513 **C 9**	52. 1243 **C 11**
5. 55 **C 5**	21. 159 **C 3**	37. 517 **C 11**	53. 2361 **C 3**
6. 57 **C 3**	22. 187 **C 11**	38. 531 **C 9**	54. 2385 **C 5**
7. 59 **P**	23. 189 **C 9**	39. 561 **C 11**	55. 2403 **C 9**
8. 91 **C 7**	24. 201 **C 3**	40. 583 **C 11**	56. 3331 **P**
9. 93 **C 3**	25. 203 **C 7**	41. 671 **C 11**	57. 3773 **C 11**
10. 97 **P**	26. 207 **C 9**	42. 693 **C 3**	58. 4011 **C 3**
11. 101 **P**	27. 209 **C 11**	43. 707 **C 7**	59. 4527 **C 9**
12. 111 **C 3**	28. 253 **C 11**	44. 781 **C 11**	60. 6111 **C 9**
13. 113 **P**	29. 313 **P**	45. 801 **C 9**	61. 7707 **C 7**
14. 119 **C 7**	30. 321 **C 3**	46. 811 **P**	62. 8541 **C 9**
15. 121 **C 11**	31. 327 **C 3**	47. 931 **C 7**	63. 9001 **P**
16. 123 **C 3**	32. 329 **C 7**	48. 957 **C 11**	64. 9647 **C 11**

Primes 2

Write the prime factorization in rows (no trees), each equivalent to the original number.
The final row must be all primes in ascending order with exponents.

1. 108 2 x 54 2 x 6 x 9 **2^2 x 3^3**	9. 500 5 x 4 x 25 **2^2 x 5^3**	17. 1260 9 x 14 x 10 9 x 2 x 7 x 2 x 5 **2^2 x 3^2 x 5 x 7**
2. 144 12 x 12 3 x 4 x 3 x 4 **2^4 x 3^2**	10. 576 6 x 96 6 x 6 x 16 **2^6 x 3^2**	18. 1625 25 x 65 5 x 5 x 5 x 13 **5^3 x 13**
3. 189 9 x 21 3 x 3 x 3 x 7 **3^3 x 7**	11. 606 6 x 101 **2 x 3 x 101**	19. 1980 18 x 11 x 10 2 x 9 x 11 x 2 x 5 **2^2 x 3^2 x 5 x 11**
4. 216 9 x 24 3 x 3 x 3 x 8 **2^3 x 3^3**	12. 621 9 x 69 9 x 3 x 23 **3^3 x 23**	20. 2025 81 x 25 9 x 9 x 5 x 5 **3^4 x 5^2**
5. 297 9 x 33 3 x 3 x 3 x 11 **3^3 x 11**	13. 675 27 x 25 **3^3 x 5^2**	21. 2106 18 x 117 2 x 9 x 9 x 13 **2 x 3^4 x 13**
6. 333 3 x 111 3 x 3 x 37 **3^2 x 37**	14. 720 8 x 9 x 10 8 x 9 x 2 x 5 **2^4 x 3^2 x 5**	22. 5400 2 x 27 x 4 x 25 **2^3 x 3^3 x 5^2**
7. 360 6 x 6 x 10 2 x 3 x 2 x 3 x 2 x 5 **2^3 x 3^2 x 5**	15. 840 4 x 21 x 10 4 x 3 x 7 x 2 x 5 **2^3 x 3 x 5 x 7**	23. 6000 6 x 10 x 10 x 10 2 x 3 x 2^3 x 5^3 **2^4 x 3 x 5^3**
8. 375 15 x 25 3 x 5 x 5 x 5 **3 x 5^3**	16. 950 5 x 19 x 10 **2 x 5^2 x 19**	24. 6600 6 x 11 x 4 x 25 2 x 3 x 2^2 x 5^2 x 11 **2^3 x 3 x 5^2 x 11**

Primes 3

Answer as indicated.

1. Find the sum of the four numbers: the least 2-digit twin primes and the greatest 2-digit twin primes. 11 + 13 + 71 + 73 = **168**	9. How many 2-digit primes may be formed by selecting two digits from {2, 3, 4, 7}? 23, 37, 43, 47, 73 **5**
2. Find the sum of the primes from 80 to 100. 83 + 89 + 97 = **269**	10. Find the sum of the primes from 100 to 120. 101 + 103 + 107 + 109 + 113 = **533**
3. How many pairs of twin primes are greater than 100 and less than 200? 101 & 103; 107 & 109; 137 & 139; 149 & 151; 179 & 181; 191 & 193; 197 & 199 **7**	11. Find the three least 3-digit numbers with exactly 3 factors each. 11^2 = **121** 13^2 = **169** 17^2 = **289**
4. Find the number of primes between 40 and 80. 41, 43, 47, 53, 59, 61, 67, 71, 73, 79 **10**	12. Find the least 3-digit prime with a prime digit sum. **101** digit sum = 2
5. Find the least positive integer divisible by 4 primes. 2 x 3 x 5 x 7 = **210**	13. Find the greatest 3-digit prime with a prime digit sum. 997 DS = 25 991 DS = 19 **991**
6. Find all 2-digit pairs of numbers such that both the number and the number with the digits reversed are primes. **13** and **31**; **17** and **71**; **37** and **73**; **79** and **97**	14. Find the product of the primes between 20 and 30. 23 x 29 = **667**
7. Find the greatest 3-digit positive integer divisible by the first 3 primes. 995 no: div by 5 but not 2 2 x 3 x 5 x 33 = **990**	15. Use the digits 2, 3, 7, and 9 each once to create two primes with the greatest product. 29 x 73 = 2117 **97 x 23 = 2231**
8. Find the sum of the prime factors of 21,945. 15 x 1463 3 x 5 x 11 x 133 sum = **45** 3 x 5 x 7 x 11 x 19	16. Find the sum of the least 2-digit prime and the greatest 2-digit prime, both with a prime digit sum. 11 + 89 = **100**

Primes 4

Answer as indicated.

1. Find the sum of the positive differences of all pairs of twin primes less than 100.
5, 7; 11, 13; 17, 19; 29, 31; 41, 43; 59, 61; 71, 73
7 x 2 = **14**

2. Find the sum x + y where x = the sum of the first 5 primes and y = the sum of the reciprocals of the first 3 primes.

2 + 3 + 5 + 7 + 11 = 28
1/2 + 1/3 + 1/5 = 31/30 $29\frac{1}{30}$

3. Selecting two of the first eight primes, find the probability of their sum equaling 16.
C(8,2) = 28 $\frac{2}{28} = \frac{1}{14}$
3+13, 5+11

4. How many ordered triples of different primes have the sum of their elements equal to 36?   **18**
(2, 3, 31)
(2, 5, 29)   six each with different
(2, 11, 23)   orderings

5. How many of the first 20 positive integers may be written as the sum of 2 distinct primes?   **14**
2+3=5; 2+5=7; 2+7=9; 2+11=13; 2+17=19
3+5=8; 3+7=10; 3+11=14; 2+13=15; 3+13=16
3+17=20; 5+7=12; 5+11=16;5+13=18

6. How many ordered triples of primes have the sum of their elements equal to 33?   **37**
3: (2, 2, 29)   6: (3, 13, 17)   3: (7, 7, 19)
6: (3, 7, 23)   3: (5, 5, 23)   3: (7, 13, 13)
6: (3, 11, 19)   6: (5, 11, 17)   1: (11, 11, 11)

7. The sum of the least three 3-digit primes is how much greater than the sum of the greatest three 2-digit primes?   101 + 103 + 107
83 + 89 + 97
18 + 14 + 10 = **42**

8. If p and p+3 are both primes, find the least composite number not divisible by either one.

odd + 3 = even, so p = 2; p + 3 = 5
9

9. Find the sum of the least 2-digit prime and the greatest 2-digit prime, both with the product of its digits prime.

13 + 71 = **84**   One digit must be 1 for the product to be prime.

10. Find five primes that form an arithmetic sequence with a common difference of 6.

5, 11, 17, 23, 29

11. When the prime 97 is written as the sum of one composite and one prime, find the least possible positive difference of the two numbers.

50 + 47 = 97
50 – 47 = **3**

12. Find the sum of the greatest 3-digit number with exactly 3 factors and the least 4-digit number with exactly 3 factors. $31^2 = 961$ **2330**
$37^2 = 1369$

13. How many of the first 20 positive integers may be written as the sum of 3 distinct primes?   **8**
2+3+5=10; 2+3+7=12; 2+5+7=14;
2+3+11=16; 2+3+13=18; 3+5+7=15;
3+5+11=19; 2+5+13=20

14. Selecting two of the first twelve primes, find the probability of their sum equaling 36.
C(12,2) = 66 $\frac{4}{66} = \frac{2}{33}$
5+31, 7+29, 13+23, 17+19

Prisms 1

Find the volume in cubic units of the rectangular prism with the given unit dimensions. Use mental math.

1.	4, 5, 8	**160**	20.	8, 8, 20	**1280**
2.	2, 7, 10	**140**	21.	7, 8, 11	**616**
3.	5, 6, 10	**300**	22.	30, 35, 40	**42,000**
4.	10, 12, 20	**2400**	23.	8, 25, 30	**6000**
5.	3, 9, 11	**297**	24.	4, 15, 15	**900**
6.	8, 9, 10	**720**	25.	5, 19, 20	**1900**
7.	10, 20, 30	**6000**	26.	20, 35, 60	**42,000**
8.	3, 4, 6	**72**	27.	7, 10, 12	**840**
9.	5, 5, 5	**125**	28.	1, 11, 17	**187**
10.	4, 6, 11	**264**	29.	50, 60, 90	**270,000**
11.	10, 12, 12	**1440**	30.	2, 7, 25	**350**
12.	5, 7, 11	**385**	31.	20, 20, 24	**9600**
13.	40, 50, 60	**120,000**	32.	2.1, 8, 10	**168**
14.	9, 9, 9	**729**	33.	1.1, 1.1, 200	**242**
15.	8, 15, 30	**3600**	34.	2.5, 4, 13	**130**
16.	2, 4, 7	**56**	35.	2, 6.1, 20	**244**
17.	2, 9, 15	**270**	36.	1.5, 4, 15	**90**
18.	2, 15, 15	**450**	37.	2.5, 3, 10	**75**
19.	4, 6, 6	**144**	38.	1.3, 1.3, 100	**169**

Prisms 2

Find the surface area of the rectangular prism in square units. Show pairwise work.

1. 11, 15, 30

   $11 \times 15 = 165$
   $15 \times 30 = 450$
   $11 \times 30 = 330$
   $\overline{}$
   $945 \times 2 = \mathbf{1890}$

9. 10, 20, 30

   $10 \times 20 = 200$
   $20 \times 30 = 600$
   $10 \times 30 = 300$
   $\overline{}$
   $1100 \times 2 = \mathbf{2200}$

2. 4.5, 11, 13

   $4.5 \times 11 \times 2 = 99$
   $11 \times 13 \times 2 = 286$
   $4.5 \times 13 \times 2 = 117$
   $\overline{}$
   $\mathbf{502}$

10. 4, 11, 11

   $4 \times 11 = 44$
   $11 \times 11 = 121$
   $4 \times 11 = 44$
   $\overline{}$
   $209 \times 2 = \mathbf{418}$

3. 5, 12, 30

   $5 \times 12 = 60$
   $12 \times 30 = 360$
   $5 \times 30 = 150$
   $\overline{}$
   $570 \times 2 = \mathbf{1140}$

11. 30, 40, 60

   $30 \times 40 = 1200$
   $40 \times 60 = 2400$
   $30 \times 60 = 1800$
   $\overline{}$
   $5400 \times 2 = \mathbf{10,800}$

4. 3.5, 6, 7

   $3.5 \times 6 \times 2 = 42$
   $6 \times 7 \times 2 = 84$
   $3.5 \times 7 \times 2 = 49$
   $\overline{}$
   $\mathbf{175}$

12. 2.5, 8, 10

   $2.5 \times 8 \times 2 = 40$
   $8 \times 10 \times 2 = 50$
   $2.5 \times 10 \times 2 = 160$
   $\overline{}$
   $\mathbf{250}$

5. 2.5, 3, 10

   $2.5 \times 3 \times 2 = 15$
   $3 \times 10 \times 2 = 60$
   $2.5 \times 10 \times 2 = 50$
   $\overline{}$
   $\mathbf{125}$

13. 1.5, 4, 15

   $1.5 \times 4 \times 2 = 12$
   $4 \times 15 \times 2 = 120$
   $1.5 \times 15 \times 2 = 45$
   $\overline{}$
   $\mathbf{177}$

6. 6, 10, 12

   $6 \times 10 = 60$
   $10 \times 12 = 120$
   $6 \times 12 = 72$
   $\overline{}$
   $252 \times 2 = \mathbf{504}$

14. 1.2, 1.5, 10

   $1.2 \times 1.5 \times 2 = 3.6$
   $1.5 \times 10 \times 2 = 30$
   $1.2 \times 10 \times 2 = 24$
   $\overline{}$
   $\mathbf{57.6}$

7. 10, 15, 50

   $10 \times 15 = 150$
   $15 \times 50 = 750$
   $10 \times 50 = 500$
   $\overline{}$
   $1400 \times 2 = \mathbf{2800}$

15. 5, 6.3, 10

   $5 \times 6.3 \times 2 = 63$
   $6.3 \times 10 \times 2 = 126$
   $5 \times 10 \times 2 = 100$
   $\overline{}$
   $\mathbf{289}$

8. 10, 12, 20

   $10 \times 12 = 120$
   $12 \times 20 = 240$
   $10 \times 20 = 200$
   $\overline{}$
   $560 \times 2 = \mathbf{1120}$

16. 2.5, 8, 9

   $2.5 \times 8 \times 2 = 40$
   $8 \times 9 \times 2 = 144$
   $2.5 \times 9 \times 2 = 45$
   $\overline{}$
   $\mathbf{229}$

Prisms 3

Find the volume in cubic units of the rectangular prism given areas of three faces in square units.

1. 24, 32, 48	7. 72, 120, 135	13. 32, 40, 80
4 x 6	8 x 9	4 x 8
4 x 8	8 x 15	4 x 10
6 x 8	9 x 15	8 x 10
4 x 6 x 8 = **192**	8 x 9 x 15 = **1080**	4 x 8 x 10 = **320**
2. 30, 40, 48	8. 35, 45, 63	14. 90, 99, 110
5 x 6	5 x 7	9 x 10
5 x 8	5 x 9	9 x 11
6 x 8	7 x 9	10 x 11
5 x 6 x 8 = **240**	5 x 7 x 9 = **315**	9 x 10 x 11 = **990**
3. 30, 35, 42	9. 88, 104, 143	15. 63, 70, 90
5 x 6	8 x 11	7 x 9
5 x 7	8 x 13	7 x 10
6 x 7	11 x 13	9 x 10
5 x 6 x 7 = **210**	8 x 11 x 13 = **1144**	7 x 9 x 10 = **630**
4. 24, 48, 72	10. 17.5, 35, 24.5	16. 54, 72, 108
4 x 6	5 x 3.5	6 x 9
4 x 12	5 x 7	6 x 12
6 x 12	3.5 x 7	9 x 12
4 x 6 x 12 = **288**	3.5 x 5 x 7 = **122.5**	6 x 9 x 12 = **648**
5. 6, 22, 33	11. 15, 51, 85	17. 22.5, 31.5, 35
2 x 3	3 x 5	4.5 x 5
2 x 11	3 x 17	4.5 x 7
3 x 11	5 x 17	5 x 7
2 x 3 x 11 = **66**	3 x 5 x 17 = **255**	4.5 x 5 x 7 = **157.5**
6. 24, 48, 128	12. 48, 54, 72	18. 88, 96, 132
3 x 8	6 x 8	8 x 11
3 x 16	6 x 9	8 x 12
8 x 16	8 x 9	11 x 12
3 x 8 x 16 = **384**	6 x 8 x 9 = **432**	8 x 11 x 12 = **1056**

Prisms 4

Find the space diagonal of the rectangular prism with the given unit dimensions.

1.	3, 4, 5	$\sqrt{9 + 16 + 25}$ $\sqrt{50}$ **$5\sqrt{2}$**	12.	2, 3, $2\sqrt{3}$	$\sqrt{4 + 9 + 12}$ $\sqrt{25}$ **5**
2.	6, 8, 10	$\sqrt{36 + 64 + 100}$ $\sqrt{200}$ **$10\sqrt{2}$**	13.	1, 5, $2\sqrt{7}$	$\sqrt{1 + 25 + 28}$ $\sqrt{54}$ **$3\sqrt{6}$**
3.	1, 5, 10	$\sqrt{1 + 25 + 100}$ $\sqrt{126}$ **$3\sqrt{14}$**	14.	4, 9, $2\sqrt{7}$	$\sqrt{16 + 81 + 28}$ $\sqrt{125}$ **$5\sqrt{5}$**
4.	5, 10, 15	$\sqrt{25 + 100 + 225}$ $\sqrt{350}$ **$5\sqrt{14}$**	15.	$\sqrt{5}$, $\sqrt{7}$, $\sqrt{13}$	$\sqrt{5 + 7 + 13}$ $\sqrt{25}$ **5**
5.	8, 9, 12	$\sqrt{64 + 81 + 144}$ $\sqrt{289}$ **17**	16.	3, 4, $2\sqrt{6}$	$\sqrt{9 + 16 + 24}$ $\sqrt{49}$ **7**
6.	3, 5, 8	$\sqrt{9 + 25 + 64}$ $\sqrt{98}$ **$7\sqrt{2}$**	17.	2, 6, $4\sqrt{2}$	$\sqrt{4 + 36 + 32}$ $\sqrt{72}$ **$6\sqrt{2}$**
7.	1, 2, 2	$\sqrt{1 + 4 + 4}$ $\sqrt{9}$ **3**	18.	1, 3, $5\sqrt{5}$	$\sqrt{1 + 9 + 125}$ $\sqrt{135}$ **$3\sqrt{15}$**
8.	2, 4, 13	$\sqrt{4 + 16 + 169}$ $\sqrt{189}$ **$3\sqrt{21}$**	19.	5, 8, $3\sqrt{7}$	$\sqrt{25 + 64 + 63}$ $\sqrt{152}$ **$2\sqrt{38}$**
9.	7, 24, 60	$\sqrt{49 + 576 + 3600}$ $\sqrt{4225}$ **65**	20.	1, 3, $5\sqrt{6}$	$\sqrt{1 + 9 + 150}$ $\sqrt{160}$ **$4\sqrt{10}$**
10.	2, 6, 9	$\sqrt{4 + 36 + 81}$ $\sqrt{121}$ **11**	21.	1, 5, $7\sqrt{2}$	$\sqrt{1 + 25 + 98}$ $\sqrt{124}$ **$2\sqrt{31}$**
11.	4, 5, 11	$\sqrt{16 + 25 + 121}$ $\sqrt{162}$ **$9\sqrt{2}$**	22.	$\sqrt{3}$, $\sqrt{7}$, $\sqrt{71}$	$\sqrt{3 + 7 + 71}$ $\sqrt{81}$ **9**

Prisms 5

Find the volume in cubic units of the rectangular prism given areas of three faces in square units.

1. 22, 24, 33

    $V = \sqrt{(22)(24)(33)}$
    $= \sqrt{(2)(11)(2)(3)(4)(3)(11)}$
    $= 2 \times 3 \times 2 \times 11$
    $= 12 \times 11$
    $= \mathbf{132}$

2. 24, 48, 200

    $V = \sqrt{(24)(48)(200)}$
    $= \sqrt{(2)(12)(4)(12)(2)(100)}$
    $= 2 \times 12 \times 2 \times 10$
    $= 48 \times 10$
    $= \mathbf{480}$

3. 17.5, 35, 50

    $V = \sqrt{(17.5)(35)(50)}$
    $= \sqrt{(17.5)(17.5)(2)(2)(25)}$
    $= 17.5 \times 2 \times 5$
    $= 35 \times 5$
    $= \mathbf{175}$

4. 20, 50, 160

    $V = \sqrt{(20)(50)(160)}$
    $= \sqrt{(2)(10)(5)(10)(2)(5)(16)}$
    $= 2 \times 5 \times 4 \times 10$
    $= 40 \times 10$
    $= \mathbf{400}$

5. 13, 26, 98

    $V = \sqrt{(13)(26)(98)}$
    $= \sqrt{(13)(13)(2)(2)(49)}$
    $= 13 \times 2 \times 7$
    $= 91 \times 2$
    $= \mathbf{182}$

6. 52, 65, 500

    $V = \sqrt{(52)(65)(500)}$
    $= \sqrt{(4)(13)(13)(5)(5)(100)}$
    $= 2 \times 13 \times 5 \times 10$
    $= 130 \times 10$
    $= \mathbf{1300}$

7. 30, 40, 48

    $V = \sqrt{(30)(40)(48)}$
    $= \sqrt{(3)(10)(4)(10)(3)(4)(4)}$
    $= 3 \times 4 \times 2 \times 10$
    $= 24 \times 10$
    $= \mathbf{240}$

8. 17, 51, 108

    $V = \sqrt{(17)(51)(108)}$
    $= \sqrt{(17)(17)(3)(3)(36)}$
    $= 17 \times 3 \times 6$
    $= 51 \times 6$
    $= \mathbf{306}$

9. 20, 70, 126

    $V = \sqrt{(20)(70)(126)}$
    $= \sqrt{(2)(10)(10)(7)(14)(9)}$
    $= 10 \times 14 \times 3$
    $= 10 \times 42$
    $= \mathbf{420}$

10. 20, 35, 112

    $V = \sqrt{(20)(35)(112)}$
    $= \sqrt{(4)(5)(5)(7)(7)(16)}$
    $= 2 \times 5 \times 7 \times 4$
    $= 10 \times 28$
    $= \mathbf{280}$

11. 12, 75, 121

    $V = \sqrt{(12)(75)(121)}$
    $= \sqrt{(4)(3)(3)(25)(11)(11)}$
    $= 2 \times 3 \times 5 \times 11$
    $= 30 \times 11$
    $= \mathbf{330}$

12. 30, 60, 72

    $V = \sqrt{(30)(60)(72)}$
    $= \sqrt{(3)(10)(6)(10)(18)(4)}$
    $= 10 \times 18 \times 2$
    $= 10 \times 36$
    $= \mathbf{360}$

13. 24, 48, 72

    $V = \sqrt{(24)(48)(72)}$
    $= \sqrt{(24)(2)(24)(2)(36)}$
    $= 24 \times 2 \times 6$
    $= 144 \times 2$
    $= \mathbf{288}$

14. 19.5, 39, 128

    $V = \sqrt{(19.5)(39)(128)}$
    $= \sqrt{(19.5)(19.5)(2)(2)(64)}$
    $= 19.5 \times 2 \times 8$
    $= 39 \times 8$
    $= \mathbf{312}$

15. 18, 34, 68

    $V = \sqrt{(18)(34)(68)}$
    $= \sqrt{(9)(2)(2)(17)(17)(4)}$
    $= 3 \times 2 \times 17 \times 2$
    $= 6 \times 34$
    $= \mathbf{204}$

16. 10, 15, 54

    $V = \sqrt{(10)(15)(54)}$
    $= \sqrt{(2)(5)(3)(5)(6)(9)}$
    $= 6 \times 5 \times 3$
    $= 6 \times 15$
    $= \mathbf{90}$

17. 20, 24, 120

    $V = \sqrt{(20)(24)(120)}$
    $= \sqrt{(5)(4)(4)(6)(30)(4)}$
    $= 30 \times 4 \times 2$
    $= 30 \times 8$
    $= \mathbf{240}$

18. 24, 30, 45

    $V = \sqrt{(24)(30)(45)}$
    $= \sqrt{(4)(6)(6)(5)(5)(9)}$
    $= 2 \times 6 \times 5 \times 3$
    $= 6 \times 30$
    $= \mathbf{180}$

Prisms 6

Find the volume of the prism in cubic units with height L of the base, area A of the base, and altitude h of the prism.

1.	triangular prism sides of base 16, 30, 34; h = 40 8, 15, 17, x 2 right A = 240 V = 240 x 40 = **9600**	8.	trapezoidal prism b = 5, B = 20, L = 18, h = 22 M = 26/2 = 12.5 A = 12.5 x 18 = 225 V = 225 x 22 = 450 x 11 = **4950**
2.	regular hexagonal prism side of base 10, h = 15 A = 6 x 10 x 10 x $\sqrt{3}$ / 4 = 150$\sqrt{3}$ V = 150$\sqrt{3}$ x 15 = **2250$\sqrt{3}$**	9.	pentagonal prism A = 42, h = 9 V = 42 x 9 = **378**
3.	rhomboidal prism diagonals of base 10 and 15, h = 8 A = (10 x 15)/2 = 75 V = 75 x 8 = **600**	10.	isosceles triangular prism sides of base 50, 50, 28; h = 100 7, 24, 25 x 2 is 14, 48, 50 right L = 48 A = 672 V = 672 x 100 = **67,200**
4.	octagonal prism A = 75, h = 11 V = 75 x 11 = **825**	11.	equilateral triangular prism side of base 18, h = 10 A = 18 x 18 x $\sqrt{3}$ / 4 = 81$\sqrt{3}$ V = 81$\sqrt{3}$ x 10 = **810$\sqrt{3}$**
5.	equilateral triangular prism side of base 22, h = 5 A = 22 x 22 x $\sqrt{3}$ / 4 = 121$\sqrt{3}$ V = 121$\sqrt{3}$ x 5 = **605$\sqrt{3}$**	12.	regular hexagonal prism side of base 14, h = 20 A = 6 x 14 x 14 x $\sqrt{3}$ / 4 = 294$\sqrt{3}$ V = 294$\sqrt{3}$ x 20 = **5880$\sqrt{3}$**
6.	isosceles triangular prism sides of base 26, 26, 20; h = 30 5, 12, 13 x 2 is 10, 24, 26 right L = 24 A = 240 V = 240 x 30 = **7200**	13.	rhomboidal prism diagonals of base 25 and 40, h = 15 A = (25 x 40)/2 = 500 V = 500 x 15 = **7500**
7.	trapezoidal prism b = 6, B = 14, L = 5, h = 11 M = 20/2 = 10 A = 10 x 5 = 50 V = 50 x 11 = **550**	14.	triangular prism sides of base 11, 60, 61; h = 20 11, 60, 61 right A = 330 V = 330 x 20 = **6600**

Prisms 7

A slice is made along the internal diagonal plane of a rectangular prism with the given dimensions to create a wedge. Find the volume and surface area of the wedge. NTS

1.
V = 7×2×3/2
V = **21**

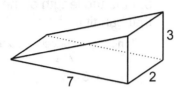

SA = 21 + 6 + 14 + 2√58
SA = 41 + 2√58

5.
V = 13×5×12/2
V = **390**

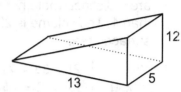

SA = 156 + 60 + 65 + 5√313
SA = 281 + 5√313

2.
V = 8×3×5/2
V = **60**

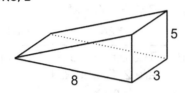

SA = 40 + 15 + 24 + 3√89
SA = 79 + 3√89

6.
V = 9×2×5/2
V = **45**

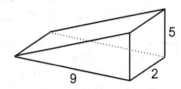

SA = 45 + 10 + 18 + 2√106
SA = 73 + 2√106

3.
V = 7×4×6/2
V = **84**

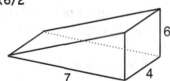

SA = 42 + 24 + 28 + 4√85
SA = 94 + 4√85

7.
V = 11×2×6/2
V = **66**

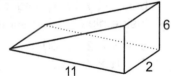

SA = 66 + 12 + 22 + 2√157
SA = 100 + 2√157

4.
V = 10×4×5/2
V = **100**

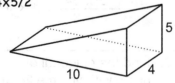

SA = 50 + 20 + 40 + 4 × 5√5
SA = 110 + 20√5

8.
V = 10×3×9/2
V = **135**

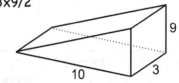

SA = 90 + 27 + 30 + 3√181
SA = 147 + 3√181

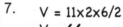

Prisms 8

Answer as indicated for the rectangular prism. Measures are units, square units, and cubic units unless otherwise specified.

1. The dimensions of a rectangular prism are 3 distinct whole numbers greater than 1. The volume is 231. Find the surface area.

11×21	$21 + 33 + 77 = 131$
$3 \times 7 \times 11$	$131 \times 2 =$ **262**

2. A rectangular box open at the top has base 10 by 20 and height 30. Find the inside plus outside surface area.

   $10 \times 20 \times 2 = \quad 400$
   $10 \times 30 \times 4 = 1200$
   $20 \times 30 \times 4 = 2400 \quad$ **4000**

3. The dimensions of a rectangular prism are 3 distinct whole numbers greater than 1. The volume is 715. Find the surface area.

5×143	$55 + 65 + 143 = 263$
$5 \times 11 \times 13$	$263 \times 2 =$ **526**

4. A rectangular box open at the top has base 5 by 6 and height 11. Find the inside plus outside surface area.

   $5 \times 6 \times 2 = \quad 60$
   $5 \times 11 \times 4 = 220$
   $6 \times 11 \times 4 = 264 \qquad$ **544**

5. The dimensions of a rectangular prism are 1, 2, and x. The space diagonal is 3. Find x.

   $9 = 1 + 4 + x^2$

   $x =$ **2**

6. Of 2 rectangular prisms, one has volume 30 and the second 105. Find the ratio of their non-congruent sides if 2 sides are congruent and all are greater than 1.

   $30 = 2 \times 3 \times 5 \qquad \dfrac{2}{7}$
   $105 = 5 \times 21 = 3 \times 5 \times 7$

7. A rectangular box measures 2 by 4 by 5. Find the length of the longest stick that can fit in the box.

   longest = space diagonal
   $25 + 16 + 4 = 45$
   $\sqrt{45} =$ **$3\sqrt{5}$**

8. The dimensions of a rectangular prism are 1, 6, and x. The space diagonal is 19. Find x.

   $361 = 1 + 36 + x^2$
   $361 - 37 = x^2$
   $x^2 = 324 \qquad x =$ **18**

9. The dimensions of a rectangular prism are 2, 5, and x. The space diagonal is 15. Find x.

   $225 = 4 + 25 + x^2$
   $225 - 29 = x^2$
   $x^2 = 196 \qquad x =$ **14**

10. Find the maximum number of rectangular blocks 3 by 5 by 2 that can fit into a box 15 by 30 by 12.

    $5 \times 6 \times 6 =$ **180**

11. Find the maximum number of rectangular blocks 6" by 4" by 8" that can fit into a box 1.5' by 1' by 2'.

    $\dfrac{18 \times 12 \times 24}{6 \times 4 \times 8} \quad 3 \times 3 \times 3 =$ **27**

12. Find the diagonal of a rectangular prism with volume 24 and height and width both $2\sqrt{2}$.

    $L = 3$
    $D = \sqrt{8 + 8 + 9} = 5$

N = # sides of base	Prisms 9	E = 3N F = N+2 V = 2N

E + F + V = 6N + 2

Given the base of a prism, find the sum of the number of edges, vertices, and faces.

Find the missing value for the prism using the formulae relating E, F, and V.

#	Base	Sum	#	Given	Answer	Type	N
1.	triangle	20	14.	15 edges	10	vertices	N = 5
2.	rectangle	26	15.	18 edges	8	faces	N = 6
3.	pentagon	32	16.	9 faces	14	vertices	N = 7
4.	hexagon	38	17.	13 faces	33	edges	N = 11
5.	heptagon	44	18.	8 vertices	12	edges	N = 4
6.	octagon	50	19.	16 vertices	10	faces	N = 8
7.	nonagon	56	20.	27 edges	18	vertices	N = 9
8.	decagon	62	21.	18 faces	48	edges	N = 16
9.	11-gon	68	22.	24 vertices	14	faces	N = 12
10.	dodecagon	74	23.	30 edges	12	faces	N = 10
11.	pentadecagon	92	24.	15 faces	26	vertices	N = 13
12.	17-gon	104	25.	30 vertices	45	edges	N = 15
13.	21-gon	128	26.	9 edges	5	faces	N = 3

Prisms 10

Answer as indicated. All tanks and trays are rectangular prisms. Trays are formed by cutting squares from corners of a rectangle and folding up the sides.

1. A tall tank partially full of water has a base 14 inches by 2.5 feet. When a rock totally sinks in the water, the water rises .3 inches. Find the volume of the rock in cubic inches.

original V = 14 x 30 x H = 420H
new V = 420(H + .3) = 420H + 126
V rock = **126**

2. A tall tank partially full of water has a base 3 feet by 3 feet. When a heavy, solid cube 1 foot on an edge totally sinks in the water, how many inches will the water level rise?

original V = 3 x 3 x H = 9H	l = 9h
V rock = 1 x 1 x 1 = 1	h = 1/9 ft
9H + 1 = 3 x 3 x (H + h)	h = **4/3** in

3. A tall tank partially full of water has a base 2.25 feet by 1.25 feet. When a rock totally sinks in the water, the water rises .6 inches. Find the volume of the rock in cubic inches.

V = (9/4)(12) x (5/4)(12) x h = 405h
405h + R = 405(h + .6)
R = 405(.6) = **243**

4. A tall tank partially full of water has a base 4 feet by 9 feet. A cube with edge 1.5 feet sinks in the water. How many inches will the water level rise?

original V = 36H	h = 3/32 ft
V rock = 27/8	h = **9/8** in
36H + 27/8 = 36(H + h)	
36h = 27/8	

5. A tank with base 3 by 6 feet is 4 feet tall and 7/8 full of water. A cube with edge 2 feet sinks in the water. Find the number of inches from the top of the water level to the top of the tank.

tot V = 3x6x4 = 72	72 – 71 = 1
(7/8)(72) = 63	6 x 3h = 1
63 + 8 = 71	h = (12)(1/18) = **2/3**

6. Two-inch squares were cut from the corners of a square with area 121 square inches. Find the volume of the tray.

original 11 x 11
11 – 2 – 2 = 7
tray 7 x 7 x 2
V = **98**

7. One-inch squares were cut from the corners of a rectangle. The tray has volume 75 cubic inches and length triple the width. Find the dimensions of the rectangle.

V = 75
tray 1 x 5 x 15
5 + 2 = 7; 15 + 2 = 17
L = 17; W = 7

8. Three-inch squares were cut from the corners of a square. Find the interior surface area in square inches of the tray with volume 432 cubic inches.

V = 432 = 3 x 144
dim = 3 x 12 x 12
bottom SA = 12 x 12 = 144
sides SA = (12 x 3) x 4 = 144   **288**

9. Four-inch squares were cut from the corners of a square. The tray has volume 256 cubic inches. Find its total surface area, inside and outside.

V = 256 = 4 x 64
dim = 4 x 8 x 8
bottom SA = (8 x 8) x 2 = 128
sides SA = (8 x 4) x 8 = 256   **384**

10. Five-inch squares were cut from the corners of a rectangle. The tray has volume 880 cubic inches and length 5 more than the width. Find the dimensions of the rectangle.

V = 880; tray 5 x 176
176 = 16 x 11
16 + 10 = 26; 11 + 10 = 21
L = 26; W = 21

Probability 1

Rolling a standard die, find the probability of:		Rolling a standard die, find the odds of:		Drawing once from a card deck, find the odds of:	
1. a 6	$\frac{1}{6}$	17. a 6	1 to 5	33. ace of spades	1 to 51
2. a prime number	$\frac{1}{2}$	18. a prime number	1 to 1	34. ace	1 to 12
3. a 7	0	19. a 7	0 to 6	35. club	1 to 3
4. an even number	$\frac{1}{2}$	20. an even number	1 to 1	36. red card	1 to 1
5. a perfect square	$\frac{1}{3}$	21. a perfect square	1 to 2	37. a perfect square	3 to 10
6. a number less than 6	$\frac{5}{6}$	22. a number less than 6	5 to 1	38. a number less than 6	5 to 8
7. a number greater than 2	$\frac{2}{3}$	23. a number greater than 2	2 to 1	39. even card	6 to 7
8. a 2-digit number	0	24. a 2-digit number	0 to 6	40. odd card	7 to 6
9. a factor of 6	$\frac{2}{3}$	25. a factor of 6	2 to 1	41. face card	3 to 10
10. a multiple of 5	$\frac{1}{6}$	26. a multiple of 5	1 to 5	42. multiple of 5	2 to 11
11. a number less than 7	1	27. a number less than 7	6 to 0	43. multiple of 4	3 to 10
12. an odd number	$\frac{1}{2}$	28. an odd number	1 to 1	44. multiple of 3	4 to 9
13. a multiple of 3	$\frac{1}{3}$	29. a multiple of 3	1 to 2	45. multiple of 2	6 to 7
14. a perfect cube	$\frac{1}{6}$	30. a perfect cube	1 to 5	46. a perfect cube	2 to 11
15. a composite number	$\frac{1}{3}$	31. a composite number	1 to 2	47. heart	1 to 3
16. a factor of 10	$\frac{1}{2}$	32. a factor of 10	1 to 1	48. queen of hearts	1 to 51

Probability 2

Find the probability of choosing a point at random from the shaded area given that the point lies inside the outer border. One box equals one square unit.

1.
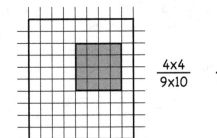
$$\frac{4\times4}{9\times10} \quad \frac{8}{45}$$

2.
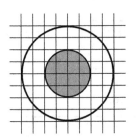
$$\frac{90-16}{6\times15} \quad \frac{37}{45}$$

3.
$$\frac{4\pi}{16\pi} \quad \frac{1}{4}$$

4.
$$\frac{36\pi-9\pi}{36\pi} \quad \frac{27\pi}{36\pi} \quad \frac{3}{4}$$

5.
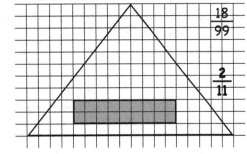
$$\frac{18}{99} \quad \frac{2}{11}$$

6.
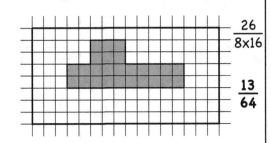
$$\frac{26}{8\times16} \quad \frac{13}{64}$$

7.
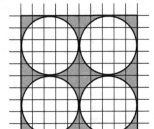
$$\frac{4}{17\times6-4-16} \quad \frac{4}{82} \quad \frac{2}{41}$$

8.
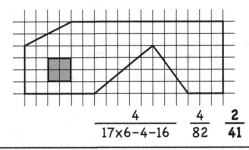
$$\frac{25\pi\times4}{4} \quad \frac{100-25\pi}{100} \quad \frac{4-\pi}{4}$$

9.
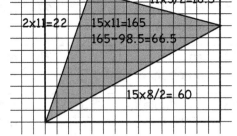
$$\frac{66.5}{165} \quad \frac{133}{330}$$
2x11=22 15x11=165 165-98.5=66.5 11x3/2=16.5 15x8/2= 60

10.
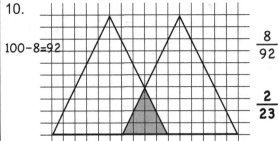
100-8=92
$$\frac{8}{92} \quad \frac{2}{23}$$

Probability 3

Form each probability by "make-a-list" over "multiplication principle." Assume solids are numbered consecutively beginning with 1.

1. Find the probability that the face-up sum is prime when rolling two dice.

1+1	1+6
1+2	6+1
2+1	2+3
1+4	3+2
4+1	

   $$\frac{9}{6 \times 6} \quad \frac{1}{4}$$

6. Rolling two regular 20-gons (1–20), find the probability that the face-up sum equals 37.

   17+20
   20+17
   18+19
   19+18

   $$\frac{4}{20 \times 20} \quad \frac{1}{100}$$

2. Rolling three octahedra (1–8), find the probability that the face-up sum equals 5.

   1+1+3
     3 in 3 places
   1+2+2
     1 in 3 places

   $$\frac{6}{8 \times 8 \times 8} \quad \frac{3}{256}$$

7. Rolling two tetrahedra (1–4), find the probability that the face-down sum is 5.

   1+4
   4+1
   3+2
   2+3

   $$\frac{4}{4 \times 4} \quad \frac{1}{4}$$

3. Rolling 2 dodecahedra (1–12), find the probability that the face-up sum is 22.

   10+12
   12+10
   11+11

   $$\frac{3}{12 \times 12} \quad \frac{1}{48}$$

8. When flipping 4 coins, find the probability of exactly 2 heads and 2 tails.

HHTT	HTTH
TTHH	THHT
THTH	
HTHT	

   $$\frac{6}{2 \times 2 \times 2 \times 2} \quad \frac{3}{8}$$

4. Find the probability that the face-up sum is a perfect cube when rolling two standard dice.

   2+6
   6+2
   3+5
   5+3
   4+4

   $$\frac{5}{6 \times 6} \quad \frac{5}{36}$$

9. When flipping 5 coins, find the probability of exactly 2 heads.

   HHTTT: HH 4 places
   HTHTT: H_H 3 places
   HTTHT: H__H 2 places
   HTTTH: 1 place

   $$\frac{10}{2 \times 2 \times 2 \times 2 \times 2} \quad \frac{5}{16}$$

5. Find the probability that the face-up sum is 16 when rolling three standard dice.

   4+6+6
     4 in 3 places
   5+5+6
     6 in 3 places

   $$\frac{6}{6 \times 6 \times 6} \quad \frac{1}{36}$$

10. Find the probability that the face-up sum is 7 when rolling two standard dice.

4+3	1+6
3+4	6+1
5+2	
2+5	

    $$\frac{6}{6 \times 6} \quad \frac{1}{6}$$

Probability 4

Find the probability when rolling two dice, following the meaning of "OR."

A bowl contains 13 blue, 11 yellow, 14 red, and 12 green marbles. Find the probability.

1+6 is also 6+1, so 2 pairs.
3+3 is only 1 pair.

1. The face-up sum equals 7 or 11.

1+6 5+6
2+5
3+4

$$\frac{6}{36} + \frac{2}{36} = \frac{4}{18} \quad \frac{2}{9}$$

9. P(blue OR yellow)

$$\frac{13}{50} + \frac{11}{50} = \frac{24}{50} = \frac{12}{25}$$

2. The face-up sum equals 4 or 6.

1+3 1+5
2+2 2+4
     3+3

$$\frac{3}{36} + \frac{5}{36} = \frac{8}{36} \quad \frac{2}{9}$$

10. P(yellow OR red)

$$\frac{11}{50} + \frac{14}{50} = \frac{25}{50} = \frac{1}{2}$$

3. The face-up sum equals an even prime or an odd composite. 2 OR 9

1+1 5+4
     6+3

$$\frac{1}{36} + \frac{4}{36} = \frac{5}{36}$$

11. P(red OR green)

$$\frac{14}{50} + \frac{12}{50} = \frac{26}{50} = \frac{13}{25}$$

4. The face-up sum equals a perfect square. 4 OR 9

2+2 3+6
     4+5

$$\frac{1}{36} + \frac{4}{36} \quad \frac{5}{36}$$

12. P(green OR blue)

$$\frac{12}{50} + \frac{13}{50} = \frac{25}{50} = \frac{1}{2}$$

5. The face-up sum is greater than 9. 10 OR 11 OR 12

4+6
5+6 5+5
     6+6

$$\frac{4}{36} + \frac{2}{36} = \frac{6}{36} \quad \frac{1}{6}$$

13. P(blue OR yellow OR red)

$$\frac{13}{50} + \frac{11}{50} + \frac{14}{50} = \frac{38}{50} = \frac{19}{25}$$

6. The face-up sum equals an odd prime. 3 OR 5 OR 7 OR 11

1+2 1+6
1+4 2+5
2+3 3+4
5+6

$$\frac{8}{36} + \frac{6}{36} = \frac{14}{36} \quad \frac{7}{18}$$

14. P(yellow OR green OR blue)

$$\frac{11}{50} + \frac{12}{50} + \frac{13}{50} = \frac{36}{50} = \frac{18}{25}$$

7. The face-up sum equals 3 or an even 2-digit number. 3 OR 10 OR 12

1+2 5+5
6+4 6+6

$$\frac{4}{36} + \frac{2}{36} = \frac{6}{36} \quad \frac{1}{6}$$

15. P(blue OR green OR red)

$$\frac{13}{50} + \frac{12}{50} + \frac{14}{50} = \frac{39}{50}$$

8. The face-up sum equals a perfect cube or an odd 2-digit number. 8 OR 11

4+4 6+5
5+3
6+2

$$\frac{5}{36} + \frac{2}{36} = \frac{7}{36}$$

16. P(red OR yellow OR green)

$$\frac{14}{50} + \frac{11}{50} + \frac{12}{50} = \frac{37}{50}$$

#13–16 may be done by 1 – P(missing color).

The "304" at top.

Probability 5

Find the probability following OR and subtracting the overlap, drawing once:

from a standard deck of cards.	*from 26 letters of the alphabet (Y is vowel).*

1. P(ace OR red)

$$\frac{4 + 26 - 2}{52} \qquad \frac{28}{52} \qquad \frac{7}{13}$$

2. P(queen OR heart)

$$\frac{4 + 13 - 1}{52} \qquad \frac{16}{52} \qquad \frac{4}{13}$$

3. P(face OR spade)

$$\frac{12 + 13 - 3}{52} \qquad \frac{22}{52} \qquad \frac{11}{26}$$

4. P (one-digit prime OR red)

$$\frac{16 + 26 - 8}{52} \qquad \frac{34}{52} \qquad \frac{17}{26}$$

5. P(8 OR king OR black)

$$\frac{4 + 4 + 26 - 4}{52} \qquad \frac{30}{52} \qquad \frac{15}{26}$$

6. P(one-digit even OR diamond)

$$\frac{16 + 13 - 4}{52} \qquad \frac{25}{52}$$

7. P(perfect square OR club)

$$\frac{12 + 13 - 3}{52} \qquad \frac{22}{52} \qquad \frac{11}{26}$$

8. P(diamond OR red)

$$\frac{13 + 26 - 13}{52} \qquad \frac{26}{52} \qquad \frac{1}{2}$$

9. P(black OR club)

$$\frac{26 + 13 - 13}{52} \qquad \frac{26}{52} \qquad \frac{1}{2}$$

10. P(king OR face card)

$$\frac{4 + 12 - 4}{52} \qquad \frac{12}{52} \qquad \frac{3}{13}$$

11. P (two-digit prime OR black)

$$\frac{8 + 26 - 4}{52} \qquad \frac{30}{52} \qquad \frac{15}{26}$$

12. P(vowel OR even ordinal letter)

$$\frac{6 + 13 - 0}{26} \qquad \frac{19}{26}$$

13. P(vowel OR odd ordinal letter)

$$\frac{6 + 13 - 6}{26} \qquad \frac{13}{26} \qquad \frac{1}{2}$$

14. P(consonant OR even ordinal letter)

$$\frac{20 + 13 - 13}{26} \qquad \frac{20}{26} \qquad \frac{10}{13}$$

15. P(consonant OR odd ordinal letter)

$$\frac{20 + 13 - 7}{26} \qquad \frac{26}{26} \qquad 1$$

16. P(vowel OR in word HOUSE)

$$\frac{6 + 5 - 3}{26} \qquad \frac{8}{26} \qquad \frac{4}{13}$$

17. P(consonant OR in word HOUSE)

$$\frac{20 + 5 - 2}{26} \qquad \frac{23}{26}$$

18. P(in word HOME OR in word HOUSE)

$$\frac{4 + 5 - 3}{26} \qquad \frac{6}{26} \qquad \frac{3}{13}$$

19. P(vowel OR in word MOVE)

$$\frac{6 + 4 - 2}{26} \qquad \frac{8}{26} \qquad \frac{4}{13}$$

20. P(consonant OR in word MOVE)

$$\frac{20 + 4 - 2}{26} \qquad \frac{22}{26} \qquad \frac{11}{13}$$

21. P(in word MOVE OR in word CARE)

$$\frac{4 + 4 - 1}{26} \qquad \frac{7}{26}$$

22. P(in word TIME OR in word GREAT)

$$\frac{4 + 5 - 2}{26} \qquad \frac{7}{26}$$

Probability 6

Answer as indicated, following the meaning of "AND" in probability.

1. The probability of rain is 3/4 and of forgetting one's umbrella is 2/3. Find P(rain AND have umbrella).

$$\frac{3}{4} \cdot \frac{1}{3} = \frac{1}{4}$$

2. Find the probability of having 3 girls as children assuming P(girl) = P(boy).

P(girl AND girl AND girl)

$$\frac{1}{2} \cdot \frac{1}{2} \cdot \frac{1}{2} = \frac{1}{8}$$

3. Find the probability of Team A losing 2 consecutive games if it has a .75 chance of winning any game.

P(lose AND lose)

$$\frac{1}{4} \cdot \frac{1}{4} = \frac{1}{16}$$

4. Find the probability of guessing correctly on four consecutive true-false questions.

$$\frac{1}{2} \cdot \frac{1}{2} \cdot \frac{1}{2} \cdot \frac{1}{2} = \frac{1}{16}$$

5. Draw one card from a standard deck and roll one die. Find P(heart AND prime).

$$\frac{1}{4} \cdot \frac{1}{2} = \frac{1}{8}$$

6. Find the probability of getting a head first and a tail second when tossing a coin twice.

$$\frac{1}{2} \cdot \frac{1}{2} = \frac{1}{4}$$

7. On 2 draws with replacement from the natural numbers less than 21, find P(odd, multiple of 4).

$$\frac{10}{20} \cdot \frac{5}{20} = \frac{1}{2} \cdot \frac{1}{4} = \frac{1}{8}$$

8. Find the probability of completing a maze on the first try if the maze has 8 forks and each fork goes right or left.

$$\left(\frac{1}{2}\right)^8 = \frac{1}{256}$$

9. Draw one card from a standard deck and roll one die. Find P(king AND odd).

$$\frac{1}{13} \cdot \frac{1}{2} = \frac{1}{26}$$

10. Find the probability of getting HTTHT when tossing a coin five times.

$$\left(\frac{1}{2}\right)^5 = \frac{1}{32}$$

11. Draw one card from a standard deck and roll one die. Find P(face card AND composite).

$$\frac{12}{52} \cdot \frac{2}{6} = \frac{3}{13} \cdot \frac{1}{3} = \frac{1}{13}$$

12. Bob's foul shot success is 70%. Find the probability of his missing his next 3 foul shots. P(miss AND miss AND miss)

$$\frac{3}{10} \cdot \frac{3}{10} \cdot \frac{3}{10} = \frac{27}{1000}$$

13. Draw one card from a standard deck and roll one die. Find P(black AND perfect square).

$$\frac{1}{2} \cdot \frac{1}{3} = \frac{1}{6}$$

14. Find the probability of sunken souffles on Geena's next four baked if the success of her souffle rising is 80%.

$$\frac{1}{5} \cdot \frac{1}{5} \cdot \frac{1}{5} \cdot \frac{1}{5} = \frac{1}{625}$$

Probability 7

A bowl contains 10 green, 14 yellow, 16 red, and 25 blue marbles. Find the specified probability for successive draws without replacement.

1. P(R, R)

$$\frac{16}{65} \cdot \frac{15}{64} = \frac{1}{13} \cdot \frac{3}{4} = \mathbf{\frac{3}{52}}$$

12. P(B, R, G)

$$\frac{25}{65} \cdot \frac{16}{64} \cdot \frac{10}{63} = \frac{5}{13} \cdot \frac{1}{2} \cdot \frac{5}{63} = \mathbf{\frac{25}{1638}}$$

2. P(B, R)

$$\frac{25}{65} \cdot \frac{16}{64} = \frac{5}{13} \cdot \frac{1}{4} = \mathbf{\frac{5}{52}}$$

13. P(NOT Y, NOT Y, NOT Y)

$$\frac{51}{65} \cdot \frac{50}{64} \cdot \frac{49}{63} = \frac{17}{13} \cdot \frac{5}{32} \cdot \frac{7}{3} = \mathbf{\frac{595}{1248}}$$

3. P(Y, Y)

$$\frac{14}{65} \cdot \frac{13}{64} = \frac{7}{5} \cdot \frac{1}{32} = \mathbf{\frac{7}{160}}$$

14. P(B OR G, R OR Y)

$$\frac{35}{65} \cdot \frac{30}{64} = \frac{7}{13} \cdot \frac{15}{32} = \mathbf{\frac{105}{416}}$$

4. P(NOT B, NOT B)

$$\frac{40}{65} \cdot \frac{39}{64} = \frac{1}{1} \cdot \frac{3}{8} = \mathbf{\frac{3}{8}}$$

15. P(B OR Y, R OR G)

$$\frac{39}{65} \cdot \frac{26}{64} = \frac{3}{5} \cdot \frac{13}{32} = \mathbf{\frac{39}{160}}$$

5. P(R, B, Y)

$$\frac{16}{65} \cdot \frac{25}{64} \cdot \frac{14}{63} = \frac{1}{13} \cdot \frac{5}{4} \cdot \frac{2}{9} = \mathbf{\frac{5}{234}}$$

16. P(NOT G, NOT G)

$$\frac{55}{65} \cdot \frac{54}{64} = \frac{11}{13} \cdot \frac{27}{32} = \mathbf{\frac{297}{416}}$$

6. P(G, G, R)

$$\frac{10}{65} \cdot \frac{9}{64} \cdot \frac{16}{63} = \frac{2}{13} \cdot \frac{1}{4} \cdot \frac{1}{7} = \mathbf{\frac{1}{182}}$$

17. P(B, NOT B, NOT B)

$$\frac{25}{65} \cdot \frac{40}{64} \cdot \frac{39}{63} = \frac{5}{1} \cdot \frac{5}{8} \cdot \frac{1}{21} = \mathbf{\frac{25}{168}}$$

7. P(B, B, Y)

$$\frac{25}{65} \cdot \frac{24}{64} \cdot \frac{14}{63} = \frac{5}{13} \cdot \frac{3}{8} \cdot \frac{2}{9} = \mathbf{\frac{5}{156}}$$

18. P(R, R, R)

$$\frac{16}{65} \cdot \frac{15}{64} \cdot \frac{14}{63} = \frac{1}{13} \cdot \frac{1}{2} \cdot \frac{1}{3} = \mathbf{\frac{1}{78}}$$

8. P(R, Y, R)

$$\frac{16}{65} \cdot \frac{14}{64} \cdot \frac{15}{63} = \frac{1}{13} \cdot \frac{2}{4} \cdot \frac{1}{3} = \mathbf{\frac{1}{78}}$$

19. P(Y, Y, Y)

$$\frac{14}{65} \cdot \frac{13}{64} \cdot \frac{12}{63} = \frac{1}{5} \cdot \frac{1}{3} \cdot \frac{1}{8} = \mathbf{\frac{1}{120}}$$

9. P(R, Y, G)

$$\frac{16}{65} \cdot \frac{14}{64} \cdot \frac{10}{63} = \frac{1}{13} \cdot \frac{2}{4} \cdot \frac{2}{9} = \mathbf{\frac{1}{117}}$$

20. P(G, G, G)

$$\frac{10}{65} \cdot \frac{9}{64} \cdot \frac{8}{63} = \frac{1}{13} \cdot \frac{1}{4} \cdot \frac{1}{7} = \mathbf{\frac{1}{364}}$$

10. P(NOT B, NOT B, NOT B)

$$\frac{40}{65} \cdot \frac{39}{64} \cdot \frac{38}{63} = \frac{1}{1} \cdot \frac{1}{4} \cdot \frac{19}{21} = \mathbf{\frac{19}{84}}$$

21. P(R, B, R)

$$\frac{16}{65} \cdot \frac{25}{64} \cdot \frac{15}{63} = \frac{1}{4} \cdot \frac{5}{13} \cdot \frac{5}{21} = \mathbf{\frac{25}{1092}}$$

11. P(Y, Y, R)

$$\frac{14}{65} \cdot \frac{13}{64} \cdot \frac{16}{63} = \frac{1}{5} \cdot \frac{1}{2} \cdot \frac{1}{9} = \mathbf{\frac{1}{90}}$$

22. P(G, G, G, G)

$$\frac{10}{65} \cdot \frac{9}{64} \cdot \frac{8}{63} \cdot \frac{7}{62} = \frac{1}{13} \cdot \frac{1}{8} \cdot \frac{1}{31} = \mathbf{\frac{1}{3224}}$$

Probability 8

For the "at least" situation, find the probability of:

1. at least one of two people born on Monday.

1 – P(not Mon, not Mon)

$$1 - \frac{6}{7} \cdot \frac{6}{7} = 1 - \frac{36}{49} \quad \textbf{13} / \textbf{49}$$

8. at least one tail when flipping a coin 3 times.

1 – P(no tail, no tail)

$$1 - \frac{1}{2} \cdot \frac{1}{2} \cdot \frac{1}{2} = 1 - \frac{1}{8} \quad \textbf{7} / \textbf{8}$$

2. at least one "4" when rolling a die 3 times.

1 – P(no 4, no 4, no 4)

$$1 - \frac{5}{6} \cdot \frac{5}{6} \cdot \frac{5}{6} = 1 - \frac{125}{216} \quad \textbf{91} / \textbf{216}$$

9. at least 1 club when drawing 2 cards from a deck without replacement.

1 – P(no club, no club)

$$1 - \frac{39}{52} \cdot \frac{38}{51} = 1 - \frac{19}{34} \quad \textbf{15} / \textbf{34}$$

3. making at least one goal in 3 tries if a player makes 1/4 of his goals.

1 – P(no goal, no goal, no goal)

$$1 - \frac{3}{4} \cdot \frac{3}{4} \cdot \frac{3}{4} = 1 - \frac{27}{64} \quad \textbf{37} / \textbf{64}$$

10. making at least one goal in 3 tries if a player makes 1/3 of his goals.

1 – P(no goal, no goal, no goal)

$$1 - \frac{2}{3} \cdot \frac{2}{3} \cdot \frac{2}{3} = 1 - \frac{8}{27} \quad \textbf{19} / \textbf{27}$$

4. at least 1 head when flipping a coin 4 times.

1 – P(no H, no H, no H, no H)

$$1 - \frac{1}{2} \cdot \frac{1}{2} \cdot \frac{1}{2} \cdot \frac{1}{2} = 1 - \frac{1}{16} \quad \textbf{15} / \textbf{16}$$

11. rain at least 1 day in the next 2 days if rain occurs 2 of 7 days.

1 – P(no rain, no rain)

$$1 - \frac{5}{7} \cdot \frac{5}{7} = 1 - \frac{35}{49} \quad \textbf{14} / \textbf{49}$$

5. at least one of 3 people born on Friday.

1 – P(not Fri, not Fri, not Fri)

$$1 - \frac{6}{7} \cdot \frac{6}{7} \cdot \frac{6}{7} = 1 - \frac{216}{343} \quad \textbf{127} / \textbf{343}$$

12. at least 1 black card when drawing 2 cards from a standard deck without replacement.   1 – P(no blk, no blk)

$$1 - \frac{26}{52} \cdot \frac{25}{51} = 1 - \frac{25}{102} \quad \textbf{77} / \textbf{102}$$

6. sun at least 1 day in the next 3 days if the sun shines 3 of 7 days.

1 – P(no sun, no sun, no sun)

$$1 - \frac{4}{7} \cdot \frac{4}{7} \cdot \frac{4}{7} = 1 - \frac{64}{343} \quad \textbf{279} / \textbf{343}$$

13. at least 1 hit in 3 tries if the probability of a hit is 3/4.

1 – P(no hits)

$$1 - \frac{1}{4} \cdot \frac{1}{4} \cdot \frac{1}{4} = 1 - \frac{1}{64} \quad \textbf{63} / \textbf{64}$$

7. at least 1 blue, selecting 2 random balls without replacement from 4 blue, 3 green, and 3 pink.   1 – P(no B, no B)

$$1 - \frac{6}{10} \cdot \frac{5}{9} = 1 - \frac{1}{3} \quad \textbf{2} / \textbf{3}$$

14. at least one red, selecting 2 random balls without replacement from 4 blue, 2 white, and 3 red. 1 – P(no R, no R)

$$1 - \frac{6}{9} \cdot \frac{5}{8} = 1 - \frac{5}{12} \quad \textbf{7} / \textbf{12}$$

Probability 9

Find the probability of starting at X, following a path, and ending at C.

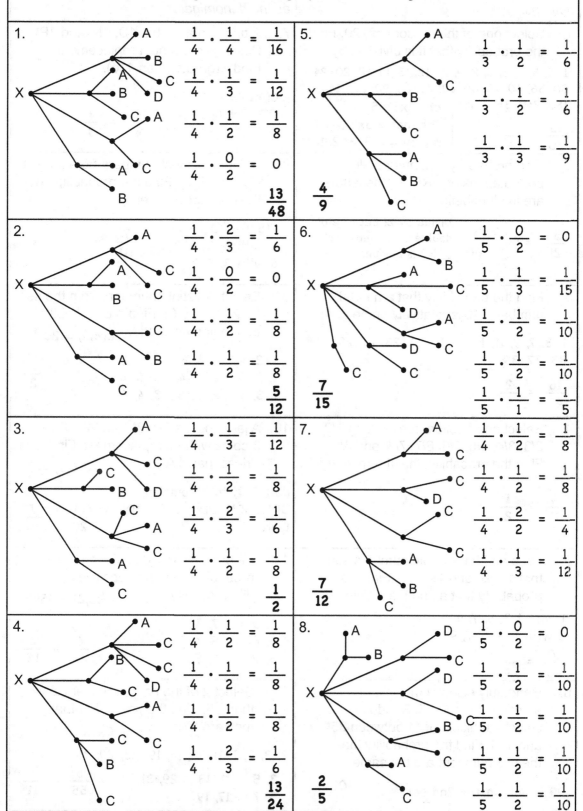

1.

$$\frac{1}{4} \cdot \frac{1}{4} = \frac{1}{16}$$

$$\frac{1}{4} \cdot \frac{1}{3} = \frac{1}{12}$$

$$\frac{1}{4} \cdot \frac{1}{2} = \frac{1}{8}$$

$$\frac{1}{4} \cdot \frac{0}{2} = 0$$

$$\frac{13}{48}$$

5.

$$\frac{1}{3} \cdot \frac{1}{2} = \frac{1}{6}$$

$$\frac{1}{3} \cdot \frac{1}{2} = \frac{1}{6}$$

$$\frac{1}{3} \cdot \frac{1}{3} = \frac{1}{9}$$

$$\frac{4}{9}$$

2.

$$\frac{1}{4} \cdot \frac{2}{3} = \frac{1}{6}$$

$$\frac{1}{4} \cdot \frac{0}{2} = 0$$

$$\frac{1}{4} \cdot \frac{1}{2} = \frac{1}{8}$$

$$\frac{1}{4} \cdot \frac{1}{2} = \frac{1}{8}$$

$$\frac{5}{12}$$

6.

$$\frac{1}{5} \cdot \frac{0}{2} = 0$$

$$\frac{1}{5} \cdot \frac{1}{3} = \frac{1}{15}$$

$$\frac{1}{5} \cdot \frac{1}{2} = \frac{1}{10}$$

$$\frac{1}{5} \cdot \frac{1}{2} = \frac{1}{10}$$

$$\frac{1}{5} \cdot \frac{1}{1} = \frac{1}{5}$$

$$\frac{7}{15}$$

3.

$$\frac{1}{4} \cdot \frac{1}{3} = \frac{1}{12}$$

$$\frac{1}{4} \cdot \frac{1}{2} = \frac{1}{8}$$

$$\frac{1}{4} \cdot \frac{2}{3} = \frac{1}{6}$$

$$\frac{1}{4} \cdot \frac{1}{2} = \frac{1}{8}$$

$$\frac{1}{2}$$

7.

$$\frac{1}{4} \cdot \frac{1}{2} = \frac{1}{8}$$

$$\frac{1}{4} \cdot \frac{1}{2} = \frac{1}{8}$$

$$\frac{1}{4} \cdot \frac{2}{2} = \frac{1}{4}$$

$$\frac{1}{4} \cdot \frac{1}{3} = \frac{1}{12}$$

$$\frac{7}{12}$$

4.

$$\frac{1}{4} \cdot \frac{1}{2} = \frac{1}{8}$$

$$\frac{1}{4} \cdot \frac{1}{2} = \frac{1}{8}$$

$$\frac{1}{4} \cdot \frac{1}{2} = \frac{1}{8}$$

$$\frac{1}{4} \cdot \frac{2}{3} = \frac{1}{6}$$

$$\frac{13}{24}$$

8.

$$\frac{1}{5} \cdot \frac{0}{2} = 0$$

$$\frac{1}{5} \cdot \frac{1}{2} = \frac{1}{10}$$

$$\frac{1}{5} \cdot \frac{1}{2} = \frac{1}{10}$$

$$\frac{1}{5} \cdot \frac{1}{2} = \frac{1}{10}$$

$$\frac{1}{5} \cdot \frac{1}{2} = \frac{1}{10}$$

$$\frac{2}{5}$$

Probability 10

Find the probability in a variety of settings.	*Find the probability using a combination as the denominator.*
1. Select one of the divisors of 720. Find the probability that it is divisible by 4. 1, 2, 3, 4, 5, 6, 8, 9, 10, 12, 15, 16, 18, 20, 24 30, 36, 40, 45, 48, 60, 72, 80, 90, 120, 144, 180, 240, 360, 720 | or $2^4×3^2×5$ $\frac{18}{30} = \frac{3}{5}$ | tot fact = 5×3×2=30 div by 4 = 3×3×2=18	7. A bag contains 1Q, 1D, 1N, and 1P. Pull 2 coins without replacement. Find P(greater than 11¢). C(4,2) = 6 QD, QN, QP, DN $\frac{4}{6} = \frac{2}{3}$
2. Find probability that among all permutations of GREED, the letters are in alphabetical order. $\frac{5!}{2!} = 60$ $\frac{1}{60}$ Divide by 2! because of double E. Alphabetical order is unique.	8. Select 2 distinct numbers from the set {3, 4, 5, 6, 7}. Find the probability that their product is even. C(5,2) = 10 4 with 3, 5, 6, 7 $\frac{7}{10}$ 6 with 3, 5, 7
3. Find the probability that a natural number ≤ 40 is relatively prime to 40. 1, 3, 7, 9, 11, 13, 17, 19, 21, 23, 27, 29, 31, 33, 37, 39 $\frac{16}{40} = \frac{2}{5}$	9. Select 3 distinct numbers from the set {1, 2, 3, 4, 5, 6}. Find the probability that their sum is odd. C(6,3) = 20 1, 2, 4 1, 4, 6 2, 4, 5 4, 5, 6 $\frac{1}{2}$ 1, 2, 6 2, 3, 4 2, 5, 6 1, 3, 5 2, 3, 6 3, 4, 6
4. Select one fraction from among 1/2, 2/3, 3/4, 4/5, 5/6, 6/7, 7/8, and 8/9. Find the probability that it terminates. $\frac{4}{8} = \frac{1}{2}$	10. A jar contains 2D, 2N, and 2P. Pull 3 coins without replacement. Find P(less than 17¢). C(6,3) = 20 do ≥ 17¢, take complement DDN, DDN, DDP, DDP, DNN, DNN $\frac{14}{20} = \frac{7}{10}$
5. Let a and b be whole numbers such that 1≤a≤4 and 1≤b≤10. Find the probability that a and b are both odd. 1 with 1 3, 5, 7, 9 3 with 1, 3, 5, 7, 9 $\frac{10}{40} = \frac{1}{4}$	11. Select 2 distinct digits. Find the probability that their positive difference is 4. C(10,2) = 45 4, 0 7, 3 5, 1 8, 4 $\frac{6}{45} = \frac{2}{15}$ 6, 2 9, 5
6. Randomly select a multiple of 9 between 0 and 75. Randomly select a multiple of 11 between 35 and 100. Find the probability that the two numbers are the same. 99 is only in the 2nd set. 0	12. Select 2 of the prime numbers less than 35. Find the probability that they are twin. C(11,2) = 55 2, 3, 5, 7, 11, 13, 17, 19, 23, 29, 31 3, 5 11, 13 29, 31 $\frac{5}{55} = \frac{1}{11}$ 5, 7 17, 19

Probability 11

Find the overall probability by adding individual probabilities.	*Find the probability with a "given" in one event (no prior dependence).*
1. Rolling a die twice, find the probability that the second roll is greater than the first. 1 with 2, 3, 4, 5, 6; 2 with 3, 4, 5, 6; etc. $\frac{5}{36} + \frac{4}{36} + \frac{3}{36} + \frac{2}{36} + \frac{1}{36} = \frac{15}{36} = \frac{5}{12}$	7. Given that when tossing 3 coins, at least 1 is a head. Find P(at least 1 tail). The given reduces the denominator from 8 to 7. HTT HHT HHH THT HTH TTH THH $\frac{6}{7}$
2. Jar A has 8 red and 2 blue marbles. Jar B has 5 blue and 7 pink marbles. Draw once from Jar A. If red, draw a 2nd from A; but if blue, draw a 2nd from B. Find P(1st any, 2nd not pink). $\frac{8}{10} \times \frac{9}{9} + \frac{2}{10} \times \frac{5}{12} = \frac{4}{5} + \frac{1}{12} = \frac{53}{60}$	8. Given that when tossing 3 coins, exactly 1 is a head. Find P(2 successive tails). The given reduces the denominator from 8 to 3. HTT THT TTH $\frac{2}{3}$
3. Rolling a die twice, find the probability that the second roll is less than or equal to the first. 1 with 1; 2 with 1,2; etc. $\frac{1}{36} + \frac{2}{36} + \frac{3}{36} + \frac{4}{36} + \frac{5}{36} + \frac{6}{36} = \frac{7}{12}$	9. A jar contains 6 red, 4 blue, 7 green, and 3 pink marbles. Draw 1 randomly. Given that it is not blue, find the probability of red. The given reduces the denominator from 20 to 16. $\frac{6}{16} = \frac{3}{8}$
4. Jar A has 8 tan and 8 red marbles. Jar B has 10 red and 20 green marbles. Draw once from Jar A. If red, draw a 2nd from A; but if tan, draw a 2nd from B. Find P(1st any, 2nd not green). $\frac{1}{2} \times \frac{15}{15} + \frac{1}{2} \times \frac{10}{30} = \frac{1}{2} + \frac{1}{6} = \frac{2}{3}$	10. Roll a red and blue die together. Given that the sum of the numbers is 10, find P(red 4). 4, 6; 5, 5; 6, 4 $\frac{1}{3}$
5. A packet contains 7 seeds: 2 produce blue flowers, 3 white, and 2 red. Select 2 seeds randomly. Find P(same color). $\frac{2}{7} \times \frac{1}{6} + \frac{3}{7} \times \frac{2}{6} + \frac{2}{7} \times \frac{1}{6} = \frac{5}{21}$	11. Roll a red and blue die together. Let P1 = P(red die even and blue die 4). Let P2 = P(red die even given blue die 4). Find P1 + P2. $P1 = \frac{1}{2} \times \frac{1}{6} = \frac{1}{12}$ $P2 = \frac{1}{2}$ $\frac{7}{12}$
6. A packet contains 9 seeds: 3 produce blue flowers, 4 white, and 2 red. Select 2 seeds randomly. Find P(same color). $\frac{3}{9} \times \frac{2}{8} + \frac{4}{9} \times \frac{3}{8} + \frac{2}{9} \times \frac{1}{8} =$ $\frac{1}{12} + \frac{1}{6} + \frac{1}{36} = \frac{10}{36} = \frac{5}{18}$	12. Roll a red and blue die together. Given that the sum of the numbers is 7, find P(red 4). 1, 6; 2, 5; 3, 4; 4, 3; 5, 3; 6, 1 $\frac{1}{6}$

Probability 12

Find the probability that 3 numbers selected randomly form the sides of an acute △.	*Find the probability that 3 numbers selected randomly form the angles of an acute △.*

1. 5, 8, 12, 13

5, 8, 12	obtuse	
5, 8, 13	not △	
5, 12, 13	right	$\dfrac{1}{4}$
8, 12, 13	acute	

9. 20, 30, 40, 50, 60, 70, and 80

30, 70, 80 $C(7,\ 3) = \dfrac{5 \times 6 \times 7}{2 \times 3}$
40, 60, 80
50, 60, 70 $\dfrac{3}{35}$

2. 3, 4, 5, 8

3, 4, 5	right	
3, 4, 8	not △	
3, 5, 8	not △	0
4, 5, 8	obtuse	

10. 35, 50, 55, 60, 65, and 80

35, 65, 80 $C(6,\ 3) = \dfrac{4 \times 5 \times 6}{2 \times 3}$
55, 60, 65 $\dfrac{2}{20}$ $\dfrac{1}{10}$

3. 5, 9, 11, 12

5, 9, 11	obtuse	
5, 9, 12	obtuse	
5, 11, 12	acute	$\dfrac{1}{2}$
9, 11, 12	acute	

11. 35, 50, 60, 70, and 85

35, 60, 85 $C(5,\ 3) = \dfrac{4 \times 5}{2}$
50, 60, 70 $\dfrac{2}{10}$ $\dfrac{1}{5}$

4. 6, 8, 9, 10

6, 8, 9	acute	
6, 8, 10	right	
6, 9, 10	acute	$\dfrac{3}{4}$
8, 9, 10	acute	

12. 20, 25, 35, 65, 70, 75, 80, and 85

20, 75, 85 $C(8,\ 3) = \dfrac{6 \times 7 \times 8}{2 \times 3}$
25, 75, 80
25, 70, 85 35, 70, 75 $\dfrac{5}{56}$
35, 65, 80

5. 5, 5, 6, 7

5, 5, 6	acute	
5, 5, 7	acute	
5, 6, 7	acute	1
5, 6, 7	acute	

13. 14, 34, 54, 56, 60, 70, 76, 80, and 86

14, 80, 86 $C(9,\ 3) = \dfrac{7 \times 8 \times 9}{2 \times 3}$
34, 70, 76
34, 60, 86 $\dfrac{4}{84}$ $\dfrac{1}{21}$
54, 56, 70

6. 7, 8, 15, 17

7, 8, 15	not △	
7, 8, 17	not △	
7, 15, 17	obtuse	0
8, 15, 17	right	

14. 27, 43, 50, 57, 67, 73, and 80

27, 73, 80 $C(7,\ 3) = \dfrac{5 \times 6 \times 7}{2 \times 3}$
50, 57, 73
43, 57, 80 $\dfrac{3}{35}$

7. 12, 14, 16, 20

12, 14, 16	acute	
12, 14, 20	obtuse	
12, 16, 20	right	$\dfrac{1}{2}$
14, 16, 20	acute	

15. 15, 25, 35, 45, 55, 60, 65, 80, and 85

15, 80, 85 $C(9,\ 3) = \dfrac{7 \times 8 \times 9}{2 \times 3}$
35, 65, 80
45, 55, 80 $\dfrac{3}{84}$ $\dfrac{1}{28}$

8. 6, 7, 8, 11

6, 7, 8	acute	
6, 7, 11	obtuse	
6, 8, 11	obtuse	$\dfrac{1}{4}$
7, 8, 11	obtuse	

16. 31, 49, 50, 51, 61, 70, 79, and 80

31, 70, 79 $C(8,\ 3) = \dfrac{6 \times 7 \times 8}{2 \times 3}$
49, 51, 80
49, 61, 70 50, 51, 79 $\dfrac{4}{56}$ $\dfrac{1}{14}$

Problem Solving 1

Solve by working backwards. | *Solve by making a chart.*

---

1. Five boys with initials A, B, C, D, and E exchanged cards. Only A had cards to start. A gave B 1/3 of his cards plus 3. B gave C 1/3 of his cards plus 3. C gave D 1/3 of his cards plus 3. D gave E 15 cards. How many cards did A have to start?

E 15	B 288	Check by going
D 36	**A 855**	forward.
C 99		

5. A 6-cup mixture is 1/4 wheat flour and 3/4 white flour. Add 4 cups of wheat flour. The new 10-cup mixture is what percent wheat flour?

	cups wheat	cups white	tot cups
Old	$\dfrac{3}{2}$	$\dfrac{9}{2}$	6
New	$\dfrac{11}{2}$	$\dfrac{9}{2}$	10

11 + 9 = 20 parts
11/20 = **55%**

---

2. Find the least natural number such that the operation cycle can occur 3 times, always yielding natural numbers: subtract 1, take 2/3 of the difference.

Try 3 left: 3, 9/2 NO
Try 6 left: 6, 9, 10, 15, 16, 24, 25 YES

25

6. Four women--Amy, Belle, Cleo, and Dee--visited Edie one day (either AM or PM). Amy visited at 8:00, Belle at 9:00, Cleo at 10:00, and Dee at 11:00. Cleo did not visit between Belle and Dee. At least one visited between Amy and Belle. Amy did not visit before Dee. Who visited last?   **Cleo**

A	B	C	D
8 PM	9 AM	10 PM	11 AM

---

3. Certain micro-organisms contained in a jar quadruple every 30 seconds. After 1.5 minutes, the jar contained 2112 of them. How many were in the jar to start?

2112	90 sec
528	60 sec
132	30 sec
33	start

7. A uniform rate of 720 ticks per cycle and 20 cycles per hour is the same rate as what number of ticks per second?

ticks	cycles	time
720	1	
720x20	20	1 hr
720x20	20	3600 sec
4		1 sec

---

4. Start with a number. Add 5, minus 8, add 3, minus 6, add 2, minus 4, and add 7. The result is 14. What was the starting number?

14 – 7 = 7	12 + 8 = 20
7 + 4 = 11	20 – 5 = **15**
11 – 2 = 9	
9 + 6 = 15	
15 – 3 = 12	

8. Saving money at a uniform weekly rate yielded $78 savings in 6 weeks. At the same rate, in how many more weeks will the savings be $208?

$	weeks	
78	6	16 – 6 = **10**
26	2	
208	16	

Problem Solving 2

Solve each problem using the Trial and Error (combined with reasoning) method.

Solve by introducing a line.

---

1. A bag contains 11 red, 6 blue, and 5 green marbles. Find the least number of blue marbles that must be added so that the probability of drawing one blue marble is greater than 1/2.

start: 6/22
add 6: 12/28
add 10: 16/32   **11**

---

6. The area of △ABC is 36 square units. Find the area of △BCD. (NTS)

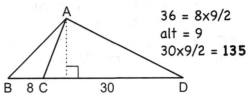

36 = 8x9/2
alt = 9
30x9/2 = **135**

---

2. A game has 5 players, each scoring from 0 to 100 inclusive. Their mean score is 80. Find the greatest number of players who can score exactly 50.

Try 3.	Try 2.
Total sum=80x5=400	2 at 50 = 100
3 at 50 = 150	400−100 = 300
400−150 = 250 (too great)	3 at 100 **2**

---

7. Find the measure of an interior angle of a regular 5-pointed star.

ext angle of reg pent =
360/5 = 72
int angle of reg pent =
180 − 72 = 108
108/3 = **36**

---

3. From the digits 3, 4, 5, and 6, form the 3-digit number and the 1-digit number that yield the greatest product. Use each digit only once.

Place one of the two greater digits in the 100s place.
643 x 5 = 3215
543 x 6 = 3258

---

8. A square with area 16 is inscribed in one quarter of a circle. Find the area of the circle.

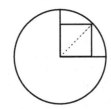

The diagonal of the square is a radius of the circle. r = 4 x (root 2)
r^2 = 32; A = **32π**

---

4. The sum of the digits of a certain two-digit number equals the square root of the number. Find the number.

2-digit perfect squares = 16, 25, 36, 49, 64, **81**
8 + 1 = 9; 9 is the square root of 81

---

9. Find the perimeter of the pentagon.

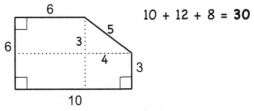

10 + 12 + 8 = **30**

---

5. Place each of 2, 3, 4, 6, and 8 in one of the circles so that the vertical and horizontal products are equal.

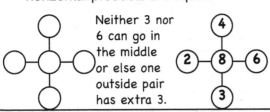

Neither 3 nor 6 can go in the middle or else one outside pair has extra 3.

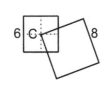

---

10. Find the area of the quadrilateral, which is the intersection of 2 squares. C is the center of the smaller square.

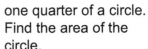

By introducing lines, one sees that the triangle could be moved to form 1/4 of the smaller square.

9

Problem Solving 3

Solve by drawing a picture.

---

1. Cut a rectanglular paper in two. Cut one of the pieces in two. Find the sum of the *different* numbers of edges from all ways of creating the 3 detached polygons.

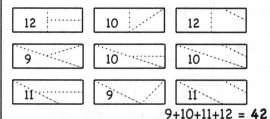

9+10+11+12 = **42**

---

5. The area of the triangle formed by connecting midpoints of adjacent sides of a square is what fractional part of the area of the square?

$\dfrac{1}{8}$

---

2. Flip the square centered at the origin 180° about the x-axis. Then flip it 180° about the y-axis. Find the final locations of the vertices.

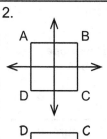

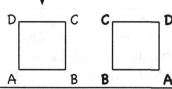

---

6. Find the greatest number of non-overlapping shaded regions when dividing the figure with exactly 2 straight lines.

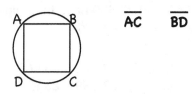

5

---

3. There are 4 pieces of chain, each 3 links long. What is the least number of opens and closes of links to join all 12 links into one complete circle?

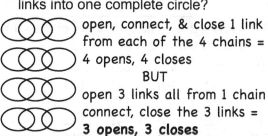

open, connect, & close 1 link from each of the 4 chains = 4 opens, 4 closes

BUT

open 3 links all from 1 chain connect, close the 3 links = **3 opens, 3 closes**

---

7. A, B, C, and D are 4 distnct points on a circle such that AB = BC = CD = AD. Name 2 diameters.

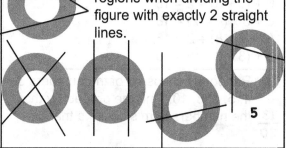

$\overline{AC}$ $\qquad$ $\overline{BD}$

---

4. A, B, and C are jars with volumes in the ratio 2:1:3. Jar A is 3/4 full. Jars B and C are empty. Pour from A to fill B. Pour the remainder from A into C. Jar C is what fractional part full?

$\dfrac{1}{6}$

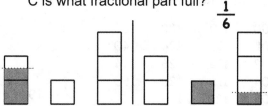

---

8. Fifteen points numbered consecutively 1 to 15 are on a circle. Cross off #3, #6, #9, and so on--every 3rd point not already crossed off. Find the number of the last point crossed off.

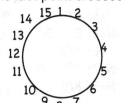

3, 6, 9, 12, 15, 4, 8, 13, 2, 10, 1, 11, 7, 14, 5

5

Problem Solving 4

Solve by making a list.

---

1. Find the number of right triangles in the square image. Assume trisection points.

1 tri : 12
2 tri: 4
2 tri, 1 sq: 8
2 tri, 2 sq: 4
6 tri, 6 sq: 4
3 tri, 3 sq: 4   **36**

---

6. Switching only pairs of adjacent digits, find the least number of switches to change 12345 to 54321.

12345	51243	54321
12354	51423	
12534	54123	
15234	54132	
51234	54312	**10**

---

2. How many distinct values are possible for 6 + 3 x 8 ÷ 2 by placing 1 pair of parentheses in the expression?

(6 + 3) x 8 ÷ 2 = 36
6 + (3 x 8) ÷ 2 = 18
6 + 3 x (8 ÷ 2) = 18
(6 + 3 x 8) ÷ 2 = 15
6 + (3 x 8 ÷ 2) = 18   **3**

---

7. Connect the points (3, 7) and (11, 23) with a segment. How many lattice points are on the segment?

(3, 7)   (8, 17)
(4, 9)   (9, 19)   follow slope
(5, 11)   (10, 21)   R 1, U 2
(6, 13)   (11, 23)
(7, 15)   **9**

---

3. Randomly choose 3 of the numbers 3, 4, 5, 6, 8, 10. Find the probability that the numbers form sides of a triangle.

$\frac{13}{20}$

3, 4, 5   **3, 5, 6**   3, 6, 10   4, 5, 10   **5, 6, 8**
3, 4, 6   3, 5, 8   **3, 8, 10**   **4, 6, 8**   **5, 6, 10**
3, 4, 8   3, 5, 10   **4, 5, 6**   4, 6, 10   **5, 8, 10**
3, 4, 10   **3, 6, 8**   **4, 5, 8**   **4, 8, 10**   **6, 8, 10**

---

8. Randomly choose 4 of the numbers 3, 4, 5, 10, 18. Find the probability that the numbers form sides of a quadrilateral. C(5, 4) = 5 $\frac{2}{5}$

3, 4, 5, 10   yes   3, 5, 10, 18   no
3, 4, 5, 18   no   4, 5, 10, 18   yes
3, 4, 10, 18 no
                      sum 3 sides > 4th side

---

4. How many whole numbers between 23,000 and 23,600 have their digits in strictly ascending order?

23,456   23,467   23,479   23,569
23,457   23,468   23,489   23,578
23,458   23,469   23,567   23,579
23,459   23,478   23,568   23,589   **16**

---

9. How many 3-digit numbers of the form AB1 are divisible by 3?

111   291   471   651   831
141   321   501   681   861
171   351   531   711   891
201   381   561   741   921
231   411   591   771   951
261   441   621   801   981   **30**

---

5. N and D are positive digits. Find the number of simplified, unique, fractions N/D that are less than 1.

N=1: 1/2 through 1/9   8
N=2: 2/3, 2/5, 2/7, 2/9   4
N=3: 3/4, 3/5, 3/7, 3/8   4
N=4: 4/5, 4/7, 4/9   3
N=5: 5/6, 5/7, 5/8, 5/9   4
N=6: 6/7   1
N=7: 7/8, 7/9   2
N=8: 8/9   1
                     27

---

10. How many 3-digit numbers from 100 to 150 inclusive are divisible by each of their nonzero digits?

100   110   122   135
101   111   124   140
102   112   126   144
104   115   128   150
105   120   132   **19**

317

Problem Solving 5

Solve by making up numbers without loss of generality (WLOG). Answers may be proven algebraically.

---

1. If the natural number n is divided by 9, the remainder is 5. Find the remainder when 3n is divided by 9.

WLOG, let n = 23.
23/9 = 2 R 5
3n = 69
69/9 = 7 R **6**

6. A flight occurred at 300 mph with the return trip on the exact route at 600 mph. Find the average speed for the round trip. Make up D.

	R	T	D
To	300	4	1200
Fr	600	2	1200
TOT	**400**	6	2400

---

2. Double the length of one side of a square and halve the other. Find the ratio of the perimeters: original square to resulting rectangle.

Sq: 10 by 10
P Sq = 40
Rec: 20 by 5
P Rec = 50   **4:5**

7. Find the percent decrease in the area of a square if each side is decreased by 10%.

Use 10 WLOG
– 10% = 9
orig area = 100
new area = 81

$\dfrac{ch}{orig} = \dfrac{19}{100}$ **19% D**

---

3. By what % is the area of a circle increased if its radius is increased by 20%?

r = 10   ch/orig = 44π/100π
a = 100π
R = 12
A = 144π   **44%**

8. The variable w increased by 10% of w is x. Then, x decreased by 50% of x is y. Finally, y increased by 20% of y is z. Find the percent that z is of w.

WLOG w = 100
x = 110
y = 55
z = 66

$\dfrac{66}{100}$ **66%**

---

4. The set X contains only 3 consecutive natural numbers. The set Y contains only the next 3 consecutive natural numbers. Find the number of different values formed by summing an element of X and an element of Y.

Let X = {1, 2, 3}   sums = 5, 6, 7, 8, 9
Let Y = {4, 5, 6}   **5**

9. Two sets of five consecutive natural numbers have exactly two numbers in common. Find the positive difference of the sums of the two sets.

WLOG
1 + 2 + 3 + 4 + 5 = 15
4 + 5 + 6 + 7 + 8 = 30
30 – 15 = **15**

---

5. AB = BC = CD = DE = DF. If a circle has a radius equal to (1/4)(AF) and point A is on the circle, then another point on the circle is between which two named points?

A   B   C   D   E   F

WLOG, AB = 4, AF = 20, 20/4 = 5
diam = 10, **C & D**

10. Find the percent change in the area of a square if one side is increased by 30% and one side is decreased by 30%.

Use 10 WLOG | original area = 100
+ 30% = 13 | new area = 91
– 30% = 7 | change = 9   **9% D**

MAVA Math: Enhanced Skills Solutions   Copyright © 2015 Marla Weiss

322

Problem Solving 6

Solve by finding a pattern.

1. Find the number of pairs of vertical angles formed when 15 lines intersect in one point.

lines	vert ang
2	2 + 0
3	3 + 3 = 3(3 − 1)
4	4 + 4 + 4 = 4(4 − 1)
15	15(15 − 1) = 15 × 14 = **210**

2. Evaluate:

$$1\frac{1}{2} \times 1\frac{1}{3} \times 1\frac{1}{4} \times \ldots \times 1\frac{1}{41}$$

$$\frac{3}{2} \times \frac{4}{3} \times \frac{5}{4} \times \ldots \times \frac{41}{40} \times \frac{42}{41}$$

= 42/2 = **21**

3. How many dots are in the 10th image in the pattern?

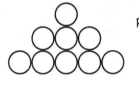

1 × 2   2 × 3   3 × 4   10 × 11 = **110**

4. Find the rightmost digit of the 10th row.

```
        1
      2 3 4
    5 6 7 8 9
10 11 12 13 14 15 16
```

row 1: 1 × 1
row 2: 2 × 2
row 3: 3 × 3
row 4: 4 × 4

row 10: 10 × 10 = **100**

5. To make TUKs, 1 TUK uses 2 TEMs, 2 TUKs use 6 TEMs, 3 TUKs use 12 TEMs, and 4 TUKs use 20 TEMs. One hundred TUKs use how many TEMs?

1:	2	1 × (1+1)
2:	6	2 × (2+1)
3:	12	3 × (3+1) 100 × 101 =
4:	20	4 × (4+1) **10,100**

6. Following the pattern, find the sum of the terms in Row 15.
Row 1: 2 + 4
Row 2: 2 + 4 + 6
Row 3: 2 + 4 + 6 + 8
  2x1 2x2 2x3 2x4

Row 15 has 16 terms. Last = 2 x 16.
(F+L)n/2 = (2 + 32)(16)/2 = 34 × 8 = **272**

7. Find the number of circles in the 15th row if the pattern continues.

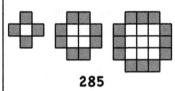

Row	Circles
1	1
2	3
3	5
R	2R−1 **29**

8. Find the number of total rectangles (including squares) in the 15th figure if the pattern continues.

Fig	1	2	3	4	5	R
Recs	1	3	6	10	15	R(R+1)/2 **120**

9. Find the number of individual small squares in the 15th figure if the pattern continues.

Fig	Squares
1	1+4=5
2	4+8=12
3	9+12=21
R	R^2+4R

285

10. Evaluate.

$$\frac{1}{2^1} + \frac{1}{2^2} + \frac{1}{2^3} + \ldots + \frac{1}{2^{14}} + \frac{1}{2^{15}}$$

$\frac{1}{2}$ $\frac{3}{4}$ $\frac{7}{8}$ Numerator is 1 less than the denominator.   **32,767** / **32,768**

MAVA Math: Enhanced Skills Solutions Copyright © 2015 Marla Weiss

Problem Solving 7

Solve by making a simpler problem.	*Solve by using logical reasoning.*

1. Find the sum of the first 12 positive perfect cubes.

OR

1, 3, 6, 10, 15, 21, 28, 36, 45, 55, 66, 78

$1^3 = 1$

$1^3 + 2^3 = 1+8 = 9$

$1^3 + 2^3 + 3^3 = 1+8+27 = 36$

$1^3 + 2^3 + 3^3 + 4^3 = 1+8+27+64 = 100$

$1 + \ldots + 12 = 78$ $78^2 = \mathbf{6084}$

---

6. Arrange the letters A, B, C, D, E, and F according to the following rules:
1. The word CAB appears.
2. D and E are at the ends in one of the two orders.
3. Two letters are beween A and E.
4. C is not next to D.

<u>E</u> <u>F</u> <u>C</u> <u>A</u> <u>B</u> <u>D</u>

---

2. How many dots form the perimeter of an array of dots 799 dots by 800 dots?

4 by 5

(5+5) + 2(4−2)

14

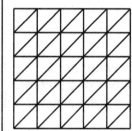

799 by 800

(800+800) + 2(799−2)

1600 + 1594 = **3194**

---

7. Evaluate. Answer in exponential form.

(994−991)(991−988)(988−985) . . . (7−4)(4−1)

Answer is 3 to a power. How many 3s are multiplied together?

(4−1) has 1 3; (4 − 1)/3 = 1

(7−4)(4−1) has 2 3s; (7 − 1)/3 = 2

(10−7)(7−4)(4−1) has 3 3s; (10 − 1)/3 = 3

(994 − 1)/3 = 331

3^{331}

---

3. How many triangles are in the figure?

1x1 sq: 5x5 sq:

25x2=50   1x2=2

2x2 sq: sum = **110**

16x2=32

3x3 sq:

9x2=18

4x4 sq:

4x2=8

---

8. Move and/or turn 2 of the short line segments shown to result in 4 of the small-size squares.

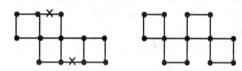

---

4. An 11 by 11 checkerboard is made up of alternating black and white squares with a black square in the bottom right corner. What fraction of the squares are black?

3 by 3

5/9

$\dfrac{61}{121}$

---

9. Find the next number in the sequence:

1, 1, 2, 3, 6, 9, 15, 135, 150, 285, ___

For each consecutive pair of adjacent terms, the pattern is: add, add, multiply, repeat. The next term is 150 x 285.

42,750

---

5. Find the value of $10^{2x} - 10^x + 1$ when x = 10.

x = 2: $10^4 - 10^2 + 1 = 9901$

x = 3: $10^6 - 10^3 + 1 = 999001$

x = 10: **99999999990000000001**

(10 9s and 9 0s)

---

10. If certain people only tell the truth on Sunday, Tuesday, and Thursday, then one of those people can say "I told the truth yesterday" on which day of the week?

Sun: no   Wed: no   Sat: yes

Mon: no   Thurs: no

Tues: no   Fri: no   **Saturday**

Problem Solving 8

Write and solve one or more equations.

1. The denominator of a fraction is 1 less than twice the numerator. Subtracting 1 from the denominator, the fraction is 2/3. Find the original fraction.

$$\frac{N}{D} = \frac{N}{2N-1} \qquad \frac{N}{2N-2} = \frac{2}{3}$$

$$4N - 4 = 3N$$
$$N = 4$$
$$D = 7$$
$$\frac{4}{7}$$

2. The units digit of a number is 4 times the tens digit. Adding 54, the digits swap places. Find the number.

$$U = 4T$$
$$10T + U + 54 = 10U + T$$
$$9T + 54 = 9U$$
$$T + 6 = U$$
$$T + 6 = 4T$$
$$T = 2$$
$$U = 8$$

28

3. The numerator of a fraction is 3 less than the denominator. Subtracting 1 from both, the fraction is 1/2. Find the original fraction.

$$\frac{N}{D} = \frac{D-3}{D} \qquad \frac{D-4}{D-1} = \frac{1}{2}$$

$$2D - 8 = D - 1$$
$$D = 7$$
$$N = 4$$
$$\frac{4}{7}$$

4. A number is 6 times the sum of its digits. The units digit is 1 less than the tens digit. Find the number.

$$10T + U = 6(T + U)$$
$$10T + U = 6T + 6U$$
$$4T = 5U$$
$$U = T - 1$$
$$4T = 5T - 5$$
$$T = 5$$
$$U = 4$$

54

5. The denominator of a fraction is 7 more than the numerator. Adding 5 to both, the fraction becomes 1/2. Find the original fraction.

$$\frac{N}{D} = \frac{N}{N+7} \qquad \frac{N+5}{N+12} = \frac{1}{2}$$

$$2N + 10 = N+12$$
$$N = 2$$
$$D = 9$$
$$\frac{2}{9}$$

6. How many ounces of pure salt must be added to 5 ounces of a 20% salt solution to yield a 25% salt solution?

$$(5) + (x) = (5 + x)$$
$$(.20)(5) + (1)(x) = (.25)(5 + x)$$
$$(20)(5) + (100)(x) = (25)(5 + x)$$
$$100 + 100x = 125 + 25x$$
$$75x = 25$$
$$x = 1/3$$
$$\frac{1}{3}$$

7. How many pounds of water must evaporate from 9 pounds of a 50% salt solution to yield a 75% salt solution?

$$(9) - (x) = (9 - x)$$
$$(.50)(9) - (0)(x) = (.75)(9 - x)$$
$$(50)(9) = (75)(9 - x)$$
$$(2)(9) = (3)(9 - x)$$
$$18 = 27 - 3x$$
$$x = 3$$

3

8. The sum of the digits of a number is 6. Subtracting 36 from the number, the digits swap places. Find the number.

$$T + U = 6$$
$$10T + U - 36 = 10U + T$$
$$9T - 36 = 9U$$
$$T - 4 = U$$
$$2T - 4 = 6$$
$$T = 5$$
$$U = 1$$

51

Properties 1

Name the property illustrated. Choose among: ClPA, APA, IdPA, InPA, CPA, ClPM, APM, IdPM, InPM, CPM, DPMA, RPE, SPE, TPE, ZPM, APE, MPE.

1.	$5 + 3 = 3 + 5$	**CPA**	20.	$5.2 \times 1.7 = 1.7 \times 5.2$	**CPM**

#			#		
1.	$5 + 3 = 3 + 5$	**CPA**	20.	$5.2 \times 1.7 = 1.7 \times 5.2$	**CPM**
2.	If $1 + 1 = 2$, then $1 + 1 + 5 = 2 + 5$	**APE**	21.	If $6 + 5 = 11$, then $11 = 6 + 5$	**SPE**
3.	$(2 \times 3) \times 8 = 2 \times (3 \times 8)$	**APM**	22.	$(7 + 2)\,5 = 35 + 10$	**DPMA**
4.	$6 \times (1/6) = 1$	**InPM**	23.	$0 + 13.2 = 13.2$	**IdPA**
5.	$126 = 126$	**RPE**	24.	2.1×3.3 is a unique real number	**ClPM**
6.	$3 + 4$ is a unique real number	**ClPA**	25.	If $4 = 9 - 5$ and $9 - 5 = 7 - 3$, then $4 = 7 - 3$	**TPE**
7.	$17 + 0 = 17$	**IdPA**	26.	$4 + (6 + 7) = (4 + 6) + 7$	**APA**
8.	$4(5 + 3) = 20 + 12$	**DPMA**	27.	$1 \times 5.43 = 5.43$	**IdPM**
9.	If $2 + 1 = 3$, then $3 = 2 + 1$	**SPE**	28.	$(7 \times 2) \times 5 = 7 \times (2 \times 5)$	**APM**
10.	2×3 is a unique real number	**ClPM**	29.	$9 + 5 = 5 + 9$	**CPA**
11.	$5 \times 19 = 19 \times 5$	**CPM**	30.	$67.4 = 67.4$	**RPE**
12.	$5 + (-5) = 0$	**InPA**	31.	$45.8 \times 0 = 0$	**ZPM**
13.	$7 + (6 + 1) = (7 + 6) + 1$	**APA**	32.	If $2 + 3 = 5$, then $4(2 + 3) = 4 \times 5$	**MPE**
14.	$32 \times 1 = 32$	**IdPM**	33.	$0 = (-3.4) + 3.4$	**InPA**
15.	If $1 = 9 - 8$ and $9 - 8 = 7 - 6$, then $1 = 7 - 6$	**TPE**	34.	$3.1 + 4.56$ is a unique real number	**ClPA**
16.	$0 \times 999 = 0$	**ZPM**	35.	$(2/3) \times (3/2) = 1$	**InPM**
17.	If $12 = 3 \times 4$, then $12 \times 5 = 3 \times 4 \times 5$.	**MPE**	36.	If $12 = 11 + 1$, then $11 + 1 = 12$.	**SPE**
18.	$6(2 + 4) = 6 \times 2 + 6 \times 4$	**DPMA**	37.	$1 \times 101 = 101$	**IdPM**
19.	If $5 \times 3 = 15$, then $15 = 5 \times 3$.	**SPE**	38.	$11 \times (7 \times 4) = (11 \times 7) \times 4$	**APM**

Properties 2

Name the property illustrated. Choose among: CIPA, APA, IdPA, InPA, CPA, CIPM, APM, IdPM, InPM, CPM, DPMA, RPE, SPE, TPE, ZPM, APE, MPE.

1. $ab + 8 = ba + 8$	**CPM**	17. $0 + mn = mn$	**IdPA**
2. If $xy = a$, then $xy + 4 = a + 4$	**APE**	18. If $w = bc$ and $bc = x$, then $w = x$	**TPE**
3. $a(b + c) = ab + ac$	**DPMA**	19. $3a \cdot 7b$ is a unique real number	**CIPM**
4. $(abc)(1) = abc$	**IdPM**	20. $4(ab) = 4(ba)$	**CPM**
5. $3x + 0 = 3x$	**IdPA**	21. $(p + g)rs = prs + grs$	**DPMA**
6. $w + m$ is a unique real number	**CIPA**	22. If $x + 1 = y$, then $y = x + 1$	**SPE**
7. $xy = xy$	**RPE**	23. $5m = 5m$	**RPE**
8. $(xy)z = x(yz)$	**APM**	24. $bc + cd = cd + bc$	**CPA**
9. rs is a unique real number	**CIPM**	25. $(9rs)(0) = 0$	**ZPM**
10. $162 = x$ implies $x = 162$	**SPE**	26. $4x + 7y$ is a unique real number	**CIPA**
11. $5y + 4x = 4x + 5y$	**CPA**	27. $g + (f + e) = (g + f) + e$	**APA**
12. $0 \cdot b^5 = 0$	**ZPM**	28. $(1)(13c^7) = 13c^7$	**IdPM**
13. $a + (y + 1) = (a + y) + 1$	**APA**	29. $3w = 5x$ implies $3wy = 5xy$	**MPE**
14. $a^3 + (-a^3) = 0$	**InPA**	30. $\dfrac{a}{c} \cdot \dfrac{c}{a} = 1$	**InPM**
15. If $10x = 7y$ and $7y = 2b$, then $10x = 2b$	**TPE**	31. $(6r)s = 6(rs)$	**APM**
16. $\dfrac{5}{3} \cdot \dfrac{3}{5} = 1$	**InPM**	32. $0 = -(5ad) + 5ad$	**InPA**

Proportions 1

Solve by simplifying first.

1. $\dfrac{45}{x} = \dfrac{35}{63}\dfrac{5}{9}$ **x = 81**	12. $\dfrac{16}{x} = \dfrac{20}{35}\dfrac{4}{7}$ **x = 28**	23. $\dfrac{35}{x} = \dfrac{25}{45}\dfrac{5}{9}$ **x = 63**
2. $\dfrac{12}{x} = \dfrac{30}{35}\dfrac{6}{7}$ **x = 14**	13. $\dfrac{3}{7}\dfrac{33}{77} = \dfrac{12}{x}$ **x = 28**	24. $\dfrac{66}{x} = \dfrac{72}{84}\dfrac{6}{7}$ **x = 77**
3. $\dfrac{3}{7}\dfrac{15}{35} = \dfrac{21}{x}$ **x = 49**	14. $\dfrac{x}{10} = \dfrac{56}{35}\dfrac{8}{5}$ **x = 16**	25. $\dfrac{7}{11}\dfrac{49}{77} = \dfrac{56}{x}$ **x = 88**
4. $\dfrac{32}{x} = \dfrac{24}{21}\dfrac{8}{7}$ **x = 28**	15. $\dfrac{x}{15} = \dfrac{26}{65}\dfrac{2}{5}$ **x = 6**	26. $\dfrac{27}{x} = \dfrac{48}{80}\dfrac{3}{5}$ **x = 45**
5. $\dfrac{33}{x} = \dfrac{55}{60}\dfrac{11}{12}$ **x = 36**	16. $\dfrac{22}{x} = \dfrac{55}{15}\dfrac{11}{3}$ **x = 6**	27. $\dfrac{36}{x} = \dfrac{81}{99}\dfrac{9}{11}$ **x = 44**
6. $\dfrac{34}{x} = \dfrac{51}{15}\dfrac{17}{5}$ **x = 10**	17. $\dfrac{x}{36} = \dfrac{95}{30}\dfrac{19}{6}$ **x = 114**	28. $\dfrac{14}{x} = \dfrac{34}{85}\dfrac{2}{5}$ **x = 35**
7. $\dfrac{x}{28} = \dfrac{55}{35}\dfrac{11}{7}$ **x = 44**	18. $\dfrac{x}{45} = \dfrac{78}{54}\dfrac{13}{9}$ **x = 65**	29. $\dfrac{x}{65} = \dfrac{57}{95}\dfrac{3}{5}$ **x = 39**
8. $\dfrac{x}{18} = \dfrac{51}{34}\dfrac{3}{2}$ **x = 27**	19. $\dfrac{3}{7}\dfrac{9}{21} = \dfrac{x}{49}$ **x = 21**	30. $\dfrac{x}{48} = \dfrac{77}{66}\dfrac{7}{6}$ **x = 56**
9. $\dfrac{32}{x} = \dfrac{52}{39}\dfrac{4}{3}$ **x = 24**	20. $\dfrac{7}{10}\dfrac{14}{20} = \dfrac{35}{x}$ **x = 50**	31. $\dfrac{55}{x} = \dfrac{99}{72}\dfrac{11}{8}$ **x = 40**
10. $\dfrac{18}{x} = \dfrac{27}{24}\dfrac{9}{8}$ **x = 16**	21. $\dfrac{x}{65} = \dfrac{64}{80}\dfrac{4}{5}$ **x = 52**	32. $\dfrac{36}{x} = \dfrac{98}{49}\dfrac{2}{1}$ **x = 18**
11. $\dfrac{x}{35} = \dfrac{88}{40}\dfrac{11}{5}$ **x = 77**	22. $\dfrac{51}{x} = \dfrac{66}{44}\dfrac{3}{2}$ **x = 34**	33. $\dfrac{x}{28} = \dfrac{54}{36}\dfrac{3}{2}$ **x = 42**

Proportions 2

Solve for x by cross multiplying. Variables are nonzero.

1. $\dfrac{6}{x} = \dfrac{5}{9}$ $5x = 54$

$x = \dfrac{54}{5}$

9. $\dfrac{h}{e} = \dfrac{c}{x}$ $hx = ce$

$x = \dfrac{ce}{h}$

17. $\dfrac{9}{x} = \dfrac{13}{11}$ $13x = 99$

$x = \dfrac{99}{13}$

2. $\dfrac{5}{x} = \dfrac{8}{15}$ $8x = 75$

$x = \dfrac{75}{8}$

10. $\dfrac{5}{x} = \dfrac{15}{4}$ $15x = 20$

$x = \dfrac{4}{3}$

18. $\dfrac{3x}{17} = \dfrac{5}{2}$ $6x = 85$

$x = \dfrac{85}{6}$

3. $\dfrac{7}{3} = \dfrac{5}{x}$ $7x = 15$

$x = \dfrac{15}{7}$

11. $\dfrac{6}{7} = \dfrac{9}{x}$ $6x = 63$

$x = \dfrac{21}{2}$

19. $\dfrac{7}{5} = \dfrac{12}{x}$ $7x = 60$

$x = \dfrac{60}{7}$

4. $\dfrac{2}{13} = \dfrac{11}{x}$ $2x = 143$

$x = \dfrac{143}{2}$

12. $\dfrac{6}{x} = \dfrac{10}{9}$ $10x = 54$

$x = \dfrac{27}{5}$

20. $\dfrac{7}{x} = \dfrac{9a}{7}$ $9ax = 49$

$x = \dfrac{49}{9a}$

5. $\dfrac{6}{5} = \dfrac{7}{x}$ $6x = 35$

$x = \dfrac{35}{6}$

13. $\dfrac{11}{x} = \dfrac{33}{7}$ $33x = 77$

$x = \dfrac{7}{3}$

21. $\dfrac{12}{x} = \dfrac{18}{5}$ $18x = 60$

$x = \dfrac{10}{3}$

6. $\dfrac{5}{x} = \dfrac{9}{4}$ $9x = 20$

$x = \dfrac{20}{9}$

14. $\dfrac{9}{2x} = \dfrac{5}{4}$ $10x = 36$

$x = \dfrac{18}{5}$

22. $\dfrac{6k}{m} = \dfrac{2a}{x}$ $6kx = 2am$

$x = \dfrac{am}{3k}$

7. $\dfrac{3}{4} = \dfrac{2}{x}$ $3x = 8$

$x = \dfrac{8}{3}$

15. $\dfrac{5a}{x} = \dfrac{2b}{y}$ $2bx = 5ay$

$x = \dfrac{5ay}{2b}$

23. $\dfrac{9}{8} = \dfrac{x}{11}$ $8x = 99$

$x = \dfrac{99}{8}$

8. $8 : x = 5 : 12$

$\dfrac{8}{x} = \dfrac{5}{12}$

$5x = 96$

$x = \dfrac{96}{5}$

16. $45 : 11 = 10 : x$

$\dfrac{45}{11} = \dfrac{10}{x}$

$9x = 22$

$x = \dfrac{22}{9}$

24. $3a : 7 = 8a : ex$

$\dfrac{3a}{7} = \dfrac{8a}{ex}$

$3ex = 56$

$x = \dfrac{56}{3e}$

Proportions 3

Write and solve a proportion, assuming a constant rate.

1. If a store needs 9 registers open for 150 customers, how many registers are needed for 200 customers?

$$\overset{3}{\underset{50}{\frac{9}{150}}} = \frac{x}{200}$$

Or by mental math, 3 registers for every 50 customers.

$$x = 12$$

7. When a four-foot child casts a 7-foot shadow, find the height of a building in feet with a 168-foot shadow.

$$\frac{4}{7} = \frac{x}{168}$$

$$x = 24 \times 4 = 96$$

2. If the tax on a $7200 item is $504, find the tax on an item costing $8000.

$$\overset{56}{\underset{800}{\frac{504}{7200}}} = \frac{x}{8000}$$

$$x = 560 \qquad \mathbf{\$560}$$

8. If 12 sheet cakes serve 26 people, how many people do 54 identical sheet cakes serve?

$$\overset{6}{\underset{13}{\frac{12}{26}}} = \frac{54}{x}$$

$$x = 117$$

3. If 8 workers load 12 trucks, how many workers are needed to load 21 trucks?

$$\overset{2}{\underset{3}{\frac{8}{12}}} = \frac{x}{21}$$

$$x = 14$$

9. A recipe needs 2 cups sugar for every 3 cups flour. How many cups of sugar are needed for 10 cups of flour?

$$\frac{2}{3} = \frac{x}{10}$$

$$3x = 20 \qquad x = 6\frac{2}{3}$$

4. Find the real estate tax on a $80,000 home if the tax on a $110,000 home is $1320.

$$\overset{12}{\underset{1000}{\frac{1320}{110,000}}} = \frac{x}{80,000}$$

$$x = 960 \qquad \mathbf{\$960}$$

10. Find the restaurant-calculated tip in dollars on a $500 bill if a $650 bill at the same restaurant had a $104 tip.

$$\overset{8}{\underset{50}{\frac{104}{650}}} = \frac{x}{500}$$

$$x = 80$$

5. 224 ounces of punch are needed to serve 35 people. How many ounces are needed to serve 50 people?

$$\overset{32}{\underset{5}{\frac{224}{35}}} = \frac{x}{50}$$

$$x = 320$$

11. The weight of 64 feet of rope is 20 pounds. What is the weight in pounds of 80 feet of the same rope?

$$\overset{16}{\underset{5}{\frac{64}{20}}} = \frac{80}{x}$$

$$x = 25$$

6. A photograph 5 by 8 inches is enlarged so that one dimension is 7 inches. Find the other new dimension.

$$\frac{5}{8} = \frac{7}{x} \qquad \begin{aligned} 5x &= 56 \\ 10x &= 112 \end{aligned}$$

$$x = 11.2 \text{ inches}$$

12. If a 13.5-foot flagpole casts a 19.5-foot shadow, what is the shadow of a nearby 54-foot tree at the same time?

$$\overset{27}{\underset{39}{\frac{13.5}{19.5}}} = \frac{54}{x}$$

$$x = 78 \text{ feet}$$

Proportions 4

Solve, using the two-column method.

1. Picking 4 bushels of peaches in 25 minutes, how many bushels can be picked in 5 hours?

bushels	time
4	25 m
4/5	5 m
48/5	1 h
48	5 h

48

2. If 7 inches of material costs $17.50, find the cost of 2 feet of the same material.

length	cost
7 in	17.50
1 in	2.50
1 ft	30.00
2 ft	60.00

$60

3. Typing 250 words in 20 minutes, how many hours are needed to type a 7500 word paper?

words	time
250	20 min
750	1 hr
7500	10 hr

10

4. A recipe requires 1.25 cups of sugar for 60 cookies. How many cups are needed for 9 dozen?

sugar	cookies
1.25	60
0.25	1 doz
2.25	9 doz

$2\frac{1}{4}$

5. If a person 79.2 inches tall casts a shadow 9 feet long, find the height of a tree in feet casting a 60-foot long shadow at the same time and place.

height	shadow
79.2 in	9 ft
8.8 in	1 ft
528 in	60 ft
44 ft	60 ft

44

6. 14 pounds of chips serve 64 people. How many ounces are needed to serve 14 people?

people	chips
64	14 lb
32/7	1 lb
2/7	1 oz
14	49 oz

49

7. If a map uses a scale of 1 cm to 150 km, find the actual distance in km if the map distance is 45 mm.

actual	scale
150 km	1 cm
15 km	1 mm
675 km	45 mm

675

8. Typing 50 words per 40 seconds, how many words can be typed in 12 minutes?

words	time
50	40 sec
150	2 min
900	12 min

900

9. A photograph 6 by 9 inches is enlarged so that the lesser dimension is 4 feet. Find the other dimension in feet.

length	width
6 in	9 in
1 ft	1.5 ft
4 ft	6 ft

6

10. The weight of 63 feet of rope is 24 pounds. What is the weight in pounds of 56 yards of the same rope?

length	weight
63 ft	24 lb
21 yd	24 lb
7 yd	8 lb
56 yd	64 lb

64

Proportions 5

Find the number.

1. A number added to 15, then divided by 11 is the same as 14 less than the number, then divided by 12.

$$\frac{x + 15}{11} = \frac{x - 14}{12}$$

$12x + 180 = 11x - 154$ **x = −334**

2. The quotient of 5 more than a number and 15 equals the quotient of 3 less than the number and 13.

$$\frac{x + 5}{15} = \frac{x - 3}{13} \quad \middle| \quad 2x = 110$$

$13x + 65 = 15x - 45$ **x = 55**

3. Half the sum of 22 and a number equals one-fifth the sum of 16 and the number.

$$\frac{x + 22}{2} = \frac{x + 16}{5} \quad \middle| \quad 3x = -78$$

$5x + 110 = 2x + 32$ **x = −26**

4. The sum of triple a number and 7, then divided by 4, equals 7 less than twice the number, then divided by 5.

$$\frac{3x + 7}{4} = \frac{2x - 7}{5} \quad \middle| \quad 7x = -63$$

$15x + 35 = 8x - 28$ **x = −9**

5. The quotient of 3 less than 4 times a number and 16 equals the quotient of 8 less than the number and 6.

$$\frac{4x - 3}{16} = \frac{x - 8}{6} \quad \middle| \quad \begin{array}{l} 12x - 9 = 8x - 64 \\ 4x = -55 \end{array}$$

$24x - 18 = 16x - 128$ **x = −13.75**

6. Half the sum of 25 and a number equals one-sixth the sum of 5 and twice the number.

$$\frac{x + 25}{2} = \frac{2x + 5}{6} \quad \middle| \quad 2x = -140$$

$6x + 150 = 4x + 10$ **x = −70**

7. A number subtracted from 8, then divided by 10, is the same as 2 less than the number, then divided by 20.

$$\frac{8 - x}{10} = \frac{x - 2}{20} \quad \middle| \quad 30x = 180$$

$160 - 20x = 10x - 20$ **x = 6**

8. Half the sum of 8 and triple a number equals one-third the sum of 7 and twice the number.

$$\frac{3x + 8}{2} = \frac{2x + 7}{3} \quad \middle| \quad 5x = -10$$

$9x + 24 = 4x + 14$ **x = −2**

9. The quotient of 7 more than a number and 30 equals the quotient of 3 less than twice the number and 50.

$$\frac{x + 7}{30} = \frac{2x - 3}{50} \quad \middle| \quad 10x = 440$$

$50x + 350 = 60x - 90$ **x = 44**

10. A number added to 50, then divided by 20, is the same as 16 less than twice the number, then divided by 11.

$$\frac{x + 50}{20} = \frac{2x - 16}{11} \quad \middle| \quad 29x = 870$$

$11x + 550 = 40x - 320$ **x = 30**

11. One-third the sum of 17 and a number equals one-sixth the sum of 16 and five times the number.

$$\frac{x + 17}{3} = \frac{5x + 16}{6} \quad \middle| \quad 9x = 54$$

$6x + 102 = 15x + 48$ **x = 6**

12. The quotient of 9 more than four times a number and 10 equals the quotient of 4 more than the number and 20.

$$\frac{4x + 9}{10} = \frac{x + 4}{20} \quad \middle| \quad 70x = -140$$

$80x + 180 = 10x + 40$ **x = −2**

One eats area, not radius. **Proportions 6**	
Answer as indicated assuming a constant rate. Radii and diameters are in units.	*Find the positive geometric mean.*

1. If a cookie with diameter 12 serves 4 people, a cookie with diameter 18 serves how many?

$a = 36\pi$
$A = 81\pi$
$\dfrac{4}{36\pi} = \dfrac{x}{81\pi}$ **9**

9. between 4 and 25

$\dfrac{4}{x} = \dfrac{x}{25}$ $x^2 = 100$ **x = 10**

2. A pizza with radius 15 serves how many people if a pizza with radius 10 serves 4 people?

$A = 225\pi$
$a = 100\pi$
$\dfrac{x}{225\pi} = \dfrac{4}{100\pi}$ **9**

10. between 8 and 50

$\dfrac{8}{x} = \dfrac{x}{50}$ $x^2 = 400$ **x = 20**

3. If a silver disk with diameter 40 makes 9 rings, a disk with radius 40 makes how many rings?

$a = 400\pi$ times 4 $\dfrac{9}{400\pi} = \dfrac{x}{1600\pi}$ **36**
$A = 1600\pi$

11. between 3 and 48

$\dfrac{3}{x} = \dfrac{x}{48}$ $x^2 = 144$ **x = 12**

4. A cake with circumference 48π serves 32 people. A cake with circumference 36π serves how many people?

$A = 576\pi$
$a = 324\pi$
$\dfrac{32}{576\pi} = \dfrac{x}{324\pi}$ **18**

12. between 18 and 50

$\dfrac{18}{x} = \dfrac{x}{50}$ $x^2 = 900$ **x = 30**

5. A pie with radius 8 serves how many people if a pie with radius 16 serves 20 people?

$a = 64\pi$
$A = 256\pi$
$\dfrac{x}{64\pi} = \dfrac{20}{256\pi}$ **5**

13. between 5 and 45

$\dfrac{5}{x} = \dfrac{x}{45}$ $x^2 = 225$ **x = 15**

6. A pizza with diameter 20 serves how many people if a pizza with radius 2 serves 5 people?

$A = 100\pi$ times 25 $\dfrac{5}{4\pi} = \dfrac{x}{100\pi}$ **125**
$a = 4\pi$

14. between 11 and 44

$\dfrac{11}{x} = \dfrac{x}{44}$ $x^2 = 484$ **x = 22**

7. If a cookie with diameter 6 serves 3 people, a cookie with diameter 24 serves how many?

$A = 144\pi$ times 16 $\dfrac{3}{9\pi} = \dfrac{x}{144\pi}$ **48**
$a = 9\pi$

15. between 6 and 54

$\dfrac{6}{x} = \dfrac{x}{54}$ $x^2 = 324$ **x = 18**

8. A cake with circumference 10π serves 7 people. A cake with circumference 30π serves how many people?

$a = 25\pi$ times 9 $\dfrac{7}{25\pi} = \dfrac{x}{225\pi}$ **63**
$A = 225\pi$

16. between 16 and 36

$\dfrac{16}{x} = \dfrac{x}{36}$ $x^2 = 576$ **x = 24**

MAVA Math: Enhanced Skills Solutions Copyright © 2015 Marla Weiss

Proportions 7

Answer as indicated, assuming inverse proportionality.	Answer as indicated, assuming direct proportionality.
1. If 24 people can do a job in 5 days, then how many people are needed to do the job in 1 day? $(24)(5) = (P)(1)$ **P = 120**	7. If the sales tax on an item costing x dollars is y dollars, find the tax on an item costing z dollars. $\dfrac{x}{y} = \dfrac{z}{T}$ $xT = yz$ $T = \dfrac{yz}{x}$
2. If the intensity of a light measures 25 at a distance of 3 units from the light, then find the intensity of the light at a distance of 15 units. $(25)(3) = (L)(15)$ **L = 5**	8. How many words can be keyed in s seconds at a rate of w words in m minutes? $\dfrac{w}{60m} = \dfrac{R}{s}$ $60mR = ws$ $R = \dfrac{ws}{60m}$
3. If 60 bees need 6 hours to produce a quantity of honey, how many bees are needed to produce the same amount of honey in 5 hours? $(60)(6) = (B)(5)$ **B = 72**	9. The weight of f feet of rope is p pounds. What is the weight in pounds of y yards of the same rope? $\dfrac{f}{p} = \dfrac{3y}{W}$ $fW = 3yp$ $W = \dfrac{3yp}{f}$
4. If 6 cats can catch 10 mice in 15 days, how many days would be needed for 10 cats to catch the same number of mice? $(6)(15) = (10)(D)$ **D = 9**	10. A recipe requires c cups of sugar for k cookies. How many cups of sugar are needed for d dozen cookies? $\dfrac{c}{k} = \dfrac{U}{12d}$ $kU = 12cd$ $U = \dfrac{12cd}{k}$
5. If manufacturing 100 items costs $520, then find the cost of manufacturing 325 of the same items. $(100)(520) = (325)(C)$ **C = 160**	11. If c pounds of chips serve p people, how many ounces of chips are needed to serve q people? $\dfrac{16c}{p} = \dfrac{Z}{q}$ $pZ = 16cq$ $Z = \dfrac{16cq}{p}$
6. If rectangular areas of land have a fixed price per square yard, then an area 32 by 45 yards is equivalent to an area 72 by how many yards? $(32)(45) = (L)(72)$ **L = 20**	12. Picking b bushels of peaches in m minutes, how many bushels can be picked in h hours ? $\dfrac{b}{m} = \dfrac{B}{60h}$ $mB = 60bh$ $B = \dfrac{60bh}{m}$

Proportions 8

Answer as indicated.

1. X varies directly as Y. X = 60 when Y = 5. Find X when Y = 11.

X = kY	X = (12)(11)
60 = 5k	**X = 132**
k = 12	

8. P varies inversely as Q. P = 75 when Q = 20. Find Q when P = 60.

PQ = k	60Q = 1500
(75)(20) = k	**Q = 25**
k = 1500	

2. A varies jointly as B and C. A = 300 when B = 5 and C = 3. Find A when B = 10 and C = 12.

A = kBC	A = (20)(10)(12)
300 = (5)(3)k	**A = 2400**
k = 20	

9. X varies directly as Y and inversely as Z. X = 231 when Y = 22 and Z = 4. Find Z when X = 35 and Y = 10.

XZ = kY	k = 42
(231)(4) = k(22)	35Z = (42)(10)
(21)(11)(2)(2) = k(22)	**Z = 12**

3. M varies inversely as N. M = 40 when N = 12. Find N when M = 24.

MN = k	24N = 480
(40)(12) = k	**N = 20**
k = 480	

10. M varies directly as the square root of N. M = 220 when N = 400. Find N when M = 165.

$M = k\sqrt{N}$	$165 = 11\sqrt{N}$
220 = 20k	$15 = \sqrt{N}$
k = 11	**N = 225**

4. M varies inversely as the cube of N. M = 8 when N = 6. Find N when M = 27.

$MN^3 = k$	$27N^3 = 1728$
(8)(216) = k	$N^3 = 64$
k = 1728	**N = 4**

11. X varies jointly as Y and Z. X = 450 when Y = 10 and Z = 15. Find Z when X = 540 and Y = 9.

X = kYZ	540 = (3)(9)(Z)
450 = k(10)(15)	**Z = 20**
k = 3	

5. R varies directly as the square of P and inversely as Q. R = 4 when P = 5 and Q = 3. Find R when P = 10 and Q = 15.

$RQ = kP^2$	15R = (12/25)(100)
(4)(3) = 25k	15R = 48
k = 12/25	**R = 16/5**

12. A varies jointly as B, C, and D. A = 360 when B = 5, C = 9, and D = 7. Find A when B = 14, C = 3, and D = 2.

A = kBCD	A = (8/7)(14)(6)
360 = (5)(9)(7)k	**A = 96**
k = 8/7	

6. X varies directly as Y. X = 57 when Y = 6. Find X when Y = 22.

X = kY	k = 19/2
57 = 6k	X = (22)(19/2)
(19)(3) = (3)(2k)	**X = 209**

13. T varies directly as the cube of S and inversely as R. T = 63 when R = 4 and S = 3. Find R when S = 3 and T = 2.

	k = 28/3
$TR = kS^3$	2R = 28 x 27/3
63 x 4 = 27k	**R = 14 x 9 = 126**

7. Y varies directly as X. X = 60 when Y = 5. Find X when Y = 11.

Y = kX	11 = X/12
5 = 60k	**X = 132**
k = 1/12	

14. R varies jointly as P and Q and inversely as the square of T. R = 12 when P = 7, Q = 6, and T = 14. Find R when T = 2, P = 3 and Q = 5.

	k = (4)(14)
$RT^2 = kPQ$	4R = (4)(14)(15)
(12)(14)(14) = (7)(6)k	**R = 210**

Pyramids 1

Find the volume of the pyramid in cubic units.

1. square base, edge 12
   altitude 8

   $V = \dfrac{12 \times 12 \times 8}{3} = \textbf{384}$

2. equilateral triangle base, edge 12
   altitude 16

   $V = \dfrac{12 \times 12 \times \sqrt{3} \times 16}{4 \times 3} = \textbf{192}\sqrt{\textbf{3}}$

3. rectangle base, length 12, width 8
   altitude 11

   $V = \dfrac{12 \times 8 \times 11}{3} = \textbf{352}$

4. square base, diagonal 18
   altitude 5

   $V = \dfrac{9\sqrt{2} \times 9\sqrt{2} \times 5}{3} = \textbf{270}$

5. equilateral triangle base, edge 6
   altitude 7

   $V = \dfrac{6 \times 6 \times \sqrt{3} \times 7}{4 \times 3} = \textbf{21}\sqrt{\textbf{3}}$

6. rectangle base, length 15, width 14
   altitude 24

   $V = \dfrac{15 \times 14 \times 24}{3} = \textbf{1680}$

7. regular hexagon base, edge 6
   altitude 20

   $V = \dfrac{6 \times 6 \times \sqrt{3} \times 6 \times 20}{4 \times 3} = \textbf{360}\sqrt{\textbf{3}}$

8. pentagon base, area 48
   altitude 11

   $V = \dfrac{48 \times 11}{3} = \textbf{176}$

9. trapezoid base, sides 5, 5, 5, and 11
   altitude 15   M = 8, h = 4

   $V = \dfrac{8 \times 4 \times 15}{3} = \textbf{160}$

10. regular hexagon base, edge 10
    altitude 6

    $V = \dfrac{10 \times 10 \times \sqrt{3} \times 6 \times 6}{4 \times 3} = \textbf{300}\sqrt{\textbf{3}}$

11. square base, edge 15
    altitude 11

    $V = \dfrac{15 \times 15 \times 11}{3} = \textbf{825}$

12. triangle base, edges 5, 5, and 6
    altitude 13

    $V = \dfrac{3 \times 4 \times 13}{3} = \textbf{52}$

13. octagon base, area 51
    altitude 10

    $V = \dfrac{51 \times 10}{3} = \textbf{170}$

14. greatest pyramid that fits in a box
    with base 5 by 5 and height 12

    $V = \dfrac{5 \times 5 \times 12}{3} = \textbf{100}$

15. triangle base, edges 25, 25, and 14
    altitude 9

    $V = \dfrac{7 \times 24 \times 9}{3} = \textbf{504}$

16. greatest pyramid that fits in a box
    with base 6 by 7 and height 11

    $V = \dfrac{6 \times 7 \times 11}{3} = \textbf{154}$

Pyramids 2

Rectanglular pyramids have 2 of each triangular face.
Find the total surface area of the pyramid in square units.

---

1. square base, edge 16
   lateral edge 10

   A base = 16x16 = 256
   A face = 16x6/2 = 48
   A 4 faces = 48x4 = 192   **448**

9. square base, edge 6
   altitude 4

   A base = 6x6 = 36
   A face = 6x5/2 = 15
   A 4 faces = 15x4 = 60   **96**

---

2. rectangle base, length 14, width 40
   lateral edge 25   **1496**

   A base = 14x40 = 560
   A face = 40x15/2 = 300
   A face = 14x24/2 = 168

10. rectangle base, length 10, width 18
    altitude 12   **564**

    A base = 10x18 = 180
    A face = 10x15/2 = 75
    A face = 18x13/2 = 117

---

3. equilateral triangle base, edge 16
   lateral edge 17
              360 + 64√3

   A base = 16 × 16 × √3 / 4
   A face = 16x15/2 = 120

11. square base, edge 24
    altitude 35

    A base = 24x24 = 576
    A face = 24x37/2 = 444
    A 4 faces = 444x4 = 1776   **2352**

---

4. square base, edge 24
   lateral edge 13

   A base = 24x24 = 576
   A face = 24x5/2 = 60
   A 4 faces = 60x4 = 240   **816**

12. square base, edge 10
    lateral edge 13

    A base = 10x10 = 100
    A face = 10x12/2 = 60
    A 4 faces = 60x4 = 240   **340**

---

5. rectangle base, length 80, width 26
   lateral edge 85   **10,264**

   A base = 80x26 = 2080
   A face = 80x75/2 = 3000
   A face = 26x84/2 = 1092

13. rectangle base, length 60, width 18
    altitude 40   **4440**

    A base = 60x18 = 1080
    A face = 50x18/2 = 450
    A face = 60x41/2 = 1230

---

6. equilateral triangle base, edge 40
   lateral edge 29
              1260 + 400√3

   A base = 40 × 40 × √3 / 4
   A face = 40x21/2 = 420

14. rectangle base, length 14, width 36
    altitude 24   **1824**

    A base = 14x36 = 504
    A face = 14x30/2 = 210
    A face = 36x25/2 = 450

---

7. square base, edge 14
   lateral edge 25

   A base = 14x14 = 196
   A face = 14x24/2 = 168
   A 4 faces = 168x4 = 672   **868**

15. square base, edge 32
    altitude 30

    A base = 32x32 = 1024
    A face = 32x34/2 = 544
    A 4 faces = 544x4 = 2176   **3200**

---

8. rectangular base, length 14, width 30
   lateral edge 25   **1356**

   A base = 14x30 = 420
   A face = 14x24/2 = 168
   A face = 30x20/2 = 300

16. rectangle base, length 22, width 50
    altitude 60   **5580**

    A base = 22x50 = 1100
    A face = 22x65/2 = 715
    A face = 50x61/2 = 1525

Pyramids 3

Find the altitude of the pyramid in units.

Find the volume of the pyramid in cubic units.

1. square base, edge 10
   volume 1000

   $1000 = \dfrac{10 \times 10 \times a}{3}$ **a = 30**

9. square base
   altitude 24, slant height 26

   $V = \dfrac{400 \times 24}{3} = \textbf{3200}$ s = 20
   A sq = 400

2. rectangle base, length 12, width 8
   volume 352

   $352 = \dfrac{12 \times 8 \times a}{3}$ **a = 11**

10. square base
    altitude 12, slant height 15

    $V = \dfrac{324 \times 12}{3} = \textbf{1296}$ s = 18
    A sq = 324

3. rhombus base with diagonals 9 and 10
   volume 225

   A rhom = 9x10/2 = 45 $225 = \dfrac{45 \times a}{3}$ **a = 15**

11. square base, area 40
    slant height 8

    $V = \dfrac{40 \times 3\sqrt{6}}{3} = \textbf{40}\sqrt{\textbf{6}}$

4. trapezoid base, edges 13, 13, 13, 23
   volume 504

   M trap = 18
   h trap = 12 $504 = \dfrac{18 \times 12 \times a}{3}$ **a = 7**

12. square base
    altitude 13, slant height 22

    $V = \dfrac{1260 \times 13}{3} = \textbf{5460}$ s = 6$\sqrt{35}$
    A sq = 1260

5. trapezoid base, edges 10, 10, 10, 26
   volume 324

   M trap = 18
   h trap = 6 $324 = \dfrac{18 \times 6 \times a}{3}$ **a = 9**

13. square base, area 72
    slant height 9

    $V = \dfrac{72 \times 3\sqrt{7}}{3} = \textbf{72}\sqrt{\textbf{7}}$

6. rhombus base with diagonals 8 and 9
   volume 156

   A rhom = 8x9/2 = 36 $156 = \dfrac{36 \times a}{3}$ **a = 13**

14. square base
    altitude 30, slant height 34

    $V = \dfrac{1024 \times 30}{3} = \textbf{10,240}$ s = 32
    A sq = 1024

7. square base, edge 12
   volume 624

   $624 = \dfrac{12 \times 12 \times a}{3}$ **a = 13**

15. square base
    altitude 24, slant height 25

    $V = \dfrac{196 \times 24}{3} = \textbf{1568}$ s = 14
    A sq = 196

8. rectangular base, length 9, width 15
   volume 765

   $765 = \dfrac{9 \times 15 \times a}{3}$ **a = 17**

16. square base
    altitude 60, slant height 61

    $V = \dfrac{484 \times 60}{3} = \textbf{9680}$ s = 22
    A sq = 484

Pyramids 4

Find the volume in cubic units of the pyramid frustrum.

1. square base with edge 3, upper edge 1, frustrum height 6

$$\frac{h}{1} = \frac{h+6}{3}$$

3h = h+6   V = 3x3x9/3 = 27
2h = 6   v = 1x1x3/3 = 1
h = 3   27 − 1 = **26**

2. triangular base with area 64, pyramid altitude 24, cut 9 up from base

$$\frac{15 \times 15}{24 \times 24} = \frac{a}{64} \quad a = 25$$

V = 64x24/3 = 512
v = 25x15/3 = 125
512 − 125 = **387**

3. square base with edge 30, pyramid altitude 25, cut 15 down from vertex

$$\frac{25}{30} = \frac{15}{b} \quad \begin{array}{l} 5b = 90 \\ b = 18 \end{array}$$

V = 30x30x25/3 = 7500
v = 18x18x15/3 = 1620
7500 − 1620 = **5880**

4. triangular base with area 49, pyramid altitude 21, cut 6 up from base

$$\frac{15 \times 15}{21 \times 21} = \frac{a}{49} \quad a = 25$$

V = 49x21/3 = 343
v = 25x15/3 = 125
343 − 125 = **218**

5. triangular base with area 100, pyramid altitude 30, frustrum altitude 9

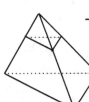

$$\frac{21 \times 21}{30 \times 30} = \frac{a}{100} \quad a = 49$$

V = 100x30/3 = 1000
v = 49x21/3 = 343
1000 − 343 = **657**

6. square base with edge 24, pyramid altitude 10, cut halfway up

$$\frac{5}{b} = \frac{10}{24} \quad b = 12$$

V = 24x24x10/3 = 1920
v = 12x12x5/3 = 240
1920 − 240 = **1680**

7. square base with edge 15, upper edge 3, frustrum height 8

$$\frac{h}{3} = \frac{h+8}{15} \quad \begin{array}{l} 5h = h+8 \\ 4h = 8 \\ h = 2 \end{array}$$

V = 15x15x10/3 = 750
v = 3x3x2/3 = 6
750 − 6 = **744**

8. square base with edge 8, upper edge 6, frustrum height 3

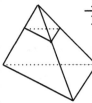

$$\frac{h}{6} = \frac{h+3}{8} \quad \begin{array}{l} 8h = 6h+18 \\ 2h = 18 \\ h = 9 \end{array}$$

V = 8x8x12/3 = 256
v = 6x6x9/3 = 108
256 − 108 = **148**

9. triangular base with area 27, pyramid altitude 15, cut 5 up from base

$$\frac{10 \times 10}{15 \times 15} = \frac{a}{27} \quad \begin{array}{l} 9a = 4 \times 27 \\ a = 12 \end{array}$$

V = 27x15/3 = 135
v = 12x10/3 = 40
135 − 40 = **95**

10. square base with edge 12, pyramid altitude 20, cut 5 up from base

$$\frac{15}{b} = \frac{20}{12} \quad b = 9$$

V = 12x12x20/3 = 960
v = 9x9x15/3 = 405
960 − 405 = **555**

Pythagorean Theorem 1

Determine whether three numbers form the sides of a right triangle (YES/NO). Give a reason.

1.	9, 40, 41	primitive	14.	24, 70, 72	by 1s digit 6 + 0 ≠ 4
	YES			**NO**	
2.	9, 12, 16	by 1s digit 1 + 4 ≠ 6	15.	20, 21, 28	by 1s digit 0 + 1 ≠ 4
	NO			**NO**	
3.	24, 32, 45	by 1s digit 6 + 4 ≠ 5	16.	1, 2.4, 2.6	×10 10, 24, 26
	NO			**YES**	÷ 2 5, 12, 13 primitive
4.	4, 5, 6	by 1s digit 6 + 5 ≠ 6	17.	10, 10.5, 14.5	×2
	NO			**YES**	20, 21, 29 primitive
5.	2, 3, $\sqrt{13}$	4 + 9 = 13	18.	4, 5, $\sqrt{43}$	16 + 25 ≠ 43
	YES			**NO**	
6.	14, 48, 50	7, 24, 25 primitive ×2	19.	6, 7, $\sqrt{85}$	36 + 49 = 85
	YES			**YES**	
7.	16, 30, 34	8, 15, 17 primitive ×2	20.	2.5, 6, 6.5	× 2
	YES			**YES**	5, 12, 13 primitive
8.	4, 4, $4\sqrt{2}$	45, 45, 90 triangle ×4	21.	5, 9, $\sqrt{106}$	25 + 81 = 106
	YES			**YES**	
9.	11, 50, 51	121 + 2500 ≠ 2601	22.	24, 32, 40	3, 4, 5 primitive ×8
	NO			**YES**	
10.	7, 7, $7\sqrt{2}$	45, 45, 90 triangle ×7	23.	3, 7, $\sqrt{59}$	9 + 49 ≠ 59
	YES			**NO**	
11.	40, 42, 48	÷ 2 20, 21, 24	24.	2, 5, $\sqrt{29}$	4 + 25 = 29
	NO	by 1s digit 0 + 1 ≠ 6		**YES**	
12.	1, 3, $\sqrt{10}$	1 + 9 = 10	25.	5, 10, $5\sqrt{5}$	÷ 5 1, 2, $\sqrt{5}$
	YES			**YES**	1 + 4 = 5
13.	1, 1, $\sqrt{2}$	1 + 1 = 2	26.	$\sqrt{5}$, 3, 4	5 + 9 ≠ 16
	YES			**NO**	

Pythagorean Theorem 2

Find the missing side (in ascending order) of the Pythagorean triplet by mental math.

1. 3, 4, __5__	17. 14, 48, __50__	33. __26__ , 168, 170
2. 7, 24 __25__	18. 2, 2, __$2\sqrt{2}$__	34. 33, 180, __183__
3. 8, __15__ , 17	19. 40, 42, __58__	35. 90, __400__ , 410
4. 5, 12, __13__	20. 5, __$5\sqrt{3}$__ , 10	36. 14, 14, __$14\sqrt{2}$__
5. 5, 5, __$5\sqrt{2}$__	21. 1, __$\sqrt{3}$__ , 2	37. 12, __16__ , 20
6. 9, 40, __41__	22. 22, __120__ , 122	38. 32, 60, __68__
7. 11, 60, __61__	23. __15__ , 36, 39	39. 77, 264, __275__
8. 12, __35__ , 37	24. 80, 150, __170__	40. 18, __24__ , 30
9. 14, __48__ , 50	25. 33, __44__ , 55	41. 24, __70__ , 74
10. 10, __$10\sqrt{3}$__ , 20	26. 16, __$16\sqrt{3}$__ , 32	42. 9, __$9\sqrt{3}$__ , 18
11. 18, 80, __82__	27. 3, 3, __$3\sqrt{2}$__	43. 25, __60__ , 65
12. 10, __24__ , 26	28. 13, 84, __85__	44. 132, 385, __407__
13. __19__ , 180, 181	29. __9__ , 12, 15	45. 24, __45__ , 51
14. 20, __21__ , 29	30. 44, 240, __244__	46. 30, 40, __50__
15. 6, 8, __10__	31. 16, 30, __34__	47. 3, __$3\sqrt{3}$__ , 6
16. 33, 33, __$33\sqrt{2}$__	32. 21, 28, __35__	48. 36, 105, __111__

Pythagorean Theorem 3

Draw a picture and calculate.

1. A tree broke 36 feet above the ground. The attached top rests on the ground 27 feet from the bottom of the tree. How many feet tall was the tree?

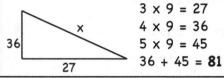

$3 \times 9 = 27$
$4 \times 9 = 36$
$5 \times 9 = 45$
$36 + 45 = $ **81**

6. A clock has a minute hand 16 cm long and an hour hand 12 cm long. How many cm apart are the ends of the hands at 3:00?

3-4-5 right △
x 4 blow-up

2. 30 steps east, 10 steps north, 10 steps east, and finally 20 steps north is how many direct steps from the starting point?

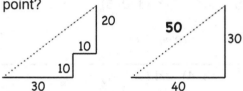

7. A car went 12 miles west, 7 miles south, and then 12 miles west. How many miles was the car from its starting point?

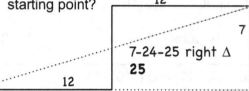

7-24-25 right △
25

3. Jo and Bo left at the same place and time. Jo walked south at 3 mph, while Bo ran east at 4 mph. How far apart were they 2 hours later?

$D = 4 \times 2 = 8$
$D = 3 \times 2 = 6$
10 miles

8. Jan and Hal left work at 6:00 PM. Jan drove north at 35 mph, while Hal sped west at 84 mph. How far apart were they at 6:30 PM?

$13 \times 7/2 = $ **45.5**
$D = 35 \times .5 = 17.5$
$35 = 5 \times 7$
$84 = 12 \times 7$
$D = 84 \times .5 = 42$
5-12-13 right △

4. If a person walks 5 miles east, 5 miles south, and then 7 miles east, how many miles would he have saved by traveling in a straight line?

$5 + 5 + 7 = 17$
5-12-13 right △
$17 - 13 = $ **4**

9. 30 steps east, 10 steps north, 6 steps west, and 3 steps south is how many steps from the starting point?

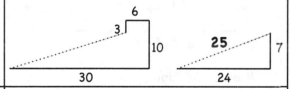

5. The front of a boat is tied by a 20-foot rope to a point on a pier that is 12 feet above the height of the boat. If 5 feet of the rope are pulled in, how many feet forward will the boat move?

$16 - 9 = $ **7**

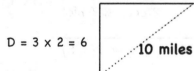

10. Two boats leave from the same dock at 11:00AM. One goes north at 60mph, while the other travels east at 80mph. At what time will they be 150 miles part?

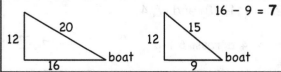

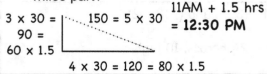

$3 \times 30 = $
$90 = $
60×1.5

$150 = 5 \times 30$

11AM + 1.5 hrs
= **12:30 PM**

$4 \times 30 = 120 = 80 \times 1.5$

Pythagorean Theorem 4

Find the distance between the pair of points.

1.	(6, –6) and (–2, 9)		12.	(7, 0) and (0, 7)
	8 across, 15 high	**17**		7 across, 7 high	$7\sqrt{2}$
2.	(5, –8) and (–5, 16)		13.	(5, 1) and (–4, –1)
	10 across, 24 high	**26**		9 across, 2 high	$\sqrt{85}$
3.	(–8, –21) and (2, 3)		14.	(2, 5) and (5, –1)
	10 across, 24 high	**26**		3 across, 6 high	$3\sqrt{5}$
4.	(–8, –2) and (16, 5)		15.	(–4, –2) and (0, 5)
	24 across, 7 high	**25**		4 across, 7 high	$\sqrt{65}$
5.	(9, –11) and (–15, –4)		16.	(3, 4) and (–1, –1)
	24 across, 7 high	**25**		4 across, 5 high	$\sqrt{41}$
6.	(10, 3) and (–14, –4)		17.	(7, 2) and (5, 9)
	24 across, 7 high	**25**		2 across, 7 high	$\sqrt{53}$
7.	(3, –8) and (–2, 4)		18.	(1, 2) and (6, 5)
	5 across, 12 high	**13**		5 across, 3 high	$\sqrt{34}$
8.	(–10, –11) and (6, 19)		19.	(5, 1) and (4, –8)
	16 across, 30 high	**34**		1 across, 9 high	$\sqrt{82}$
9.	(–1, –11) and (7, –5)		20.	(–2, 1) and (–1, 2)
	8 across, 6 high	**10**		1 across, 1 high	$\sqrt{2}$
10.	(4, –8) and (–4, 7)		21.	(3, 10) and (–1, 6)
	8 across, 15 high	**17**		4 across, 4 high	$4\sqrt{2}$
11.	(6, 8) and (–18, –2)		22.	(11, 10) and (7, 4)
	24 across, 10 high	**26**		4 across, 6 high	$2\sqrt{13}$

Pythagorean Theorem 5

Given measures for 3 sides, indicate whether the triangle is acute or obtuse.

1.	6, 12, 13 **acute** $36 + 144 = 180$ $169 < 180$	**12.**	5, 12, 14 **obtuse** $25 + 144 = 169$ $196 > 169$	**23.**	2, 4, 5 **obtuse** $4 + 16 = 20$ $25 > 20$
2.	4, 5, 7 **obtuse** $16 + 25 = 41$ $49 > 41$	**13.**	4, 5, 6 **acute** $16 + 25 = 41$ $36 < 41$	**24.**	3, 6, 8 **obtuse** $9 + 36 = 45$ $64 > 45$
3.	8, 11, 13 **acute** $64 + 121 = 185$ $169 < 185$	**14.**	2, 3, 4 **obuse** $4 + 9 = 13$ $16 > 13$	**25.**	4, 9, 10 **acute** $36 + 81 = 117$ $100 < 117$
4.	2, 4, 5 **obtuse** $4 + 16 = 20$ $25 > 20$	**15.**	5, 7, 11 **obtuse** $25 + 49 = 74$ $121 > 74$	**26.**	10, 15, 18 **acute** $100 + 225 = 325$ $324 < 325$
5.	10, 11, 15 **obtuse** $100 + 121 = 221$ $225 > 221$	**16.**	7, 8, 9 **acute** $49 + 64 = 113$ $81 < 113$	**27.**	5, 7, 9 **obtuse** $25 + 49 = 74$ $81 > 74$
6.	5, 10, 12 **obtuse** $25 + 100 = 125$ $144 > 125$	**17.**	6, 10, 15 **obtuse** $36 + 100 = 136$ $225 > 136$	**28.**	4, 11, 13 **obtuse** $16 + 121 = 137$ $169 > 137$
7.	7, 9, 12 **obtuse** $49 + 81 = 130$ $144 > 130$	**18.**	8, 10, 11 **acute** $64 + 100 = 164$ $121 < 164$	**29.**	2, 6, 7 **obtuse** $4 + 36 = 40$ $49 > 40$
8.	5, 6, 8 **obtuse** $25 + 36 = 61$ $64 > 61$	**19.**	10, 11, 12 **acute** $100 + 121 = 221$ $144 < 221$	**30.**	12, 15, 19 **acute** $144 + 225 = 369$ $361 < 369$
9.	5, 8, 11 **obtuse** $25 + 64 = 89$ $121 > 89$	**20.**	5, 6, 7 **acute** $25 + 36 = 61$ $49 < 61$	**31.**	9, 10, 15 **obtuse** $81 + 100 = 181$ $225 > 181$
10.	6, 7, 8 **acute** $36 + 49 = 85$ $64 < 85$	**21.**	3, 9, 10 **obtuse** $9 + 81 = 90$ $100 > 90$	**32.**	8, 9, 10 **acute** $64 + 81 = 145$ $100 < 145$
11.	10, 13, 15 **acute** $100 + 169 = 269$ $225 < 269$	**22.**	10, 20, 25 **obtuse** $100 + 400 = 500$ $625 > 500$	**33.**	6, 8, 11 **obtuse** $36 + 64 = 100$ $121 > 100$

Pythagorean Theorem 6

Answer as indicated.

---

1. Find the sum of the perimeter and area of the square if the legs of the triangle are 6 and 8.

P = 4 x 10 = **40**
A = 10 x 10 = **100**

---

6. Find AB.

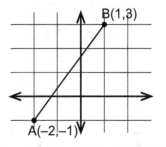

AB = **5**

---

2. Find AB if the 3 congruent squares each have area 9.

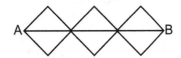

side = 3
AB = 3(3√2) = **9√2**

---

7. Find FH if EG = 42, EF = 26, and FG = 40.

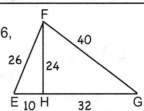

FH = **24**

Split 42 into 2 right triangle compatible numbers.

---

3. Find AC if BC = 18, CD = 14, and AD = 40.

3, 4, 5, x8 = 24, 32, 40
3, 4, 5, x6 = 18, 24, 30
AC = **30**

---

8. The areas of the 3 squares are 4, 9, and 16. Find the length of the diagonal line shown.

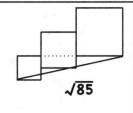

√85

sides = 2, 3, and 4
diagonal is hypotenuse of 2 by 9 Δ

---

4. Find AC if AB = 10, BD = 21, and AD = 17.

AC = **8**

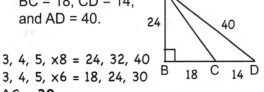

---

9. Find the perimeter of the parallelogram.

14 + 12√2

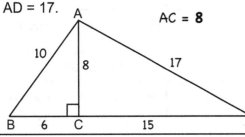

---

5. Find the sum of all of the diagonals of the three squares if the largest square has area 16.

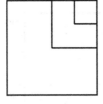

2(4√2 + 2√2 + √2)   **14√2**

10.

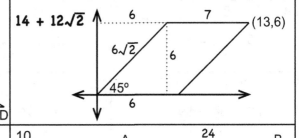

ABCD is a 16 by 24 rectangle. F is a midpoint. BE = 7. Find the perimeter of ΔAEF.

P = 15 + 20 + 25 = **60**

Pythagorean Theorem 7

Find the area of the rectangle in square units.	*Find the perimeter of the isosceles right triangle in units.*
1. width 2, diagonal 5 $L^2 + 4 = 25$ $L = \sqrt{21}$ $A = 2\sqrt{21}$	10. leg 6 $P = 12 + 6\sqrt{2}$
2. length 10, diagonal 14 Use lowercase w for the "reduced" width. Use uppercase W for the actual width. $w^2 + 25 = 49$ $W = 2\sqrt{24} = 4\sqrt{6}$ $A = 40\sqrt{6}$	11. hypotenuse $\sqrt{6}$ $L = \sqrt{3}$ $P = 2\sqrt{3} + \sqrt{6}$
3. width 6, diagonal 25 $L^2 + 36 = 625$ $L = \sqrt{589}$ $A = 6\sqrt{589}$	12. hypotenuse 6 Divide by 2. Tack on root 2. $L = 3\sqrt{2}$ $P = 6 + 6\sqrt{2}$
4. length 18, diagonal 20 $w^2 + 81 = 100$ $W = 2\sqrt{19}$ $A = 36\sqrt{19}$	13. leg $\sqrt{6}$ $H = (\sqrt{6})(\sqrt{2}) = 2\sqrt{3}$ $P = 2\sqrt{6} + 2\sqrt{3}$
5. length 14, diagonal 24 $w^2 + 49 = 144$ $W = 2\sqrt{95}$ $A = 28\sqrt{95}$	14. hypotenuse 10 $L = 5\sqrt{2}$ $P = 10 + 10\sqrt{2}$
6. length 30, diagonal 33 $w^2 + 100 = 121$ $W = 3\sqrt{21}$ $A = 90\sqrt{21}$	15. leg 10 $P = 10 + 10 + 10\sqrt{2} = 20 + 10\sqrt{2}$
7. length 15, diagonal 35 $w^2 + 9 = 49$ $W = 5\sqrt{40} = 10\sqrt{10}$ $A = 150\sqrt{10}$	16. leg $\sqrt{10}$ $H = (\sqrt{10})(\sqrt{2}) = 2\sqrt{5}$ $P = 2\sqrt{10} + 2\sqrt{5}$
8. length 16, diagonal 44 $w^2 + 16 = 121$ $W = 4\sqrt{105}$ $A = 64\sqrt{105}$	17. hypotenuse $\sqrt{10}$ $L = \sqrt{5}$ $P = 2\sqrt{5} + \sqrt{10}$
9. width 8, diagonal 9 $L^2 + 64 = 81$ $L = \sqrt{17}$ $A = 8\sqrt{17}$	18. hypotenuse $\sqrt{2}$ $L = 1$ $P = 2 + \sqrt{2}$

Factors of the sides are bold.	

Pythagorean Theorem 8

Find the number of right triangles with whole number sides that have one side as given.

1. 5	**2**	**8.** 16	**3**	**15.** 40	**7**
3, 4, **5** **5**, 12, 13		3, 4, 5 **8**, 15, 17 **16**, 63, 65		3, **4**, 5 **5**, 12, 13 **8**, 15, 17 9, **40**, 41	20, 21, 29 20, 99, 101
2. 8	**2**	**9.** 20	**5**	**16.** 48	**9**
3, 4, 5 **8**, 15, 17		3, 4, **5** **5**, 12, 13 **20**, 21, 29 **20**, 99, 101		3, 4, 5 5, **12**, 13 7, **24**, 25 8, 15, 17	**12**, 35, 37 **16**, 63, 65 **24**, 143, 145 **48**, 55, 73
3. 9	**2**	**10.** 24	**7**	**17.** 60	**10**
3, 4, 5 **9**, 40, 41		3, 4, 5 5, **12**, 13 7, **24**, 25 **8**, 15, 17	**12**, 35, 37 **24**, 143, 145	3, 4, 5 5, **12**, 13 8, **15**, 17 11, **60**, 61	20, 21, 29 20, 99, 101 **60**, 63, 87
4. 10	**2**	**11.** 25	**3**	**18.** 100	**6**
3, 4, **5** **5**, 12, 13		3, 4, **5** **5**, 12, 13 7, 24, **25**		3, 4, **5** **5**, 12, 13 7, 24, **25** 20, 21, 29	20, 99, 101
5. 12	**4**	**12.** 30	**4**	**19.** 144	**11**
3, 4, 5 5, **12**, 13 **12**, 35, 37		3, 4, 5 **5**, 12, 13 8, **15**, 17		3, **4**, 5 5, **12**, 13 7, **24**, 25 8, 15, 17 9, 40, 41	**12**, 35, 37 **16**, 63, 65 17, **144**, 145 **24**, 143, 145 65, **72**, 97
6. 14	**1**	**13.** 35	**5**	**20.** 160	**7**
7, 24, 25		3, 4, **5** **5**, 12, 13 **7**, 24, 25 12, **35**, 37	35, 612, 613	3, 4, **5** 5, 12, 13 8, 15, 17 9, **40**, 41	20, 21, 29 20, 99, 101
7. 15	**4**	**14.** 36	**6**	**21.** 200	**8**
3, 4, 5 5, 12, 13 8, **15**, 17		**3**, 4, 5 5, **12**, 13 **9**, 40, 41 **12**, 35, 37	36, 77, 85	3, 4, **5** **5**, 12, 13 7, 24, **25** **8**, 15, 17	9, **40**, 41 20, 21, 29 **20**, 99, 101

Radicals 1

Simplify the radical.

1.	$\sqrt{1}$	**1**	17.	$\sqrt{(47)^2}$	**47**	33.	$\sqrt{\dfrac{400}{361}}$ **$\dfrac{20}{19}$**
2.	$\sqrt{49}$	**7**	18.	$\sqrt{64}$	**8**	34.	$\sqrt{121,000,000}$ **11,000**
3.	$\sqrt{256}$	**16**	19.	$\sqrt{900}$	**30**	35.	$\sqrt{324}$ **18**
4.	$\sqrt{4900}$	**70**	20.	$\sqrt{(91)^2}$	**91**	36.	$\sqrt{961}$ **31**
5.	$\sqrt{(39)^2}$	**39**	21.	$\sqrt{\dfrac{16}{49}}$	**$\dfrac{4}{7}$**	37.	$\sqrt{(111)^2}$ **111**
6.	$\sqrt{225}$	**15**	22.	$\sqrt{441}$	**21**	38.	$\sqrt{\dfrac{4900}{22,500}}$ **$\dfrac{7}{15}$**
7.	$\sqrt{\dfrac{81}{25}}$	**$\dfrac{9}{5}$**	23.	$\sqrt{486^2}$	**486**	39.	$\sqrt{8100}$ **90**
8.	$\sqrt{6400}$	**80**	24.	$\sqrt{1600}$	**40**	40.	$\sqrt{151^2}$ **151**
9.	$\sqrt{360,000}$	**600**	25.	$\sqrt{28,900}$	**170**	41.	$\sqrt{1,440,000}$ **1200**
10.	$\sqrt{121}$	**11**	26.	$\sqrt{625}$	**25**	42.	$\sqrt{\dfrac{324}{289}}$ **$\dfrac{18}{17}$**
11.	$\sqrt{10,000}$	**100**	27.	$\sqrt{32,400}$	**180**	43.	$\sqrt{160,000}$ **400**
12.	$\sqrt{12,100}$	**110**	28.	$\sqrt{617^2}$	**617**	44.	$\sqrt{1,000,000}$ **1000**
13.	$\sqrt{\dfrac{1}{9}}$	**$\dfrac{1}{3}$**	29.	$\sqrt{\dfrac{121}{625}}$	**$\dfrac{11}{25}$**	45.	$\sqrt{\dfrac{169}{289}}$ **$\dfrac{13}{17}$**
14.	$\sqrt{\dfrac{90}{40}}$	**$\dfrac{3}{2}$**	30.	$\sqrt{361}$	**19**	46.	$\sqrt{2,560,000}$ **1600**
15.	$\sqrt{144}$	**12**	31.	$\sqrt{\dfrac{49}{64}}$	**$\dfrac{7}{8}$**	47.	$\sqrt{810,000}$ **900**
16.	$\sqrt{400}$	**20**	32.	$\sqrt{\dfrac{441}{961}}$	**$\dfrac{21}{31}$**	48.	$\sqrt{\dfrac{361}{196}}$ **$\dfrac{19}{14}$**

Radicals 2

Simplify the radical.

1. $\sqrt{18}$ $\sqrt{9}\sqrt{2}$
   $3\sqrt{2}$

2. $\sqrt{27}$ $\sqrt{9}\sqrt{3}$
   $3\sqrt{3}$

3. $\sqrt{40}$ $\sqrt{4}\sqrt{10}$
   $2\sqrt{10}$

4. $\sqrt{48}$ $\sqrt{16}\sqrt{3}$
   $4\sqrt{3}$

5. $\sqrt{50}$ $\sqrt{25}\sqrt{2}$
   $5\sqrt{2}$

6. $\sqrt{54}$ $\sqrt{9}\sqrt{6}$
   $3\sqrt{6}$

7. $\sqrt{72}$ $\sqrt{36}\sqrt{2}$
   $6\sqrt{2}$

8. $\sqrt{75}$ $\sqrt{25}\sqrt{3}$
   $5\sqrt{3}$

9. $\sqrt{80}$ $\sqrt{16}\sqrt{5}$
   $4\sqrt{5}$

10. $\sqrt{90}$ $\sqrt{9}\sqrt{10}$
    $3\sqrt{10}$

11. $\sqrt{98}$ $\sqrt{49}\sqrt{2}$
    $7\sqrt{2}$

12. $\sqrt{108}$ $\sqrt{36}\sqrt{3}$
    $6\sqrt{3}$

13. $\sqrt{125}$ $\sqrt{25}\sqrt{5}$
    $5\sqrt{5}$

14. $\sqrt{128}$ $\sqrt{64}\sqrt{2}$
    $8\sqrt{2}$

15. $\sqrt{147}$ $\sqrt{49}\sqrt{3}$
    $7\sqrt{3}$

16. $\sqrt{150}$ $\sqrt{25}\sqrt{6}$
    $5\sqrt{6}$

17. $\sqrt{162}$ $\sqrt{81}\sqrt{2}$
    $9\sqrt{2}$

18. $\sqrt{175}$ $\sqrt{25}\sqrt{7}$
    $5\sqrt{7}$

19. $\sqrt{180}$ $\sqrt{36}\sqrt{5}$
    $6\sqrt{5}$

20. $\sqrt{192}$ $\sqrt{64}\sqrt{3}$
    $8\sqrt{3}$

21. $\sqrt{242}$ $\sqrt{121}\sqrt{2}$
    $11\sqrt{2}$

22. $\sqrt{243}$ $\sqrt{81}\sqrt{3}$
    $9\sqrt{3}$

23. $\sqrt{245}$ $\sqrt{49}\sqrt{5}$
    $7\sqrt{5}$

24. $\sqrt{275}$ $\sqrt{25}\sqrt{11}$
    $5\sqrt{11}$

25. $\sqrt{320}$ $\sqrt{64}\sqrt{5}$
    $8\sqrt{5}$

26. $\sqrt{360}$ $\sqrt{36}\sqrt{10}$
    $6\sqrt{10}$

27. $\sqrt{363}$ $\sqrt{121}\sqrt{3}$
    $11\sqrt{3}$

28. $\sqrt{432}$ $\sqrt{144}\sqrt{3}$
    $12\sqrt{3}$

29. $\sqrt{448}$ $\sqrt{64}\sqrt{7}$
    $8\sqrt{7}$

30. $\sqrt{567}$ $\sqrt{81}\sqrt{7}$
    $9\sqrt{7}$

31. $\sqrt{675}$ $\sqrt{225}\sqrt{3}$
    $15\sqrt{3}$

32. $\sqrt{800}$ $\sqrt{400}\sqrt{2}$
    $20\sqrt{2}$

33. $\sqrt{847}$ $\sqrt{121}\sqrt{7}$
    $11\sqrt{7}$

Radicals 3

Simplify and multiply.

1. $\sqrt{10} \cdot \sqrt{15} \cdot \sqrt{6}$ $\sqrt{5}\ \sqrt{2}\ \sqrt{5}\ \sqrt{3}\ \sqrt{3}\ \sqrt{2}$ $5 \times 3 \times 2 = \mathbf{30}$	12. $\sqrt{18} \cdot \sqrt{14} \cdot \sqrt{21}$ $3\sqrt{2}\ \sqrt{7}\ \sqrt{2}\ \sqrt{7}\ \sqrt{3}$ $3 \times 7 \times 2 \times \sqrt{3} = \mathbf{42\sqrt{3}}$	23. $5\sqrt{35} \cdot 6\sqrt{15}$ $5\sqrt{5}\ \sqrt{7}\ 6\sqrt{3}\ \sqrt{5}$ $5 \times 6 \times 5 \times \sqrt{21} = \mathbf{150\sqrt{21}}$
2. $\sqrt{26} \cdot \sqrt{39} \cdot \sqrt{6}$ $\sqrt{13}\ \sqrt{2}\ \sqrt{13}\ \sqrt{3}\ \sqrt{3}\ \sqrt{2}$ $13 \times 3 \times 2 = \mathbf{78}$	13. $\sqrt{20} \cdot \sqrt{21} \cdot \sqrt{35}$ $2\sqrt{5}\ \sqrt{7}\ \sqrt{3}\ \sqrt{7}\ \sqrt{5}$ $2 \times 5 \times 7 \times \sqrt{3} = \mathbf{70\sqrt{3}}$	24. $2\sqrt{8} \cdot 4\sqrt{14}$ $2\ 2\sqrt{2}\ 4\sqrt{2}\ \sqrt{7}$ $2 \times 2 \times 2 \times 4\sqrt{7} = \mathbf{32\sqrt{7}}$
3. $\sqrt{35} \cdot \sqrt{21} \cdot \sqrt{15}$ $\sqrt{7}\ \sqrt{5}\ \sqrt{7}\ \sqrt{3}\ \sqrt{5}\ \sqrt{3}$ $7 \times 5 \times 3 = \mathbf{105}$	14. $\sqrt{36} \cdot \sqrt{35} \cdot \sqrt{28}$ $6\sqrt{5}\ \sqrt{7}\ 2\sqrt{7}$ $6 \times 2 \times 7 \times \sqrt{5} = \mathbf{84\sqrt{5}}$	25. $3\sqrt{6} \cdot 2\sqrt{10}$ $3\sqrt{3}\ \sqrt{2}\ 2\sqrt{2}\ \sqrt{5}$ $3 \times 2 \times 2 \times \sqrt{15} = \mathbf{12\sqrt{15}}$
4. $\sqrt{34} \cdot \sqrt{51} \cdot \sqrt{6}$ $\sqrt{17}\ \sqrt{2}\ \sqrt{17}\ \sqrt{3}\ \sqrt{3}\ \sqrt{2}$ $17 \times 3 \times 2 = \mathbf{102}$	15. $\sqrt{12} \cdot \sqrt{33} \cdot \sqrt{55}$ $2\sqrt{3}\ \sqrt{3}\ \sqrt{11}\ \sqrt{5}\ \sqrt{11}$ $2 \times 3 \times 11 \times \sqrt{5} = \mathbf{66\sqrt{5}}$	26. $4\sqrt{24} \cdot 5\sqrt{20}$ $4\ 2\sqrt{6}\ 5\ 2\sqrt{5}$ $4 \times 2 \times 5 \times 2\sqrt{30} = \mathbf{80\sqrt{30}}$
5. $\sqrt{98} \cdot \sqrt{25} \cdot \sqrt{32}$ $7\sqrt{2}\ 5\ 4\sqrt{2}$ $7 \times 2 \times 5 \times 4 = \mathbf{280}$	16. $\sqrt{18} \cdot \sqrt{10} \cdot \sqrt{6}$ $3\sqrt{2}\ \sqrt{5}\ \sqrt{2}\ \sqrt{3}\ \sqrt{2}$ $3 \times 2 \times \sqrt{30} = \mathbf{6\sqrt{30}}$	27. $5\sqrt{27} \cdot 2\sqrt{18}$ $5\ 3\sqrt{3}\ 2\ 3\sqrt{2}$ $5 \times 3 \times 2 \times 3\sqrt{6} = \mathbf{90\sqrt{6}}$
6. $\sqrt{36} \cdot \sqrt{28} \cdot \sqrt{7}$ $6\ 2\sqrt{7}\ \sqrt{7}$ $6 \times 2 \times 7 = \mathbf{84}$	17. $\sqrt{72} \cdot \sqrt{99} \cdot \sqrt{44}$ $6\sqrt{2}\ 3\sqrt{11}\ 2\sqrt{11}$ $6 \times 6 \times 11 \times \sqrt{2} = \mathbf{396\sqrt{2}}$	28. $5\sqrt{39} \cdot 2\sqrt{26}$ $5\sqrt{3}\ \sqrt{13}\ 2\sqrt{13}\ \sqrt{2}$ $5 \times 2 \times 13 \times \sqrt{6} = \mathbf{130\sqrt{6}}$
7. $\sqrt{21} \cdot \sqrt{14} \cdot \sqrt{6}$ $\sqrt{7}\ \sqrt{3}\ \sqrt{7}\ \sqrt{2}\ \sqrt{2}\ \sqrt{3}$ $7 \times 3 \times 2 = \mathbf{42}$	18. $\sqrt{75} \cdot \sqrt{48} \cdot \sqrt{18}$ $5\sqrt{3}\ 4\sqrt{3}\ 3\sqrt{2}$ $5 \times 4 \times 3 \times 3\sqrt{2} = \mathbf{180\sqrt{2}}$	29. $3\sqrt{60} \cdot 3\sqrt{45}$ $3\ 2\sqrt{3}\ \sqrt{5}\ 3\ 3\sqrt{5}$ $3 \times 2 \times 5 \times 9\sqrt{3} = \mathbf{270\sqrt{3}}$
8. $\sqrt{27} \cdot \sqrt{15} \cdot \sqrt{5}$ $3\sqrt{3}\ \sqrt{3}\ \sqrt{5}\ \sqrt{5}$ $3 \times 3 \times 5 = \mathbf{45}$	19. $\sqrt{54} \cdot \sqrt{21} \cdot \sqrt{3}$ $3\sqrt{6}\ \sqrt{3}\ \sqrt{7}\ \sqrt{3}$ $3 \times 3 \times \sqrt{42} = \mathbf{9\sqrt{42}}$	30. $3\sqrt{55} \cdot 5\sqrt{33}$ $3\sqrt{5}\ \sqrt{11}\ 5\sqrt{11}\ \sqrt{3}$ $3 \times 5 \times 11\sqrt{15} = \mathbf{165\sqrt{15}}$
9. $\sqrt{54} \cdot \sqrt{24} \cdot \sqrt{64}$ $3\sqrt{6}\ 2\sqrt{6}\ 8$ $3 \times 2 \times 8 \times 6 = \mathbf{288}$	20. $\sqrt{40} \cdot \sqrt{50} \cdot \sqrt{60}$ $2\sqrt{5}\ \sqrt{2}\ 5\sqrt{2}\ 2\sqrt{5}\ \sqrt{3}$ $10 \times 2 \times 5 \times 2\sqrt{3} = \mathbf{200\sqrt{3}}$	31. $3\sqrt{22} \cdot 2\sqrt{44}$ $3\sqrt{2}\ \sqrt{11}\ 2\ 2\sqrt{11}$ $3 \times 2 \times 2 \times 11\sqrt{2} = \mathbf{132\sqrt{2}}$
10. $\sqrt{28} \cdot \sqrt{80} \cdot \sqrt{35}$ $2\sqrt{7}\ 4\sqrt{5}\ \sqrt{5}\ \sqrt{7}$ $2 \times 4 \times 5 \times 7 = \mathbf{280}$	21. $\sqrt{24} \cdot \sqrt{25} \cdot \sqrt{27}$ $2\sqrt{3}\ \sqrt{2}\ 5\ 3\sqrt{3}$ $2 \times 5 \times 3 \times 3\sqrt{2} = \mathbf{90\sqrt{2}}$	32. $7\sqrt{48} \cdot 8\sqrt{12}$ $7\ 4\sqrt{3}\ 8\ 2\sqrt{3}$ $7 \times 4 \times 8 \times 2 \times 3 = \mathbf{1344}$
11. $\sqrt{48} \cdot \sqrt{63} \cdot \sqrt{21}$ $4\sqrt{3}\ 3\sqrt{7}\ \sqrt{3}\ \sqrt{7}$ $4 \times 3 \times 3 \times 7 = \mathbf{252}$	22. $\sqrt{35} \cdot \sqrt{14} \cdot \sqrt{81}$ $\sqrt{5}\ \sqrt{7}\ \sqrt{7}\ \sqrt{2}\ 9$ $9 \times 7 \times \sqrt{10} = \mathbf{63\sqrt{10}}$	33. $9\sqrt{42} \cdot 4\sqrt{28}$ $9\sqrt{6}\ \sqrt{7}\ 4\ 2\sqrt{7}$ $9 \times 4 \times 2 \times 7\sqrt{6} = \mathbf{504\sqrt{6}}$

Radicals 4

Simplify and add or subtract.

1. $5\sqrt{28} - 2\sqrt{63}$

$10\sqrt{7} - 6\sqrt{7}$

$\mathbf{4\sqrt{7}}$

2. $2\sqrt{48} + 4\sqrt{147}$

$8\sqrt{3} + 28\sqrt{3}$

$\mathbf{36\sqrt{3}}$

3. $6\sqrt{18} + 5\sqrt{72}$

$18\sqrt{2} + 30\sqrt{2}$

$\mathbf{48\sqrt{2}}$

4. $\sqrt{175} + 3\sqrt{63}$

$5\sqrt{7} + 9\sqrt{7}$

$\mathbf{14\sqrt{7}}$

5. $5\sqrt{75} - 2\sqrt{300}$

$25\sqrt{3} - 20\sqrt{3}$

$\mathbf{5\sqrt{3}}$

6. $\sqrt{288} + 2\sqrt{800}$

$12\sqrt{2} + 40\sqrt{2}$

$\mathbf{52\sqrt{2}}$

7. $\sqrt{720} + 5\sqrt{245}$

$12\sqrt{5} + 35\sqrt{5}$

$\mathbf{47\sqrt{5}}$

8. $\sqrt{252} + 2\sqrt{448}$

$6\sqrt{7} + 16\sqrt{7}$

$\mathbf{22\sqrt{7}}$

9. $5\sqrt{12} - 2\sqrt{27}$

$10\sqrt{3} - 6\sqrt{3}$

$\mathbf{4\sqrt{3}}$

10. $6\sqrt{50} - 2\sqrt{8}$

$30\sqrt{2} - 4\sqrt{2}$

$\mathbf{26\sqrt{2}}$

11. $5\sqrt{20} - \sqrt{125}$

$10\sqrt{5} - 5\sqrt{5}$

$\mathbf{5\sqrt{5}}$

12. $\sqrt{150} + 2\sqrt{54}$

$5\sqrt{6} + 6\sqrt{6}$

$\mathbf{11\sqrt{6}}$

13. $5\sqrt{99} - \sqrt{275}$

$15\sqrt{11} - 5\sqrt{11}$

$\mathbf{10\sqrt{11}}$

14. $\sqrt{147} + 2\sqrt{108}$

$7\sqrt{3} + 12\sqrt{3}$

$\mathbf{19\sqrt{3}}$

15. $\sqrt{363} + 2\sqrt{75}$

$11\sqrt{3} + 10\sqrt{3}$

$\mathbf{21\sqrt{3}}$

16. $\sqrt{600} - 4\sqrt{24}$

$10\sqrt{6} - 8\sqrt{6}$

$\mathbf{2\sqrt{6}}$

17. $7\sqrt{20} + 6\sqrt{45}$

$14\sqrt{5} + 18\sqrt{5}$

$\mathbf{32\sqrt{5}}$

18. $\sqrt{128} + 2\sqrt{98}$

$8\sqrt{2} + 14\sqrt{2}$

$\mathbf{22\sqrt{2}}$

19. $\sqrt{200} - 2\sqrt{32}$

$10\sqrt{2} - 8\sqrt{2}$

$\mathbf{2\sqrt{2}}$

20. $\sqrt{176} + 3\sqrt{44}$

$4\sqrt{11} + 6\sqrt{11}$

$\mathbf{10\sqrt{11}}$

21. $\sqrt{500} - 2\sqrt{80}$

$10\sqrt{5} - 8\sqrt{5}$

$\mathbf{2\sqrt{5}}$

22. $\sqrt{171} + 3\sqrt{76}$

$3\sqrt{19} + 6\sqrt{19}$

$\mathbf{9\sqrt{19}}$

23. $6\sqrt{52} - 2\sqrt{117}$

$12\sqrt{13} - 6\sqrt{13}$

$\mathbf{6\sqrt{13}}$

24. $\sqrt{112} + 2\sqrt{343}$

$4\sqrt{7} + 14\sqrt{7}$

$\mathbf{18\sqrt{7}}$

Radicals 5

Round the radical as specified. Use a calculator.	Simplify the radical. Use mental math.
1. $\sqrt{37}$ to the nearest thousandth **6.083**	14. $\sqrt{2.56}$ **1.6**
2. $\sqrt{59}$ to the nearest tenth **7.7**	15. $\sqrt{0.000064}$ **0.008**
3. $\sqrt{91}$ to the nearest hundredth **9.54**	16. $\sqrt{6.25}$ **2.5**
4. $\sqrt{67}$ to the nearest ten-thousandth **8.1854**	17. $\sqrt{0.0049}$ **0.07**
5. $\sqrt{11}$ to the nearest thousandth **3.317**	18. $\sqrt{.81}$ **.9**
6. $\sqrt{29}$ to the nearest hundredth **5.39**	19. $\sqrt{0.0361}$ **0.19**
7. $\sqrt{63}$ to the nearest tenth **7.9**	20. $\sqrt{.36}$ **.6**
8. $\sqrt{95}$ to the nearest ten-thousandth **9.7468**	21. $\sqrt{0.0256}$ **0.16**
9. $\sqrt{55}$ to the nearest ten-thousandth **7.4162**	22. $\sqrt{(1.8)^2}$ **1.8**
10. $\sqrt{97}$ to the nearest hundredth **9.85**	23. $\sqrt{0.000016}$ **0.004**
11. $\sqrt{77}$ to the nearest thousandth **8.775**	24. $\sqrt{1.21}$ **1.1**
12. $\sqrt{31}$ to the nearest hundredth **5.57**	25. $\sqrt{0.0324}$ **0.18**
13. $\sqrt{23}$ to the nearest thousandth **4.796**	26. $\sqrt{1.69}$ **1.3**

Radicals 6

Rationalize the denominator. Simplify.

1. $\dfrac{20}{\sqrt{5}} \cdot \dfrac{\sqrt{5}}{\sqrt{5}}$ **4√5**

2. $\dfrac{75}{\sqrt{3}} \cdot \dfrac{\sqrt{3}}{\sqrt{3}}$ **25√3**

3. $\dfrac{14}{\sqrt{2}} \cdot \dfrac{\sqrt{2}}{\sqrt{2}}$ **7√2**

4. $\dfrac{63}{\sqrt{7}} \cdot \dfrac{\sqrt{7}}{\sqrt{7}}$ **9√7**

5. $\dfrac{66}{\sqrt{11}} \cdot \dfrac{\sqrt{11}}{\sqrt{11}}$ **6√11**

6. $\dfrac{42}{\sqrt{6}} \cdot \dfrac{\sqrt{6}}{\sqrt{6}}$ **7√6**

7. $\dfrac{80}{\sqrt{10}} \cdot \dfrac{\sqrt{10}}{\sqrt{10}}$ **8√10**

8. $\dfrac{60}{\sqrt{15}} \cdot \dfrac{\sqrt{15}}{\sqrt{15}}$ **4√15**

9. $\dfrac{20}{\sqrt{5}+3} \cdot \dfrac{\sqrt{5}-3}{\sqrt{5}-3}$ −4   **−5√5 + 15**

10. $\dfrac{25}{\sqrt{6}+1} \cdot \dfrac{\sqrt{6}-1}{\sqrt{6}-1}$ 5   **5√6 − 5**

11. $\dfrac{12}{\sqrt{7}-2} \cdot \dfrac{\sqrt{7}+2}{\sqrt{7}+2}$ 3   **4√7 + 8**

12. $\dfrac{42}{\sqrt{3}-3} \cdot \dfrac{\sqrt{3}+3}{\sqrt{3}+3}$ −6   **−7√3 − 21**

13. $\dfrac{16}{\sqrt{9}+5} \cdot \dfrac{\sqrt{9}-5}{\sqrt{9}-5}$ −16   **−√9 + 5**

14. $\dfrac{18}{\sqrt{6}-3} \cdot \dfrac{\sqrt{6}+3}{\sqrt{6}+3}$ −3   **−6√6 − 18**

15. $\dfrac{28}{\sqrt{2}-4} \cdot \dfrac{\sqrt{2}+4}{\sqrt{2}+4}$ −14   **−2√2 − 8**

16. $\dfrac{56}{\sqrt{8}-1} \cdot \dfrac{\sqrt{8}+1}{\sqrt{8}+1}$ 7   **8√8 + 8**

17. $\dfrac{57}{2\sqrt{5}+1} \cdot \dfrac{2\sqrt{5}-1}{2\sqrt{5}-1}$ 19   **6√5 − 3**

18. $\dfrac{55}{3\sqrt{4}-5} \cdot \dfrac{3\sqrt{4}+5}{3\sqrt{4}+5}$ 11   **15√4 + 25**

19. $\dfrac{42}{5\sqrt{2}+6} \cdot \dfrac{5\sqrt{2}-6}{5\sqrt{2}-6}$ 14   **15√2 − 18**

20. $\dfrac{-13}{4\sqrt{3}+7} \cdot \dfrac{4\sqrt{3}-7}{4\sqrt{3}-7}$ −1   **52√3 − 91**

21. $\dfrac{60}{2\sqrt{6}-3} \cdot \dfrac{2\sqrt{6}+3}{2\sqrt{6}+3}$ 15   **8√6 + 12**

22. $\dfrac{58}{3\sqrt{5}+4} \cdot \dfrac{3\sqrt{5}-4}{3\sqrt{5}-4}$ 29   **6√5 − 8**

23. $\dfrac{45}{4\sqrt{4}+7} \cdot \dfrac{4\sqrt{4}-7}{4\sqrt{4}-7}$ 15   **12√4 − 21**

24. $\dfrac{44}{5\sqrt{3}-8} \cdot \dfrac{5\sqrt{3}+8}{5\sqrt{3}+8}$ 11   **20√3 + 32**

Ratios 1

Write each ratio first as identifying letters and then as a simplified fraction.

1. number of letters to vowels in the word SCHOOLHOUSE	$\frac{L}{V} = \frac{11}{5}$	
2. number of vowels to consonants in the word SCHOOLHOUSE	$\frac{V}{C} = \frac{5}{6}$	
3. number of leap years to years from 2000 through 2009	$\frac{L}{Y} = \frac{3}{10}$	
4. number of months of the year to months with 31 days	$\frac{M}{T} = \frac{12}{7}$	
5. number of primes to composites from 1 through 30	$\frac{P}{C} = \frac{10}{19}$	
6. number of prime digits to composite digits in 7,535,197,228	$\frac{P}{C} = \frac{7}{2}$	
7. number of percent signs to dollar signs in the string ??$#@$%*?%#??+@$?&	$\frac{P}{D} = \frac{2}{3}$	
8. number of blues to greens in the list Navy, Grass, Sky, Forest, Lime	$\frac{B}{G} = \frac{2}{3}$	
9. number of even digits to odd digits in 5,347,190,943	$\frac{V}{D} = \frac{3}{7}$	
10. number of vowels (without Y) to letters in the alphabet	$\frac{V}{L} = \frac{5}{26}$	
11. number of face cards to total cards in a standard deck Do 1 suit directly.	$\frac{F}{C} = \frac{3}{13}$	

12. number of letters to vowels in the word BOOKFAIR	$\frac{L}{V} = \frac{8}{4} = \frac{2}{1}$
13. number of consonants to vowels to in the word WORKSHOP	$\frac{C}{V} = \frac{6}{2} = \frac{3}{1}$
14. number of leap years to years from 1900 through 1999 not 1900	$\frac{L}{Y} = \frac{24}{100} = \frac{6}{25}$
15. number of months with 30 days to months of the year	$\frac{T}{M} = \frac{4}{12} = \frac{1}{3}$
16. number of primes to total numbers from 1 through 20	$\frac{P}{N} = \frac{8}{20} = \frac{2}{5}$
17. number of prime digits to composite digits in 6,035,831,396,491	$\frac{P}{C} = \frac{4}{6} = \frac{2}{3}$
18. number of asterisks to percents in the string *%*#&*%*?*#%&+%*$	$\frac{A}{P} = \frac{6}{4} = \frac{3}{2}$
19. number of tans to pinks in the list Beige, Hot, Magenta, Taupe	$\frac{T}{P} = \frac{2}{2} = \frac{1}{1}$
20. number of odd digits to even digits in 7,815,240,968	$\frac{D}{V} = \frac{4}{6} = \frac{2}{3}$
21. number of vowels (with Y) to letters in the alphabet	$\frac{V}{L} = \frac{6}{26} = \frac{3}{13}$
22. number of face cards to even cards (with Q) in a standard deck	$\frac{F}{E} = \frac{3}{6} = \frac{1}{2}$

Ratios 2

Answer as indicated.

1.	The sum of two whole numbers is 36. If the numbers are in the ratio 5:4, find their product. 5+4 = 9 36/9 = 4 20 × 16 = **320** 20:16	8.	In a class if the ratio of the number of girls to boys is 3:5, find the ratio of the number of boys to students. G = 3x, B = 5x, S = 8x B:S = **5:8**
2.	If $27,000 were divided among three people in the ratio 2:3:5, find the value of the greatest share. 2+3+5 = 10 27,000/10 = 2700 5×2700 = **13,500**	9.	A recipe has the ratio of cups of flour to sugar to butter as 4:2:1. If the recipe is doubled, find the new ratio of the three ingredients. **4:2:1**
3.	If $30,000 were divided among three people in the ratio 3:5:7, find the value of the least share. 3+5+7 = 15 30,000/15 = 2000 3×2000 = **6,000**	10.	If the ratio of a to b is 5 to 8 and the ratio of b to c is 4 to 3, find the ratio of a to c. a:b = 5:8 b:c = 4:3 = 8:6 a:c = **5:6**
4.	The sum of two whole numbers is 60. If the numbers are in the ratio 7:5, find their product. 7+5 = 12 60/12 = 5 35 × 25 = **875** 35:25	11.	In a class if the ratio of the number of students to boys is 11:5, find the ratio of the number of boys to girls. S = 11x, B = 5x, G = 6x B:G = **5:6**
5.	Three whole numbers have a sum of 80 and are in the ratio 2:5:9. Find the least of the numbers. 2+5+9 = 16 80/16 = 5 2×5= **10**	12.	The ratio of x to 50 equals the ratio of 18 to 30. Find x. x:50 = 18:30 x:50 = 6:10 x:50 = 30:50 x = **30**
6.	If 600 items were divided four ways in the ratio 3:2:8:2, find the median value. 3+2+8+2 = 15 600/15 = 40 80, 80, 120, 320 median = **100**	13.	The ratio of Jo's age to Tom's age is 2:3. The ratio of Bo's age to Tom's age is 8:7. Find the ratio of Bo's age to Jo's age. J:T = 2:3 = 14:21 B:T = 8:7 = 24:21 B:J =24:14 = **12:7**
7.	Three whole numbers have a sum of 72 and are in the ratio 1:5:6. Find the greatest of the numbers. 1+5+6 = 12 72/12 = 6 6 × 6 = **36**	14.	If the ratio of c to d is 7 to 9 and the ratio of d to e is 27 to 8, find the ratio of e to c. c:d = 7:9 = 21:27 d:e = 27:8 e:c = **8:21**

Ratios 3

Find the unit rate.	*Find the cost for the number specified, assuming a constant rate.*
1. $55 for 25 candles $11 for 5 **$2.20 for 1**	12. 7 if 3 cost $111 1 costs $37 7 cost **$259**
2. $70 for 42 pens $5 for 3 **$1.67 for 1**	13. 9 if 4 cost $236 1 costs $59 9 cost **$531**
3. $165 for 132 pencils $15 for 12 **$1.25 for 1**	14. 11 if 5 cost $425 1 costs $85 11 cost **$935**
4. $130 for 78 tags $10 for 6 **$1.67 for 1**	15. 8 if 6 cost $522 2 cost $174 8 cost **$696**
5. $105 for 90 buttons $7 for 6 **$1.17 for 1**	16. 3 if 11 cost $572 1 costs $52 3 cost **$156**
6. 572 apples for 44 crates 52 for 4 **13 for 1**	17. 3 if 10 cost $13.20 1 costs $1.32 3 cost **$3.96**
7. 204 people for 51 cars 68 for 17 **4 for 1**	18. 13 if 8 cost $16.40 1 costs $2.05 13 cost **$26.65**
8. 336 students for 21 teachers 112 for 7 **16 for 1**	19. 7 if 15 cost $360 1 costs $24 7 cost **$168**
9. 88 pies for 132 people 8 for 12 **2/3 for 1**	20. 10 if 6 cost $30.90 2 cost $10.30 10 cost **$51.50**
10. 600 pounds for 25 boxes 120 for 5 **24 for 1**	21. 12 if 7 cost $21.28 1 costs $3.04 12 cost **$36.48**
11. 209 pounds for 22 cartons 19 for 2 **9.5 for 1**	22. 5 if 12 cost $72.24 1 costs $6.02 5 cost **$30.10**

Ratios 4

Find the better buy using mental math.	*Find the scale or dimensions by mental math.*
1. 2 for $9.60 or 3 for $14.25 4.80 v. round $15; 5 - .25 = 4.75 **3**	12. actual dimensions 605 m by 990 m drawn dimensions 11 cm by 18 cm scale: **1 cm = 55 m**
2. 3 for $13.65 or 4 for $18.24 4.55 v. 4.56 **3**	13. actual dimensions 99 yd by 176 yd drawn dimensions 9 in by 16 in scale: **1 in = 11 yd**
3. 5 for $23.55 or 6 for $28.50 4.71 v. 4.75 **5**	14. actual dimensions 120 yd by 165 yd drawn dimensions 8 in by 11 in scale: **1 in = 15 yd**
4. 11 for $171.60 or 9 for $139.50 15.60 v. 15.50 **9**	15. actual dimensions 98 km by 126 km drawn dimensions 7 cm by 9 cm scale: **1 cm = 14 km**
5. 6 for $5.88 or 7 for $6.93 0.98 v. 0.99 **6**	16. actual dimensions 324 yd by 244 yd scale 1 in = 8 yd drawn dimensions: **40.5 in by 30.5 in**
6. 5 for $178.25 or 2 for $71.34 35.65 v. 35.67 **5**	17. drawn dimensions 5.5 cm by 7 cm scale 1 cm = 26 m actual dimensions: **143 m by 182 m**
7. 8 for $9.20 or 12 for $13.92 1.15 v. 1.16 **8**	18. actual dimensions 110 ft by 84 ft scale 1 in = 20 ft drawn dimensions: **5.5 in by 4.2 in**
8. 15 for $32.25 or 19 for $39.90 2.15 v. 2.10 **19**	19. drawn dimensions 3.5 cm by 2.1 cm scale 1 cm = 30 m actual dimensions: **105 m by 63 m**
9. 13 for $97.50 or 11 for $83.60 7.50 v. 7.60 **13**	20. actual dimensions 65 yd by 117 yd drawn dimensions 10 in by 18 in scale: **1 in = 6.5 yd**
10. 6 for $14.40 or 8 for $19.28 2.40 v. 2.41 **6**	21. drawn dimensions 2.5 cm by 9 cm scale 1 cm = 2.5 m actual dimensions: **6.25 m by 22.5 m**
11. 10 for $29.50 or 9 for 25.20 2.95 v. 2.80 **9**	22. actual dimensions 420 ft by 115.5 ft scale 1 in = 10.5 ft drawn dimensions: **40 in by 11 in**

Ratios 5

Find the number that can be added to both values of the 1st ratio to get the 2nd ratio.	*A, B, C, D, and E are consecutive points on a line. Find the specified ratio.*
1. 1st ratio 3:7 2nd ratio 3:4 $\dfrac{3+x}{7+x} = \dfrac{3}{4}$ $4x+12 = 3x+21$ $\dfrac{12}{16} = \dfrac{3}{4}$ $x = \mathbf{9}$	**8.** AB:BC = 1:2 **3:10** BC:CD = 2:3 CD:DE = 3:4 Find AC:AE. AC = AB + BC = 3 AE = AB + BC + CD + DE = 1+2+3+4 = 10
2. 1st ratio 2:5 2nd ratio 8:9 $\dfrac{2+x}{5+x} = \dfrac{8}{9}$ $9x+18 = 8x+40$ $\dfrac{24}{27} = \dfrac{8}{9}$ $x = \mathbf{22}$	**9.** AB:BC = 1:1 AB:BC = 2:2 **2:21** BC:CD = 1:3 BC:CD = 2:6 CD:DE = 2:5 CD:DE = 6:15 Find AB:CE. AB = 2 CE = CD + DE = 6+15 = 21
3. 1st ratio 5:12 2nd ratio 11:12 $\dfrac{5+x}{12+x} = \dfrac{11}{12}$ $12x+60 = 11x+132$ $\dfrac{77}{84} = \dfrac{11}{12}$ $x = \mathbf{72}$	**10.** AB:BC = 1:4 AB:BC = 1:4 **5:36** BC:CD = 1:2 BC:CD = 4:8 CD:DE = 1:3 CD:DE = 8:24 Find AC:BE. AC = AB + BC = 1+4 = 5 BE = BC + CD + DE = 4+8+24 = 36
4. 1st ratio 2:9 2nd ratio 4:5 $\dfrac{2+x}{9+x} = \dfrac{4}{5}$ $5x+10 = 4x+36$ $\dfrac{28}{35} = \dfrac{4}{5}$ $x = \mathbf{26}$	**11.** AB:BC = 2:3 AB:BC = 2:3 **4:15** BC:CD = 3:7 BC:CD = 3:7 CD:DE = 1:5 CD:DE = 7:35 Find AD:BE. AD = AB + BC + CD = 2+3+7 = 12 BE = BC + CD + DE = 3+7+35 = 45
5. 1st ratio 2:13 2nd ratio 7:8 $\dfrac{2+x}{13+x} = \dfrac{7}{8}$ $8x+16 = 7x+91$ $\dfrac{77}{88} = \dfrac{7}{8}$ $x = \mathbf{75}$	**12.** AC:CD = 3:5 AC:CD = 6:10 **46:15** BC:CD = 1:2 BC:CD = 5:10 CD:DE = 1:3 CD:DE = 10:30 Find AE:BD. AE = AC + CD + DE = 6+10+30 = 46 BD = BC + CD = 5+10 = 15
6. 1st ratio 3:8 2nd ratio 5:6 $\dfrac{3+x}{8+x} = \dfrac{5}{6}$ $6x+18 = 5x+40$ $\dfrac{25}{30} = \dfrac{5}{6}$ $x = \mathbf{22}$	**13.** AB:BD = 1:10 AB:BD = 1:10 **4:7** BD:DE = 2:3 BD:DE = 10:15 BC:DE = 2:5 BC:DE = 6:15 Find CD:AC. CD = BD − BC = 10−6 = 4 AC = AB + BC = 1+6 = 7
7. 1st ratio 2:9 2nd ratio 6:7 $\dfrac{2+x}{9+x} = \dfrac{6}{7}$ $7x+14 = 6x+54$ $\dfrac{42}{49} = \dfrac{6}{7}$ $x = \mathbf{40}$	**14.** AC:CD = 3:2 AC:CD = 9:6 **7:16** BC:CD = 1:3 BC:CD = 2:6 DE:CD = 5:3 DE:CD = 10:6 Find AB:CE. AB = AC − BC = 9−2 = 7 CE = CD + DE = 6+10 = 16

Ratios 6

Answer as indicated for the plane geometry ratios.

---

1. Find the ratio of a side of an equilateral triangle to its perimeter.

    1:3

8. The ratio of a rectangle's length to width is 4:3. If the width is 48, find the perimeter.

    W = 3x16 = 48
    L = 4x16 = 64
    P = 2x112 = **224**

---

2. Given two equilateral triangles, one with side 6 and one with side 20. Find the ratio of the perimeters, larger triangle to smaller.

    10:3

9. Find the ratio of the area of the hexagon to the area of the 8 by 8 square. The legs of the two isosceles right triangles are 2.

    $\dfrac{64 - 4}{64} = \dfrac{15}{16}$

    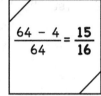

---

3. The ratio of a rectangle's length to width is 5:2. If the width is 8, find the area.

    W = 2x4 = 8
    L = 5x4 = 20
    A = **160**

10. Find the ratio of the sides, smaller to larger, of equilateral triangles with perimeters 36 and 90.

    $\dfrac{12}{30} = \dfrac{2}{5}$

---

4. The measures of three angles of a triangle are in the ratio 2:3:4. Find the measure of the least angle.

    2+3+4 = 9
    180/9 = 20
    2x20 = **40**

11. Find the ratio of Area I to Area II in the rectangle.

    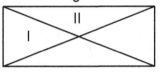

    $\dfrac{(H)(B/2)(.5)}{(B)(H/2)(.5)}$

    1:1

---

5. Find the ratio of a side of a regular hexagon to its perimeter.

    1:6

12. A rectangle has sides in the ratio 2:9. If the perimeter is 528, find the length of the shorter side.

    P = 528 | 11x24 = 264
    SP = 264 | blow-up = 24
    2+9 = 11 | sides = **48**, 216

---

6. A rectangle has sides in the ratio 3:5. If the perimeter is 160, find the length of the longer side.

    P = 160
    SP = 80
    sides = 30, **50**

13. The radii of the smaller circles are 2 and 3. Find the ratio of the areas, smallest circle to largest circle.   **4:25**

    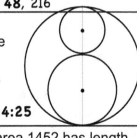

---

7. The measures of the angles of a quadrilateral are in the ratio 1:2:3:4. Find the sum of the measures of the two greatest angles.

    1+2+3+4 = 10
    360/10 = 36
    7x36 = **252**

14. A rectangle with area 1452 has length and width in the ratio 3:4. Find the diagonal.

    1452 = 132x11 = 12x11x11
    L = 33, W = 44
    diagonal = **55**

Ratios 7

Find the specified ratio.

1.	A:B = 5:3 C:B = 9:7 Find 2A:B.	35:21 (x7) 27:21 (x3) 70:21 **10:3**	9.	S:A:T = 5:9:12 N:O:T = 6:11:15 Find A:N.	25:45:60 (x5) 24:44:60 (x4) A:N = 45:24 **15:8**
2.	X:Y = 5:11 Y:Z = 3:7 Find (X+Y):(X+Z).	15:33 (x3) 33:77 (x11) 48:92 **12:23**	10.	F:U:N = 7:22:5 S:U:B = 8:33:14 Find B:F.	21:66:15 (x3) 16:66:28 (x2) B:F = 28:21 **4:3**
3.	A:B = 6:7 B:C = 4:3 Find (A + 2B):B.	24:28 (x4) 28:21 (x7) 80:28 **20:7**	11.	C:A:N = 11:8:9 N:O:T= 15:12:13 Find O:A.	55:40:45 (x5) 45:36:39 (x3) O:A = 36:40 **9:10**
4.	D:O:G = 5:4:9 G:H:J = 7:6:8 Find O:J.	35:28:63 (x7) 63:54:72 (x9) O:J = 28:72 **7:18**	12.	S:U:N = 9:10:8 T:A:N = 12:11:6 Find T:S.	27:30:24 (x3) 48:44:24 (x4) T:S = 48:27 **16:9**
5.	P:A:N = 35:5:12 T:O:P = 9:8:15 Find A:T.	105:15:36 (x3) 63:56:105 (x7) A:T = 15:63 **5:21**	13.	A:B:C = 6:5:14 H:A:T = 10:7:15 Find T:B.	42:35:98 (x7) 60:42:90 (x6) T:B = 90:35 **18:7**
6.	A:B:C:D = 3:2:5:7 C:P:Q:R = 4:7:8:9 Find R:A.	12:8:20:28 (x4) 20:35:40:45 (x5) R:A = 45:12 **15:4**	14.	C:A:K:E = 9:4:6:8 S:E:N:D = 2:3:5:7 Find A:D.	27:12:18:24 (x3) 16:24:40:56 (x8) A:D = 12:56 **3:14**
7.	B:I:G = 8:4:3 B:U:S = 12:5:4 Find U:G.	24:12:9 (x3) 24:10:8 (x2) U:G = 10:9 **10:9**	15.	E:F = 2:9 D:F = 7:10 Find 4D:3E.	20:90 (x10) 63:90 (x9) (4x63):(3x20) **21:5**
8.	H:O:T = 7:8:20 T:I:P = 25:9:21 Find H:P.	35:40:100 (x5) 100:36:84 (x4) H:P = 35:84 **5:12**	16.	G:H = 9:5 E:G = 4:11 Find (E + H):7E.	99:55 (x11) 36:99 (x9) (91):(7x36) **13:36**

Ratios 8

Find the ratio of an angle to its complement or supplement as indicated.	*Answer as indicated.*
1. angle to complement if the ratio of the angle to its supplement is 2:7 A:S = 2:7 (x20) A = 40, S = 140 A:C = 40:50 **4:5**	8. For a cube with edge e, find the ratio of the volume (in cubic units) to the surface area (in square units). $$\frac{V}{SA} = \frac{e^3}{6e^2} = \frac{e}{6}$$
2. angle to supplement if the ratio of the angle to its complement is 1:8 A:C = 1:8 (x10) A = 10, C = 80 A:S = 10:170 **1:17**	9. The radius of circle A equals the diameter of circle B. Find the ratio of the area of circle A to the area of circle B. $$\frac{area\ A}{area\ B} = \frac{\pi(2r)^2}{\pi r^2} = \frac{4}{1}$$
3. angle to complement if the ratio of the angle to its supplement is 3:7 A:S = 3:7 (x18) A = 54, S = 126 A:C = 54:36 **3:2**	10. For a square with side s, find the ratio of the area (in square units) to the perimeter (in units)? $$\frac{A}{P} = \frac{s^2}{4s} = \frac{s}{4}$$
4. angle to complement if the ratio of the angle to its supplement is 4:5 A:S = 4:5 (x20) A = 80, S = 100 A:C = 80:10 **8:1**	11. A sphere "just fits" in a cylinder. Find the ratio of the volume of the sphere to the volume of the cylinder. $$\frac{v\ S}{v\ C} = \frac{4\pi r^3}{3} \div \pi r^2(2r) = \frac{2}{3}$$
5. angle to complement if the ratio of the angle to its supplement is 7:8 A:S = 7:8 (x12) A = 84, S = 96 A:C = 84:6 **14:1**	12. The ratio of the diameters of two circles is 1:4. Find the ratio of their radii. $$\frac{1}{4}$$
6. angle to supplement if the ratio of the angle to its complement is 1:9 A:C = 1:9 (x9) A = 9, C = 81 A:S = 9:171 **1:19**	13. A square and a circle have the same area. Find the ratio of a diagonal of the square to the radius of the circle. $$\frac{d^2}{2} = \pi r^2 \quad\bigg|\quad \frac{d^2}{r^2} = 2\pi \quad\bigg|\quad \frac{d}{r} = \frac{\sqrt{2\pi}}{1}$$ $$d^2 = 2\pi r^2$$
7. angle to complement if the ratio of the angle to its supplement is 5:7 A:S = 5:7 (x15) A = 75, S = 105 A:C = 75:15 **5:1**	14. Two similar triangles have areas 75 and 108. Find the ratio of their perimeters. $$\frac{75}{108} = \frac{25}{36} \qquad \frac{5}{6}$$

Area of square (rhombus) is half the product of the diagonals.

Ratíos 9

Find the ratio of the areas as indicated. Inscribed polygons or circles are drawn to midpoints when apparent.

1. inner square to outer square
  $$\frac{4}{8} = \frac{1}{2}$$
	7. inner triangle to outer equilateral triangle
	  $$\frac{1}{4}$$

2. inner square to outer square $$\frac{1}{4}$$	8. isosceles right triangle to circle with triangle drawn on diameter 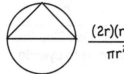 $$\frac{(2r)(r)/2}{\pi r^2} = \frac{1}{\pi}$$
3. circle to square 2r $$\frac{\pi r^2}{4r^2} = \frac{\pi}{4}$$	9. equilateral triangle to circle 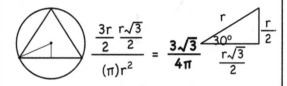 $$\frac{\frac{3r}{2}\frac{r\sqrt{3}}{2}}{(\pi)r^2} = \frac{3\sqrt{3}}{4\pi}$$

4. square to circle
  $$\frac{(2r)^2/2}{\pi r^2} = \frac{2}{\pi}$$
	10. regular hexagon to circle
	  $$\frac{(6\sqrt{3})r^2/4}{\pi r^2} = \frac{3\sqrt{3}}{2\pi}$$

5. inner circle to outer circle $$\frac{\pi r^2}{\pi(r\sqrt{2})^2} = \frac{1}{2}$$	11. smaller circle to larger concentric circle 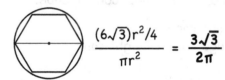 $$\frac{\pi r^2}{\pi(2r)^2} = \frac{1}{4}$$
6. inner square to outer square $$\frac{(2r)^2/2}{4r^2} = \frac{1}{2}$$	12. smaller circle to larger circle 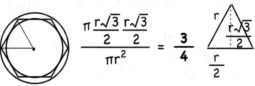 $$\frac{\pi \frac{r\sqrt{3}}{2}\frac{r\sqrt{3}}{2}}{\pi r^2} = \frac{3}{4}$$

Ratios 10

Find the ratio of the volumes of the space figures that share a base.

1. A cone just fits inside a hemisphere. Find the ratio of the volumes, cone to hemisphere.

    r = radius cone = radius hemisphere
    r = height cone = height hemisphere

    $$V = \dfrac{\frac{1}{3}(\pi)r^2(r)}{\frac{1}{2}\frac{4}{3}(\pi)r^3} = \frac{1}{3} \times \frac{3}{2} = \frac{1}{2}$$

2. A cone just fits in a cylinder. Find the ratio of the volumes, cone to cylinder.

    r = radius cylinder = radius cone

    $$V = \dfrac{\frac{1}{3}\pi r^2 h}{\pi r^2 h} = \frac{1}{3}$$

3. A cone just fits in a cube. Find the ratio of the volumes, cone to cube.

    r = radius cone
    2r = edge cube = height cone

    $$V = \dfrac{\frac{1}{3}(\pi)r^2(2r)}{(2r)^3} = \frac{2\pi}{24} = \frac{\pi}{12}$$

4. A square pyramid just fits inside a cube. Find the ratio of the volumes, pyramid to cube.

    e = cube edge
    e = base edge pyramid = height pyramid

    $$V = \dfrac{\frac{1}{3}e^2(e)}{e^3} = \frac{1}{3}$$

5. A sphere just fits in a cube. Find the ratio of the volumes, sphere to cube.

    r = radius sphere
    2r = edge cube

    $$V = \dfrac{\frac{4}{3}(\pi)r^3}{(2r)^3} = \frac{4\pi}{24} = \frac{\pi}{6}$$

6. A sphere just fits in a cylinder. Find the ratio of the volumes, sphere to cylinder.

    r = radius sphere = radius cylinder
    2r = height cylinder

    $$V = \dfrac{\frac{4}{3}(\pi)r^3}{(\pi)r^2(2r)} = \frac{4}{6} = \frac{2}{3}$$

7. A hemisphere just fits in a rectangular prism. Find the ratio of the volumes, hemisphere to prism.

    r = radius hemisphere
    2r by 2r = base prism; r = height prism

    $$V = \dfrac{\frac{1}{2}\frac{4}{3}(\pi)r^3}{(2r)(2r)(r)} = \frac{2\pi}{12} = \frac{\pi}{6}$$

8. A hemisphere just fits in a cylinder. Find the ratio of the volumes, hemisphere to cylinder.

    r = radius hemisphere = radius cylinder
    r = height cylinder

    $$V = \dfrac{\frac{1}{2}\frac{4}{3}(\pi)r^3}{(\pi)r^2(r)} = \frac{2}{3}$$

Rectangles 1

Find the area and perimeter of the rectangle by mental math.

	BASE	HEIGHT	AREA	PERIMETER		BASE	HEIGHT	AREA	PERIMETER
1.	36	11	396	94	20.	13	13	169	52
2.	14	14	196	56	21.	74	11	814	170
3.	5	16	80	42	22.	52	10	520	124
4.	51	9	459	120	23.	9	25	225	68
5.	3	17	51	40	24.	30	4.5	135	69
6.	20	5.5	110	51	25.	5	12	60	34
7.	12	1.5	18	27	26.	3.5	22	77	51
8.	30	50	1500	160	27.	40	25	1000	130
9.	8	20	160	56	28.	11	27	297	76
10.	11	43	473	108	29.	8	31	248	78
11.	12	3.5	42	31	30.	50	90	4500	280
12.	6	18	108	48	31.	7	51	357	116
13.	19	3	57	44	32.	13	100	1300	226
14.	7	21	147	56	33.	8	15	120	46
15.	9	13	117	44	34.	20	60	1200	160
16.	10	16	160	52	35.	25	16	400	82
17.	16	4.25	68	40.5	36.	10	6.75	67.5	33.5
18.	20	5.75	115	51.5	37.	14	2.25	31.5	32.5
19.	11	55	605	132	38.	11	38	418	98

Rectangles 2

Answer as indicated.

1. Find the perimeter of a new square formed when an 8 by 8 square has its area reduced by 15 square units.

old A = 64
new A = 64 − 15 = 49
new side = 7
new P = 7 x 4 = **28**

7. Find the perimeter of a new square formed when a 10 by 10 square has its area reduced by 19 square units.

old A = 100
new A = 100 − 19 = 81
new side = 9
new P = 9 x 4 = **36**

2. Square A has one-fourth the area of square B. Find the perimeter of square B if the perimeter of square A is 40.

side A = 10 | side B = 20
area A = 100 | perim B = **80**
area B = 400 |

8. A square with area 20.5 and a rectangle with whole sides have congruent diagonals. Find the rectangle's area and perimeter.

for sq d^2 = 41 | rec is 5 by 4
d = root 41 | A rec = **20**
41 = 25 + 16 | P rec = **18**

3. Find the maximum area of a rectangle with diagonal 10.

max area is square
side = $5\sqrt{2}$
area = **50**

9. Find the sum of the numeric values of the area and perimeter of a square if they are equal.

(S)(S) = 4S
S = 4
A = P = 16
A + P = **32**

4. A square with area 36 and a rectangle with area 72 have one common congruent side. Find the perimeter of the rectangle.

side = 6
6 x 12 = 72
P = 18 + 18 = **36**

10. Find the area of a rectangle if the area equals the perimeter and the length is twice the width.

LW = 2L + 2W | W^2 = 3W
L = 2W | W = 3
(2W)W = 2(2W) + 2W | L = 6
| A = **18**

5. The square is divided into 5 congruent rectangles, each with perimeter 30. Find the perimeter of the square.

2S + 2S/5 = 30 | 12S = 150 | P = **50**
10S + 2S = 150 | S = 12.5

11. A square with diagonal 6 and a rectangle with length 9 have equal areas. Find the diagonal of the rectangle.

A sq = 36/2 = 18 | diag rec = $\sqrt{85}$
W rec = 2 |

6. A rectangle measures 3x + 5 by 6. Find 2 other rectangles with the same area in terms of x.

area = 18x + 30
6x + 10 by 3
9x + 15 by 2

12. Three rectangles are shown. Find the perimeter of the greatest. NTS

68 | 60 | 75
8 x 4 | 15 x 3

3,4,5 and 8,15,17 blow-ups
32 + 45 = 77
120 + 154 = **274**

Rectangles 3

Complete the chart for the rectangle. | *Complete the chart for the square.*

	L	W	A	P	D		SIDE	AREA	PERIM	DIAG
1.	11	**5**	**55**	32	$\sqrt{146}$	20. **13**	169	52	$13\sqrt{2}$
2.	**9**	3	27	**24**	$3\sqrt{10}$	21. **9**	81	**36**	$9\sqrt{2}$
3.	**24**	10	**240**	**68**	26	22. **1**	1	**4**	$\sqrt{2}$
4.	24	**7**	**168**	**62**	25	23. **$\sqrt{2}$**	2	**$4\sqrt{2}$**	2
5.	13	**7**	**91**	40	$\sqrt{218}$	24. **$\sqrt{6}$**	6	**$4\sqrt{6}$**	$2\sqrt{3}$
6.	16	**30**	**480**	**92**	34	25. **1.2**	1.44	**4.8**	$1.2\sqrt{2}$
7.	**11**	60	**660**	**142**	61	26. **$\sqrt{30}$**	30	**$4\sqrt{30}$**	$2\sqrt{15}$
8.	**12**	35	420	**94**	**37**	27. **$10\sqrt{2}$**	200	**$40\sqrt{2}$**	20
9.	**15**	36	**540**	102	**39**	28. **$6\sqrt{2}$**	72	**$24\sqrt{2}$**	12
10.	**12**	15	**180**	54	**$3\sqrt{41}$**	29. **12**	144	48	$12\sqrt{2}$
11.	12	**20**	**240**	64	**$4\sqrt{34}$**	30. **$5\sqrt{2}$**	50	**$20\sqrt{2}$**	10
12.	4	**8**	32	**24**	**$4\sqrt{5}$**	31. **24**	576	96	$24\sqrt{2}$
13.	18	**2**	36	**40**	**$2\sqrt{82}$**	32. **1.5**	2.25	**6**	$1.5\sqrt{2}$
14.	**9**	13	117	**44**	**$5\sqrt{10}$**	33. **$4\sqrt{2}$**	32	**$16\sqrt{2}$**	8
15.	**6**	12	72	**36**	**$6\sqrt{5}$**	34. 4	**16**	**16**	$4\sqrt{2}$
16.	21	**20**	**420**	82	29	35. **20**	400	80	$20\sqrt{2}$
17.	8	**10**	**80**	36	**$2\sqrt{41}$**	36. **11**	121	44	$11\sqrt{2}$
18.	**12**	16	**192**	**56**	20	37. **$8\sqrt{2}$**	128	**$32\sqrt{2}$**	16
19.	5	**10**	**50**	30	**$5\sqrt{5}$**	38. **1.3**	1.69	**5.2**	$1.3\sqrt{2}$

Rectangles 4

Find the area in square units algebraically.	*Find the perimeter in units algebraically.*
1. The length of a rectangle is twice its width. Find the area if the perimeter is 120.	7. The length of a rectangle is twice its width. Find the perimeter if the area is 72.

1.
L = 2W
L + W = 60
3W = 60
W = 20
| L = 40
A = (40)(20) = **800**

7.
L = 2W
LW = 72
(2W)(W) = 72
(W)(W) = 36
| W = 6
L = 12
P = (2)(18) = **36**

2. The width of a rectangle is 6 more than its length. Find the area if the perimeter is 100.
W = L + 6
L + W = 50
2L + 6 = 50
L = 22
| W = 28
A = (22)(28) = (11)(56)
A = **616**

8. The width of a rectangle is 3 more than its length. Find the perimeter if the area is 340.
W = L + 3
LW = 340
L(L + 3) = 340
17 × 20 = 340
| L = 17
W = 20
P = (2)(37) = **74**

3. The length of a rectangle is 4 more than triple the width. If the perimeter is 264, find the area.
L = 4 + 3W
L + W = 132
4W + 4 = 132
4W = 128
| W = 32
L = 100
A = (100)(32) = **3200**

9. The width of a rectangle is 5 more than its length. Find the perimeter if the area is 150.
W = L + 5
LW = 150
L(L + 5) = 150
10 × 15 = 150
| L = 10
W = 15
P = (2)(25) = **50**

4. The width of a rectangle is 7 more than twice the length. Find the area if the perimeter is 146.
W = 2L + 7
L + W = 73
3L + 7 = 73
3L = 66
| L = 22
W = 51
A = (22)(51) = (2)(561)
A = **1122**

10. The length of a rectangle is triple its width. Find the perimeter if the area is 243.
L = 3W
LW = 243
(3W)(W) = 243
(W)(W) = 81
| W = 9
L = 27
P = (2)(36) = **72**

5. The length of a rectangle is five times its width. If the perimeter is 360, find the area.
L = 5W
L + W = 180
6W = 180
W = 30
| L = 150
A = (150)(30) = **4500**

11. The length of a rectangle is one more than twice its width. Find the perimeter if the area is 253.
L = 2W + 1
LW = 253
W(2W + 1) = 253
11 × 23 = 253
| W = 11
L = 23
P = (2)(34) = **68**

6. The length of a rectangle is triple its width. If the perimeter is 264, find the area.
L = 3W
L + W = 132
4W = 132
W = 33
| L = 99
A = (99)(33) = (9)(363)
A = **3267**

12. The length of a rectangle is two less than triple its width. Find the perimeter if the area is 341.
L = 3W − 2
LW = 341
W(3W − 2) = 341
11 × 31 = 341
| W = 11
L = 31
P = (2)(42) = **84**

Rhombus 1

Find the area of the rhombus in square units.

Find the height of the rhombus.

---

1. diagonals 14 and 35

$$\frac{14 \times 35}{2}$$ A = 7 × 35 = **245**

8. diagonals 18 and 24

A = 18 × 24 / 2 = 216
each right tri is 9, 12, 15
A = BH = 15H
15H = 216   H = **14.4**

---

2. perimeter 80, one angle 60°

30-60-90 △
diagonals 20 and 20√3̄
A = **200√3̄**

9. perimeter 44, area 165

side = 11
A = BH
165 = 11H
H = **15**

---

3. perimeter 52, shorter diagonal 10

5-12-13 right △
diagonals 10 and 24
A = 10 × 24 / 2 = **120**

10. diagonals 20 and 48 $$\frac{20 \times 48}{2}$$

A = 20 × 48 / 2 = 480
each right tri is 10, 24, 26
A = BH = 26H
26H = 480   H = **240/13**

---

4. diagonals 22 and 47

$$\frac{22 \times 47}{2}$$ A = 11 × 47 = **517**

11. shorter diagonal 10, one angle 120°

30-60-90 △
sides 5, 5√3̄, and 10
A = (10)(10√3̄)/2 = 50√3̄
A = BH   B = 10
H = **5√3̄**

---

5. perimeter 60, longer diagonal 24

9-12-15 right △
diagonals 18 and 24
A = 18 × 24 / 2 = **216**

12. perimeter 100, area 375

side = 25
25H = 375
H = **15**

---

6. perimeter 20, longer diagonal 8

diagonals 6 and 8
side = 5
3-4-5 right △
A = 6 × 8 / 2 = **24**

13. diagonals 14 and 48

A = 14 × 48 / 2 = 336
each right tri is 7, 24, 25
A = BH = 25H
25H = 336   H = **13.44**

---

7. diagonals 22 and 38

$$\frac{22 \times 38}{2}$$ A = 11 × 38 = **418**

14. one angle 45°, perimeter 4√2̄

isos △
side = √2̄
H = **1**

Rhombus 2

Find the perimeter of the rhombus. | *Find the diagonal(s) of the rhombus.*

1.   diagonals 24 and 32

each right $\triangle$ is 12, 16, 20
   (3,4,5 x4)
side = 20
P = 20 x 4 = **80**

7.   perimeter 60, longer diagonal 24

side = 15
half longer diagonal = 12
9-12-15 right $\triangle$
shorter diagonal = 9 + 9 = **18**

2.   longer diagonal 18, one angle 60°

30-60-90 $\triangle$
$3\sqrt{3}$, 9, $6\sqrt{3}$
P = **$24\sqrt{3}$**

8.   perimeter 32, one diagonal 10

side = 8
half one diagonal = 5
25 + x^2 = 64
x^2 = 39
diagonal = **$2\sqrt{39}$**

3.   area 1320, shorter diagonal 22

$\dfrac{22D}{2}$ = 1320

D = 120

half diagonals = 11, 60
11, 60, 61 right $\triangle$
side = 61
P = 4 x 61 = **244**

9.   area 720, perimeter 164

side = 41
try 9-40-41 right $\triangle$
diagonals = **18, 80**
area = 18 x 80 / 2 = 720

4.   diagonals 30 and 72

each right $\triangle$ is 15, 36, 39
   (5,12,13 x3)
side = 39
P = 39 x 4 = **156**

10. one diagonal $8\sqrt{2}$, perimeter $16\sqrt{5}$

side = $4\sqrt{5}$
half diagonal = $4\sqrt{2}$
32 + x^2 = 80
x^2 = 48
x = $4\sqrt{3}$, diagonal = **$8\sqrt{3}$**

5.   diagonals 10 and 14

each right $\triangle$ is 5, 7, $\sqrt{74}$
P = **$4\sqrt{74}$**

11. area 840, perimeter 116

side = 29
try 20-21-29 right $\triangle$
diagonals = **40, 42**
area = 40 x 42 / 2 = 840

6.   one angle 135°, area $72\sqrt{2}$

one angle 45°, isos right $\triangle$
consider factors of 72
BH = 12 x $6\sqrt{2}$
side = 12
P = 12 x 4 = **48**

12. perimeter 20, ratio of diagonals 2:1

side = 5
diagonals = 2x, x
$4x^2 + x^2 = 5x^2$ = 25
x = $\sqrt{5}$
diagonals = **$4\sqrt{5}$, $2\sqrt{5}$**

Scientific Notation 1

Write in scientific notation.

1. 5468 5.468×10^3	17. 54,931 5.4931×10^4	33. $.4545 \times 10^{-4}$ 4.545×10^{-5}
2. 29 2.9×10^1	18. 44×10^3 4.4×10^4	34. .0078 7.8×10^{-3}
3. 0.932 9.32×10^{-1}	19. .000071 7.1×10^{-5}	35. $5^2 \times 3^2$ 2.25×10^2
4. 80,022 8.0022×10^4	20. .4378 4.378×10^{-1}	36. 12 x 400 4.8×10^3
5. 321 3.21×10^2	21. 12,968 1.2968×10^4	37. $58,932 \div 100$ 5.8932×10^2
6. 56 x 10 5.6×10^2	22. 22.675×10^4 2.2675×10^5	38. 10,001 1.0001×10^4
7. 895.2×10^6 8.952×10^8	23. $76.45 \div 100$ 7.645×10^{-1}	39. 157.9×10^7 1.579×10^9
8. 553×10^{-3} 5.53×10^{-1}	24. .0045 4.5×10^{-3}	40. 36 x 11 3.96×10^2
9. $321 \div 1000$ 3.21×10^{-1}	25. 36×10^{-4} 3.6×10^{-3}	41. .000422 4.22×10^{-4}
10. $.72 \times 10^3$ 7.2×10^2	26. 623×10^7 6.23×10^9	42. 85×10^2 8.5×10^3
11. 50 x 15 7.5×10^2	27. 387.93 3.8793×10^2	43. 675.8×10^5 6.758×10^7
12. 0.99 9.9×10^{-1}	28. 1911.7 1.9117×10^3	44. 903.6 9.036×10^2
13. 10,854 1.0854×10^4	29. 12 x 12 1.44×10^2	45. 20 x 30 x 40 2.4×10^4
14. $6519 \div 100$ 6.519×10^1	30. 78,935 7.8935×10^4	46. $.97 \times 10^{-3}$ 9.7×10^{-4}
15. 42.35×10^4 4.235×10^5	31. 25×10^5 2.5×10^6	47. .0265 2.65×10^{-2}
16. 300 x 14 4.2×10^3	32. $5370 \div 10$ 5.37×10^2	48. $4^2 \times 2^5$ 5.12×10^2

Scientific Notation 2

Write in scientific notation.

1. $(7 \times 10^{-6}) \div (8 \times 10^{-4})$

   0.875×10^{-2}

   8.75×10^{-3}

11. $(2.5 \times 10^{10}) \times (5 \times 10^{-3})$

    12.5×10^{7}

    1.25×10^{8}

2. $(5 \times 10^{-8}) \times (9 \times 10^{2})$

   45×10^{-6}

   4.5×10^{-5}

12. $(451 \times 10^{6}) \div (11 \times 10^{3})$

    41×10^{3}

    4.1×10^{4}

3. $(42 \times 10^{15}) \div (4 \times 10^{13})$

   10.5×10^{2}

   1.05×10^{3}

13. $(25 \times 10^{11}) \div (2 \times 10^{8})$

    12.5×10^{3}

    1.25×10^{4}

4. $(2 \times 10^{9}) \div (8 \times 10^{3})$

   0.25×10^{6}

   2.5×10^{5}

14. $(3 \times 10^{9}) \div (4 \times 10^{-3})$

    0.75×10^{12}

    7.5×10^{11}

5. $(9 \times 10^{-3}) \times (9 \times 10^{11})$

   81×10^{8}

   8.1×10^{9}

15. $(5 \times 10^{-2}) \times (7 \times 10^{13})$

    35×10^{11}

    3.5×10^{12}

6. $(64 \times 10^{3}) \div (5 \times 10^{6})$

   12.8×10^{-3}

   1.28×10^{-2}

16. $(253 \times 10^{-7}) \div (1.1 \times 10^{5})$

    230×10^{-12}

    2.3×10^{-10}

7. $(3 \times 10^{5}) \div (8 \times 10^{-5})$

   0.375×10^{10}

   3.75×10^{9}

17. $(1 \times 10^{5}) \div (8 \times 10^{-1})$

    0.125×10^{6}

    1.25×10^{5}

8. $(6 \times 10^{-3}) \times (6 \times 10^{-7})$

   36×10^{-10}

   3.6×10^{-9}

18. $(11 \times 10^{2}) \times (36 \times 10^{-17})$

    396×10^{-15}

    3.96×10^{-13}

9. $(5 \times 10^{4}) \div (8 \times 10^{-3})$

   0.625×10^{7}

   6.25×10^{6}

19. $(2.5 \times 10^{-2}) \div (10 \times 10^{5})$

    $.25 \times 10^{-7}$

    2.5×10^{-8}

10. $(3 \times 10^{2}) \div (12 \times 10^{-9})$

    0.25×10^{11}

    2.5×10^{10}

20. $(4.5 \times 10^{9}) \div (9 \times 10^{-1})$

    0.5×10^{10}

    5.0×10^{9}

Sequences 1

Find the next term, write the key, and circle A for arithmetic, G for geometric, or N for neither.

1. 6, 9, 12, 15, 18, __21__

   (A) G N Key: __+3__

2. 3, 12, 48, __192__

   A (G) N Key: __x4__

3. 5, –1, –7, –13, –19, __–25__

   (A) G N Key: __–6__

4. 1, 4, 9, 16, 25, __36__

   A G (N) Key: __n^2__

5. 2, 3, 5, 7, 11, 13, 17, __19__

   A G (N) Key: __primes__

6. 3, 10, 31, 94, __283__

   A G (N) Key: __x3 +1__

7. 1, 3, 6, 10, 15, 21, 28, __36__

   A G (N) Key: __+2, 3, 4, . . .__

8. 300, 150, 75, 37.5, __18.75__

   A (G) N Key: __x(1/2)__

9. 6, –12, 24, –48, 96, __–192__

   A (G) N Key: __x(–2)__

10. 1, 1, 2, 3, 5, 8, 13, 21, 34, __55__

    A G (N) Key: __Fibonacci: After 1, 1, add 2 previous.__

11. 3, 14, 58, 234, __938__

    A G (N) Key: __x4 +2__

12. 1, 8, 27, 64, 125, __216__

    A G (N) Key: __n^3__

13. 1, 4, 16, 64, __256__

    A (G) N Key: __x4__

14. 3, 7, 15, 31, 63, __127__

    A G (N) Key: __x2 +1__

15. 5, 15, 45, 135, __405__

    A (G) N Key: __x3__

16. 1, 2, 6, 24, 120, __720__

    A G (N) Key: __x2, 3, 4, 5, 6__

17. 500, 100, 20, 4, 0.8, __0.16__

    A (G) N Key: __x(1/5)__

18. 3.3, 4.5, 5.7, 6.9, __8.1__

    (A) G N Key: __+1.2__

19. 13, 16, 22, 31, 43, __58__

    A G (N) Key: __+3, 6, 9, 12, 15__

20. 13, 4, –5, –14, –23, __–32__

    (A) G N Key: __–9__

21. 6, 11, 21, 41, __81__ or +5, 10, 20, 40

    A G (N) Key: __x2 –1__

22. 20, 30, 50, 70, 110, 130, __170__

    A G (N) Key: __primes x 10__

Sequences 2

Find the specified ordinal term.

CL is "cycle length"
a | b is "a divides b"
R is "remainder"

1.　1900th of ABCDEABCDE . . . CL = 5 5 | 1900　　R0　　　E	12.　1466th of 76543217654321 . . . CL = 7 7 | 1463　　R3　　　5
2.　942nd of ABCDABCD . . . CL = 4 4 | 940　　R2　　　B	13.　767th of +&*@!%$+&*@!%$ . . . CL = 7 7 | 763　　R4　　　@
3.　1663rd of DEFGHDEFGH . . . CL = 5 5 | 1660　　R3　　　F	14.　598th of BOPHBOPH . . . CL = 4 4 | 596　　R2　　　O
4.　389th of @#$%&@#$%& . . . CL = 5 5 | 385　　R4　　　%	15.　5392nd of SWAESWAE . . . CL = 4 4 | 5392　　R0　　　E
5.　355th of ?+¢?+¢ . . . CL = 3 3 | 354　　R1　　　?	16.　1937th of B9X5B9X5 . . . CL = 4 4 | 1936　　R1　　　B
6.　667th of ABWXYZABWXYZ . . . CL = 6 6 | 666　　R1　　　A	17.　6431st of KLHGTKLHGT . . . CL = 5 5 | 6430　　R1　　　K
7.　557th of N8M9PN8M9P . . . CL = 5 5 | 555　　R2　　　8	18.　742nd of ≠Δ^≠Δ^≠Δ^ . . . CL = 3 3 | 741　　R1　　　≠
8.　843rd of 1234567812345678 . . . CL = 8 8 | 840　　R3　　　3	19.　797th of ABABAB . . . CL = 2 2 | 796　　R1　　　A
9.　758th of abcdefgabcdefg . . . CL = 7 7 | 756　　R2　　　b	20.　8613rd of @%$@%$@%$ . . . CL = 3 3 | 8613　　R0　　　$
10.　9572nd of ßπ∂ßπ∂ßπ∂ . . . CL = 3 3 | 9570　　R2　　　π	21.　2146th of mnopqswmnopqsw . . . CL = 7 7 | 2142　　R4　　　p
11.　6583rd of ©®∑§©®∑§ . . . CL = 4 4 | 6580　　R3　　　∑	22.　7898th of SEDRTSEDRT . . . CL = 5 5 | 7895　　R3　　　D

Sequences 3

Find the ones digit.

1. 97^{241} 7, 9, 3, 1 241/4 R1 **7**	12. 264^{265} 4, 6 265/2 R1 **4**	23. 293^{255} 3, 9, 7, 1 255/4 R3 **7**
2. 2^{391} 2, 4, 8, 6 391/4 R3 **8**	13. 624^{326} 4, 6 326/2 R0 **6**	24. 143^{512} 3, 9, 7, 1 512/4 R0 **1**
3. 293^{615} 3, 9, 7, 1 615/4 R3 **7**	14. 358^{170} 8, 4, 2, 6 170/4 R2 **4**	25. 17^{902} 7, 9, 3, 1 902/4 R2 **9**
4. 84^{781} 4, 6 781/2 R1 **4**	15. 172^{379} 2, 4, 8, 6 379/4 R3 **8**	26. 118^{987} 8, 4, 2, 6 987/4 R3 **2**
5. 37^{362} 7, 9, 3, 1 362/4 R2 **9**	16. 437^{272} 7, 9, 3, 1 272/4 R0 **1**	27. 355^{255} 5 only **5**
6. 325^{506} 5 only **5**	17. 93^{762} 3, 9, 7, 1 762/4 R2 **9**	28. 552^{988} 2, 4, 8, 6 988/4 R0 **6**
7. 189^{137} 9, 1 137/2 R1 **9**	18. 397^{175} 7, 9, 3, 1 175/4 R3 **3**	29. 559^{154} 9, 1 154/2 R0 **1**
8. 78^{765} 8, 4, 2, 6 765/4 R1 **8**	19. 532^{197} 2, 4, 8, 6 197/4 R1 **2**	30. 128^{521} 8, 4, 2, 6 521/4 R1 **8**
9. 237^{236} 7, 9, 3, 1 236/4 R0 **1**	20. 106^{234} 6 only **6**	31. 337^{333} 7, 9, 3, 1 333/4 R1 **7**
10. 112^{137} 2, 4, 8, 6 137/4 R1 **2**	21. 268^{268} 8, 4, 2, 6 268/4 R0 **6**	32. 353^{422} 3, 9, 7, 1 422/4 R2 **9**
11. 117^{183} 7, 9, 3, 1 183/4 R3 **3**	22. 802^{650} 2, 4, 8, 6 650/4 R2 **4**	33. 651^{156} 1 only **1**

Sequences 4

Find the specified term of the arithmetic sequence.

1. 4, 7, 10, 13, . . . Find 20th term. 4 + 3 x 19 = 4 + 57 = **61**	12. 40, 50, 60, 70, . . . Find 111th term. 40 + 10 x 110 = 40 + 1100 = **1140**
2. 4, 4.5, 5, 5.5, . . . Find 20th term. 4 + .5 x 19 = 4 + 9.5 = **13.5**	13. 2, 7, 12, 17, . . . Find 45th term. 2 + 5 x 44 = 2 + 220 = **222**
3. 8, 10, 12, 14, . . . Find 57th term. 8 + 2 x 56 = 8 + 112 = **120**	14. 15, 30, 45, 60, . . . Find 41st term. 15 + 15 x 40 = 15 + 600 = **615**
4. 50, 45, 40, 35, . . . Find 19th term. 50 + −5 x 18 = 50 − 90 = **−40**	15. 17, 11, 5, −1, . . . Find 81st term. 17 − 6 x 80 = 17 − 480 = **−463**
5. 80, 40, 0, −40, . . . Find 11th term. 80 − 40 x 10 = 80 − 400 = **−320**	16. −8, −5, −2, 1, . . . Find 17th term. −8 + 3 x 16 = −8 + 48 = **40**
6. 33, 44, 55, 66, . . . Find 26th term. 33 + 11 x 25 = 33 + 275 = **308**	17. 9, 20, 31, 42, . . . Find 28th term. 9 + 11 x 27 = 9 + 297 = **306**
7. 40, 36, 32, 28, . . . Find 43rd term. 40 + −4 x 42 = 40 − 168 = **−128**	18. 60, 25, −10, −45, . . . Find 11th term. 60 − 35 x 10 = 60 − 350 = **−290**
8. 11, 17, 23, 29, . . . Find 51st term. 11 + 6 x 50 = 11 + 300 = **311**	19. 12, 37, 62, 87, . . . Find 10th term. 12 + 25 x 9 = 12 + 225 = **237**
9. 13, 21, 29, 37, . . . Find 26th term. 13 + 8 x 25 = 13 + 200 = **213**	20. 40, 29, 18, 7, . . . Find 16th term. 40 − 11 x 15 = 40 − 165 = **−125**
10. 28, 25, 22, 19, . . . Find 32nd term. 28 − 3 x 31 = 28 − 93 = **−65**	21. −20, −10, 0, 10, . . . Find 14th term. −20 + 10 x 13 = −20 + 130 = **110**
11. 30, 38, 46, 54, . . . Find 71st term. 30 + 8 x 70 = 30 + 560 = **590**	22. −7, −3, 1, 5, . . . Find 21st term. −7 + 4 x 20 = −7 + 80 = **73**

Sequences 5

Find the next term in each sequence by successive differences.

1.

7		7		11		20		37		69		129		238		**427**
	0		4		9		17		32		60		109		189		
		4		5		8		15		28		49		80				
			1		3		7		13		21		31					
				2		4		6		8		10						

2.

0		12		25		41		62		90		127		**175**
	12		13		16		21		28		37		48		
		1		3		5		7		9		11			

3.

5		6		14		32		64		115		191		**299**
	1		8		18		32		51		76		108		
		7		10		14		19		25		32			
			3		4		5		6		7				

4.

6		11		20		35		60		101		166		265		**410**
	5		9		15		25		41		65		99		145		
		4		6		10		16		24		34		46			
			2		4		6		8		10		12				

5.

11		21		39		69		116		188		298		466		**721**
	10		18		30		47		72		110		168		255		
		8		12		17		25		38		58		87			
			4		5		8		13		20		29				
				1		3		5		7		9					

6.

11		21		40		69		112		176		271		410		**609**
	10		19		29		43		64		95		139		199		
		9		10		14		21		31		44		60			
			1		4		7		10		13		16				

7.

18		28		40		56		78		108		148		200		**266**
	10		12		16		22		30		40		52		66		
		2		4		6		8		10		12		14			

Sequences 6

For competition, memorize 6-digit pattern for 1/7.

Answer as indicated.

Find the specified digit.

1. 90, A, 76, 69, 62, B, 48, C, 34 Find A + B + C. −7 $\qquad$ 83 + 55 + 41 = **179**	**12.** 18th decimal digit of 3/11 $\dfrac{3}{11} = \dfrac{27}{99}$ $\quad$ $.\overline{27}$ $\qquad$ **7**
2. 3, A, 19, B, 35, 43, 51 Find AB. +8 $\qquad$ 11 × 27 = **297**	**13.** 37th decimal digit of 1/7 $\dfrac{1}{7}$ $\qquad$ $.\overline{142857}$ $\qquad$ **1**
3. 3, 15, 75, A, 1875, B Find B − A. $\qquad$ 5 × 1875 − 5 × 75 = ×5 $\qquad$ 5 × 1800 = **9000**	**14.** 64th decimal digit of 2/7 $\;$ Same 6 digits rotate for all 7ths. $\dfrac{2}{7}$ $\qquad$ $.\overline{285714}$ $\qquad$ **7**
4. 5, A, 17, 23, B, C, 41, 47 Find A + B + C. +6 $\qquad$ 11 + 29 + 35 = **75**	**15.** 85th decimal digit of 1/18 $\dfrac{1}{18} = \dfrac{.5}{9}$ $\quad$ $.0\overline{5}$ $\qquad$ **5**
5. 8, 20, 50, A, B, 781.25 Find A + B. ×2.5 $\qquad$ 125 + 312.5 = **437.5**	**16.** 44th decimal digit of 3/7 $\dfrac{3}{7}$ $\qquad$ $.\overline{428571}$ $\qquad$ **2**
6. 8, 19, A, 41, 52, B, C, 85 Find C(B − A). $\quad$ 74(63 − 30) = $\qquad$ 74 × 33 = +11 $\qquad$ 222 × 11 = **2442**	**17.** 98th decimal digit of 1/37 $\dfrac{1}{37}$ $\qquad$ $.\overline{027}$ $\qquad$ **2**
7. A, 1600, 1200, 800, B Find B − A. −400 $\qquad$ 400 − 2000 = **−1600**	**18.** 54th decimal digit of 2/37 $\;$ 2/37 is double 1/37 $\dfrac{2}{37}$ $\qquad$ $.\overline{054}$ $\qquad$ **4**
8. 60, 55.5, 51, A, B, 37.5, 33, C Find C − (A + B). 28.5 − (46.5 + 42) = $\qquad$ 28.5 − 88.5 = −4.5 $\qquad$ **−60**	**19.** 76th decimal digit of 4/37 $\;$ 54+54 = 108 see #18 $\dfrac{4}{37}$ $\qquad$ $.\overline{108}$ $\qquad$ **1**
9. 2, 6, 18, 54, A, B, 1458 Find B − A. ×3 $\qquad$ 486 − 162 = **324**	**20.** 35th decimal digit of 5/7 $\dfrac{5}{7}$ $\qquad$ $.\overline{714285}$ $\qquad$ **8**
10. 1000, 200, 40, A, 1.6, B Find A + B. ÷5 $\qquad$ 8 + .32 = **8.32**	**21.** 64th decimal digit of 6/7 $\dfrac{6}{7}$ $\qquad$ $.\overline{857142}$ $\qquad$ **1**
11. 6, 9.5, A, B, 20, 23.5, 27, C Find A + B + C. +3.5 $\qquad$ 13 + 16.5 + 30.5 = **60**	**22.** 100th decimal digit of 1/36 $\dfrac{1}{36}$ $\qquad$ $.02\overline{7}$ $\qquad$ **7**

Sequences 7

List the first 5 elements of the linear sequence for whole number input.	*Find the formula in functional notation with whole number input for the sequence.*
1. $f(n) = 3n - 4$ **-4, -1, 2, 5, 8**	14. $-2, 3, 8, 13, 18, \ldots$ **f(n) = 5n - 2**
2. $f(n) = -5n + 6$ **6, 1, -4, -9, -14**	15. $-7, -3, 1, 5, 9, \ldots$ **f(n) = 4n - 7**
3. $f(n) = 2n - 7$ **-7, -5, -3, -1, 1**	16. $-1, 5, 11, 17, 23, \ldots$ **f(n) = 6n - 1**
4. $f(n) = -4n - 2$ **-2, -6, -10, -14, -18**	17. $-3, -1, 1, 3, 5, \ldots$ **f(n) = 2n - 3**
5. $f(n) = 6n + 1$ **1, 7, 13, 19, 25**	18. $-8, -5, -2, 1, 4, \ldots$ **f(n) = 3n - 8**
6. $f(n) = 7n + 4$ **4, 11, 18, 25, 32**	19. $-4, 3, 10, 17, 24, \ldots$ **f(n) = 7n - 4**
7. $f(n) = 8n - 5$ **-5, 3, 11, 19, 27**	20. $6, 10, 14, 18, 22, \ldots$ **f(n) = 4n + 6**
8. $f(n) = 5n - 6$ **-6, -1, 4, 9, 14**	21. $12, 6, 0, -6, -12, \ldots$ **f(n) = -6n + 12**
9. $f(n) = 9n + 3$ **3, 12, 21, 30, 39**	22. $8, 11, 14, 17, 20, \ldots$ **f(n) = 3n + 8**
10. $f(n) = -3n + 7$ **7, 4, 1, -2, -5**	23. $10, 6, 2, -2, -6, \ldots$ **f(n) = -4n + 10**
11. $f(n) = -9n - 8$ **-8, -17, -26, -35, -44**	24. $5, -2, -9, -16, -23, \ldots$ **f(n) = -7n + 5**
12. $f(n) = -8n + 9$ **9, 1, -7, -15, -23**	25. $-5, -9, -13, -17, -21, \ldots$ **f(n) = -4n - 5**
13. $f(n) = -2n - 1$ **-1, -3, -5, -7, -9**	26. $-6, 3, 12, 21, 30, \ldots$ **f(n) = 9n - 6**

Add d n−1 times to the first term to reach the nth term.	Sequences 8	F (first) is the same as a_1 in the formula (F+L)n/2.

Find the first term a_1 of the arithmetic sequence given two terms. | | *Find the first term a_1 and the common difference d of the arithmetic sequence.*

1.
$a_6 = 24$
$a_{20} = 80$

24 + 14d = 80
14d = 56
d = 4
24 − (4 × 5) = **4**

11. sum of 1st 20 terms = 1280
$a_{20} = 121$ $\dfrac{(F+121)(20)}{2}$ F + 121 = 128
F = 7
d = 6

2.
$a_5 = −13$
$a_{30} = 62$

−13 + 25d = 62
25d = 75
d = 3
−13 − (3 × 4) = **−25**

12. sum of 1st 25 terms = 950
$a_{25} = 92$ $\dfrac{(F+92)(25)}{2}$ F + 92 = 76
F = −16
d = 4.5

3.
$a_8 = −10$
$a_{41} = 221$

−10 + 33d = 221
33d = 231 = 11 × 21
d = 7
−10 − (7 × 7) = **−59**

13. sum of 1st 37 terms = 5439
$a_{37} = 282$ $\dfrac{(F+282)(37)}{2}$ F + 282 = 294
F = 12
d = 7.5

4.
$a_7 = 9$
$a_{35} = 233$

9 + 28d = 233
28d = 224
d = 8
9 − (8 × 6) = **−39**

14. sum of 1st 31 terms = 1798
$a_{31} = 166$ $\dfrac{(F+166)(31)}{2}$ F + 166 = 116
F = −50
d = 7.2

5.
$a_{11} = 44$
$a_{45} = −58$

44 + 34d = −58
34d = −102
d = −3
44 − (−3 × 10) = **74**

15. sum of 1st 19 terms = 2356
$a_{19} = 232$ $\dfrac{(F+232)(19)}{2}$ F + 232 = 248
F = 16
d = 12

6.
$a_9 = 31$
$a_{53} = 97$

31 + 44d = 97
44d = 66
d = 1.5
31 − (1.5 × 8) = **19**

16. sum of 1st 17 terms = 1700
$a_{17} = 156$ $\dfrac{(F+156)(17)}{2}$ F + 156 = 200
F = 44
d = 7

7.
$a_{12} = −12$
$a_{42} = 60$

−12 + 30d = 60
30d = 72
d = 12/5 = 2.4
−12 − (2.4 × 11) = **−38.4**

17. sum of 1st 21 terms = 1281
$a_{21} = 111$ $\dfrac{(F+111)(21)}{2}$ F + 111 = 122
F = 11
d = 5

8.
$a_{13} = 10$
$a_{33} = 80$

10 + 20d = 80
20d = 70
d = 7/2 = 3.5
10 − (3.5 × 12) = **−32**

18. sum of 1st 15 terms = 1245
$a_{15} = 146$ $\dfrac{(F+146)(15)}{2}$ F + 146 = 166
F = 20
d = 9

9.
$a_{10} = 195$
$a_{55} = 420$

195 + 45d = 420
45d = 225
d = 5
195 − (5 × 9) = **150**

19. sum of 1st 45 terms = 3195
$a_{45} = 137$ $\dfrac{(F+137)(45)}{2}$ F + 137 = 142
F = 5
d = 3

10.
$a_{20} = 150$
$a_{58} = 378$

150 + 38d = 378
38d = 228
d = 6
150 − (6 × 19) = **36**

20. sum of 1st 13 terms = 806
$a_{13} = 110$ $\dfrac{(F+110)(13)}{2}$ F + 110 = 124
F = 14
d = 8

Series 1

Find the sum by the formula $(F + L)n/2$.

If consecutive, can also subtract L–F and add 1 to get the number of numbers.

1.　$20 + 21 + 22 + \ldots + 78 + 79 + 80$ $\dfrac{(20 + 80)\,(61)}{2}$　$50 \times 61 = \mathbf{3050}$ 　　　　　　$80 \text{ nos} - 19 \text{ nos} = 61 \text{ nos}$	9.　$11 + 13 + 15 + \ldots + 87 + 89 + 91$ $\dfrac{(11 + 91)\,(41)}{2}$　$51 \times 41 = \mathbf{2091}$ 　　　　　$46 \text{ odds} - 5 \text{ odds} = 41 \text{ nos}$
2.　$16 + 20 + 24 + \ldots + 64 + 68 + 72$ $\dfrac{(16 + 72)\,(15)}{2}$　$4 \times 11 \times 15 = \mathbf{660}$ 　　　　$4\text{x}4, 4\text{x}5, \ldots, 4\text{x}18$　$18{-}3 = 15 \text{ nos}$	10.　$15 + 20 + 25 + \ldots + 75 + 80 + 85$ $\dfrac{(15 + 85)\,(15)}{2}$　$50 \times 15 = \mathbf{750}$ 　　　　$5\text{x}3, 5\text{x}4, \ldots, 5\text{x}17$　$17{-}2 = 15 \text{ nos}$
3.　$18 + 24 + 30 + \ldots + 84 + 90 + 96$ $\dfrac{(18 + 96)\,(14)}{2}$　$114 \times 7 = \mathbf{798}$ 　　　　$6\text{x}3, 6\text{x}4, \ldots, 6\text{x}16$　$16{-}2 = 14 \text{ nos}$	11.　$31 + 38 + 45 + 52 + 59 + 66 + 73$ $\dfrac{(31 + 73)\,(7)}{2}$　$52 \times 7 = \mathbf{364}$
4.　$100 + 105 + 110 + \ldots + 245 + 250$ $\dfrac{(100 + 250)\,(31)}{2}$　$175 \times 31 = \mathbf{5425}$ 　$5\text{x}20, 5\text{x}21, \ldots, 5\text{x}50$　$50{-}19 = 31 \text{ nos}$	12.　$300 + 301 + \ldots + 499 + 500$ $\dfrac{(300 + 500)\,(201)}{2}$　$400 \times 201 = \mathbf{80{,}400}$ 　　　　$500 \text{ nos} - 299 \text{ nos} = 201 \text{ nos}$
5.　$32 + 33 + 34 + \ldots + 76 + 77 + 78$ $\dfrac{(32 + 78)\,(47)}{2}$　$5 \times 11 \times 47 = \mathbf{2585}$ 　　　　　$78 \text{ nos} - 31 \text{ nos} = 47 \text{ nos}$	13.　$44 + 50 + 56 + \ldots + 86 + 92 + 98$ $\dfrac{(44 + 98)\,(10)}{2}$　$71 \times 10 = \mathbf{710}$
6.　$28 + 29 + 30 + \ldots + 84 + 85 + 86$ $\dfrac{(28 + 86)\,(59)}{2}$　$57 \times 59 = \mathbf{3363}$ 　　　　　$86 \text{ nos} - 27 \text{ nos} = 59 \text{ nos}$	14.　$6 + 9 + 12 + \ldots + 66 + 69 + 72$ $\dfrac{(6 + 72)\,(23)}{2}$　$\begin{array}{l} 39 \times 23 = 780 + 117 = \\ \mathbf{897} \end{array}$ 　$3\text{x}2, 3\text{x}3, \ldots, 3\text{x}24$　$24{-}1 = 23 \text{ nos}$
7.　$2 + 4 + 6 + \ldots + 64 + 66 + 68$ $\dfrac{(2 + 68)\,(34)}{2}$　$\begin{array}{l} 35 \times 34 = 1050 + 140 = \\ \mathbf{1190} \end{array}$ 　　$2\text{x}1, 2\text{x}2, \ldots, 2\text{x}34$　34 nos	15.　$1 + 8 + 15 + \ldots + 78 + 85 + 92$ $\dfrac{(1 + 92)\,(14)}{2}$　$93 \times 7 = \mathbf{651}$
8.　$50 + 52 + 54 + \ldots + 94 + 96 + 98$ $\dfrac{(50 + 98)\,(25)}{2}$　$74 \times 25 = \mathbf{1850}$ 　$2\text{x}25, 2\text{x}26, \ldots, 2\text{x}49$　$49{-}24 = 25 \text{ nos}$	16.　$1 + 2 + 3 + \ldots + 198 + 199 + 200$ $\dfrac{(1 + 200)\,(200)}{2}$　$201 \times 100 = \mathbf{20{,}100}$

Series 2

Find the sum by the formula for consecutive odds starting at one.

$$1$$
$$1 + 3 = 4$$
$$1 + 3 + 5 = 9$$
$$1 + 3 + 5 + 7 = 16$$

1. the first 7 odd numbers $7^2 = \textbf{49}$	11. $1 + 3 + 5 + 7 + \ldots + 49$ 1 to 50 is 50 nos. $\quad 25^2 = \textbf{625}$ 25 evens, 25 odds
2. the first 11 odd numbers $11^2 = \textbf{121}$	12. $1 + 3 + 5 + 7 + \ldots + 79$ 1 to 80 is 80 nos. $\quad 40^2 = \textbf{1600}$ 40 evens, 40 odds
3. the first 15 odd numbers $15^2 = \textbf{225}$	13. $1 + 3 + 5 + 7 + \ldots + 119$ 1 to 120 is 120 nos. $\quad 60^2 = \textbf{3600}$ 60 evens, 60 odds
4. the first 20 odd numbers $20^2 = \textbf{400}$	14. $1 + 3 + 5 + 7 + \ldots + 149$ 1 to 150 is 150 nos. $\quad 75^2 = \textbf{5625}$ 75 evens, 75 odds
5. the first 49 odd numbers $49^2 = \textbf{2401}$ $\quad$ $50^2 = 2500$ 49th odd is 99 $2500 - 99 = 2401$	15. $1 + 3 + 5 + 7 + \ldots + 179$ 1 to 180 is 180 nos. $\quad 90^2 = \textbf{8100}$ 90 evens, 90 odds
6. the first 70 odd numbers $70^2 = \textbf{4900}$	16. $1 + 3 + 5 + 7 + \ldots + 199$ 1 to 200 is 200 nos. $\quad 100^2 = \textbf{10,000}$ 100 evens, 100 odds
7. the first 80 odd numbers $80^2 = \textbf{6400}$	17. $1 + 3 + 5 + 7 + \ldots + 249$ 1 to 250 is 250 nos. $\quad 125^2 = \textbf{15,625}$ 125 evens, 125 odds
8. the first 91 odd numbers $91^2 = \textbf{8281}$	18. $1 + 3 + 5 + 7 + \ldots + 299$ 1 to 300 is 300 nos. $\quad 150^2 = \textbf{22,500}$ 150 evens, 150 odds
9. the first 100 odd numbers $100^2 = \textbf{10,000}$	19. $1 + 3 + 5 + 7 + \ldots + 399$ 1 to 400 is 400 nos. $\quad 200^2 = \textbf{40,000}$ 200 evens, 200 odds
10. the first 200 odd numbers $200^2 = \textbf{40,000}$	20. $1 + 3 + 5 + 7 + \ldots + 499$ 1 to 500 is 500 nos. $\quad 250^2 = \textbf{62,500}$ 250 evens, 250 odds

Sets 1

Check for yes and dash for no if the number belongs to the set.

1.

	25	1.5	$\frac{2}{7}$	$|{-2}|$	0	$\sqrt{16}$	$\frac{-6}{3}$	$\sqrt{-36}$	-17	$9.\overline{4}$	$9\frac{3}{11}$
N	✔	–	–	✔	–	✔	–	–	–	–	–
W	✔	–	–	✔	✔	✔	–	–	–	–	–
Z	✔	–	–	✔	✔	✔	✔	–	✔	–	–
Q	✔	✔	✔	✔	✔	✔	✔	–	✔	✔	✔
R	✔	✔	✔	✔	✔	✔	✔	–	✔	✔	✔

2.

	$\frac{3}{5}$	π	$\sqrt{81}$	64	$2.\overline{14}$	-0.6	$-|{-8}|$	-1	$27\frac{4}{5}$	$\frac{56}{7}$	$\sqrt{-4}$
N	–	–	✔	✔	–	–	–	–	–	✔	–
W	–	–	✔	✔	–	–	–	–	–	✔	–
Z	–	–	✔	✔	–	–	✔	✔	–	✔	–
Q	✔	–	✔	✔	✔	✔	✔	✔	✔	✔	–
R	✔	✔	✔	✔	✔	✔	✔	✔	✔	✔	–

3.

	-8	$12\frac{3}{8}$	$\frac{63}{9}$	$37.\overline{11}$	$\sqrt{-9}$	121	$\frac{1}{4}$	$\sqrt{49}$	.053	4π	$|2-3|$
N	–	–	✔	–	–	✔	–	✔	–	–	✔
W	–	–	✔	–	–	✔	–	✔	–	–	✔
Z	✔	–	✔	–	–	✔	–	✔	–	–	✔
Q	✔	✔	✔	✔	–	✔	✔	✔	✔	–	✔
R	✔	✔	✔	✔	–	✔	✔	✔	✔	✔	✔

4.

	12.8	$|{-45}|$	$\sqrt{-26}$	-81	$\sqrt{144}$	3.888	$\frac{35}{7}$	$10.\overline{64}$	100	$\frac{4}{9}$	$33\frac{1}{3}$
N	–	✔	–	–	✔	–	✔	–	✔	–	–
W	–	✔	–	–	✔	–	✔	–	✔	–	–
Z	–	✔	–	✔	✔	–	✔	–	✔	–	–
Q	✔	✔	–	✔	✔	✔	✔	✔	✔	✔	✔
R	✔	✔	–	✔	✔	✔	✔	✔	✔	✔	✔

Sets 2

Complete with the set name and/or operation(s). Also name the property.	Find the intersection.
1. $A \cup B = B \cup$ __A__ **CPU**	17. {primes} ∩ {multiples of 11} **{11}**
2. $B \cap C =$ __C__ $\cap B$ **CPI**	18. {zero} ∩ {evens} **{0}**
3. $(A \cup F) \cup H = A \cup (F$ __U__ $H)$ **APU**	19. {odds} ∩ {evens} **{ }**
4. $M \cup ($ __N__ ∩ __T__ $) = (M \cup N) \cap (M \cup T)$ **DPUI**	20. {primes} ∩ {factors of 30} **{2, 3, 5}**
5. F __∩__ $(C \cap D) = (F \cap C)$ __∩__ D **API**	21. {factors of 36} ∩ {multiples of 4} **{4, 12, 36}**
6. $G \cup B = B$ __U__ G **CPU**	22. {2-digit product of 2 primes} ∩ {factors of 330} **{10, 15, 22, 33, 55}**
7. $B \cap (C \cap D) = (B \cap C) \cap$ __D__ **APU**	23. {composites} ∩ {factors of 20} **{4, 10, 20}**
8. $Y \cap$ __X__ $= X \cap Y$ **CPI**	24. {perfect squares} ∩ {factors of 500} **{1, 4, 25, 100}**
9. __T__ $\cup S = S \cup T$ **CPU**	25. {perfect squares} ∩ {2-digit perfect cubes} **{64}**
10. $(D$ __U__ $F) \cup S = D \cup (F$ __U__ $S)$ **APU**	26. {primes} ∩ {evens} **{2}**
11. $K \cap (L \cup M) = (K$ __∩__ $L) \cup (K$ __∩__ $M)$ **DPIU**	27. {x-axis ordered pairs} ∩ {y-axis ordered pairs} **{ (0,0) }**
12. $($ __J__ $\cap H) \cap K = J \cap (H \cap K)$ **API**	28. $\{x \mid x \ \varepsilon \ N, 2x < 8 \} \cap \{x \mid x \ \varepsilon \ N, 3 \le 3x \le 9\}$ **{1, 2, 3}**
13. $(B \cup C) \cap D = (B \cap D) \cup (C \cap$ __D__ $)$ **DPIU**	29. $\{x \mid x \ \varepsilon \ Z, 4x < 28 \} \cap \{x \mid x \ \varepsilon \ Z, 3x > 9\}$ **{4, 5, 6}**
14. __B__ $\cup (C \cup E) = (B \cup C) \cup E$ **APU**	30. $\{x \mid x \ \varepsilon \ W, 3x < 9 \} \cap \{x \mid x \ \varepsilon \ W, -2 < 7x < 20\}$ **{0, 1, 2}**
15. $X \cup (S$ __∩__ $T) = (X \cup S)$ __∩__ $(X \cup T)$ **DPUI**	31. $\{x \mid x \ \varepsilon \ Z, 5x \ge 25 \} \cap \{x \mid x \ \varepsilon \ Z, -5 \le 3x < 25\}$ **{5, 6, 7, 8}**
16. F __∩__ $(G \cup H) = (F$ __∩__ $G) \cup ($ __F__ $\cap H)$ **DPIU**	32. $\{x \mid x \ \varepsilon \ N, 4x \le 9 \} \cap \{x \mid x \ \varepsilon \ N, 3x + 4 > 9\}$ **{2}**

Sets 3

Operate as indicated.

| A = {1, 2, 5, 7}   C = {1, 3, 5, 7}
B = {2, 4, 7, 8}   D = {2, 4, 6, 8}
U = {1, 2, 3, 4, 5, 6, 7, 8, 9} | A = {1, 2, 3, 5, 6}   C = {1, 3, 5, 7}
B = {2, 4, 5, 8}   D = {1, 3, 9}
U = {1, 2, 3, 4, 5, 6, 7, 8, 9, 10} |

1. A∪C

   {1, 2, 3, 5, 7}

10. (A∪B)∪C

    {1, 2, 3, 4, 5, 6, 7, 8}

2. C∩D

   { }

11. (A∩B)∩C

    {2, 5} ∩ {1, 3, 5, 7}
    {5}

3. A∩C

   {1, 5, 7}

12. A∪(B∩C)

    A∪{5}
    A   or {1, 2, 3, 5, 6}

4. C′

   {2, 4, 6, 8, 9}

13. A∩(B∪D)

    {1, 2, 3, 5, 6} ∩ {1, 2, 3, 4, 5, 8, 9}
    {1, 2, 3, 5}

5. B′∪C

   {1, 3, 5, 6, 9} ∪ {1, 3, 5, 7}
   {1, 3, 5, 6, 7, 9}

14. (A∪D)∪B′

    {1, 2, 3, 5, 6, 9} ∪ {1, 3, 6, 7, 9, 10}
    {1, 2, 3, 5, 6, 7, 9, 10}

6. B∪(D′∩A)

   {2, 4, 7, 8} ∪ {1, 5, 7}
   {1, 2, 4, 5, 7, 8}

15. (A∪C)∩D′

    {1, 2, 3, 5, 6, 7} ∩ {2, 4, 5, 6, 7, 8, 10}
    {2, 5, 6, 7}

7. (A∪B)∩(C∪D)

   {1, 2, 4, 5, 7, 8} ∩ {1, 2, 3, 4, 5, 6, 7, 8}
   {1, 2, 4, 5, 7, 8}

16. (C′∪B)∪A

    {2, 4, 5, 6, 8, 9, 10} ∪ {1, 2, 3, 5, 6}
    {1, 2, 3, 4, 5, 6, 8, 9, 10}

8. A∩(D′∪C)

   {1, 2, 5, 7} ∩ {1, 3, 5, 7, 9}
   {1, 5, 7}

17. (B∪D)∪A′

    {1, 2, 3, 4, 5, 8, 9} ∪ {4, 7, 8, 9, 10}
    {1, 2, 3, 4, 5, 7, 8, 9, 10}

9. (B′∩C)∪(A′∩D)

   {1, 3, 5} ∪ {4, 6, 8}
   {1, 3, 4, 5, 6, 8}

18. D∩(A∪C′)

    {1, 3, 9} ∩ {1, 2, 3, 4, 5, 6, 8, 9, 10}
    {1, 3, 9} or D

Sets 4

Draw a Venn diagram of the sets to show their relationship.

1. A = {2, 4, 6, 8}   B = {2, 3, 4, 5}

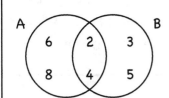

2. A = {1, 2, 3, 4}   B = {5, 6, 7}

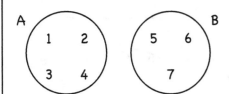

3. A = {1, 2, 3, 4, 5}   B = {2, 4, 5}

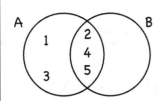

4. A = {1, 2, 3, 4, 5}   B = {3, 5, 7}
   C = {6, 7, 8, 9}

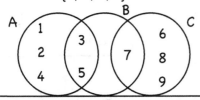

5. A = {0, 3, 4, 7}   B = {1, 3, 4}
   C = {2, 5}

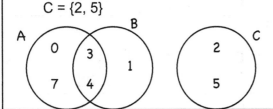

6. A = {2, 4, 6, 8}
   B = {3, 4, 8, 9}
   C = {1, 6, 7, 8, 9}

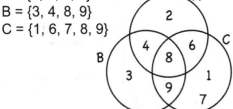

7. A = {1, 2, 3, 4}
   B = {5, 6}
   C = {1, 2, 5, 6, 7, 8}

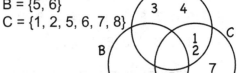

8. A = {2, 8, 9}   B = {1, 4, 5, 7}
   C = {0, 4, 5, 6}

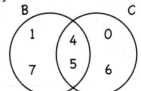

9. A = {0, 1, 5}   B = {3, 7, 8}
   C = {0, 1, 2, 4, 5, 6}

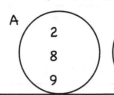

10. A = {1, 2, 3, 6, 7}
    B = {1, 2}
    C = {1, 2, 3, 4, 5}

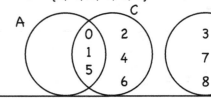

11. A = {1, 2, 9}   B = {3, 4, 5}
    C = {0, 6, 7, 8}

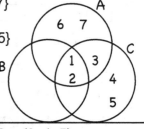

12. A = {1, 3, 4, 6, 9}
    B = {0, 1, 2, 8}
    C = {1, 2, 6, 9}

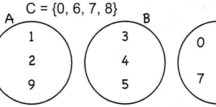

MAVA Math: Enhanced Skills Solutions   Copyright © 2015 Marla Weiss

Sets 5

Operate on the set of points in the figures.

Note that answers for rays and lines may vary due to several ways to name them. Also, the order of segment endpoints may reverse.

1. $\overline{AB} \cap \overline{BC}$ **B**	14. $\overline{AD} \cap \overline{GH}$ **$\overline{GH}$**	27. $\overleftrightarrow{AB} \cap \overline{BC}$ **$\overline{BC}$**
2. $\overline{AC} \cap \overline{BD}$ **E**	15. $\overline{AD} \cup \overline{GH}$ **$\overline{AD}$**	28. $\overline{AB} \cap \overline{DC}$ **∅**
3. $\overline{AE} \cup \overline{EC}$ **$\overline{AC}$**	16. $\overline{EH} \cap \overline{HC}$ **H**	29. $\overrightarrow{BA} \cap \overrightarrow{BC}$ **B**
4. $\overrightarrow{DA} \cup \overrightarrow{DC}$ **∠ADC**	17. $\overline{AE} \cup \overline{EH} \cup \overline{AH}$ **△AEH**	30. $\overline{AB} \cup \overline{BC}$ **$\overline{AC}$**
5. $\overline{BE} \cap \overline{AD}$ **∅**	18. $\overrightarrow{GH} \cup \overrightarrow{AH}$ **$\overrightarrow{AH}$**	31. $\overrightarrow{BC} \cup \overrightarrow{CD}$ **$\overrightarrow{BD}$**
6. $\overline{AB} \cup \overline{BE} \cup \overline{AE}$ **△ABE**	19. $\overline{AG} \cup \overline{GH} \cup \overline{HD}$ **$\overline{AD}$**	32. $\overline{AD} \cap \overline{AB}$ **$\overline{AB}$**
7. $\overline{BC} \cap \overline{CD}$ **C**	20. $\overline{AB} \cap \overline{CD}$ **∅**	33. $\overrightarrow{AC} \cap \overrightarrow{DC}$ **$\overline{AD}$**
8. $\overline{AB} \cap \overline{CD}$ **∅**	21. $\overrightarrow{CF} \cup \overrightarrow{CE}$ **∠FCE**	34. $\overline{AC} \cup \overline{CD}$ **$\overline{AD}$**
9. $\overrightarrow{EB} \cup \overrightarrow{EC}$ **∠BEC**	22. $\overline{AG} \cup \overline{GH}$ **$\overline{AH}$**	35. $\overline{AB} \cap \overline{BD}$ **$\overline{BD}$**
10. $\overrightarrow{DB} \cap \overrightarrow{EB}$ **$\overrightarrow{EB}$**	23. $\overline{AE} \cap \overline{AF}$ **$\overline{AF}$**	36. $\overline{CD} \cap \overline{AD}$ **$\overline{CD}$**
11. $\overrightarrow{EC} \cup \overrightarrow{EA}$ **$\overleftrightarrow{AC}$**	24. $\overrightarrow{FG} \cap \overrightarrow{GC}$ **$\overrightarrow{GC}$**	37. $\overrightarrow{BA} \cup \overrightarrow{BC}$ **$\overleftrightarrow{AC}$**
12. $\overline{BD} \cap \overline{CD}$ **D**	25. $\overline{GH} \cup \overline{HC} \cup \overline{GC}$ **△GCH**	38. $\overrightarrow{AB} \cup \overrightarrow{CD}$ **$\overrightarrow{AD}$**
13. $\overline{BC} \cup \overline{BD} \cup \overline{CD}$ **△BCD**	26. $\overrightarrow{GA} \cap \overrightarrow{HD}$ **∅**	39. $\overline{AD} \cap \overline{BC}$ **$\overline{BC}$**

Sets 6

2^n where n is the number of elements

Operate on the infinite sets.

Find the number of subsets.

1. Z ∪ W = __z__	17. {Bob, Joe}	**4**	
2. N ∩ W = __N__	18. {0, 1, 2, 3, 4, 5, 6, 7, 8}	**512**	
3. W ∩ Q = __W__	19. {H, E, L, L, O} omit one L	**16**	
4. N ∪ Z = __z__	20. {10, 20, 30}	**8**	
5. Q ∩ Z = __z__	21. {n, u, m, b, e, r}	**64**	
6. R ∩ Q = __Q__	22. {c, l, a, p} ∪ {h, a, n, d, s} omit one a	**256**	
7. N ∪ Q = __Q__	23. {c, o, u, n, t}	**32**	
8. Z ∪ (Q ∪ N) = __Q__	24. {s, e, t} ∪ {t, h, e, o, r, y} omit one e, t	**128**	
9. R ∩ (N ∪ W) = __W__	25. {M, A, T, H}	**16**	
10. Q ∩ (N ∪ Z) = __z__	26. {A, L, G, E, B, R, A} omit one A	**64**	
11. (Q ∪ Z) ∩ R = __Q__	27. {h, o, u, s, e} ∪ {b, o, a, t} omit one o	**256**	
12. (Z ∪ R) ∪ W = __R__	28. {1, 3, 5, 7} ∪ {2, 4, 6, 8}	**256**	
13. Z ∩ (Q ∪ R) = __z__	29. {9, 8, 7, 6, 5, 4, 3, 2, 1, 0}	**1024**	
14. W ∩ (Q ∪ N) = __W__	30. {3, 4, 5, 6} ∪ {6, 7, 8, 9} omit one 6	**128**	
15. Z ∪ (W ∩ R) = __z__	31. {v, a, l, u, e}	**32**	
16. W ∪ (Q ∩ N) = __W__	32. {23, 24, 24, . . . , 33} subtract and add 1 for the number of numbers	**2048**	

Spheres 1

Find the volume and surface area of the sphere with the specified dimension.

---

1. r = 3

$V = \dfrac{4\pi(3\times3\times3)}{3} = \mathbf{36\pi}$ SA = 4π(3×3) = **36π**

9. d = 8

$V = \dfrac{4\pi(4\times4\times4)}{3} = \dfrac{\mathbf{256\pi}}{\mathbf{3}}$ SA = 4π(4×4) = **64π**

---

2. d = 18

$V = \dfrac{4\pi(9\times9\times9)}{3} = \mathbf{972\pi}$ SA = 4π(9×9) = **324π**

10. r = 3.5

$V = \dfrac{4\pi(7\times7\times7)}{3\times2\times2\times2} = \dfrac{\mathbf{343\pi}}{\mathbf{6}}$ $SA = \dfrac{4\pi(7\times7)}{2\times2} = \mathbf{49\pi}$

---

3. r = 1.5

$V = \dfrac{4\pi(3\times3\times3)}{3\times2\times2\times2} = \dfrac{\mathbf{9\pi}}{\mathbf{2}}$ $SA = \dfrac{4\pi(3\times3)}{2\times2} = \mathbf{9\pi}$

11. r = 5

$V = \dfrac{4\pi(5\times5\times5)}{3} = \dfrac{\mathbf{500\pi}}{\mathbf{3}}$ SA = 4π(5×5) = **100π**

---

4. d = 5

$V = \dfrac{4\pi(5\times5\times5)}{3\times2\times2\times2} = \dfrac{\mathbf{125\pi}}{\mathbf{6}}$ $SA = \dfrac{4\pi(5\times5)}{2\times2} = \mathbf{25\pi}$

12. r = 1/3

$V = \dfrac{4\pi(1\times1\times1)}{3\times3\times3\times3} = \dfrac{\mathbf{4\pi}}{\mathbf{81}}$ $SA = \dfrac{4\pi(1\times1)}{3\times3} = \dfrac{\mathbf{4\pi}}{\mathbf{9}}$

---

5. r = 6

$V = \dfrac{4\pi(6\times6\times6)}{3} = \mathbf{288\pi}$ SA = 4π(6×6) = **144π**

13. d = 30

$V = \dfrac{4\pi(15\times15\times15)}{3} = \mathbf{4500\pi}$ SA = 4π(15×15) = **900π**

---

6. r = 1

$V = \dfrac{4\pi(1\times1\times1)}{3} = \dfrac{\mathbf{4\pi}}{\mathbf{3}}$ SA = 4π(1×1) = **4π**

14. r = .25

$V = \dfrac{4\pi(1\times1\times1)}{3\times4\times4\times4} = \dfrac{\boldsymbol{\pi}}{\mathbf{48}}$ $SA = \dfrac{4\pi(1\times1)}{4\times4} = \dfrac{\boldsymbol{\pi}}{\mathbf{4}}$

---

7. r = 2

$V = \dfrac{4\pi(2\times2\times2)}{3} = \dfrac{\mathbf{32\pi}}{\mathbf{3}}$ SA = 4π(2×2) = **16π**

15. d = 24

$V = \dfrac{4\pi(12\times12\times12)}{3} = \mathbf{2304\pi}$ SA = 4π(12×12) = **576π**

---

8. d = 1

$V = \dfrac{4\pi(1\times1\times1)}{3\times2\times2\times2} = \dfrac{\boldsymbol{\pi}}{\mathbf{6}}$ $SA = \dfrac{4\pi(1\times1)}{2\times2} = \boldsymbol{\pi}$

16. d = 9

$V = \dfrac{4\pi(9\times9\times9)}{3\times2\times2\times2} = \dfrac{\mathbf{243\pi}}{\mathbf{2}}$ $SA = \dfrac{4\pi(9\times9)}{2\times2} = \mathbf{81\pi}$

Spheres 2

Find the ratio.

The 4π or 4π/3 in the formulae simplify so do not need to be written.

---

1. Find the ratio of the volumes of two spheres whose surface areas are 36 and 64.

$$\frac{sa}{SA} = \frac{r \times r}{R \times R} = \frac{36}{64} \qquad \begin{array}{l} r = 6 \\ R = 8 \end{array}$$

$$\frac{v}{V} = \frac{r \times r \times r}{R \times R \times R} = \frac{3 \times 3 \times 3}{4 \times 4 \times 4} = \frac{\mathbf{27}}{\mathbf{64}}$$

6. Find the ratio of the volumes of two spheres whose surface areas are 20 and 245.

$$\frac{sa}{SA} = \frac{r \times r}{R \times R} = \frac{20}{245} \qquad \begin{array}{l} r = 2\sqrt{5} \\ R = 7\sqrt{5} \end{array}$$

$$\frac{v}{V} = \frac{r \times r \times r}{R \times R \times R} = \frac{\mathbf{8}}{\mathbf{343}} \qquad \text{The root 5s simplify.}$$

---

2. Find the ratio of the volumes of two spheres whose surface areas are 12 and 27.

$$\frac{sa}{SA} = \frac{r \times r}{R \times R} = \frac{12}{27} \qquad \begin{array}{l} r = 2\sqrt{3} \\ R = 3\sqrt{3} \end{array}$$

$$\frac{v}{V} = \frac{r \times r \times r}{R \times R \times R} = \frac{\mathbf{8}}{\mathbf{27}} \qquad \text{The root 3s simplify.}$$

7. Find the ratio of the surface areas of 2 spheres whose volumes are in the ratio 24:81.

$$\frac{v}{V} = \frac{r \times r \times r}{R \times R \times R} = \frac{24}{81} = \frac{8}{27} \qquad \begin{array}{l} r = 2 \\ R = 3 \end{array}$$

$$\frac{sa}{SA} = \frac{r \times r}{R \times R} = \frac{2 \times 2}{3 \times 3} = \frac{\mathbf{4}}{\mathbf{9}}$$

---

3. Find the ratio of the surface areas of two spheres with volumes 64 and 729.

$$\frac{v}{V} = \frac{r \times r \times r}{R \times R \times R} = \frac{64}{729} \qquad \begin{array}{l} r = 4 \\ R = 9 \end{array}$$

$$\frac{sa}{SA} = \frac{r \times r}{R \times R} = \frac{4 \times 4}{9 \times 9} = \frac{\mathbf{16}}{\mathbf{81}}$$

8. Find the ratio of the volumes of two spheres whose surface areas are in the ratio 12:75.

$$\frac{sa}{SA} = \frac{r \times r}{R \times R} = \frac{12}{75} = \frac{4}{25} \qquad \begin{array}{l} r = 2 \\ R = 5 \end{array}$$

$$\frac{v}{V} = \frac{r \times r \times r}{R \times R \times R} = \frac{\mathbf{8}}{\mathbf{125}}$$

---

4. Find the ratio of the surface areas of 2 spheres whose volumes are in the ratio 512:1331.

$$\frac{v}{V} = \frac{r \times r \times r}{R \times R \times R} = \frac{512}{1331} \qquad \begin{array}{l} r = 8 \\ R = 11 \end{array}$$

$$\frac{sa}{SA} = \frac{r \times r}{R \times R} = \frac{8 \times 8}{11 \times 11} = \frac{\mathbf{64}}{\mathbf{121}}$$

9. Find the ratio of the surface areas of 2 spheres whose volumes are in the ratio 8:243.

$$\frac{v}{V} = \frac{r \times r \times r}{R \times R \times R} = \frac{8}{243} \qquad \begin{array}{l} r = 2 \\ R = 7 \end{array}$$

$$\frac{sa}{SA} = \frac{r \times r}{R \times R} = \frac{\mathbf{4}}{\mathbf{49}}$$

---

5. Find the ratio of the surface areas of 2 spheres whose volumes are in the ratio 343:1728.

$$\frac{v}{V} = \frac{r \times r \times r}{R \times R \times R} = \frac{343}{1728} \qquad \begin{array}{l} r = 7 \\ R = 12 \end{array}$$

$$\frac{sa}{SA} = \frac{r \times r}{R \times R} = \frac{7 \times 7}{12 \times 12} = \frac{\mathbf{49}}{\mathbf{144}}$$

10. Find the ratio of the volumes of two spheres whose surface areas are in the ratio 81:121.

$$\frac{sa}{SA} = \frac{r \times r}{R \times R} = \frac{81}{121} \qquad \begin{array}{l} r = 9 \\ R = 11 \end{array}$$

$$\frac{v}{V} = \frac{r \times r \times r}{R \times R \times R} = \frac{\mathbf{729}}{\mathbf{1331}}$$

Spheres 3

Answer as indicated about inscribed spheres.	Answer as indicated, converting volume and surface area of spheres.
1. Find the volume of a sphere inscribed in a cube with edge 6. $r = 3$ $V = \dfrac{4\pi\, r^3}{3} = \mathbf{36\pi}$	7. A sphere has surface area 36π. Find its volume. $SA = 36\pi$ $4\pi r^2 = 36\pi$ $r = 3$ $V = (4/3) \cdot 3^3 \cdot \pi = 4 \cdot 9\pi = \mathbf{36\pi}$
2. Find the volume of a sphere inscribed in a cube with edge 11. $r = 11/2$ $V = \dfrac{(4\pi)11^3}{3 \times 8} = \dfrac{\mathbf{1331\pi}}{\mathbf{6}}$	8. A sphere has surface area 324π. Find its volume. $SA = 324\pi$ $4\pi r^2 = 324\pi$ $r = 9$ $V = (4/3) \cdot 9^3 \cdot \pi = 4 \cdot 243\pi = \mathbf{972\pi}$
3. A sphere just fits in a cylinder with volume 54π. Find the surface area of the cylinder. $V_{cyl} = 54\pi = \pi r^2 h = \pi r^2(2r)$ $r = 3,\ d = 6,\ h = 6$ $SA\ cyl = 9\pi + 9\pi + 6\pi(6) = \mathbf{54\pi}$	9. A sphere has volume 288π. Find its surface area. $(4/3)\pi r^3 = 288\pi$ $r^3 = (3/4)(288) = 3 \cdot 72 = 216$ $r = 6$ $SA = 4 \cdot 6^2 \cdot \pi = 4 \cdot 36\pi = \mathbf{144\pi}$
4. A cylinder circumscribed about a sphere has volume 128π. Find the volume of the sphere. $V_{cyl} = 128\pi = \pi r^2 h = \pi r^2(2r)$ $r = 4$ $V\ sp = (4/3)64\pi = \dfrac{\mathbf{256\pi}}{\mathbf{3}}$	10. A sphere has volume $36{,}000\pi$. Find its surface area. $(4/3)\pi r^3 = 36{,}000\pi$ $r^3 = (3/4)(36{,}000) = 3 \cdot 9000 = 27{,}000$ $r = 30$ $SA = 4 \cdot 30^2 \cdot \pi = 4 \cdot 900\pi = \mathbf{3600\pi}$
5. Three spheres each with radius 10 just fit inside a cylinder. Find the volume and surface area of the cylinder. $r = 10,\ d = 20,\ h = 60$ $V_{cyl} = 100\pi \cdot 60 = \mathbf{6000\pi}$ $SA\ cyl = 100\pi + 100\pi + 20\pi \cdot 60 = \mathbf{1400\pi}$	11. A sphere has surface area 576π. Find its volume. $4\pi r^2 = 576\pi$ $r^2 = 144$ $r = 12$ $V = (4/3) \cdot 12^3 \cdot \pi = 4 \cdot 576\pi = \mathbf{2304\pi}$
6. A sphere with radius 6 just fits in a cylinder. Find the surface area of the cylinder. $r = 6,\ d = 12,\ h = 12$ $SA\ cyl = 36\pi + 36\pi + 12\pi(12)$ $ = 72\pi + 144\pi = \mathbf{216\pi}$	12. A sphere has volume 4500π. Find its surface area. $(4/3)\pi r^3 = 4500\pi$ $r^3 = (3/4)(4500) = 3 \cdot 1125 = 3375$ $r = 15$ $SA = 4 \cdot 15^2 \cdot \pi = 4 \cdot 225\pi = \mathbf{900\pi}$

Spheres 4

Find the total volume or surface area of the shape, attaching base to base.	*Answer about reshaped spheres.*

1. A hemisphere is attached to one base of a cylinder with radius 3 and height 8. Find the volume.

$V_{cyl} = \pi \cdot 3^2 \cdot 8 = 72\pi$

$V_{hem} = (2/3)\pi \cdot 3^3 = 18\pi$

$V = \mathbf{90\pi}$

---

6. A meatball with a 3-inch radius can be reformed into how many meatballs with a 1-inch radius?

$$V = \frac{(4\pi)(3)^3}{3} = \frac{(4\pi)(1)^3(3)^3}{3}$$

$3 \times 3 \times 3 = \mathbf{27}$

---

2. Use the same shape as #1. Find the surface area.

SA cyl base = 9π

SA cyl lat = 48π

SA hemi = 18π

SA = $\mathbf{75\pi}$

---

7. Eight metal balls each with radius 1 are melted and reshaped into one ball. Find the new radius.

$$V = \frac{8(4\pi)(1)^3}{3} = \frac{(4\pi)(2)^3}{3}$$

$\mathbf{2}$

---

3. A hemisphere is attached to the base of a cone with height 11 and radius 5. Find the volume.

$V_{cone} = (1/3)\pi \cdot 5^2 \cdot 11 = 275\pi/3$

$V_{hem} = (2/3)\pi \cdot 5^3 = 250\pi/3$

$V = 525\pi/3 = \mathbf{175\pi}$

---

8. How many spherical snowballs with a 1-inch radius can be made from a 6-inch radius spherical snowball?

$$V = \frac{(4\pi)(6)^3}{3} = \frac{(4\pi)(1)^3(6)^3}{3}$$

$6 \times 6 \times 6 = \mathbf{216}$

---

4. A hemisphere attached to each base of a cylinder. The completed shape has diameter 12 and total length 26. Find the volume.

hemi r = 6

26 – 12 = 14

cyl h = 14

$V_{cyl} = \pi \cdot 6^2 \cdot 14 = 504\pi$

$V_{sph} = (4/3)\pi \cdot 6^3 = 288\pi$

$V = \mathbf{792\pi}$

---

9. How many spherical snowballs with a 3-inch radius can be made from a 1-foot radius spherical snowball?

$$V = \frac{(4\pi)(12)^3}{3} = \frac{(4\pi)(3)^3(4)^3}{3}$$

$4 \times 4 \times 4 = \mathbf{64}$

---

5. Use the same shape as #4. Find the surface area.

SA cyl lat = $\pi dh = 12\pi \cdot 14 = 168\pi$

SA sph = $4\pi \cdot 36 = 144\pi$

SA = $\mathbf{312\pi}$

---

10. How many spherical snowballs with a 3-inch radius can be made from a 2-foot radius spherical snowball?

$$V = \frac{(4\pi)(24)^3}{3} = \frac{(4\pi)(3)^3(8)^3}{3}$$

$8 \times 8 \times 8 = \mathbf{512}$

Statistics 1

Answer as indicated.

1. Of 5 test scores from 0 to 100, the mode is 80, the median is 75, the mean is 69, and the range is 35. Find the next to the least.

45	**65**	75	80	80

    80 – L = 35   69 x 5 = 345
    L = 45   345 – (120 + 160) = 65

6. Of a set of seven 2-digit positive multiples of 10, 30 is the median, 20 and 40 are the modes, and 110 is the range. Find the mean.

10	20	20	30	40	40	120

    (10+40+30+80+120)/7 = 280/7 = **40**

2. In a set of 5 numbers, the mode is less than the median. If four of the numbers are 15, 35, 45, and 65, find the missing number.

15	**15**	35	45	65

    15, 35, 35, 45, 65 no
    15, 35, 45, 45, 65 no

7. Of 5 numbers from 10 to 50, the mean is 33, the median is 30, the mode is 48, and the range is 37. Find the second least.

11	**28**	30	48	48

    33x5 = 165
    165–(11+30+48+48) = 165–137 = 28

3. In a set of 5 numbers, the mean, median, and mode are all equal. If 4 of the numbers are 50, 60, 40, and 10, find the missing number.

10	40	**40**	50	60

    (10+40+50+50+60)/5 = 210/5 = 42
    (10+40+40+50+60)/5 = 200/5 = 40

8. Of 6 test scores from 0 to 100, the modes are 60 and 80, the median is 65, and the range is 70. Find the mean.

10	60	60	70	80	80

    (130+70+160)/6 = 360/6 = **60**

4. Five natural numbers have mean 5, median 5, and mode 8. Find the range.

x	y	5	8	8

    21 + x + y = 5 x 5 = 25
    x + y = 4 both not 2 or else 2nd mode
    x = 1, y = 3   range = **7**

9. The mode of x, $4x$, -12, 18, and $10x$ is 18. If the median is 7.2, find the mean.

–12	x	4x	18	10x
–12	1.8	7.2	18	18

    36 + 9 – 12 = 33
    33/5 = **6.6**

5. Of a set of seven 2-digit multiples of 5, the range is 85, the median is 25, the modes are 10 and 25, and the mean is 40. Find the sum of the 3 greatest.

10	10	25	25	x	y	95

    40 x 7 = 280
    280 – (10+10+25+25+95) = 115
    115 + 95 = **210**

10. In a set of 5 numbers, the mean, median, and mode are all equal. If 4 of the numbers are 18, 29, 21, and 16, find the missing number.

16	18	21	**21**	29

    16+18+21+29=84 | 105/5 = 21
    x = 18 or 21 |
    84+21 = 105 |

Statistics 2

Find the mean of:	Answer as indicated.
1. $5x + 1$, $3x - 2$, $6x + 7$, and $2x + 10$. $$\frac{16x + 16}{4} = 4x + 4$$	8. Find the mean, median, and range of the factors of 24. 1, 2, 3, 4, 6, 8, 12, 24 range = **23** median = **5** mean = 60/8 = **7.5**
2. $-3x + 5$, $-x - 7$, $11x + 6$, $2x - 12$, $4x$, and $8x + 17$. $$\frac{21x + 9}{6} = \frac{7x + 3}{2}$$	9. Find the mean, median, mode, and range of 10 numbers: the 1st 5 squares and the 1st 5 cubes. 1, 4, 9, 16, 25 mode = **1** 1, 8, 27, 64, 125 median = (9+16)/2 = **12.5** range = **124** mean = 280/10 = **28**
3. $6x + 4$, $4x - 3$, $7x + 8$, $x - 11$, and $2x + 17$. $$\frac{20x + 15}{5} = 4x + 3$$	10. Find the mean, median, and range of the factors of 36. 1, 2, 3, 4, 6, 9, 12, 18, 36 range = **35** median = **6** mean = 91/7 = **13**
4. $-8x + 6$, $-4x - 5$, $10x + 9$, $3x - 17$, $8x + 1$, $3x$, and $9x + 6$. $$\frac{21x}{7} = 3x$$	11. Find the mean, median, and range of the arithmetic sequence with F = 2, L = 20, and d = 3. 2, 5, 8, 11, 14, 17, 20 mean = **11** range = **18** median = mean for median = **11** arithmetic seq
5. $-2x + 8$, $-7x - 9$, $12x + 4$, $4x - 10$, $5x + 5$, $9x$, $x + 1$, and $-8x + 13$. $$\frac{14x + 12}{8} = \frac{7x + 6}{4}$$	12. Find the mean, median, and range of the arithmetic sequence with F = 2, L = 200, and d = 2. 2, 4, 6, . . . , 200 100, median, 102 range = **198** median = **101** N = 100 mean = **101**
6. $-4x + 6$, $-2x - 9$, $13x + 4$, $2x - 11$, $6x$, $5x + 1$, $2x - 6$, $-x - 8$, and $9x + 14$. $$\frac{30x - 9}{9} = \frac{10x - 3}{3}$$	13. Find the mean, median, and range of the first 8 prime numbers. 2, 3, 5, 7, 11, 13, 17, 19 range = **17** median = **9** mean = 77/8 = 72 5/8 = **9.625**
7. $-x + 8$, $-5x - 9$, $14x + 5$, $3x - 10$, $8x + 3$, and $7x + 7$. $$\frac{26x + 4}{6} = \frac{13x + 2}{3}$$	14. Find the mean, median, and range of the 3-digit multiples of 50. 100, 150, . . , 950 median = **525** range = **850** mean = **525** N = 18 (arithmetic seq)

Statistics 3

Find the positive difference of the median and mode.

1.
stem	leaf
2	1 1 3 4 7 7
3	0 6 6 6 8
5	3 3 6 6 8 9
6	1 2 4 4
8	0 1 3
9	8

median = 53
mode = 36

17

2.
stem	leaf
1	0 0 0 6 7 9 9
2	1 1 2 3 4 5 6
4	2 2 3 7 7
6	3 3 3 4 5 6
7	0 1
8	3 3 5 5 5 5

median = 43
mode = 85

42

3.
stem	leaf
1	1 2 2 4 4 6 8
2	0 0
4	0 1 2 3 3 5
6	2 8 9
7	3 3 3 5 7
8	0 0 0
9	1 2 2 2 2 5 7 8 9

median = 69
mode = 92

23

4.
stem	leaf
1	0 0 1 1 2
2	2 3 3
3	3 4 4 4
4	5 6 6 7 8
5	5 5
6	1 1 2 2 5 5 5 5 6
7	0 3 6

median = 47
mode = 65

18

5.
stem	leaf
3	1 1
4	3 4 5 6
5	0 0 9 9 9 9
6	0 2 2 4 4 5 5 6 6
7	0 0 0 0 0 2 3 4 5 6
8	1 1 1 1 1 1 2 2 2 4 5
9	5 5

median = 70
mode = 81

11

6.

```
X
X         X       X
X      X  X       X
X  X  X  X        X
X  X  X  X  X  X
X  X  X  X  X  X
3  4  5  6  7  8
```

median = 5
mode = 3

2

7.

```
X
X          X        X  X
X     X  X          X  X
X  X  X  X          X  X
X  X  X  X  X  X  X
X  X  X  X  X  X  X
0  2  4  6  8  10 12
```

median = 6
mode = 0

6

8.
score	frequency
100	5
90	10
80	15
70	20
60	5
50	15
40	15

median = 70
mode = 70

0

9.
score	frequency
70	12
60	22
50	18
40	20
30	12
20	10
10	8

median = 50
mode = 60

10

10.
score	frequency
100	20
110	30
120	11
130	16
140	19
150	40
160	10

median = 130
mode = 150

20

Statistics 4

Answer as indicated.

---

1. The median of 11, 42, 65, *x*, and 16 is 8 less than the mean. Find *x* if *x* is negative.

x 11 16 42 65
median = 16
mean = 24
24 · 5 = 120
11+16+42+65 = 134   x = **−14**

---

6. The mean of a set of 3 test scores is 90, the median is 88, and the greatest is 97. Find the least score.

x 88 97
mean = 90
sum = 90 · 3 = 270
185 + x = 270
x = **85**

---

2. The median of 18, 32, *x*, 12, 50, and 15 is 15 less than the mean. Find *x* if *x* is 3-digit and positive.

12 15 18 32 50 x
median = 25
mean = 40
40 · 6 = 240
127 + x = 240   x = **113**

---

7. The mean of a set of 3 test scores is 85, the median is 82, and the least is 77. Find the greatest score.

77 82 x
mean = 85
sum = 85 · 3 = 255
159 + x = 255
x = **96**

---

3. The median of 30, 12, *x*, 15, 40, and 17 is 10 more than the mean. Find *x* if *x* is negative.

x 12 15 17 30 40
median = 16
mean = 6
6 · 6 = 36
114 + x = 36   x = **−78**

---

8. Of 4 test scores, the mean is 81, the median is 80, the mode is 73, and the least is 73. Find the greatest score.

73 73 87 x
deduce 87 from median
mean = 81
sum = 81 · 4 = 324
73 + 73 + 87 = 233

233 + x = 324
x = **91**

---

4. The median of 9, 2, *x*, 4, 3, 9, 7, and 5 is 9 less than the mean. Find *x* if *x* is 2-digit and positive.

2 3 4 5 7 9 9 x
median = 6
mean = 15
15 · 8 = 120
39 + x = 120   x = **81**

---

9. Of 4 test scores, the mean is 78, the median is 79, the mode is 86, and the greatest is 86. Find the least score.

x 72 86 86
deduce 72 from median
mean = 78
sum = 78 · 4 = 312
72 + 86 + 86 = 244

244 + x = 312
x = **68**

---

5. The median and mode of 8, *x*, 2.5*x*, 3*x*, and 10 are both 10. Find the sum of possible means if all five numbers are natural.

If x = 10, 2.5x = 25 and 3x = 30. Mean = 83/5 = 16.6. If 2.5x = 10, x = 4. 3x = 12. Mean = 44/5 = 8.8. Median = 10 works for both. If 3x = 10, x is not natural. Sum = 16.6 + 8.8 = **25.4**

---

10. The median and mode of 7, *x*, 1.5*x*, 2*x*, and 8 are both 8. Find the sum of possible means if all five numbers are natural.

If x = 8, 1.5x = 12 and 2x = 16. Mean = 51/5 = 10.2. If 2x = 8, x = 4. 1.5x = 6. Median is not 8. If 1.5x = 8, x is not natural. Only mean = **10.2**

Transformations 1

Draw the reflection over the specified line. One box equals one unit.

1. over the *y*-axis

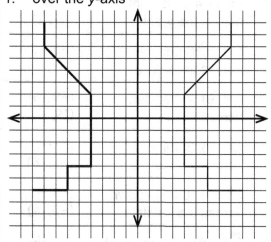

4. over the line *x* = 2

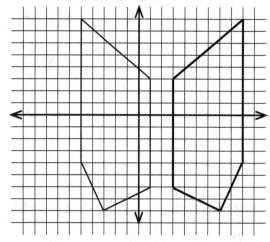

2. over the *x*-axis

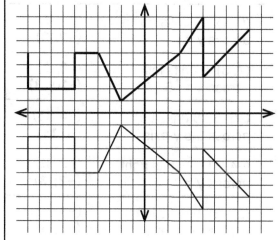

5. over the line x = –1

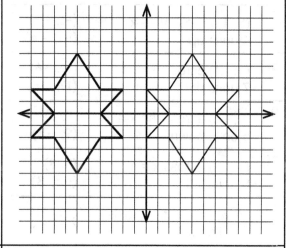

3. over the line *y* = –1

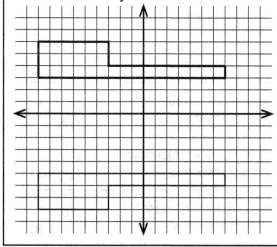

6. over the *x*-axis

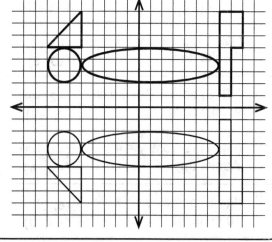

Transformations 2

Find the coordinates of the point after the successive reflection and translation.		Find the coordinates of the point after the successive translation and reflection.	
1. (1, 4) across the x-axis 2 units left	(1, –4) **(–1, –4)**	17. (3, 1) 4 units right across the y-axis	(7, 1) **(–7, 1)**
2. (2, 7) across the y-axis 9 units down	(–2, 7) **(–2, –2)**	18. (–3, 5) 6 units up across the x-axis	(–3, 11) **(–3, –11)**
3. (–1, 5) across y = 2 5 units right	(–1, –1) **(4, –1)**	19. (–4, 6) 7 units left across x = 4	(–11, 6) **(19, 6)**
4. (–7, –2) across y = x 4 units up	(–2, –7) **(–2, –3)**	20. (8, –9) 3 units up across y = –x	(8, –6) **(6, –8)**
5. (4, –5) across the y-axis 8 units down	(–4, –5) **(–4, –13)**	21. (–5, 6) 10 units down across the x-axis	(–5, –4) **(–5, 4)**
6. (1, 3) across x = 8 14 units left	(15, 3) **(1, 3)**	22. (–1, –8) 12 units right across y = –3	(11, –8) **(11, 2)**
7. (8, 1) across y = –x 4 units up	(–1, –8) **(–1, –4)**	23. (0, 4) 11 units down across y = x	(0, –7) **(–7, 0)**
8. (10, 4) across y = –1 7 units right	(10, –6) **(17, –6)**	24. (7, 3) 9 units left across x = –6	(–2, 3) **(–10, 3)**
9. (9, –2) across the x-axis 5 units down	(9, 2) **(9, –3)**	25. (–6, 5) 8 units up across the y-axis	(–6, 13) **(6, 13)**
10. (–3, 8) across x = 1 8 units left	(5, 8) **(–3, 8)**	26. (–4, 11) 2 units right across y = 7	(–2, 11) **(–2, 3)**
11. (7, 3) across the y-axis 2 units right	(–7, 3) **(–7, 5)**	27. (2, –3) 5 units left across the x-axis	(–3, –3) **(–3, 3)**
12. (–5, 6) across y = 3 12 units up	(–5, 0) **(–5, 12)**	28. (–3, 12) 15 units down across x = 5	(–3, –3) **(13, –3)**
13. (12, 10) across y = x 6 units down	(10, 12) **(10, 6)**	29. (9, 8) 14 units up across y = –x	(9, 22) **(–22, –9)**
14. (–2, 9) across the x-axis 9 units right	(–2, –9) **(7, –9)**	30. (–7, 8) 6 units left across the y-axis	(–13, 8) **(13, 8)**
15. (–4, 11) across y = 4 7 units up	(–4, –3) **(–4, 4)**	31. (–7, 0) 4 units down across x = 2	(–7, –4) **(11, –4)**
16. (0, –6) across y = –x 10 units left	(6, 0) **(–4, 0)**	32. (–5, –1) 13 units right across y = x	(8, –1) **(–1, 8)**

Transformations 3

Draw the reflection of each shape over the origin.

Reflect the vertices first.

1.

4.

2.

5.

3.

6.

Transformations 4

Name the fewest transformation(s)–rotation, reflection, and/or translation–done to the left figure to obtain the right figure.

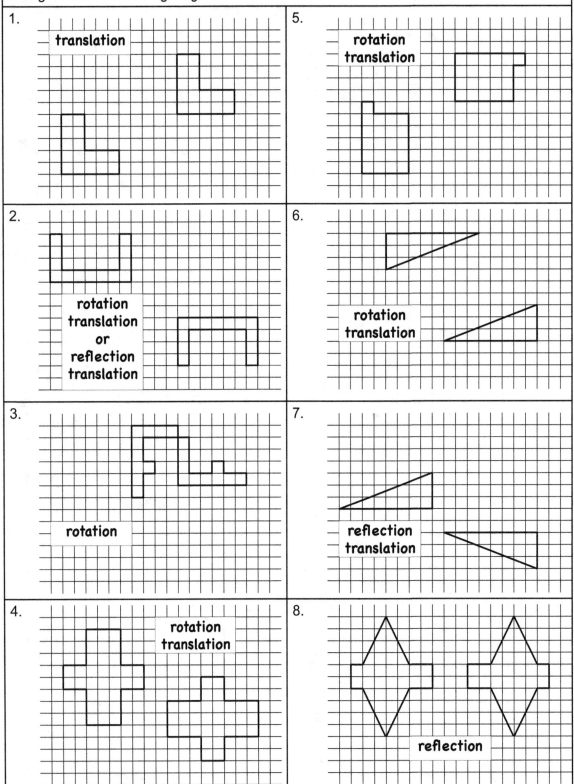

1. translation

2. rotation translation or reflection translation

3. rotation

4. rotation translation

5. rotation translation

6. rotation translation

7. reflection translation

8. reflection

Trapezoids 1

Find the midline or base of the trapezoid by mental math.

	b	M	B		b	M	B		b	M	B
1.	10	**15**	20	17.	6.25	18.5	**30.75**	33.	**33**	40	47
2.	8	**16**	24	18.	7	15	**23**	34.	**17.9**	22	26.1
3.	8.2	**9.25**	10.3	19.	9	29.5	**50**	35.	**32.6**	35.6	38.6
4.	11	**12**	13	20.	64	64.5	**65**	36.	**73**	84	95
5.	29	**35**	41	21.	35	40	**45**	37.	**6.2**	11	15.8
6.	10	**12.5**	15	22.	2.3	4	**5.7**	38.	**13.6**	13.8	14
7.	9.5	**12.5**	15.5	23.	12.4	24	**35.6**	39.	**51**	55	59
8.	36	**44**	52	24.	12.5	15.3	**18.1**	40.	**57**	70	83
9.	19	**30.5**	42	25.	22.8	23.5	**24.2**	41.	**46**	63	90
10.	11	**15.5**	20	26.	18	24.2	**30.4**	42.	**15**	18.4	21.8
11.	16.5	**20**	23.5	27.	53	58.5	**64**	43.	**15**	29	43
12.	13	**19**	25	28.	11.1	17.2	**23.3**	44.	**43**	51	59
13.	15	**20**	25	29.	30.4	35.4	**40.4**	45.	**1.5**	14.5	27.5
14.	12.5	**15.6**	18.7	30.	55	72.5	**90**	46.	**28.8**	33	37.2
15.	4.1	**6**	7.9	31.	16.3	20.4	**24.5**	47.	**15.1**	16.2	17.3
16.	22.6	**27.5**	32.4	32.	5.7	5.75	**5.8**	48.	**7**	20	33

Trapezoids 2

Find the area of the trapezoid in square units by mental math given the bases and height.

1.	b = 12, B = 18 H = 9	M = 15 **A = 135**	17.	b = 59, B = 65 H = 11	M = 62 **A = 682**	
2.	b = 6.5, B = 9.5 H = 7	M = 8 **A = 56**	18.	b = 22, B = 66 H = 5	M = 44 **A = 220**	
3.	b = 7, B = 15 H = 11	M = 11 **A = 121**	19.	b = 17, B = 30 H = 20	M = 23.5 **A = 470**	
4.	b = 6.25, B = 11.75 H = 7	M = 9 **A = 63**	20.	b = 11.2, B = 18.8 H = 15	M = 15 **A = 225**	
5.	b = 11, B = 14 H = 10	M = 12.5 **A = 125**	21.	b = 18, B = 34 H = 25	M = 26 **A = 650**	
6.	b = 13, B = 19 H = 15	M = 16 **A = 240**	22.	b = 13, B = 19 H = 15	M = 16 **A = 240**	
7.	b = 10.1, B = 11.9 H = 13	M = 11 **A = 143**	23.	b = 9.25, B = 10.75 H = 31	M = 10 **A = 310**	
8.	b = 16, B = 17 H = 20	M = 16.5 **A = 330**	24.	b = 27, B = 51 H = 20	M = 39 **A = 780**	
9.	b = 21.4, B = 28.6 H = 12	M = 25 **A = 300**	25.	b = 28, B = 42 H = 35	M = 35 **A = 1225**	
10.	b = 2.2, B = 8.8 H = 8	M = 5.5 **A = 44**	26.	b = 4, B = 14 H = 9	M = 9 **A = 81**	
11.	b = 34, B = 46 H = 11	M = 40 **A = 440**	27.	b = 11, B = 89 H = 23	M = 50 **A = 1150**	
12.	b = 14, B = 30 H = 17	M = 22 **A = 374**	28.	b = 14, B = 44 H = 30	M = 29 **A = 870**	
13.	b = 25, B = 43 H = 30	M = 34 **A = 1020**	29.	b = 16, B = 34 H = 25	M = 25 **A = 625**	
14.	b = 11.3, B = 15.7 H = 40	M = 13.5 **A = 540**	30.	b = 15, B = 45 H = 17	M = 30 **A = 510**	
15.	b = 9.9, B = 16.1 H = 12	M = 13 **A = 156**	31.	b = 15.3, B = 26.7 H = 21	M = 21 **A = 441**	
16.	b = 27, B = 33 H = 14	M = 30 **A = 420**	32.	b = 3, B = 25 H = 14	M = 14 **A = 196**	

Trapezoids 3

Find the base or height of the trapezoid as indicated.

1. The area of a trapezoid is 72, the height is 9, and one base is 5. Find the other base.

    72 = 9M
    M = 8
    b = 5
    B = 11

2. The area of a trapezoid is 96, and the bases are 7 and 17. Find the height.

    M = 12
    96 = 12H
    H = 8

3. The area of a trapezoid is 42, the height is 6, and one base is 5. Find the other base.

    42 = 6M
    M = 7
    b = 5
    B = 9

4. The area of a trapezoid is 135, and the bases are 8 and 22. Find the height.

    M = 15
    135 = 15H
    H = 9

5. The area of a trapezoid is 230, the height is 10, and one base is 16. Find the other base.

    230 = 10M
    M = 23
    b = 16
    B = 30

6. The area of a trapezoid is 108, one base is 12, and the height is 8. Find the other base.

    108 = 8M
    M = 13.5
    b = 12
    B = 15

7. The area of a trapezoid is 165, the height is 10, and one base is 23. Find the other base.

    165 = 10M
    M = 16.5
    B = 23
    b = 10

8. The area of a trapezoid is 90, and the bases are 9 and 36. Find the height.

    M = 22.5
    90 = 22.5H
    H = 4

9. The area of a trapezoid is 290, the height is 20, and one base is 12. Find the other base.

    290 = 20M
    M = 14.5
    b = 12
    B = 17

10. The area of a trapezoid is 625, the height is 25, and one base is 25.3. Find the other base.

    625 = 25M
    M = 25
    B = 25.3
    b = 24.7

11. The area of a trapezoid is 286, and the bases are 10.2 and 11.8. Find the height.

    M = 11
    286 = 11H
    H = 26

12. The area of a trapezoid is 167, and the bases are 13.5 and 6.5. Find the height.

    M = 10
    167 = 10H
    H = 16.7

Trapezoids 4

Find the perimeter in units and the area in square units of the isosceles trapezoid.

1. sides 17, 17, 17, and 33

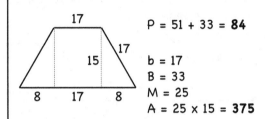

P = 51 + 33 = **84**

b = 17
B = 33
M = 25
A = 25 x 15 = **375**

6. bases 6 and 18; legs $4\sqrt{3}$

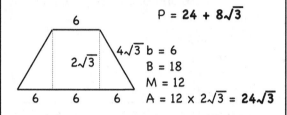

P = **24 + 8√3**

b = 6
B = 18
M = 12
A = 12 x 2√3 = **24√3**

2. bases 2 and 4; one angle 135°

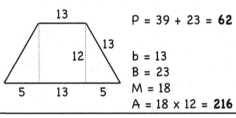

P = **6 + 2√2**

b = 2
B = 4
M = 3
A = 3 x 1 = **3**

7. bases 8 and 18; one angle 120°

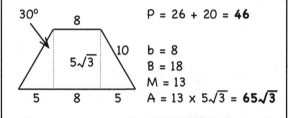

P = 26 + 20 = **46**

b = 8
B = 18
M = 13
A = 13 x 5√3 = **65√3**

3. sides 13, 13, 13, 23

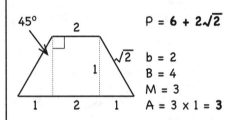

P = 39 + 23 = **62**

b = 13
B = 23
M = 18
A = 18 x 12 = **216**

8. bases 25 and 35; one angle 45°

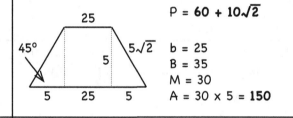

P = **60 + 10√2**

b = 25
B = 35
M = 30
A = 30 x 5 = **150**

4. sides 25, 25, 25, and 39

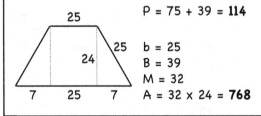

P = 75 + 39 = **114**

b = 25
B = 39
M = 32
A = 32 x 24 = **768**

9. sides 25, 25, 30, and 60

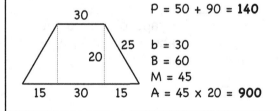

P = 50 + 90 = **140**

b = 30
B = 60
M = 45
A = 45 x 20 = **900**

5. parallel sides 9 and 15; non-parallel sides 5

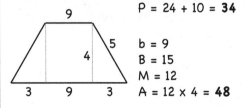

P = 24 + 10 = **34**

b = 9
B = 15
M = 12
A = 12 x 4 = **48**

10. bases 6 and 12; one angle 60°

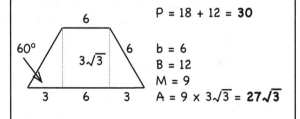

P = 18 + 12 = **30**

b = 6
B = 12
M = 9
A = 9 x 3√3 = **27√3**

Triangles 1

Given two angles of a triangle, find the third angle by mental math.

1. 30, 60 **90**	17. 18, 100 **62**	33. 13, 99 **68**	49. 55.5, 72.5 **52**
2. 39, 51 **90**	18. 12, 12 **156**	34. 35, 53 **92**	50. 60.3, 60.6 **59.1**
3. 50, 50 **80**	19. 15, 115 **50**	35. 45, 85 **50**	51. 29.1, 31.9 **119**
4. 26, 90 **64**	20. 35, 45 **100**	36. 27, 57 **96**	52. 26, 134.7 **19.3**
5. 45, 90 **45**	21. 21, 29 **130**	37. 29, 86 **65**	53. 32.1, 43.2 **104.7**
6. 6, 102 **72**	22. 26, 70 **84**	38. 16, 61 **103**	54. 15.2, 91.8 **73**
7. 12, 86 **82**	23. 25, 25 **130**	39. 73, 74 **33**	55. 16.25, 83 **80.75**
8. 18, 36 **126**	24. 51, 52 **77**	40. 10, 55 **115**	56. 11.1, 101.1 **67.8**
9. 63, 100 **17**	25. 16, 101 **63**	41. 14, 84 **82**	57. 33.3, 44.4 **102.3**
10. 47, 90 **43**	26. 59, 60 **61**	42. 22, 75 **83**	58. 13.5, 111 **55.5**
11. 12, 43 **125**	27. 21, 46 **113**	43. 39, 39 **102**	59. 80.5, 81.5 **18**
12. 22, 73 **85**	28. 11, 53 **116**	44. 80, 90 **10**	60. 42.5, 49.5 **88**
13. 42, 69 **69**	29. 15, 72 **93**	45. 52, 62 **66**	61. 31.4, 52.6 **96**
14. 70, 80 **30**	30. 24, 82 **74**	46. 46, 64 **70**	62. 25.9, 75.4 **78.7**
15. 20, 86 **74**	31. 38, 64 **78**	47. 77, 78 **25**	63. 34.7, 43.6 **101.7**
16. 31, 41 **108**	32. 20, 21 **139**	48. 39, 40 **101**	64. 65.5, 69.5 **45**

Triangles 2

Identify △ABC with as many labels as possible among: Isosceles, Equilateral, Scalene, Right, Acute, Obtuse, Equiangular.

1. AC = BC **Isosceles**	**12.** m∠A = 100º, m∠B = 40º **Isosceles, Obtuse** 3rd angle 40
2. AC = 11, BC = 11 **Isosceles**	**13.** AB = BC = AC **Equilateral, Isosceles, Equiangular, Acute**
3. m∠C = 90º **Right**	**14.** m∠A = 33º, m∠B = 33º **Isosceles, Obtuse** 3rd angle 114
4. m∠A = 21º, m∠B = 69º **Right, Scalene**	**15.** m∠A = 51º, m∠B = 51º **Isosceles, Acute** 3rd angle 78
5. m∠A = 60º **None**	**16.** AB = 10, BC = 11 **None**
6. m∠A = 41º, m∠B = 49º **Right, Scalene**	**17.** m∠A = 90º, AB = 12, AC = 12 **Isosceles, Right**
7. m∠A = 30º, m∠B = 120º **Isosceles, Obtuse** 3rd angle 30	**18.** m∠A = 40º, AB = 9, BC = 9 **Isosceles, Obtuse** vertex angle 100
8. m∠A = 90.01º **Obtuse**	**19.** m∠A = 40º, AB = 9, AC = 9 **Isosceles, Acute** base angles 70
9. m∠A = 60º, m∠C = 60º **Equilateral, Isosceles, Equiangular, Acute**	**20.** m∠A = 90º, m∠B = 45º **Isosceles, Right**
10. m∠A = 41º, m∠B = 52º **Acute, Scalene**	**21.** AB = 8, AC = 8, BC = $8\sqrt{2}$ **Isosceles, Right**
11. AB = 10, BC = 10, AC = 10 **Equilateral, Isosceles, Equiangular, Acute**	**22.** AB = 6, BC = 7, AC = 8 **Scalene, Acute** Close to 7, 7, 7

Medians, angle bisectors, and altitudes
(if interior) meet at 1 pont.

Triangles 3

Use paper corner to
check for right angle.

Draw 3 medians, 3 angle bisectors, and 3 altitudes in the triangles.

1.	Medians	Angle Bisectors	Altitudes

1.

2.

3.   altitude   altitude

4.

5.

6.

7.

Triangles 4

Find the base, height, or area in square units of each triangle by mental math.

#	BASE	HEIGHT	AREA	#	BASE	HEIGHT	AREA	#	BASE	HEIGHT	AREA
1.	11	36	**198**	17.	48	220	**5280**	33.	19	**38**	361
2.	29	**10**	145	18.	200	13.5	**1350**	34.	80	12.5	**500**
3.	**50**	60	1500	19.	25	180	**2250**	35.	17	170	**1445**
4.	28	14	**196**	20.	**100**	11.1	555	36.	70	35	**1225**
5.	**10**	37	185	21.	220	35	**3850**	37.	**12**	24	144
6.	50	**50**	1250	22.	40	143	**2860**	38.	**32**	16	256
7.	12	97	**582**	23.	14	**25**	175	39.	13.5	80	**540**
8.	20	50	**500**	24.	22	59	**649**	40.	25	**54**	675
9.	15	**30**	225	25.	45	440	**9900**	41.	45	90	**2025**
10.	**12**	16	96	26.	**60**	15	450	42.	26	26	**338**
11.	14	42	**294**	27.	120	140	**8400**	43.	70	75	**2625**
12.	11	78	**429**	28.	240	**11**	1320	44.	**54**	22	594
13.	40	**40**	800	29.	**86**	200	8600	45.	20.2	30	**303**
14.	**8**	15	60	30.	80	55	**2200**	46.	24.5	40	**490**
15.	18	18	**162**	31.	**40**	121	2420	47.	**72**	25	900
16.	**22**	36	396	32.	22	**60**	660	48.	**60**	65	1950

Triangles 5

Find the area. One box equals one square unit.

Draw recs around tris as in #1.

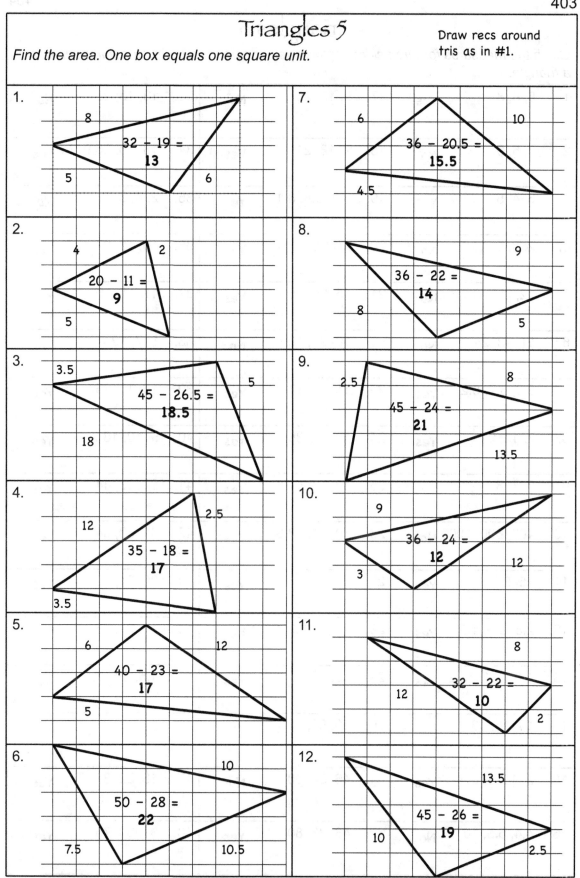

1.
8
$32 - 19 =$
13
5
6

2.
4
2
$20 - 11 =$
9
5

3.
3.5
$45 - 26.5 =$
18.5
5
18

4.
12
2.5
$35 - 18 =$
17
3.5

5.
6
12
$40 - 23 =$
17
5

6.
10
$50 - 28 =$
22
7.5
10.5

7.
6
10
$36 - 20.5 =$
15.5
4.5

8.
9
$36 - 22 =$
14
8
5

9.
2.5
8
$45 - 24 =$
21
13.5

10.
9
$36 - 24 =$
12
3
12

11.
8
$32 - 22 =$
10
12
2

12.
13.5
$45 - 26 =$
19
10
2.5

MAVA Math: Enhanced Skills Solutions   Copyright © 2015 Marla Weiss

Triangles 6

Given the measures for three sides, answer yes or no as to whether the numbers form a triangle.

1.	8, 8, 8	Yes	17.	9, 12, 25	No	33.	13, 13, 28	No
2.	3, 5, 9	No	18.	7, 18, 21	Yes	34.	18, 18, 35.9	Yes
3.	3, 4, 5	Yes	19.	3.2, 5.8, 8.1	Yes	35.	2, 4, 8	No
4.	2, 3, 5	No	20.	.01, .01, .01	Yes	36.	16, 18, 24	Yes
5.	12, 25, 40	No	21.	15, 20, 25	Yes	37.	11, 11, 11	Yes
6.	11, 13, 24	No	22.	16, 23, 39	No	38.	6, 7, 12	Yes
7.	8.2, 8.2, 8.2	Yes	23.	11, 12, 13	Yes	39.	15, 15, 25	Yes
8.	7, 15, 21	Yes	24.	19, 22, 35	Yes	40.	9, 10, 18.7	Yes
9.	6, 7, 8	Yes	25.	13.1, 22, 35	Yes	41.	22, 23, 44	Yes
10.	400, 600, 1000	No	26.	10, 14, 23.9	Yes	42.	31, 33, 54	Yes
11.	15, 20, 32	Yes	27.	33, 34, 43	Yes	43.	20, 24, 40	Yes
12.	2.5, 3.5, 4.5	Yes	28.	13.1, 13.2, 26.3	No	44.	17, 21, 36	Yes
13.	13, 13.1, 13.2	Yes	29.	19, 33, 51	Yes	45.	1.5, 5.5, 7.5	No
14.	14, 16, 30	No	30.	50, 50, 51	Yes	46.	1, 1, 3	No
15.	2, 10, 11	Yes	31.	46, 34, 90	No	47.	18, 22, 37	Yes
16.	4.4, 5.5, 10	No	32.	48, 33, 80	Yes	48.	37, 39, 75	Yes

Triangles 7

Given two sides of a triangle,	write an inequality for the value of the 3rd side s.	write an inequality for the value of the perimeter P.
1. 3, 12	$9 < s < 15$	$3+12=15$ $15+9=24, 15+15 = 30$ $\quad$ $24 < P < 30$
2. 9, 13	$4 < s < 22$	$9+13=22$ $22+4=26, 22+22 = 44$ $\quad$ $26 < P < 44$
3. 6, 15	$9 < s < 21$	$6+15=21$ $21+9=30, 21+21 = 42$ $\quad$ $30 < P < 42$
4. 8, 20	$12 < s < 28$	$8+20=28$ $28+12=40, 28+28 = 56$ $\quad$ $40 < P < 56$
5. 12, 13	$1 < s < 25$	$12+13=25$ $25+1=26, 25+25 = 50$ $\quad$ $26 < P < 50$
6. 8, 18	$10 < s < 26$	$8+18=26$ $26+10=36, 26+26 = 52$ $\quad$ $36 < P < 52$
7. 3, 4	$1 < s < 7$	$3+4=7$ $7+1=8, 7+7 = 14$ $\quad$ $8 < P < 14$
8. 12, 15	$3 < s < 27$	$12+15=27$ $27+3=30, 27+27 = 54$ $\quad$ $30 < P < 54$
9. 11, 22	$11 < s < 33$	$11+22=33$ $33+11=44, 33+33 = 66$ $\quad$ $44 < P < 66$
10. 5, 17	$12 < s < 22$	$5+17=22$ $22+12=34, 22+22 = 44$ $\quad$ $34 < P < 44$
11. 15, 32	$17 < s < 47$	$15+32=47$ $47+17=64, 47+47 = 94$ $\quad$ $64 < P < 94$
12. 14, 19	$5 < s < 33$	$14+19=33$ $33+5=38, 33+33 = 66$ $\quad$ $38 < P < 66$
13. 31, 38	$7 < s < 69$	$31+38=69$ $69+7=76, 69+69 = 138$ $\quad$ $76 < P < 138$
14. 22, 45	$23 < s < 67$	$22+45=67$ $67+23=90, 67+67 = 134$ $\quad$ $90 < P < 134$
15. 20, 43	$23 < s < 63$	$20+43=63$ $63+23=86, 63+63 = 126$ $\quad$ $86 < P < 126$
16. 18, 21	$3 < s < 39$	$18+21=39$ $39+3=42, 39+39 = 78$ $\quad$ $42 < P < 78$
17. 10, 31	$21 < s < 41$	$10+31=41$ $41+21=62, 41+41 = 82$ $\quad$ $62 < P < 82$
18. 16, 30	$14 < s < 46$	$16+30=46$ $46+14=60, 46+46 = 92$ $\quad$ $60 < P < 92$
19. a, b; $a < b$	$b - a < s < a + b$	$2b < P < 2a + 2b$

Triangles 8

Answer as indicated, making a list when necessary.	*Answer as indicated. Linear measures are units; area measures are square units.*

1. How many triangles with whole number sides have longest side 5?

5, 5, 5	5, 5, 2	5, 4, 3
5, 5, 4	5, 5, 1	5, 4, 2
5, 5, 3	5, 4, 4	5, 3, 3

9

7. Find the area of an isosceles right triangle with hypotenuse 10.

leg = $5\sqrt{2}$

area = $\dfrac{(5\sqrt{2})(5\sqrt{2})}{2}$ = **25**

2. How many triangles with whole number sides have longest side 6?

6, 6, 6	6, 5, 5	6, 4, 4
through	through	6, 4, 3
6, 6, 1	6, 5, 2	

12

8. Find the leg of an isosceles right triangle with area 32.

A square = 64
leg = **8**

3. Given two sticks with lengths 3 and 11. Find all possible integral lengths of a third stick that could be used to construct a triangle.

8 < s < 14
9, 10, 11, 12, 13

9. The area of a rectangle 9 by 6 equals the area of a triangle with base 27. Find the corresponding altitude of the triangle.

9 x 6 = 27 x H / 2
2 x 54 = 27H
H = **4**

4. A triangle has one side 3 and another side 7. Find the product of the least and greatest possible integral values of the perimeter.

4 < s < 10
14 < P < 20
15 x 19 = **285**

10. Find the leg of an isosceles right triangle with area 50.

A square = 100
leg = **10**

5. A triangle has one side 4 and another side 13. Find the sum of the least and greatest possible integral values of the perimeter.

9 < s < 17
26 < P < 34
27 + 33 = **60**

11. If the area of a square with side 18 equals the area of a triangle with altitude 8, find the corresponding base of the triangle.

18 x 18 = 8 x b / 2   Keep factored.
2 x 18 x 18 = 8b   Multiply last!
b = **81**

6. Given two sticks with lengths 5 and 12. How many possible integral lengths of a third stick could be used to construct a triangle?

7 < s < 17
8 through 16
16 − 8 + 1 = 9

9

12. Find the area of an isosceles triangle with perimeter 36 and altitude to the base 12.

h = 12
b = 10
A = **60**

Triangles 9

Answer as indicated. NTS

---

1. Find x.

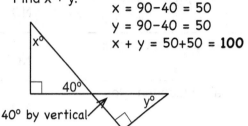

51+33 = 84
180−84 = 96
x = 96/2 = **48**

6. Find x + y.

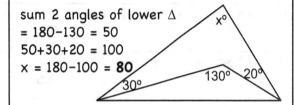

x = 90−40 = 50
y = 90−40 = 50
x + y = 50+50 = **100**

40° by vertical

---

2. Find x + y.

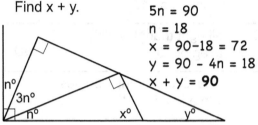

5n = 90
n = 18
x = 90−18 = 72
y = 90 − 4n = 18
x + y = **90**

7. Find x.

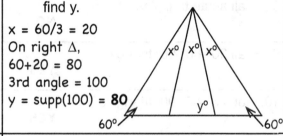

sum 2 angles of lower △
= 180−130 = 50
50+30+20 = 100
x = 180−100 = **80**

---

3. Find x.

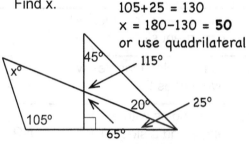

105+25 = 130
x = 180−130 = **50**
or use quadrilateral

8. If the largest triangle is equilateral, find y.

x = 60/3 = 20
On right △,
60+20 = 80
3rd angle = 100
y = supp(100) = **80**

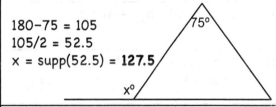

---

4. Find x.

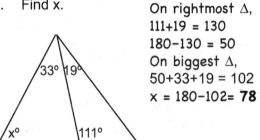

On rightmost △,
111+19 = 130
180−130 = 50
On biggest △,
50+33+19 = 102
x = 180−102= **78**

9. Find x if the triangle is isosceles with a 75° vertex angle.

180−75 = 105
105/2 = 52.5
x = supp(52.5) = **127.5**

---

5. Find x given the smaller bottom triangle is isosceles with a 37° base angle.

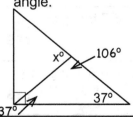

x = 180−106 = **74**
Or, for a △, an external angle is the sum of the two remote interior angles.

10. Find x.

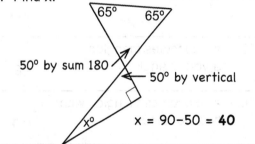

50° by sum 180

50° by vertical

x = 90−50 = **40**

MAVA Math: Enhanced Skills Solutions Copyright © 2015 Marla Weiss

Triangles 10

Answer YES or NO as to whether the triangles are similar.

#			#		
1.	all right triangles	NO	17.	all 20°-60°-100° triangles	YES
2.	all scalene triangles	NO	18.	all isosceles triangles with base 5	NO
3.	all obtuse triangles	NO	19.	all isosceles triangles with base angles 20°	YES
4.	all equilateral triangles	YES	20.	all right triangles with hypotenuse 10	NO
5.	all isosceles right triangles	YES	21.	all right triangles with a leg of 10	NO
6.	all equiangular triangles	YES	22.	all isosceles right triangles with a leg of 10	YES
7.	all isosceles triangles	NO	23.	all rotations of 6-6-7 triangles	YES
8.	all acute triangles	NO	24.	all isosceles triangles with congruent sides 15	NO
9.	all 30°-60°-90° triangles	YES	25.	all isosceles triangles with two angles 60°	YES
10.	all 45°-45°-90° triangles	YES	26.	all triangles with a 70° angle and a side 10	NO
11.	all triangles with a 100° angle	NO	27.	all 25°-40°-115° triangles	YES
12.	all obtuse triangles with a 120° angle	NO	28.	all dilations of 5-5-5 triangles	YES
13.	all right triangles with a 10° angle	YES	29.	all scalene triangles with 2 sides 10 and 12	NO
14.	all dilations of 3-4-5 triangles	YES	30.	all acute triangles with a 40° angle	NO
15.	all isosceles triangles with a vertex angle 20°	YES	31.	all dilations of 5-6-7 triangles	YES
16.	all isosceles triangles with two sides 10	NO	32.	all reflections of 4-6-8 triangles	YES

MAVA Math: Enhanced Skills Solutions Copyright © 2015 Marla Weiss

Triangles 11

Complete the chart for 2 similar triangles.

GIVEN	RATIO OF SIDES	RATIO OF ALTITUDES	RATIO OF PERIMETERS	RATIO OF AREAS
1. Perimeters 5 and 9	5:9	5:9	5:9	25:81
2. Altitudes 3 and 7	3:7	3:7	3:7	9:49
3. Sides 4, 8, 10 and 6, 12, 15	2:3	2:3	2:3	4:9
4. Greatest sides 12 and 15	4:5	4:5	4:5	16:25
5. Sides 3, 4, 5 and 18, 24, 30	1:6	1:6	1:6	1:36
6. Greatest sides 15 and 20	3:4	3:4	3:4	9:16
7. Altitudes 4 and 12	1:3	1:3	1:3	1:9
8. Perimeters 12 and 20	3:5	3:5	3:5	9:25
9. Least sides 16 and 40	2:5	2:5	2:5	4:25
10. Sides 5, 12, 13 and 10, 24, 26	1:2	1:2	1:2	1:4
11. Perimeters 14 and 22	7:11	7:11	7:11	49:121
12. Least sides 18 and 21	6:7	6:7	6:7	36:49
13. Greatest sides 30 and 42	5:7	5:7	5:7	25:49
14. Altitudes 11 and 44	1:4	1:4	1:4	1:16
15. Perimeters 10 and 45	2:9	2:9	2:9	4:81
16. Least sides 12 and 44	3:11	3:11	3:11	9:121
17. Altitudes 24 and 27	8:9	8:9	8:9	64:81
18. Greatest sides 21 and 36	7:12	7:12	7:12	49:144
19. Perimeters 12 and 45	4:15	4:15	4:15	16:225

Triangles 12

△ABC ~ △DEF. Find the lengths of the two sides not given using mental math. NTS

1.   AB = 5, BC = 3, AC = 4, ED = 10

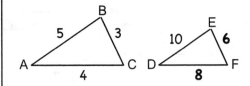

2.   ED = 8, EF = 6, DF = 4, AC = 6

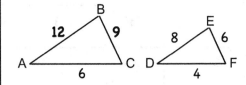

3.   AB = 6, BC = 9, AC = 12, EF = 3

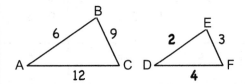

4.   ED = 6, EF = 9, DF = 9, AB = 8

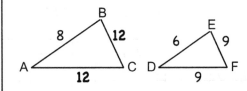

5.   AB = 9, BC = 18, AC = 12, DF = 8

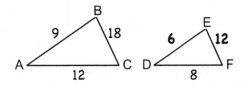

6.   AB = 8, AC = 12, BC = 10, DE = 6

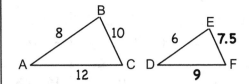

7.   AB = 6, AC = 9, BC = 8, EF = 12

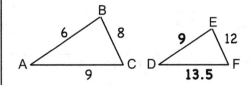

8.   AB = 3, BC = 4, AC = 5, DF = 20

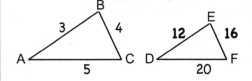

9.   AC = 10, BC = 6, DE = 6, DF = 12

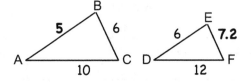

10.   AB = 9, AC = 12, DE = 15, EF = 30

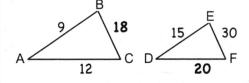

11.   AB = 14, DF = 12, DE = 6, EF = 9

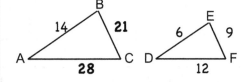

12.   DE = 6, EF = 18, AB = 8, AC = 20

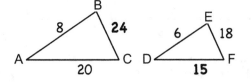

Triangles 13

Find x. NTS	Find the area of the triangle by Hero.
1. $x = 2$	8. 4, 6, 8 semi-perimeter = (4 + 6 + 8)/2 = 9 $A = \sqrt{(9)(9-4)(9-6)(9-8)}$ $= \sqrt{9 \times 5 \times 3} = \mathbf{3\sqrt{15}}$
2. $x = \sqrt{30}$	9. 3, 6, 7 semi-perimeter = (3 + 6 + 7)/2 = 8 $A = \sqrt{(8)(8-3)(8-6)(8-7)}$ $= \sqrt{8 \times 5 \times 2} = \mathbf{4\sqrt{5}}$
3. $x = \sqrt{84} = 2\sqrt{21}$	10. 5, 7, 10 semi-perimeter = (5 + 7 + 10)/2 = 11 $A = \sqrt{(11)(11-5)(11-7)(11-10)}$ $= \sqrt{11 \times 6 \times 4} = \mathbf{2\sqrt{66}}$
4. $x = \sqrt{54} = 3\sqrt{6}$	11. 4, 4, 6 semi-perimeter = (4 + 4 + 6)/2 = 7 $A = \sqrt{(7)(7-4)(7-4)(7-6)}$ $= \sqrt{7 \times 3 \times 3} = \mathbf{3\sqrt{7}}$
5. $x = \sqrt{99} = 3\sqrt{11}$	12. 3, 7, 8 semi-perimeter = (3 + 7 + 8)/2 = 9 $A = \sqrt{(9)(9-3)(9-7)(9-8)}$ $= \sqrt{9 \times 6 \times 2} = \mathbf{6\sqrt{3}}$
6. $x = \sqrt{147} = 7\sqrt{3}$	13. 13, 14, 15 semi-perimeter = (13 + 14 + 15)/2 = 21 $A = \sqrt{(21)(21-13)(21-14)(21-15)}$ $= \sqrt{21 \times 8 \times 7 \times 6} = \sqrt{3 \times 7 \times 4 \times 2 \times 7 \times 2 \times 3}$ $= 3 \times 7 \times 2 \times 2 = \mathbf{84}$
7. $x = \sqrt{135} = 3\sqrt{15}$	14. 5, 6, 7 semi-perimeter = (5 + 6 + 7)/2 = 9 $A = \sqrt{(9)(9-5)(9-6)(9-7)}$ $= \sqrt{9 \times 4 \times 3 \times 2} = \mathbf{6\sqrt{6}}$

412

Triangles 14

Find the missing sides, area, and perimeter. NTS
m ∠ A = 30°. | m ∠ A = 45°.

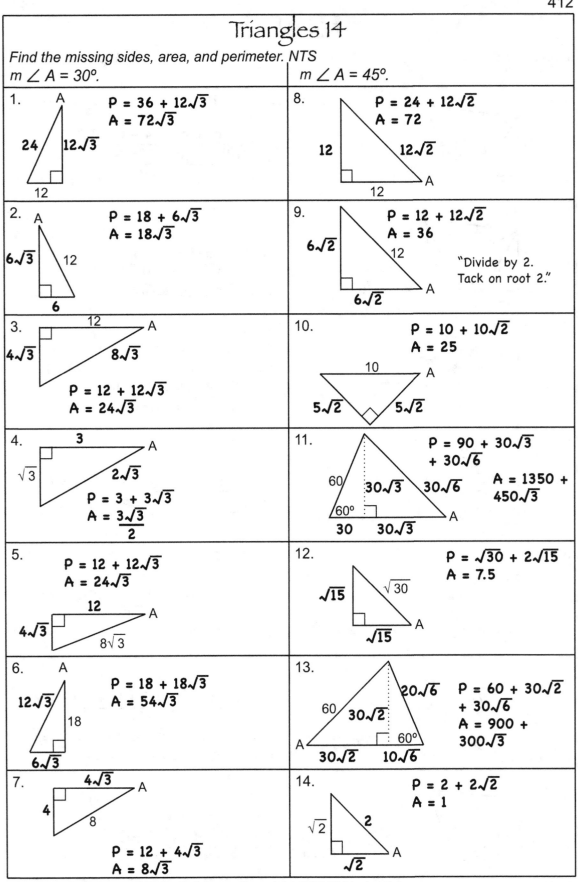

1. 24, $12\sqrt{3}$, 12
P = $36 + 12\sqrt{3}$
A = $72\sqrt{3}$

2. $6\sqrt{3}$, 12, 6
P = $18 + 6\sqrt{3}$
A = $18\sqrt{3}$

3. 12, $4\sqrt{3}$, $8\sqrt{3}$
P = $12 + 12\sqrt{3}$
A = $24\sqrt{3}$

4. 3, $\sqrt{3}$, $2\sqrt{3}$
P = $3 + 3\sqrt{3}$
A = $\dfrac{3\sqrt{3}}{2}$

5. P = $12 + 12\sqrt{3}$
A = $24\sqrt{3}$
12, $4\sqrt{3}$, $8\sqrt{3}$

6. $12\sqrt{3}$, 18, $6\sqrt{3}$
P = $18 + 18\sqrt{3}$
A = $54\sqrt{3}$

7. $4\sqrt{3}$, 4, 8
P = $12 + 4\sqrt{3}$
A = $8\sqrt{3}$

8. 12, $12\sqrt{2}$, 12
P = $24 + 12\sqrt{2}$
A = 72

9. $6\sqrt{2}$, 12, $6\sqrt{2}$
P = $12 + 12\sqrt{2}$
A = 36
"Divide by 2. Tack on root 2."

10. 10, $5\sqrt{2}$, $5\sqrt{2}$
P = $10 + 10\sqrt{2}$
A = 25

11. 60, $30\sqrt{3}$, $30\sqrt{6}$, 60°, 30, $30\sqrt{3}$
P = $90 + 30\sqrt{3} + 30\sqrt{6}$
A = $1350 + 450\sqrt{3}$

12. $\sqrt{15}$, $\sqrt{30}$, $\sqrt{15}$
P = $\sqrt{30} + 2\sqrt{15}$
A = 7.5

13. $20\sqrt{6}$, 60, $30\sqrt{2}$, 60°, $30\sqrt{2}$, $10\sqrt{6}$
P = $60 + 30\sqrt{2} + 30\sqrt{6}$
A = $900 + 300\sqrt{3}$

14. $\sqrt{2}$, 2, $\sqrt{2}$
P = $2 + 2\sqrt{2}$
A = 1

MAVA Math: Enhanced Skills Solutions Copyright © 2015 Marla Weiss

Triangles 15

Find the area in square units of the isosceles triangle with the given sides.

1. 5, 5, 6   **12**
5, 4, 3

2. 5, 5, 8   **12**
5, 3, 4

3. 14, 25, 25   **168**
25, 24, 7

4. 10, 13, 13   **60**
13, 12, 5

5. 13, 13, 24   **60**
13, 5, 12

6. 16, 17, 17   **120**
17, 15, 8

7. 24, 37, 37   **420**
37, 35, 12

8. 17, 17, 30   **120**
17, 8, 15

9. 18, 41, 41   **360**
41, 40, 9

10. 25, 25, 48   **168**
25, 7, 24

11. 29, 29, 42   **420**
29, 20, 21

12. 22, 61, 61   **660**
61, 60, 11

13. 4, 4, 6   **$3\sqrt{5}$**
4, ×, 3   16−9 = 5

14. 10, 12, 12   **$5\sqrt{119}$**
12, ×, 5   144−25 = 119

15. 10, 10, 18   **$9\sqrt{19}$**
10, ×, 9   100−81 = 19

16. 8, 8, 14   **$7\sqrt{15}$**
8, ×, 7   64−49 = 15

17. 8, 9, 9   **$4\sqrt{65}$**
9, ×, 4   81−16 = 65

18. 11, 11, 12   **$6\sqrt{85}$**
11, ×, 6   121−36 = 85

MAVA Math: Enhanced Skills Solutions   Copyright © 2015 Marla Weiss

A △ has 1 area but 3 base and height pairs.	Triangles 16	See the pattern after 2 examples.
Find the altitude to the hypotenuse.		*Find the median to the hypotenuse.*

1. sides 3, 4, 5 bh = BH 3 x 4 = 5 x H H = 12/5 = **2.4**	9. sides 3, 4, 5 $2^2 + 1.5^2 = M^2$ $4 + 2.25 = M^2$ M = **2.5**
2. sides 5, 12, 13 bh = BH 5 x 12 = 13 x H H = $\dfrac{60}{13}$	10. sides 5, 12, 13 $6^2 + 2.5^2 = M^2$ $36 + 6.25 = M^2$ M = **6.5**
3. sides 7, 24, 25 bh = BH 7 x 24 = 25 x H H = $\dfrac{168}{25}$	11. sides 7, 24, 25 M = **12.5**
4. sides 8, 15, 17 bh = BH 8 x 15 = 17 x H H = $\dfrac{120}{17}$	12. sides 8, 15, 17 M = **8.5**
5. sides 9, 40, 41 bh = BH 9 x 40 = 41 x H H = $\dfrac{360}{41}$	13. sides 9, 40, 41 M = **20.5**
6. sides 11, 60, 61 bh = BH 11 x 60 = 61 x H H = $\dfrac{660}{61}$	14. sides 11, 60, 61 M = **30.5**
7. sides 12, 35, 37 bh = BH 12 x 35 = 37 x H H = $\dfrac{420}{37}$	15. sides 12, 35, 37 M = **18.5**
8. sides 20, 21, 29 bh = BH 20 x 21 = 29 x H H = $\dfrac{420}{29}$	16. sides 20, 21, 29 M = **14.5**

Triangles 17

Find the area of the equilateral triangle with the given side.

1. $s = 2$ $\dfrac{4\sqrt{3}}{4}$ $\sqrt{3}$

2. $s = 4$ $\dfrac{16\sqrt{3}}{4}$ $4\sqrt{3}$

3. $s = 5$ $\dfrac{25\sqrt{3}}{4}$

4. $s = 8$ $\dfrac{64\sqrt{3}}{4}$ $16\sqrt{3}$

5. $s = 10$ $\dfrac{100\sqrt{3}}{4}$ $25\sqrt{3}$

6. $s = 12$ $\dfrac{12 \times 12\sqrt{3}}{4}$ $36\sqrt{3}$

7. $s = 14$ $\dfrac{14 \times 14\sqrt{3}}{4}$ $49\sqrt{3}$

8. $s = 16$ $\dfrac{16 \times 16\sqrt{3}}{4}$ $64\sqrt{3}$

9. $s = 20$ $\dfrac{20 \times 20\sqrt{3}}{4}$ $100\sqrt{3}$

10. $s = 24$ $\dfrac{24 \times 24\sqrt{3}}{4}$ $144\sqrt{3}$

11. $s = 32$ $\dfrac{32 \times 32\sqrt{3}}{4}$ $256\sqrt{3}$

12. $s = 40$ $\dfrac{40 \times 40\sqrt{3}}{4}$ $400\sqrt{3}$

13. $s = 44$ $\dfrac{44 \times 44\sqrt{3}}{4}$ $484\sqrt{3}$

Find the area of the regular hexagon with the given side.

14. $s = 3$ $\dfrac{9 \times 6\sqrt{3}}{4}$ $\dfrac{27\sqrt{3}}{2}$

15. $s = 6$ $\dfrac{36 \times 6\sqrt{3}}{4}$ $54\sqrt{3}$

16. $s = 7$ $\dfrac{49 \times 6\sqrt{3}}{4}$ $\dfrac{147\sqrt{3}}{2}$

17. $s = 9$ $\dfrac{9 \times 9 \times 6\sqrt{3}}{4}$ $\dfrac{243\sqrt{3}}{2}$

18. $s = 11$ $\dfrac{11 \times 11 \times 6\sqrt{3}}{4}$ $\dfrac{363\sqrt{3}}{2}$

19. $s = 18$ $\dfrac{18 \times 18 \times 6\sqrt{3}}{4}$ $486\sqrt{3}$

20. $s = 22$ $\dfrac{22 \times 22 \times 6\sqrt{3}}{4}$ $726\sqrt{3}$

21. $s = 26$ $\dfrac{26 \times 26 \times 6\sqrt{3}}{4}$ $1014\sqrt{3}$

22. $s = 28$ $\dfrac{28 \times 28 \times 6\sqrt{3}}{4}$ $1176\sqrt{3}$

23. $s = 30$ $\dfrac{30 \times 30 \times 6\sqrt{3}}{4}$ $1350\sqrt{3}$

24. $s = 36$ $\dfrac{36 \times 36 \times 6\sqrt{3}}{4}$ $1944\sqrt{3}$

25. $s = 42$ $\dfrac{42 \times 42 \times 6\sqrt{3}}{4}$ $2646\sqrt{3}$

26. $s = 48$ $\dfrac{48 \times 48 \times 6\sqrt{3}}{4}$ $3456\sqrt{3}$

Triangles 18

Answer as indicated for the congruent triangles. NTS

Answer as indicated for adjacent or overlapping triangles. NTS

1. △ACB ≅ △ACD. Find the perimeter of quadrilateral ABCD.

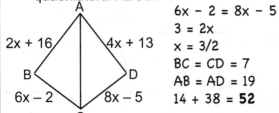

6x – 2 = 8x – 5
3 = 2x
x = 3/2
BC = CD = 7
AB = AD = 19
14 + 38 = **52**

6. Find the area and perimeter of quadrilateral ABCD.

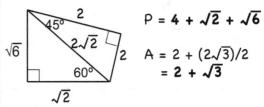

P = 4 + √2 + √6

A = 2 + (2√3)/2
= **2 + √3**

2. △ACB ≅ △DCE. AC = 2x + 9. BC = 15. CD = 4x – 1. Find AE.

2x + 9 = 4x – 1
10 = 2x
x = 5
AC = 19
CE = BC = 15
AE = **4**

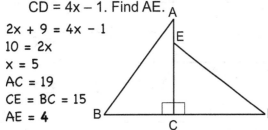

7. Given AE = 9, AC = 12, F is one of the two trisection points of $\overline{AE}$ closer to E, and B is the midpoint of $\overline{AC}$. Find the sum of the areas of △ACF and △ABE.

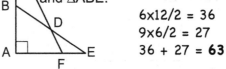

6x12/2 = 36
9x6/2 = 27
36 + 27 = **63**

3. △ACB ≅ △DBC. Find the greatest possible whole sum of the perimeters of the 2 triangles. **410**

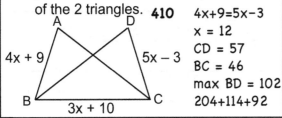

4x+9=5x–3
x = 12
CD = 57
BC = 46
max BD = 102
204+114+92

8. Find the perimeter of the pentagon.

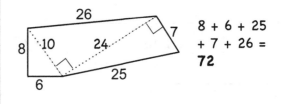

8 + 6 + 25
+ 7 + 26 =
72

4. △ABC ≅ △DEF. Given A(–6,0), B(0,–5), C(0,0), D(2,4), and F(2,10). Find the sum of all x and y values of E(x,y).

Given	Then	E = (7, 10) or
AC = 6	EF = 5	E = (–3, 10)
BC = 5		sum = **24**
DF = 6		

9. If all small squares are congruent, the area of the total shaded regions is what fractional part of the area of the greatest square?

$\frac{4}{9}$

5. △ABC ≅ △DEF. AB = 3x – 7. BC = 8. AC = 5z + 2. DE = 11. DF = 12. EF = 3y + 11. Find x – y + z.

3x – 7 = 11	3y + 11 = 8	5z + 2 = 12
3x = 18	3y = –3	5z = 10
x = 6	y = –1	z = 2

x – y + z = 6 + 1 + 2 = **9**

10. If the area of △ABC is 48, what is the area of △ACD?

A △ABD = 18
48 – 18 = **30**

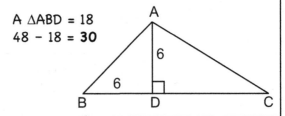

MAVA Math: Enhanced Skills Solutions Copyright © 2015 Marla Weiss

Triangles 19

Answer as indicated for the similar triangles. NTS

---

1. The area of the larger triangle is 96. Find the horizontal base of the smaller triangle.

ratio = 1/4

96 = 8B/2
B = 24
b = **6**

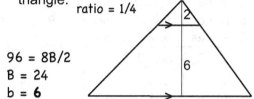

6. Find the area of the right trapezoid.

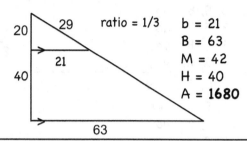

ratio = 1/3

b = 21
B = 63
M = 42
H = 40
A = **1680**

---

2. Find the sum of the perimeters of the 2 triangles.

14 + 10 + 8 = 32
21 + 15 + 12 = 48
sum = **80**

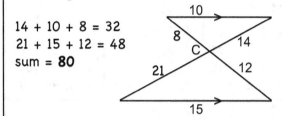

7. A, B, C, D are midpoints of sides of a parallelogram with diagonals 20 and 30. Find the perimeter of ABCD.

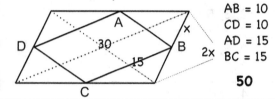

AB = 10
CD = 10
AD = 15
BC = 15

50

---

3. Find x.

$$\frac{10}{14} = \frac{5}{7} = \frac{15}{21}$$

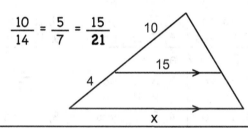

8. $\triangle ABC \sim \triangle EBD$.
BC = 30. BD = 20.
BE = 30. Find AE.

AB:BC:AC = BE:BD:DE
AB:30:AC = 30:20:DE
45:30:AC = 30:20:DE
AE = AB−BE =
45−30 = **15**

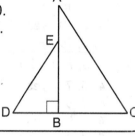

---

4. Find the perimeter of the larger triangle.

$$\frac{9}{x+10} = \frac{7}{x+6}$$

9x+54=7x+70
2x = 16
x = 8

18 + 26 + 14 = **58**

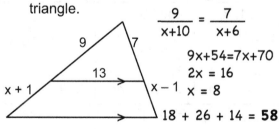

9. Find the perimeter of the smallest triangle.

10 + 2√13

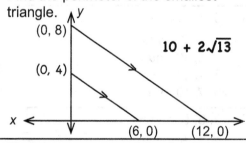

---

5. The horizontal lines are parallel. Find x, y, and z if their sum is 25.

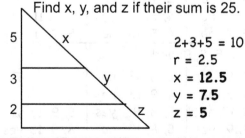

2+3+5 = 10
r = 2.5
x = **12.5**
y = **7.5**
z = **5**

10. Find the measures of the 2 missing sides.

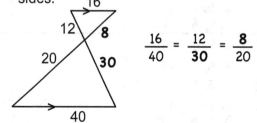

$$\frac{16}{40} = \frac{12}{30} = \frac{8}{20}$$

MAVA Math: Enhanced Skills Solutions Copyright © 2015 Marla Weiss

C(4,3) = 4	Triangles 20	C(5,3) = 10

Find the probability that 3 numbers selected randomly from a list of numbers form the sides of a triangle.

1. 3, 5, 8, and 10	9. 2, 6, 8, 9, and 11

1. 3, 5, 8, and 10

3, 5, 8 no	
3, 5, 10 no	
3, 8, 10 yes	$\frac{1}{2}$
5, 8, 10 yes	

9. 2, 6, 8, 9, and 11 — $\frac{1}{2}$

2, 6, 8 no	2, 8, 11 no	
2, 6, 9 no	2, 9, 11 no	
2, 6, 11 no	6, 8, 9 yes	6, 9, 11 yes
2, 8, 9 yes	6, 8, 11 yes	8, 9, 11 yes

2. 3, 5, 7, and 9

3, 5, 7 yes	
3, 5, 9 no	
3, 7, 9 yes	$\frac{3}{4}$
5, 7, 9 yes	

10. 2, 3, 5, 8, and 9 — $\frac{3}{10}$

2, 3, 5 no	2, 5, 9 no	
2, 3, 8 no	2, 8, 9 yes	
2, 3, 9 no	3, 5, 8 no	3, 8, 9 yes
2, 5, 8 no	3, 5, 9 no	5, 8, 9 yes

3. 2, 3, 5, and 7

2, 3, 5 no	
2, 3, 7 no	
2, 5, 7 no	$\frac{1}{4}$
3, 5, 7 yes	

11. 4, 6, 7, 8, and 9 — 1

4, 6, 7 yes	4, 7, 9 yes	
4, 6, 8 yes	4, 8, 9 yes	
4, 6, 9 yes	6, 7, 8 yes	6, 8, 9 yes
4, 7, 8 yes	6, 7, 9 yes	7, 8, 9 yes

4. 4, 6, 8, and 10

4, 6, 8 yes	
4, 6, 10 no	
4, 8, 10 yes	$\frac{3}{4}$
6, 8, 10 yes	

12. 3, 6, 7, 9, and 15 — $\frac{2}{5}$

3, 6, 7 yes	3, 7, 15 no	
3, 6, 9 no	3, 9, 15 no	
3, 6, 15 no	6, 7, 9 yes	6, 9, 15 no
3, 7, 9 yes	6, 7, 15 no	7, 9, 15 yes

5. 4, 5, 7, and 9

4, 5, 7 yes	
4, 5, 9 no	
4, 7, 9 yes	$\frac{3}{4}$
5, 7, 9 yes	

13. 3, 4, 5, 8, and 11 — $\frac{2}{5}$

3, 4, 5 yes	3, 5, 11 no	
3, 4, 8 no	3, 8, 11 no	
3, 4, 11 no	4, 5, 8 yes	4, 8, 11 yes
3, 5, 8 no	4, 5, 11 no	5, 8, 11 yes

6. 3, 5, 7, and 10

3, 5, 7 yes	
3, 5, 10 no	
3, 7, 10 no	$\frac{1}{2}$
5, 7, 10 yes	

14. 1, 3, 5, 9, and 13 — $\frac{1}{10}$

1, 3, 5 no	1, 5, 13 no	
1, 3, 9 no	1, 9, 13 no	
1, 3, 13 no	3, 5, 9 no	3, 9, 13 no
1, 5, 9 no	3, 5, 13 no	5, 9, 13 yes

7. 2, 3, 5, and 6

2, 3, 5 no	
2, 3, 6 no	
2, 5, 6 yes	$\frac{1}{2}$
3, 5, 6 yes	

15. 2, 4, 5, 7, and 11 — $\frac{3}{10}$

2, 4, 5 yes	2, 5, 11 no	
2, 4, 7 no	2, 7, 11 no	
2, 4, 11 no	4, 5, 7 yes	4, 7, 11 no
2, 5, 7 no	4, 5, 11 no	5, 7, 11 yes

8. 3, 5, 6, and 8

3, 5, 6 yes	
3, 5, 8 no	
3, 6, 8 yes	$\frac{3}{4}$
5, 6, 8 yes	

16. 4, 5, 8, 9, and 13 — $\frac{1}{2}$

4, 5, 8 yes	4, 8, 13 no	
4, 5, 9 no	4, 9, 13 no	
4, 5, 13 no	5, 8, 9 yes	5, 9, 13 yes
4, 8, 9 yes	5, 8, 13 no	8, 9, 13 yes

Triangles 21

Given triangles with sides: 3, 4, 5; 3, 6, 8; 6, 6, 6; 6, 7, 7; 6, 7, 8; 6, 8, 10; and 8, 8, 8.
Selecting two of the triangles, find the probability that:

WORK SPACE

$$C(7, 2) = \frac{6 \times 7}{2} = 21$$

3, 4, 5 right, scalene   6, 6, 6 acute, equilateral   6, 7, 8 acute, scalene
3, 6, 8 obtuse, scalene   6, 7, 7 acute, isosceles   6, 8, 10 right, scalene
                                                              8, 8, 8 acute, equilateral

---

1. the triangles are similar.

3, 4, 5 with 6, 8, 10
6, 6, 6 with 8, 8, 8
$\frac{2}{21}$

8. the triangles are both isosceles.

6, 6, 6 with 6, 7, 7   Equilateral is
6, 6, 6 with 8, 8, 8   isosceles by
6, 7, 7 with 8, 8, 8   definition.
$\frac{1}{7}$

---

2. the triangles are both right.

3, 4, 5 with 6, 8, 10
$\frac{1}{21}$

9. the triangles are congruent.

0

---

3. the triangles are both acute.

6, 6, 6 with 6, 7, 7   6, 7, 7 with 8, 8, 8
6, 6, 6 with 6, 7, 8   6, 7, 8 with 8, 8, 8
6, 6, 6 with 8, 8, 8
6, 7, 7 with 6, 7, 8
$\frac{2}{7}$

10. all 6 altitudes are in the interiors of the triangles.

equivalent condition to both acute
$\frac{2}{7}$

---

4. the triangles are both obtuse.

Only 1 of the triangles is obtuse.

0

11. the triangles are both scalene.

3, 4, 5 with 3, 6, 8   3, 6, 8 with 6, 8, 10
3, 4, 5 with 6, 7, 8   6, 7, 8 with 6, 8, 10
3, 4, 5 with 6, 8, 10
3, 6, 8 with 6, 7, 8
$\frac{2}{7}$

---

5. the triangles are both equilateral.

6, 6, 6 with 8, 8, 8
$\frac{1}{21}$

12. exactly 4 of the 6 altitudes are the sides.

equivalent condition to both right
$\frac{1}{21}$

---

6. one triangle is right and one is acute.

2 right, 4 acute
by multiplication principle,
2 x 4 = 8
$\frac{8}{21}$

13. one triangle is equilateral and one is scalene.

2 equilateral, 4 scalene
by multiplication principle,
2 x 4 = 8
$\frac{8}{21}$

---

7. one triangle is right and one is obtuse.

2 right, 1 obtuse
by multiplication principle,
2 x 1 = 2
$\frac{2}{21}$

14. one triangle is acute and one is obtuse.

4 acute, 1 obtuse
by multiplication principle,
4 x 1 = 4
$\frac{4}{21}$

Triangles 22

Find the probability that 3 numbers selected randomly form the sides of a right Δ.	Find the probability that 3 numbers selected randomly form the angles of a right triangle.

---

1. 3 to 13 inclusive

3, 4, 5 $C(11, 3) = \dfrac{9 \times 10 \times 11}{2 \times 3}$

6, 8, 10

5, 12, 13 $\dfrac{3}{165}$ $\dfrac{1}{\mathbf{55}}$

9. 10, 30, 40, 50, 60, 80, and 90

10, 80, 90 $C(7, 3) = \dfrac{5 \times 6 \times 7}{2 \times 3}$

30, 60, 90

40, 50, 90 $\dfrac{3}{\mathbf{35}}$

---

2. 7 to 26 inclusive

9, 12, 15 $C(20, 3) = \dfrac{18 \times 19 \times 20}{2 \times 3}$

12, 16, 20

15, 20, 25   7, 24, 25 $\dfrac{6}{57 \times 20}$ $\dfrac{1}{\mathbf{190}}$

10, 24, 26   8, 15, 17

10. 15, 20, 40, 50, 75, and 90

15, 75, 90 $C(6, 3) = \dfrac{4 \times 5 \times 6}{2 \times 3}$

40, 50, 90

$\dfrac{2}{20}$ $\dfrac{1}{\mathbf{10}}$

---

3. 5 to 20 inclusive

6, 8, 10 $C(16, 3) = \dfrac{14 \times 15 \times 16}{2 \times 3}$

9, 12, 15

12, 16, 20

5, 12, 13   8, 15, 17 $\dfrac{5}{14 \times 40}$ $\dfrac{1}{\mathbf{112}}$

11. 25, 35, 55, 65, and 90

25, 65, 90 $C(5, 3) = \dfrac{4 \times 5}{2}$

35, 55, 90

$\dfrac{2}{10}$ $\dfrac{1}{\mathbf{5}}$

---

4. 6 to 17 inclusive

6, 8, 10 $C(12, 3) = \dfrac{10 \times 11 \times 12}{2 \times 3}$

9, 12, 15

5, 12, 13 $\dfrac{4}{10 \times 22}$ $\dfrac{1}{\mathbf{55}}$

8, 15, 17

12. 5, 15, 20, 60, 70, 75, 85, and 90

5, 85, 90 $C(8, 3) = \dfrac{6 \times 7 \times 8}{2 \times 3}$

15, 75, 90

20, 70, 90 $\dfrac{3}{\mathbf{56}}$

---

5. 15 to 29 inclusive

15, 20, 25 $C(15, 3) = \dfrac{13 \times 14 \times 15}{2 \times 3}$

20, 21, 29

$\dfrac{2}{13 \times 35}$ $\dfrac{2}{\mathbf{455}}$

13. 5, 10, 15, 20, 70, 75, 80, 85, and 90

5, 85, 90 $C(9, 3) = \dfrac{7 \times 8 \times 9}{2 \times 3}$

10, 80, 90

15, 75, 90 $\dfrac{4}{84}$ $\dfrac{1}{\mathbf{21}}$

20, 70, 90

---

6. 12 to 37 inclusive

12, 16, 20 $C(26, 3) = \dfrac{24 \times 25 \times 26}{2 \times 3}$

15, 20, 25

12, 35, 37   18, 24, 30 $\dfrac{7}{26 \times 100}$ $\dfrac{7}{\mathbf{2600}}$

16, 30, 34   21, 28, 35

20, 21, 29

14. 12, 33, 39, 57, 61, 78, and 90

12, 78, 90 $C(7, 3) = \dfrac{5 \times 6 \times 7}{2 \times 3}$

33, 57, 90

$\dfrac{2}{\mathbf{35}}$

---

7. 20 to 29 inclusive

20, 21, 29 $C(10, 3) = \dfrac{8 \times 9 \times 10}{2 \times 3}$

$\dfrac{1}{\mathbf{120}}$

15. 3, 16, 17, 18, 63, 72, 74, 87, and 90

3, 87, 90 $C(9, 3) = \dfrac{7 \times 8 \times 9}{2 \times 3}$

16, 74, 90

18, 72, 90 $\dfrac{3}{84}$ $\dfrac{1}{\mathbf{28}}$

---

8. 5 to 25 inclusive

6, 8, 10 $C(21, 3) = \dfrac{19 \times 20 \times 21}{2 \times 3}$

9, 12, 15   5, 12, 13

12, 16, 20   7, 24, 25 $\dfrac{7}{19 \times 70}$ $\dfrac{1}{\mathbf{190}}$

15, 20, 25   8, 15, 17

16. 4, 13, 21, 46, 59, 77, 86, and 90

4, 86, 90 $C(8, 3) = \dfrac{6 \times 7 \times 8}{2 \times 3}$

13, 77, 90

$\dfrac{2}{56}$ $\dfrac{1}{\mathbf{28}}$

Triplets 1

Work rows may vary.

Answer using the "keep 1–change 2" chart method. Assume uniform rates.

1. If 6 cats can catch 6 mice in 6 minutes, how many cats are needed to catch 100 mice in 100 minutes?

cats	mice	min
6	6	6
6	2	2
6	100	100

2. Five boys can pack 60 boxes in 3 days. How many boxes can 2 boys pack in 4 days?

boys	boxes	days
5	60	3
1	12	3
2	24	3
2	8	1
2	**32**	4

3. Four girls can mow 30 lawns in 5 days. Eight girls can mow how many lawns in a day-and-a half?

girls	lawns	days
4	30	5
8	30	2.5
8	6	0.5
8	**18**	1.5

4. 2000 bees need one year to make 7 pounds of honey. How many years do 5000 bees need to make 70 pounds of honey?

bees	years	honey
2000	1	7
2000	10	70
1000	20	70
5000	**4**	70

5. Ten movers can pack 15 studios in 2 days. How many movers are needed to pack 42 studios in one week?

movers	studios	days
10	15	2
2	3	2
28	42	2
56	42	1
8	42	7

6. Nine chefs can prepare 51 dinners in 2 hours. 24 chefs can prepare how many dinners in 1 day?

chefs	dinners	hours
9	51	2
3	17	2
24	136	2
24	**1632**	24

7. If 8 cats can catch 12 mice in 20 minutes, how many cats are needed to catch 36 mice in 2 hours?

cats	mice	min
8	12	20
8	36	60
4	36	120

8. If a chicken-and-a-half lays an egg-and-a-half in a day-and-a-half, how many eggs will 12 chickens lay in 12 days?

chickens	eggs	days
1.5	1.5	1.5
1	1	1.5
12	12	1.5
12	**96**	12

MAVA Math: Enhanced Skills Solutions Copyright © 2015 Marla Weiss

Rows may vary.

Triplets 2

Answer using the "keep 1–change 2" method. Assume uniform rates and positive, integral variables.

1. If x students can drink y cans of soda in z days, how many cans of soda can w students drink in 1 week?

	students	cans	days
	x	y	z
	xz	y	1
	1	$\frac{y}{xz}$	1
	1	$\frac{7y}{xz}$	7
	w	$\frac{7wy}{xz}$	7

5. If m men can mow w lawns in 1 week, n men can mow how many lawns in 52 weeks?

	men	lawns	weeks
	m	w	1
	m	52w	52
	1	$\frac{52w}{m}$	52
	n	$\frac{52nw}{m}$	52

2. If c cats can catch m mice in a half-hour, how many cats are needed to catch e mice in 1 week?

	cats	mice	hours
	c	m	.5
	c	2m	1
	$\frac{c}{2m}$	1	1
	$\frac{ce}{2m}$	e	1
	$\frac{ce}{336m}$	e	168

6. If e chefs can prepare m meals in 1 week, f chefs can prepare how many meals in 1 hour?

	chefs	meals	hours
	e	m	7x24
	1	$\frac{m}{e}$	7x24
	f	$\frac{mf}{e}$	7x24
	f	$\frac{mf}{168e}$	1

3. If c chickens lay e eggs in d days, how many eggs will k chickens lay in June?

	chicks	eggs	days
	c	e	d
	dc	e	1
	1	$\frac{e}{dc}$	1
	k	$\frac{ek}{dc}$	1
	k	$\frac{30ek}{cd}$	30

7. If v movers can pack r rooms in d days, how many movers are needed to pack m rooms in January?

	movers	rooms	days
	v	r	d
	dv	r	1
	$\frac{dv}{31}$	r	31
	$\frac{dv}{31r}$	1	31
	$\frac{dvm}{31r}$	m	31

4. If b boys can pack x boxes in h hours, how many boxes can y boys pack in one day?

	boys	boxes	hours
	b	x	h
	bh	x	1
	1	$\frac{x}{bh}$	1
	y	$\frac{xy}{bh}$	1
	y	$\frac{24xy}{bh}$	24

8. If b bees need one year to make p pounds of honey, e bees will make how many pounds of honey in one century?

	bees	years	pounds
	b	1	p
	1	1	$\frac{p}{b}$
	e	1	$\frac{ep}{b}$
	e	100	$\frac{100ep}{b}$

Vocabulary 1

Complete the blank.

1. In a plane, two lines that are __perpendicular__ form four right angles.

2. The x-axis and y-axis divide the coordinate plane into four __quadrants__.

3. A line that crosses two parallel lines is called a __transversal__.

4. If a=b and b=c, then a=c by the property __transitivity__ of equality.

5. An angle with its vertex at the center of a circle is called a __central__ angle.

6. A ratio equal to another ratio is called a __proportion__.

7. A point that bisects a line segment is called its __midpoint__.

8. A __complex__ fraction has fractions in its numerator and/or denominator.

9. The __intersection__ of sets is the set of elements common to the given sets.

10. A synonym for element of a set is __member__.

11. The more common name for multiplicative inverse is __reciprocal__.

12. An angle measuring 180° is called a __straight angle__.

13. The longest chord of a circle is a __diameter__.

14. The union of two __rays__ sharing a common endpoint is an angle.

15. A regular quadrilateral is called a __square__.

16. The more common name for arithmetic mean is __average__.

17. The tilt of a line is called its __slope__.

18. On the coordinate plane, a __lattice point__ has integers as both coordinates.

19. A synonym for height of a triangle is __altitude__.

Vocabulary 2

Complete the blanks.

1. An isosceles triangle has one ____vertex____ angle and two ____base____ angles.

2. "Positive, ____negative____, or zero" is an example of a ____trichotomy____.

3. 2 angles are __complementary__ if their degree sum is 90° but __supplementary__ if 180°.

4. The distributive field property uses the operations __multiplication__ and ____addition____.

5. 5 and 7 are _____twin_____ primes; 7 and 11 are ____consecutive____ primes.

6. A __composite__ number has 3 or more factors; a ____prime____ number has exactly 2 different factors.

7. Formed by parallel lines and a transversal, alternate angles may be ____interior____ or ____exterior____.

8. A triangle with 3 congruent sides is both ____isosceles____ and ____equilateral____.

9. Dividing by ____zero____ and raising ____zero____ to the 0th power are undefined.

10. 1, 2, 3, 4, . . . are called ____counting____ or ____natural____ numbers.

11. Circumference is to ____circle____ as ____perimeter____ is to polygon.

12. The identity element for multiplication is ____one____ and for addition is ____zero____.

13. ____Mean____, ____median____, and mode are three measures of central tendency.

14. A set of ordered pairs is a ____relation____, which may or may not be a ____function____.

15. In a parallelogram, ____adjacent____ angles are supplementary, and ____opposite____ angles are congruent.

16. The sides of a right triangle are called the ____hypotenuse____ and the ____legs____.

17. Three set operations are ____union____, ____intersection____, and complementation.

18. 1, 4, 9, 16, . . . are __perfect squares__; 1, 8, 27, 64, . . . are __perfect cubes__.

19. The 2 parts of a fraction are its ____numerator____ and ____denominator____.

Rate and Time are reciprocals.	# Work Problems 1

Find the time needed for the people to complete the job working together.

1.   Person A can do the job in 1 hour.
     Person B can do the job in 2 hours.
 	
A rate = 1
B rate = 1/2
together rate = 3/2
together time = **2/3 hour** or **40 min**
	8. Person A can do the job in 5 hours.
	     Person B can do the job in 7 hours.
	
	A rate = 1/5
	B rate = 1/7
	together rate = 12/35 **2 hr**
	together time = **35/12 hour** or **55 min**

2.   Person A can do the job in 1 hour.
     Person B can do the job in 3 hours.
 	
A rate = 1
B rate = 1/3
together rate = 4/3
together time = **3/4 hour** or **45 min**
	9. Person A can do the job in 7 hours.
	     Person B can do the job in 8 hours.
	
	A rate = 1/7
	B rate = 1/8
	together rate = 15/56 **3 hr**
	together time = **56/15 hour** or **44 min**

3.   Person A can do the job in 2 hours.
     Person B can do the job in 3 hours.
 	
A rate = 1/2
B rate = 1/3
together rate = 5/6
together time = **6/5 hour** or **72 min**
	10. Person A can do the job in 6 hours.
	      Person B can do the job in 9 hours.
	
	A rate = 1/6
	B rate = 1/9
	together rate = 5/18 **3 hr**
	together time = **18/5 hour** or **36 min**

4.   Person A can do the job in 2 hours.
     Person B can do the job in 4 hours.
 	
A rate = 1/2
B rate = 1/4
together rate = 3/4
together time = **4/3 hour** or **80 min**
	11. Person A can do the job in 4 hours.
	      Person B can do the job in 6 hours.
	
	A rate = 1/4
	B rate = 1/6
	together rate = 5/12 **2 hr**
	together time = **12/5 hour** or **24 min**

5.   Person A can do the job in 1 hour.
     Person B can do the job in 4 hours.
 	
A rate = 1
B rate = 1/4
together rate = 5/4
together time = **4/5 hour** or **48 min**
	12. Person A can do the job in 1 hour.
	      Person B can do the job in 5 hours.
	
	A rate = 1
	B rate = 1/5
	together rate = 6/5
	together time = **5/6 hour** or **50 min**

6.   Person A can do the job in 2 hours.
     Person B can do the job in 6 hours.
 	
A rate = 1/2
B rate = 1/6
together rate = 4/6 = 2/3
together time = **3/2 hour** or **90 min**
	13. Person A can do the job in 5 hours.
	      Person B can do the job in 5 hours.
	
	A rate = 1/5
	B rate = 1/5
	together rate = 2/5 **2 hr**
	together time = **2.5 hours** or **30 min**

7.   Person A can do the job in 3 hours.
     Person B can do the job in 6 hours.
A rate = 1/3
B rate = 1/6
together rate = 3/6 = 1/2
together time = **2 hours** or **120 min**
	14. Person A can do the job in 4 hours.
	      Person B can do the job in 8 hours.
	A rate = 1/4
	B rate = 1/8
	together rate = 3/8 **2 hr**
	together time = **8/3 hour** or **40 min**

Work Problems 2

Find the time for Person B alone to do the job given the time for A alone to do the job and their time together.

1.   Time for Person A alone = 4 hours
       Time together to finish = 3 hours
 	
A rate = 1/4
tog rate = 1/3
B rate = 1/12
B time = **12 hours**
	7. Time for Person A alone = 4 hours
	       Time together to finish = 1.5 hours
	
	A rate = 1/4
	tog rate = 2/3
	B rate = 5/12
	B time = **12/5 hours** or **2 hrs 24 min**

2.   Time for Person A alone = 4 hours
       Time together to finish = 2 hours
 	
A rate = 1/4
tog rate = 1/2
B rate = 1/4
B time = **4 hours**
	8. Time for Person A alone = 5.5 hours
	       Time together to finish = 3 hours
	
	A rate = 2/11
	tog rate = 1/3
	B rate = 5/33
	B time = **33/5 hours** or **6 hrs 36 min**

3.   Time for Person A alone = 3 hours
       Time together to finish = 1 hour
 	
A rate = 1/3
tog rate = 1
B rate = 2/3
B time = **1.5 hours**
	9. Time for Person A alone = 4 hours
	       Time together to finish = 1 hour
	
	A rate = 1/4
	tog rate = 1
	B rate = 3/4
	B time = **4/3 hours** or **1 hr 20 min**

4.   Time for Person A alone = 4 hours
       Time together to finish = 2.5 hours
 	
A rate = 1/4
tog rate = 2/5
B rate = 3/20
B time = **20/3 hours** or **6 hr 40 min**
	10. Time for Person A alone = 5 hours
	       Time together to finish = 3.5 hours
	
	A rate = 1/5
	tog rate = 2/7
	B rate = 3/35
	B time = **35/3 hours** or **11 hrs 40 min**

5.   Time for Person A alone = 3.5 hours
       Time together to finish = 2.5 hours
 	
A rate = 2/7
tog rate = 2/5
B rate = 4/35
B time = **35/4 hours** or **8 hr 45 min**
	11. Time for Person A alone = 6 hours
	       Time together to finish = 4.5 hours
	
	A rate = 1/6
	tog rate = 2/9
	B rate = 1/18
	B time = **18 hours**

6.   Time for Person A alone = 5 hours
       Time together to finish = 3 hours
 	
A rate = 1/5
tog rate = 1/3
B rate = 2/15
B time = **15/2 hours** or **7 hr 30 min**
	12. Time for Person A alone = 7 hours
	       Time together to finish = 3 hours
	
	A rate = 1/7
	tog rate = 1/3
	B rate = 4/21
	B time = **21/4 hours** or **5 hr 15 min**

ABBREVIATIONS

Use common sense in decoding abbreviations. For example, in division problems, R means remainder, yet R means radius in geometry and common ratio in sequences.

General

CBD	can't be done
DS	digit sum
GCF	greatest common factor
LCM	least common multiple
NTS	not to scale
T&E	trial and error
WP	word problem
WLOG	without loss of generality

Coordinate Plane

D	down
L	left
Q	quadrant
R	right
U	up

Logic

T	true
F	false

Sets

N	Natural numbers
W	Whole numbers
Z	Integers
Q	Rational numbers
R	Real numbers
C	Complex numbers

Geometry

A	area
B	base
C	circumference
D	diameter
E	edge
H	height
L	length
M	midline of a trapezoid
P	perimeter
R	radius
S	side
SA	surface area
V	volume
W	width

Functions

ABS	absolute value
SGN	1 for > 0, −1 for < 0, 0 for = 0
TRUNC	truncate decimal part
INT	greatest integer contained in
SIGMA	number of divisors
SQR	square
SQRT	square root

Money

D	dime
H	half dollar
N	nickel
P	penny
Q	quarter

Percent Change

D	decrease
I	increase

Triangles

A	acute
E	equilateral
I	isosceles
O	obtuse
R	right
S	scalene

Solids

E	number of edges
F	number of faces
V	number of vertices

Sequences

A	arithmetic
d	common difference
G	geometric
r	common ratio

Word Problems

D	distance
R	rate
T	time

Arithmetic

R	remainder

MAVA Math: Enhanced Skills Solutions   Copyright © 2015 Marla Weiss

ABBREVIATIONS (continued)

Properties

ClPA	Closure Property of Addition
ClPM	Closure Property of Multiplication
APA	Associative Property of Addition
APM	Associative Property of Multiplication
CPA	Commutative Property of Addition
CPM	Commutative Property of Multiplication
IdPA	Identity Property of Addition
IdPM	Identity Property of Multiplication
InPA	Inverse Property of Addition
InPM	Inverse Property of Multiplication
DPMA	Distributive Property of Multiplication over Addition
ZPM	Zero Property of Multiplication
RPE	Reflexive Property of Equality
SPE	Symmetric Property of Equality
TPE	Transitive Property of Equality
APE	Addition Property of Equality
MPE	Multiplication Property of Equality
APU	Associative Property of Union
CPU	Commutative Property of Union
API	Associative Property of Intersection
CPI	Commutative Property of Intersection
DPUI	Distributive Property of Union over Intersection
DPIU	Distributive Property of Intersection over Union
RPC	Reflexive Property of Congruence
SPC	Symmetric Property of Congruence
TPC	Transitive Property of Congruence
APC	Addition Property of Congruence
MPC	Multiplication Property of Congruence

NOTES

FORMULAE

Area square s^2	Diagonal square $s\sqrt{2}$
Area rectangle bh	Diagonal cube $e\sqrt{3}$
Area rhombus $Dd/2$	Diagonal rectangular prism $\sqrt{a^2 + b^2 + c^2}$
Area parallelogram bh	sum of the angles of an n-gon $180(n-2)$
Area trapezoid $h(b + B)/2$	number of diagonals of an n-gon $n(n-3)/2$
Area triangle $bh/2$	interior angle of a regular n-gon $180(n-2)/n$
Area equilateral triangle $\dfrac{s^2\sqrt{3}}{4}$	exterior angle of a regular n-gon $360/n$
Area circle πr^2	midpoint of (x, y) and (w, z) $((x+w)/2,(y+z)/2)$
Perimeter square $4s$	Pythagorean Theorem $a^2 + b^2 = c^2$
Perimeter rectangle $2(L+W)$	GCF(m,n) x LCM(m,n) = mn
Circumference circle πd	distance = rate x time
Surface area cube $6e^2$	simple interest = principal x rate x time
Surface area rectangular prism $2(ab+bc+ac)$	value after compound interest $P(1+ r)^T$
Surface area sphere $4\pi r^2$	number of subsets of a set with n elements 2^n
Surface area cylinder $2\pi r^2 + \pi dh$	
Surface area cone $\pi r^2 + \pi r l$	number of squares in an n by n grid $1^2 + 2^2 + 3^2 + \ldots + n^2$
Volume cube e^3	number of rectangles in an n by n grid $(1 + 2 + 3 + \ldots + n)^2$
Volume rectangular prism lwh	E + F + V = 6N + 2 for prisms:
Volume right prism Bh	E = 3N N = # sides of base

Volume pyramid $Bh/3$	E = # edges
	 F = N + 2 V = # vertices
	 F = # faces

Volume cylinder $\pi r^2 h$	V = 2N
Volume cone $\pi r^2 h/3$	Hero: Area of Triangle $\sqrt{S(S-a)(S-b)(S-c)}$ S = semi-perimeter
Volume sphere $4\pi r^3/3$	Pick: Area on Geoboard $B/2 + I - 1$ B = # border points, I = # interior pts

48 PYTHAGOREAN TRIPLES

3, 4, 5
5, 12, 13
7, 24, 25
8, 15, 17
9, 40, 41
11, 60, 61
12, 35, 37
13, 84, 85
15, 112, 113
16, 63, 65
17, 144, 145
19, 180, 181
20, 21, 29
20, 99, 101
21, 220, 221
23, 264, 265
24, 143, 145
28, 45, 53
28, 195, 197
32, 255, 257
33, 56, 65
35, 612, 613
36, 77, 85
39, 80, 89
44, 117, 125
48, 55, 73
51, 140, 149
52, 165, 173
57, 176, 185
60, 91, 109
60, 221, 229
65, 72, 97
68, 285, 293
69, 260, 269
84, 187, 205
85, 132, 157
88, 105, 137
95, 168, 193
96, 247, 265
104, 153, 185
105, 208, 233
115, 252, 277
119, 120, 169
120, 209, 241
133, 156, 205
140, 171, 221
160, 231, 281
161, 240, 289

PRIME NUMBERS
(Less Than 1070)

2	199	467	769
3	211	479	773
5	223	487	787
7	227	491	797
11	229	499	809
13	233	503	811
17	239	509	821
19	241	521	823
23	251	523	827
29	257	541	829
31	263	547	839
37	269	557	853
41	271	563	857
43	277	569	859
47	281	571	863
53	283	577	877
59	293	587	881
61	307	593	883
67	311	599	887
71	313	601	907
73	317	607	911
79	331	613	919
83	337	617	929
89	347	619	937
97	349	631	941
101	353	641	947
103	359	643	953
107	367	647	967
109	373	653	971
113	379	659	977
127	383	661	983
131	389	673	991
137	397	677	997
139	401	683	1009
149	409	691	1013
151	419	701	1019
157	421	709	1021
163	431	719	1031
167	433	727	1033
173	439	733	1039
179	443	739	1049
181	449	743	1051
191	457	751	1061
193	461	757	1063
197	463	761	1069

Middle School Math Vocabulary

absolute value	calculator
abundant number	calendar problem
acute angle	center
acute triangle	centimeter
add	central angle
addition	chart
Addition Property of Equality	chord
additive inverse	cipher
adjacent	circle
adjacent angles	circle graph
adjacent sides	circumference
age problem	circumscribed
algebra	Closure Property of Addition
algebraic expression	Closure Property of Multiplication
algorithm	coefficient
alternate exterior angles	coin problem
alternate interior angles	collinear
altitude	combination
AND	common denominator
angle	common difference
apothem	common element
approximation	common factor
arc	common fraction
area	common multiple
arithmetic	common ratio
arithmetic mean	Commutative Property of Addition
arithmetic sequence	Commutative Property of Multiplication
array	compass
ascending order	complement
Associative Property of Addition	complementary angles
Associative Property of Multiplication	complementation
at least	complex fraction
at most	complex number
average	composite number
average rate	composition
axis (axes)	compound interest
bar graph	concatenation
base	concave
billion	concentric
binary	cone
bisect	congruent
bisection	consecutive
boundary problem	consecutive even

consecutive odd
constant
convex
coordinate plane
coordinates
coplanar
corresponding angles
corresponding sides
counterexample
counting number
cross multiply
cube
cube root
cylinder
data
decagon
decimal
decimeter
decrease
decrement
definition
degree
denominator
dependent events
descending order
diagonal
diameter
dichotomy
difference
digit
dimension
direct variation
directed graph
discount
discrete sets
disjoint sets
distance
distance-rate-time problem
distinct
Distributive Property
dividend
divisibility rule
divisible
division

divisor
dodecagon
dodecahedron
domain
double
edge
element
empty set
endpoint
equal
equality
equation
equiangular
equidistant
equilateral
equilateral triangle
equivalent
equivalent fractions
estimation
Euler Graph
evaluate
even number
event
expanded notation
exponent
exponential form
exponentiation
expression
exterior
exterior angle
externally tangent
face
factor
factorial
fence post problem
Fibonacci Sequence
fictitious operation
field
First-Plus-Last Method
flip
foot (feet)
formula
fraction
fractional part

frequency distribution	inscribed
frequency table	insufficient information
function	integer
geometric mean	intercept
geometric sequence	interest
geometry	interior
Goldbach's Conjecture	interior angle
graph	internally tangent
greater	interpolate
greater than	intersect
greater than or equal to	intersecting lines
greatest	intersection
greatest common divisor	inverse
greatest common factor	inverse operation
half	Inverse Property of Addition
halve	Inverse Property of Multiplication
height	inverse variation
hemisphere	irrational number
heptagon	isosceles right triangle
hexagon	isosceles trapezoid
hexagonal prism	isosceles triangle
hexagonal pyramid	kilometer
histogram	kite
How Many? problem	lateral area
hundred	lateral edge
hundredth	lattice point
hypotenuse	least
i	least common denominator
icosahedron	least common multiple
identity element	leg
Identity Property of Addition	length
Identity Property of Multiplication	less
imaginary number	less than
improper fraction	less than or equal to
in terms of	like terms
inch	line
increase	line of symmetry
increment	line segment
independent events	linear equation
inequality	linear function
infinite	long division
infinite sequence	lowest common denominator
infinite series	lowest terms
infinity	magic square

mathematical maturity	number line
mathematician	number theory
mathematics	numeral
matrix (matrices)	numerator
matrix addition	obtuse angle
matrix multiplication	obtuse triangle
maximum	octagon
mean	octahedron
measure	odd number
measurement	odds
median	ones digit
member	ones place
mental math	operation
meter	opposite
metric system	opposite angles
midline	opposite sides
midpoint	OR
millimeter	order
million	ordered pair
minimum	organized list
mixed number	origin
mixture problem	original cost
mode	outcome
modulo	outlier
multiple	palindrome
multiplication	parabola
Multiplication Principle	parallel
Multiplication Property of Equality	parallelepiped
multiplicative inverse	parallelogram
mutually exclusive	parenthesis (parentheses)
natural number	pattern
negate	pentagon
negative number	pentagonal prism
negative sign	pentagonal pyramid
net	percent
nonagon	percent change
noncollinear	perfect cube
nonoverlapping	perfect number
nonzero	perfect square
NOT	perimeter
not to scale	permutation
nth term	perpendicular
number	perpendicular bisector
number base	perpendicular lines

perpendicular planes	ratio
pi	rational
place	ray
place value	real number
plane	reciprocal
point	rectangle
polygon	rectangular prism
polyhedron	rectangular pyramid
positive number	rectangular solid
power	reflection
power of ten	Reflexive Property of Equality
power of two	region
preceding term	region bounded by
prime factorization	regular
prime number	regular polygon
principal	relation
prism	relatively prime
probability	remainder
problem solving	repeating decimal
product	rhombus
profit problem	right angle
proof	right cylinder
proper fraction	right prism
proper subset	right pyramid
property	right triangle
proportion	rigid motion
protractor	root
prove	rotation
pyramid	sale price
Pythagorean primitive or triplet	sales tax
Pythagorean Theorem	scalar multiplication
Quadrant I, II, III, and IV	scalene trapezoid
quadratic equation	scalene triangle
quadratic formula	scientific notation
quadrilateral	sector
quadruple	segment
quantity	semicircle
quarter	semiperimeter
quotient	sequence
radical	series
radius	set
random number	shaded area
range	short division
rate	side

similar	term
similar polygons	terminating decimal
similar triangles	tetrahedron
simple interest	thousand
simplest form	thousandth
simplify	transformation
skew lines	Transitive Property of Equality
slant height	transitivity
slide	translation
slope	transversal
solid	trapezoid
solution	triangle
solution set	trichotomy
solve	trillion
space	trillions place
special right triangle	triple
sphere	triplet problem
square	trisect
square root	trisection
standard form	twin primes
standard notation	undefined
statistics	uniform border problem
stem-and-leaf plot	union
straight angle	unit
straight edge	unit fraction
subset	units digit
substitution	units place
subtraction	universal set
succeeding term	value
successive terms	variable
sum	Venn Diagram
sum of the digits	vertex (vertices)
supplement	vertical angles
supplementary angles	volume
surface area	whole number
symbol	width
symmetry	work problem
symmetric about	x-axis
Symmetric Property of Equality	x-coordinate
system of equations	yard
tangent	y-axis
ten billions place	y-coordinate
ten millions place	zero
tens place	Zero Property of Multiplication

Printed in the United States
By Bookmasters